VISUALIZING

EARTH HISTORY

VISUALIZING
EARTH HISTORY

Loren E. Babcock
The Ohio State University

WILEY

In collaboration with

THE NATIONAL GEOGRAPHIC SOCIETY

CREDITS

VP AND PUBLISHER Jay O'Callaghan
MANAGING DIRECTOR Helen McInnis
EXECUTIVE EDITOR Ryan Flahive
DIRECTOR OF DEVELOPMENT Barbara Heaney
MANAGER, PRODUCT DEVELOPMENT Nancy Perry
DEVELOPMENT EDITOR Charity Robey
ASSISTANT EDITOR Courtney Nelson
EDITORIAL ASSISTANT Erin Grattan
EXECUTIVE MARKETING MANAGER Jeffrey Rucker
MARKETING MANAGER Danielle Torio
PRODUCTION MANAGER Micheline Frederick
SR. PRODUCTION EDITOR Sujin Hong
MEDIA EDITOR Lynn Pearlman, Bridget O'Lavin
CREATIVE DIRECTOR Harry Nolan
COVER DESIGNER Harry Nolan
INTERIOR DESIGN Vertigo Design
PHOTO EDITOR Hilary Newman
PHOTO RESEARCHERS Teri Stratford/Stacey Gold,
National Geographic Society
SR. ILLUSTRATION EDITOR Sandra Rigby
PRODUCTION SERVICES Jeanine Furino/
GGS Book Services PMG

Top cover photo: © John Cancalosi/NG Image Collection
Bottom inset cover photos (left to right): Courtesy Loren Babcock;
Courtesy Alycia L. Stigall; © C. D. Winters/Photo Researchers, Inc.;
© O. Louis Mazzatenta/NG Image Collection; © Peter Carsten/
NG Image Collection

Page ii: © Marc Adamus/Getty Images

This book was set in Times New Roman by GGS Book Services PMG,
printed and bound by Quebecor World. The cover was printed by
Phoenix Color.

To order books or for customer service please, call
1-800-CALL WILEY (225-5945)

ISBN-13: 978-0-471-72490-2
BRV ISBN: 978-0470-41845-1

Printed in the United States of America

10 9 8 7 6 5 4 3 2 1

VISUALIZING
EARTH HISTORY

Loren E. Babcock
The Ohio State University

WILEY

In collaboration with
THE NATIONAL GEOGRAPHIC SOCIETY

CREDITS

VP AND PUBLISHER Jay O'Callaghan
MANAGING DIRECTOR Helen McInnis
EXECUTIVE EDITOR Ryan Flahive
DIRECTOR OF DEVELOPMENT Barbara Heaney
MANAGER, PRODUCT DEVELOPMENT Nancy Perry
DEVELOPMENT EDITOR Charity Robey
ASSISTANT EDITOR Courtney Nelson
EDITORIAL ASSISTANT Erin Grattan
EXECUTIVE MARKETING MANAGER Jeffrey Rucker
MARKETING MANAGER Danielle Torio
PRODUCTION MANAGER Micheline Frederick
SR. PRODUCTION EDITOR Sujin Hong
MEDIA EDITOR Lynn Pearlman, Bridget O'Lavin
CREATIVE DIRECTOR Harry Nolan
COVER DESIGNER Harry Nolan
INTERIOR DESIGN Vertigo Design
PHOTO EDITOR Hilary Newman
PHOTO RESEARCHERS Teri Stratford/Stacey Gold,
National Geographic Society
SR. ILLUSTRATION EDITOR Sandra Rigby
PRODUCTION SERVICES Jeanine Furino/
GGS Book Services PMG

Top cover photo: © John Cancalosi/NG Image Collection
Bottom inset cover photos (left to right): Courtesy Loren Babcock;
Courtesy Alycia L. Stigall; © C. D. Winters/Photo Researchers, Inc.;
© O. Louis Mazzatenta/NG Image Collection; © Peter Carsten/
NG Image Collection

Page ii: © Marc Adamus/Getty Images

This book was set in Times New Roman by GGS Book Services PMG,
printed and bound by Quebecor World. The cover was printed by
Phoenix Color.

To order books or for customer service please, call
1-800-CALL WILEY (225-5945)

ISBN-13: 978-0-471-72490-2
BRV ISBN: 978-0470-41845-1

Printed in the United States of America

10 9 8 7 6 5 4 3 2 1

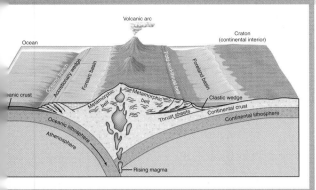

Visualizing Earth History is designed to help your students learn effectively. Created in collaboration with the National Geographic Society and our Wiley Visualizing consulting editor, Professor Jan Plass of New York University, *Visualizing Earth History* integrates rich visuals and media with text to direct students' attention to important information. This approach illustrates complex processes, organizes related pieces of information, and integrates information in clear, concise representations. Beautifully illustrated, *Visualizing Earth History* shows your students what the discipline is all about—its main concepts and applications—while also instilling an appreciation and excitement about the richness of the subject.

Visuals, as used throughout this text, are instructional components that display facts, concepts, processes, or principles. The visuals include diagrams, graphs, maps, photographs, illustrations, and schematics, and together they create the foundation for the text.

Why should a textbook based on visuals be effective? Research shows that we learn better from integrated text and visuals than from either medium separately. Beginners in a subject benefit most from reading about the topic, attending class, and studying well-designed and integrated visuals. A visual, with good accompanying discussion, really can be worth a thousand words!

Well-designed visuals can also improve the efficiency with which a learner processes information. The more effectively we process information, the more likely it is that we will learn. This process takes place in our working memory. As we learn, we integrate new information in our working memory with existing knowledge in our long-term memory.

Have you ever read a paragraph or a page in a book, stopped, and said to yourself: "I don't remember one thing I just read"? This may happen when your working memory has been overloaded, and the text you read was not successfully integrated into long-term memory. Visuals don't automatically solve the problem of overload, but well-designed visuals can reduce the number of elements that working memory must process, thus aiding learning.

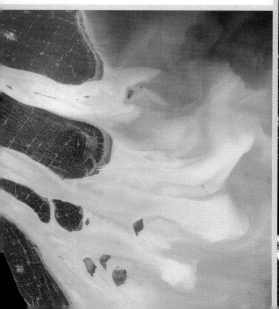

Habitat in North America	South America	North America	Eurasia	Africa
Holocene 0.01 Ma — Prairie			■ Wild asses	■ Zebras
Pleistocene 2.6 Ma	■ Hippidion			
Pliocene 5.3 Ma — Savanna		■ Equus ■ Hipparion group	■ Hipparion	
Miocene 23.0 Ma — Woodland	■ Pliohippus group	■ Dinohippus ■ Merychippus ■ Hypohippus ■ Archeohippus	■ Anchitherium	
Oligocene 33.9 Ma		■ Parahippus ■ Miohippus ■ Mesohippus		
Eocene 55.8 Ma — Forest		■ Epihippus ■ Orohippus ■ Hyracotherium		

You, as an instructor, facilitate your student's learning. Well-designed visuals used in class can help you in that effort. Here are six methods for using the visuals in *Visualizing Earth History* in classroom instruction:

1. **Assign students to study visuals in addition to reading the text.**
   It is important to make sure your students know that the visuals are just as essential as the text.

2. **Use visuals during class discussions or presentations.**
   By pointing out important information as the students look at the visuals during class discussions, you can help focus students' attention on key elements of the visuals and help them begin to organize the information and develop an integrated model of understanding.

3. **Use visuals to review content knowledge.**
   Students can review key concepts, principles, processes, vocabulary, and relationships displayed visually. Better understanding results when new information in working memory is linked to prior knowledge.

4. **Use visuals for assignments or when assessing learning.**
   Visuals can be used for comprehension activities or assessments. For example, students could be asked to identify examples of concepts portrayed in visuals. Visuals can be very useful for drawing inferences, for predicting, and for problem solving.

5. **Use visuals to situate learning in authentic contexts.**
   Learning is made more meaningful when a learner can apply facts, concepts, and principles to realistic situations or examples. Visuals can provide that realistic context.

6. **Use visuals to encourage collaboration.**
   Collaborative groups are often required to practice interactive processes. These interactive, face-to-face processes provide the information needed to build a verbal mental model. Learners also benefit from collaboration in many instances, such as when making decisions or solving problems.

Visualizing Earth History not only aids student learning with extraordinary use of visuals but also offers an array of remarkable photos from the National Geographic Society collections. National Geographic has also performed an invaluable service in fact-checking *Visualizing Earth History:* it has verified every fact in the book with two outside sources, ensuring the accuracy and currency of the text.

Given all its strengths and resources, *Visualizing Earth History* will immerse your students in the discipline—its main concepts and applications— while also instilling an appreciation for and excitement about the subject area.

Additional information on learning tools and instructional design is provided electronically, including an *Instructor's Manual* that provides guidelines and suggestions on using the text and visuals most effectively. Other supplementary materials include the Test Bank, with visuals used in assessment; PowerPoints; the Image Gallery, to provide you with the same visuals used in the text; and web-based learning materials for homework and assessment, including images, video, and media resources from National Geographic.

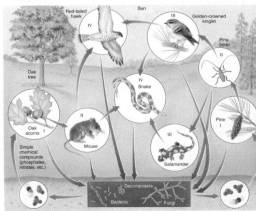

PREFACE

The goal of *Visualizing Earth History* is to introduce students to the excitement of discovering Earth's past, understanding its present, and predicting its future, using an Earth systems approach combined with an innovative program of visuals. Applying the visual learning method is the hallmark of the Wiley Visualizing series, and through it, students will learn not only how to decipher the influence of physical, chemical, and biologic processes on Earth's development but also why knowing about Earth's fascinating history is relevant in our modern world. *Visualizing Earth History* has been developed around the theme of visual learning, with spectacular photographs and pedagogically useful line art providing a seamless extension of the textual material. The illustrations, many of them from the archives of the National Geographic Society, transport students to points of interest around the globe, helping them experience these wonders through photographs that are the next best thing to being there and helping them explore concepts by stepping through the discovery process or examining phenomena from different scales, from different vantage points, or through different observational tools. The book builds on two great organizing concepts, plate tectonics and biologic evolution, which unite many disparate observations into coherent patterns.

Visualizing Earth History is organized into two major sections to maximize its value in teaching. Chapters 1 through 7 provide the principles and organizing concepts used in interpreting Earth history. These chapters are richly illustrated with examples of the types of questions geologists, paleontologists, evolutionary biolo gists, and other scientists ask and the ways that we go about seeking answers to those questions. Chapters 8 through 12 lay out a concise trip through geologic time, placing major events in their correct time context, integrating the principles and organizing concepts developed earlier in the text, and emphasizing the interactions among elements of the Earth systems through geologic time.

ILLUSTRATED BOOK TOUR

A number of pedagogical features using visuals have been developed specifically for *Visualizing Earth History*. Presenting the highly varied and often technical concepts woven throughout Earth history raises challenges for reader and instructor alike. The **Illustrated Book Tour** on the following pages provides a guide to the diverse features contributing to *Visualizing Earth History*'s pedagogical plan.

CHAPTER INTRODUCTIONS illustrate certain concepts in the chapter, using narratives featured alongside striking photographs. The chapter openers also include illustrated **CHAPTER OUTLINES** that use thumbnails of illustrations from the chapter to refer visually to the content.

VISUALIZING features are specially designed multipart visual spreads that focus on a key concept or topic in the chapter, exploring it in detail or in broader context, using a combination of photos and figures.

PROCESS DIAGRAMS present a series of figures or a combination of figures and photos that describe and depict a complex process, helping students to observe, follow, and understand the process.

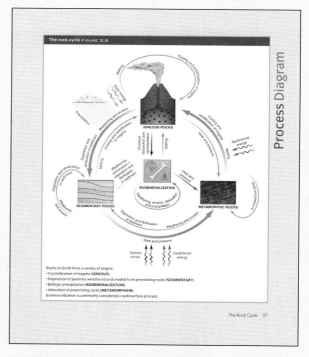

BOOK TOUR

WHAT A GEOLOGIST SEES features highlight a concept or phenomenon, using photos and figures that would stand out to a professional in the field and helping students to develop observational skills.

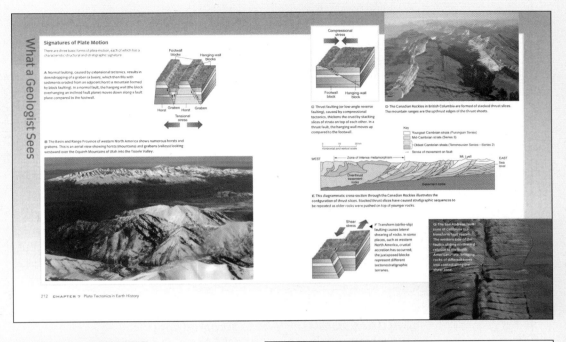

WHAT IS HAPPENING IN THESE PICTURES? is an end-of-chapter feature that presents photographs that are relevant to chapter topics but that illustrate situations students are not likely to have encountered previously. The photographs are paired with questions designed to stimulate creative thinking.

The **AMAZING PLACES** section takes the student to a location in the world that provides a vivid illustration of a theme in the chapter. Students could visit most of the Amazing Places someday and so continue their geologic education after they finish this book.

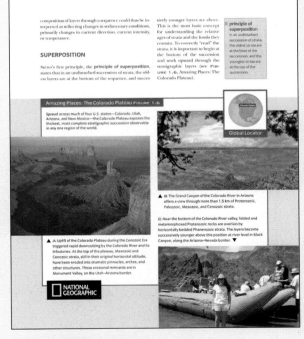

OTHER PEDAGOGICAL FEATURES

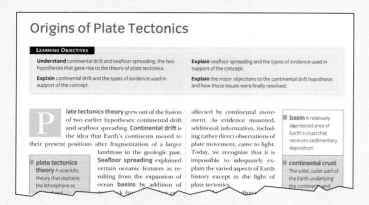

Origins of Plate Tectonics

LEARNING OBJECTIVES

Understand continental drift and seafloor spreading, the two hypotheses that gave rise to the theory of plate tectonics.

Explain continental drift and the types of evidence used in support of the concept.

Explain seafloor spreading and the types of evidence used in support of the concept.

Explain the major objections to the continental drift hypothesis and how those issues were finally resolved.

P late tectonics theory grew out of the fusion of two earlier hypotheses: continental drift and seafloor spreading. **Continental drift** is the idea that Earth's continents moved to their present positions after fragmentation of a larger landmass in the geologic past. **Seafloor spreading** explained certain oceanic features as resulting from the expansion of ocean **basins** by addition of

affected by continental movement. As evidence mounted, additional information, including rather direct observations of plate movement, came to light. Today, we recognize that it is impossible to adequately explain the varied aspects of Earth history except in the light of plate tectonics.

plate tectonics theory A scientific theory that explains the lithosphere as

basin A relatively depressed area of Earth's crust that receives sedimentary deposition.

continental crust The solid, outer part of the Earth underlying the continents and

LEARNING OBJECTIVES at the beginning of each section indicate what the student must be able to do to demonstrate mastery of the material in the chapter.

preservation to occur, the impact of predators, scavengers, and microbial decay must be minimized. Anoxia, salinity stress, desiccation, and rapid burial are some of the conditions under which the activity of biodegraders can be suspended. Exceptional preservation often involves precipitating fine mineral coatings over organic tissues, and the process apparently takes place within a few weeks of death. In other words, exceptionally preserved bodies become lithified long before the sediments do. Some of the microbes involved in decay can play a role in the precipitation of minerals that coat the nonmineralized tissues (Figure 4.17D–F on page 111).

CONCEPT CHECK STOP

What is a fossil?

What are the three major categories of fossils?

What kinds of information about past life do body fossils, trace fossils, and biomarkers provide?

How are body fossils preserved?

What are the types of preservation of body fossils?

What conditions favor fossilization?

What is exceptional fossil preservation, and how does it differ from normal fossil preservation?

CONCEPT CHECK questions at the end of each section give students the opportunity to test their comprehension of the learning objectives.

GLOBAL LOCATOR MAPS, prepared specifically for this book by The National Geographic Society, accompany figures that address matters encountered in a particular geographic region. These maps help students visualize where the areas discussed are situated.

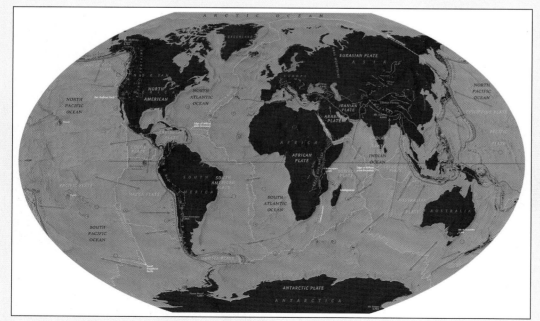

NATIONAL GEOGRAPHIC MAPS appear in various chapters of the book.

crust The outermost layer of Earth, defined by density, composition, and seismic velocity differences from the underlying mantle.

asthenosphere The layer within the upper mantle and below the lithosphere where rocks are relatively ductile and easily deformed.

MARGINAL GLOSSARY TERMS (in green boldface) introduce each chapter's most important terms. The second most important terms appear in **black boldface** and are defined in the text.

ILLUSTRATIONS AND PHOTOS support concepts covered in the text, elaborate on relevant issues, and add visual detail. Many of the photos originate from National Geographic's rich sources.

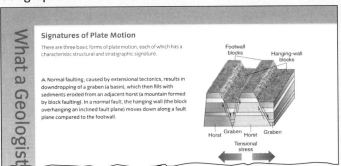

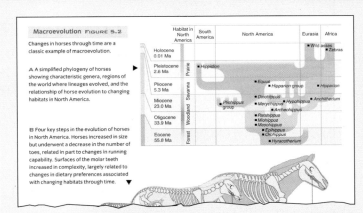

TABLES AND GRAPHS, with data sources cited at the end of the text, summarize and organize important information.

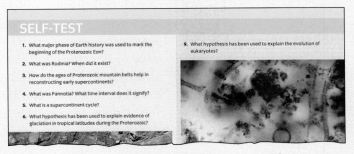

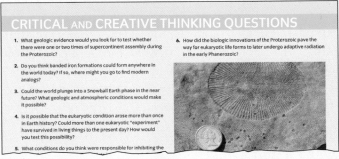

CRITICAL AND CREATIVE THINKING QUESTIONS encourage critical, incisive thinking and highlight each chapter's important concepts and applications.

A **SELF-TEST** at the end of each chapter provides a series of questions, many of which incorporate visuals, that review major concepts from the chapter. Answers to the Self-Tests are available on the John Wiley & Sons website.

The **CHAPTER SUMMARY** revisits each learning objective and redefines marginal glossary terms. In addition, a list of **KEY TERMS** points to the use of each term in the text. Students are thus able to study vocabulary words in the context of related concepts. Each portion of the Chapter Summary is illustrated with a relevant photo from its respective chapter section.

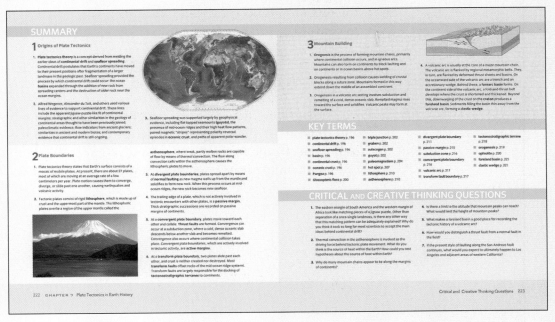

MEDIA AND SUPPLEMENTS

Visualizing Earth History is accompanied by media and supplements that incorporate the visuals from the textbook to form a pedagogically cohesive package. For example, a Process Diagram from the book appears in the Instructor's Manual, with suggestions on using it as a PowerPoint in the classroom; it may be the subject of a short video; and it may also appear with questions in the Test Bank as part of the chapter review.

INSTRUCTOR SUPPLEMENTS

VIDEOS

A rich collection of videos, some from the award-winning National Geographic Film Collection, have been selected by Arthur Lee of Roane State Community College to accompany and enrich the text. Each chapter includes at least one video clip, available online as digitized streaming video, that illustrate and expand on a concept or topic to aid student understanding. Accompanying each of the videos is contextualized commentary and questions that can further develop student understanding. The videos are available at **www.wiley.com/college/Babcock**.

POWERPOINT PRESENTATIONS

A complete set of highly visual PowerPoint presentations by Guillermo Rocha of Brooklyn College is available online to enhance classroom presentations. Tailored to the text's topical coverage and learning objectives, these presentations are designed to convey key text concepts, illustrated by embedded text art.

IMAGE GALLERY

All photographs, figures, maps, and other visuals from the text are available online, and you can use them as you wish in the classroom. You can easily incorporate these online electronic files into your PowerPoint presentations as you choose, and you can use them to create your own overhead transparencies and handouts.

TEST BANK (AVAILABLE IN ELECTRONIC FORMAT)

The visuals from the textbook are also included in the Test Bank, by Donald Thieme of Valdosta State University. The Test Bank contains approximately 1200 test items, at least 25% of which incorporate visuals from the book. The test items include multiple-choice and essay questions that test a variety of comprehension levels. The Test Bank is distributed as Microsoft Word files and as a Computerized Test Bank in both Mac and Windows formats via the secure Instructor's Resources section of the Book Companion Site (www.wiley.com/college/babcock). The easy-to-use test-generation program fully supports graphics, printed tests, student answer sheets, and answer keys. The software's advanced features allow you to create an exam to your exact specifications.

INSTRUCTOR'S MANUAL (AVAILABLE IN ELECTRONIC FORMAT)

The Instructor's Manual begins with the special introduction *Using Visuals in the Classroom*, prepared by Matthew Leavitt of the Arizona State University, that provides guidelines and suggestions on how to use the visuals in teaching the course. Each chapter includes suggestions and directions for using web-based learning modules in the classroom and for homework assignments, as well as creative ideas for in-class activities.

BOOK COMPANION SITE (WWW.WILEY.COM/COLLEGE/BABCOCK)

Instructor Resources on the book companion site include the Test Bank, Instructor's Manual, all illustrations and photos in the textbook in jpeg format, and selected Geo-discoveries animations and videos for use in classroom presentation. Answers to all the self-test questions in the book are found at the book companion site.

The book companion site also includes an extended glossary, listing all the terms defined in the book, and over 300 additional terms students may encounter as they study.

ACKNOWLEDGMENTS

PROFESSIONAL FEEDBACK

Throughout the process of writing and developing this text and the visual pedagogy, I benefited from the comments and constructive criticism provided by the instructors and colleagues listed here. I offer my sincere appreciation to these individuals for their helpful reviews:

Mead Allison
Tulane University

William Bartels
Albion College

Robert Benson
Adams State College

Brian Bodenbender
Hope College

Danita Brandt
Michigan State University

Danny Burns
University of Hawaii

Erik Burtis
Northern Virginia Community College—Woodridge

James Carew
College of Charleston

Eric Carson
San Jacinto College

Robert Cicerone
Bridgewater State College

Mitchell Colgan
College of Charleston

Cynthia Coron
Southern Connecticut State University

Chris Dewey
Mississippi State University

Frank Florence
Jefferson Community College

Jeffrey Hanor
Louisiana State University—Baton Rouge

Stephen Hasiotis
University of Kansas

John Haynes
Kent State University

Daniel Hembree
Ohio University

David Johnson
New Mexico Institute of Mining and Technology

Amanda Julson
Blinn College

Lawrence Krissek
The Ohio State University

Tim Kroeger
Bemidji State University

Mark Kulp
University of New Orleans

Arthur Lee
University of Nebraska

Lindsey Leighton
San Diego State University

Steve LoDuca
Eastern Michigan University

Steve Lundblad
Coastal Georgia Community College

Steven Mojzsis
University of Colorado—Bolder

Bruce Monger
Cornell University

Anton Olenik
Florida Atlantic University

Mark Ouimette
Hardin-Simmons University

Sarah Powell
University of Alaska—Fairbanks

Hermann Pfefferkorn
University of Pennsylvania

Ignacio Pujana
University of Texas—Dallas

Kenneth Rasmussen
Northern Virginia Community College—Annandale

Beth Rinard
Tarleton State University

Steven Schimmrich
SUNY Ulster County Community College

Todd Schneider
Morgan Community College

Alycia Stigall
Ohio University

Leif Tapanila
Idaho State University—Pocatello

Donald Theime
Georgia Perimeter College

Sarah Tindall
Kutztown University

Heyo Van Iten
Hanover College

Mari Vice
University of Wisconsin—Platteville

David Watkins
University of Nebraska

SPECIAL THANKS

I am extremely grateful to the many members of the editorial and production staff at John Wiley & Sons who guided me through the challenging steps of developing this book. Their tireless enthusiasm, professional assistance, and endless patience smoothed the path to publication. I thank, in particular, Anne Smith, vice-president of John Wiley & Sons, who expertly launched this process; Ryan Flahive, who guided this process through its many stages; Helen McInnis, managing director, Wiley Visualizing, who oversaw the concept of the book; and Jeffrey Rucker, executive marketing manager for Wiley Visualizing, who superbly represents the Wiley Visualizing imprint. I am deeply thankful for the skilled editing and processing of the many versions of the text provided by Nancy Perry and Charity Robey, development editors. Christine Moore also provided considerable assistance in manuscript development. Hilary Newman and Teri Stratford worked tirelessly to obtain photos for the text. Stacy Gold, research editor and account executive at the National Geographic Image Collection, was of invaluable assistance in selecting NGS photos. Micheline Frederick, production manager, and Sujin Hong, production editor, oversaw the scheduling of segments in the production process, and Jeanine Furino of GGS Book Services PMG supervised final layout of the text and artwork.

Many other individuals at National Geographic offered their expertise and assistance in developing this book: Francis Downey, vice president and publisher, and Richard Easby, supervising editor, National Geographic School Division; Mimi Dornack, Sales Manager, and Lori Franklin, assistant account executive, National Geographic Image Collection; Dierdre Bevington-Attardi, project manager, and Kevin Allen, director of map services, National Geographic Maps. I appreciate their contributions and support.

PERSONAL ACKNOWLEDGMENTS

The writing of *Visualizing Earth History* has been a wonderful experience for me. With the creative assistance of all the amazing people at John Wiley and National Geographic, I have developed a different, more engaging text that provides not only the essential background necessary for understanding our planet's past and present but also the relevance of that information for predicting our planet's future. The text boldly reveals a sense of the excitement that accompanies modern scientific study of Earth systems. This outlook and the impressive new vision expressed in this book is in no small way attributable to the remarkable talents of the editors, photo researchers, graphics artists, layout team, fact checkers, and many others.

It's impossible to name all the colleagues and friends who have assisted or inspired me and in so doing helped make this book a reality. Many who provided illustrations are thanked elsewhere in this book. Others, over a span of many years, have led or accompanied me to field localities or provided ideas, support, encouragement, or inspiration of diverse kinds. Among these many superb people I'd especially like to acknowledge Soo-Yeun Ahn, William I. Ausich, Gordon C. Baird, Alyssa M. Bancroft, Stig M.Bergström, Philip S. Borkow, Carlton E. Brett, Derek E. G. Briggs, James W. Collinson, Simon Conway Morris, Alan H. Coogan, Donald J. Crowley, Jerry F. Downhower, David H. Elliot, Adam M. English, Mats E. Eriksson, Rodney M. Feldmann, Dale M. Gnidovec, Lloyd and Val Gunther, Laura W. Hellstrom, J. Stewart and Mary Hollingsworth, John L. Isbell, Shelley A. Judge, the late Roger L. Kaesler, Sadie A. Kingsbury, Stephen A. Leslie, Jih-Pai Lin, Scott C. McKenzie, Margaret L. Matsui, James F. Miller, Molly F. Miller, Allison R. Palmer, Shanchi Peng, Margaret N. Rees, Richard A. Robison, James M. St. John, John S. Peel, Matthew R. Saltzman, Sharon K. Schafer, Ethan S. Skinner, Alycia L. Stigall, Walter C. Sweet, Kate E. Tierney, Gregory J. Wasserman, Marilyn D. Wegweiser, Wentang Zhang, and Maoyan Zhu. Margaret T. Wilson and our daughter Kathleen were of immense support through the writing of this book. My parents, sisters, and brother provided a lifetime of support and encouragement. Preya Nandwani provided the spark in my life that I needed as this project was drawing to a close, and without her, this work might never have come to fruition.

ABOUT THE AUTHOR

Loren Babcock earned his Ph.D. degree in Geology from the University of Kansas in 1990, and is presently Professor of Earth Sciences at The Ohio State University. He has published more than 125 articles in the refereed scientific literature, mostly on early and mid-Paleozoic time, stratigraphy, and fossils; on the processes of fossilization; and on topics in evolutionary paleobiology such as the role of predation in evolution, and the evolutionary history of biologic asymmetry. In 2001, he was awarded the Charles Schuchert Award for Excellence and Promise in Paleontology from The Paleontological Society. He is a Fellow of the Geological Society of America and The Paleontological Society, and currently serves as Secretary of the International Subcommission on Cambrian Stratigraphy (part of the International Commission on Stratigraphy).

CONTENTS *in Brief*

CONTENTS

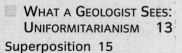

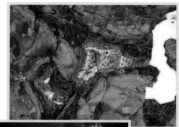

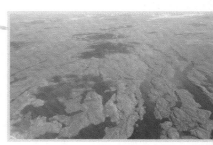

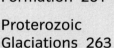
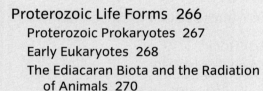

VISUALIZING FEATURES

Multi-part visual presentations that focus on a key concept or topic in the chapter

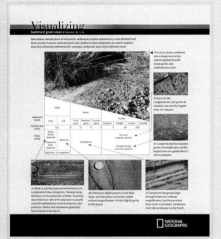

PROCESS DIAGRAMS

A series or combination of figures and photos that describe and depict a complex process

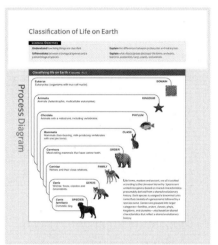

Take Your Students on a Journey of Discovery

SPECIAL ☐ EDITION

NATIONAL GEOGRAPHIC

EarthPulse

WILEY

THE ESSENTIAL VISUAL REPORT ON GLOBAL TRENDS

Exclusively from National Geographic Society and John Wiley & Sons

www.wiley.com/college/earthpulse

VISUALIZING
EARTH HISTORY

Introduction to Earth System History

From outer space, Earth appears as a surprisingly fragile bluish, mostly water-covered sphere blanketed by a thin, delicate, gaseous layer. Beneath this veneer, though, lies the record of an amazingly dynamic history recorded in rocks you walk over daily.

As you read on, you will learn how to interpret Earth's history. You will go on a remarkable journey—a trip through Earth's 4.5 billion-year development, during which you will encounter evidence of this planet's often violent growing pains. You will witness the evidence for dramatic changes in this planet's land and sea relationships, mountain ranges, river systems, glaciers, and life forms. As you journey through Earth's history, you will begin to appreciate the interaction among the processes—physical, chemical, and biologic—that work to shape Earth's surface.

You will learn techniques and tools used to interpret Earth's development and will discover the critical role of the natural sciences in modern civilization. Many things you depend on every day have a deep antiquity, and geologists play a central role in the exploration and production of their raw materials. Gasoline, electricity, plastics, cosmetics, steel, glass, concrete, fertilizer, and food additives are all derivative of such raw Earth materials as oil, coal, asphalt, iron, silica, limestone, phosphate, and water.

As we try to predict what the future holds for our planet, our best source of guidance is the long-term record of change locked in Earth's stratigraphic layers.

What Is Earth History?

Geology is the science of Earth. It is the study of all aspects of the planet, including its composition, its structure, the physical and chemical processes that act upon it, its origin, its life forms, and its history.

The insight that geologists develop comes from an array of techniques, data sources, and instrumentation. Understanding how Earth systems operate requires assimilation and integration of information from diverse fields of science. Geology, therefore, is really an interdisciplinary science. To understand the processes responsible for shaping Earth through time, geologists study Earth in many ways (**FIGURE 1.1**). They may study strata in a roadside outcrop or a remote mountain range, they may examine individual crystals visible microscopically in a rock, they may analyze the chemical composition of a fossil or backtrack through Earth's changing temperature regimes by studying cores drilled through glacial ice, and they may study movement along fractures in Earth from seismographs anchored to bedrock or from satellites orbiting the planet. Increasingly over recent decades, geologists have also studied other planets and moons in the Solar System to draw parallels with, and to interpret the context of, the evolution of our planet. Geologists work to understand Earth's past in order to, among other things, recover and responsibly utilize the planet's natural resources,

geology The science of Earth, including its composition, its structure, its origin, its life forms, the physical and chemical processes that affect it, and its history.

Geologists at work FIGURE 1.1

Geologists study Earth's history in many ways. Their interpretations are applied to the recovery of natural resources and predictions about Earth's future.

A A geologist studying the inclination of strata in a limestone quarry hopes to find clues to the tectonic history of western Brazil.

B Geologists core a frozen lake in Antarctica for a climate record.

understand what conditions have led to current conditions on the planet's surface and avoid or minimize the effects of natural disasters by separating human civilization from areas historically subject to such problems, and preserve key habitats for Earth's life forms.

Geologists are concerned with figuring out both the long-term history of evolutionary processes in animals, plants, and other life forms and the evolutionary relationships of those organisms. Information about biologic relationships has medical and agricultural applications, among other uses. In addition to deciphering Earth's past conditions and studying its present conditions, scientists want to successfully predict the future of this planet. To understand how perturbations in Earth systems are likely to change the conditions on Earth in the future, our best guide is to study Earth's ancient history as recorded in rocks, sediments, glacial ice, or other repositories of information.

■ **Earth history**
Study of the origin and development of Earth, including its life forms, through time. Also known as historical geology.

The science of geology is commonly divided into two broad but overlapping sub-disciplines referred to as **Earth dynamics** (or **physical geology**) and **Earth history** (or **historical geology**). Earth dynamics comprises the geologic processes operating on and within the Earth (including physical and chemical weathering, sedimentation, tectonic plate movement, and other processes) and the study of Earth materials and features resulting from those processes. Earth materials include rocks, minerals, sediments, water, and ice, and geologic features include mountains, valleys, streams, lakes, oceans, glaciers, faults, and many others. Earth history includes study of the origin of the planet and its physical, chemical, and biologic changes through time. Interpretation of our planet's history depends on knowledge of both geologic and biologic materials, as well as on processes and reactions affecting them. Some key concepts of Earth dynamics and biology are reviewed in this book to help you understand the context of Earth's evolution.

C A diver examines living corals and sponges forming carbonate reef rocks in the Mediterranean Sea off the coast of France.

D An oil rig in Libya produces petroleum needed for fuel, lubricants, and chemical products such as plastics, textiles, and cosmetics.

SCIENTIFIC THEORIES

Geology, like the other natural sciences, is based on facts or observations about nature. It is not a **belief system**, which is what constitutes the basis of a religion. Key characteristics of science are predictability and testability. This means that science is a self-correcting process. If ideas are demonstrated to be incorrect, they are abandoned, and new ideas, ones that square better with our observations about natural systems, replace them. In contrast, infallibility is the cornerstone of a belief system.

All scientific hypotheses, or ideas used to explain how or why something happened, are subject to testing. If a hypothesis fails to explain an outcome properly, it is abandoned or modified, and a new idea is formulated. If a hypothesis passes testing, it is provisionally accepted. This is an iterative process, and future discoveries can result in modification of scientific ideas over time because of the self-correcting aspect of **scientific methodology**. In physics and chemistry, a hypothesis can usually be tested using the outcome of an experiment. In other words, the hypothesis has a predictive component to it. Unlike physics and chemistry, however, geology is, for the most part, a historical science. Geologists look backward into Earth's deep time. When postulating tests of hypotheses, geologists must employ a technique known as **retrodiction**, which involves "prediction" in a backward direction in time. The outcome is known, so scientists must test alternative models to determine what circumstances led to the outcome. Scientists working in other historical sciences, such as biology, anthropology, and astronomy, also commonly use retrodiction in their hypothesis testing.

A **scientific theory** is a unifying idea that incorporates a number of provisionally accepted hypotheses, explains a large number of facts, has withstood much testing, and is regarded as tantamount to fact. In contrast to a scientific theory is a "theory" in the colloquial sense. The popular meaning of **theory** is speculative, an idea in need of testing, and one that is not necessarily widely accepted by scientists. This makes a popular theory similar to a scientific hypothesis. A scientific theory should not be dismissed simply because "it is just a theory." This phraseology is misleading, and it trivializes the great importance placed by scientists in the central concepts of science by confusing scientific theories with speculative popular theories.

Two important scientific theories are central to an understanding of Earth history: plate tectonics and biologic evolution. **Evolutionary theory** (FIGURE 1.2) is the great unifying idea of biology. It is the process by which biologic species give rise to other species by way of genetic changes and natural selection. **Plate tectonics theory** (FIGURE 1.3 pages 8–9) is the great unifying idea of geology. It states that Earth's outer shell (the **lithosphere**), which consists of the **crust** and upper mantle, is cracked and composed of pieces that "float" on a hot, deformable **asthenosphere**. The pieces move in various directions and may slowly spread apart, collide, or slip past one another. Both scientific theories provide a sound, consistent framework within which to interpret the past and present, and both ideas are well tested and supported by an

---

scientific methodology
A general term for a scientific investigation involving an iterative process of empirical observation, hypothesis building (with a predictive or retrodictive component), and testing.

scientific theory
A scientific concept that is tantamount to fact.

evolutionary theory The scientific theory that explains processes by which biologic species give rise to other species, principally by way of genetic changes and natural selection.

plate tectonics theory The scientific theory that Earth's outer shell, or lithosphere, is composed of pieces that interact with each other as they "float" on a hot, deformable asthenosphere.

lithosphere The outer, relatively rigid layer of Earth, approximately 100 km thick, overlying the asthenosphere. It includes the entire crust plus the upper mantle.

crust The outermost layer of Earth, defined by density, composition, and seismic velocity differences from the underlying mantle.

asthenosphere The layer within the upper mantle and below the lithosphere where rocks are relatively ductile and easily deformed.

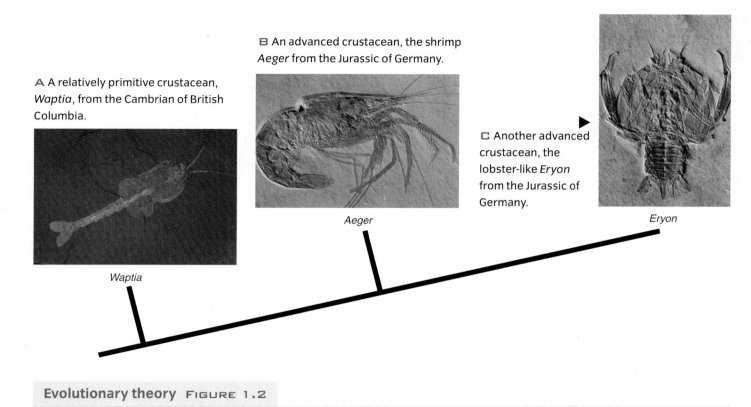

A A relatively primitive crustacean, *Waptia*, from the Cambrian of British Columbia.

B An advanced crustacean, the shrimp *Aeger* from the Jurassic of Germany.

C Another advanced crustacean, the lobster-like *Eryon* from the Jurassic of Germany.

Waptia

Aeger

Eryon

Evolutionary theory FIGURE 1.2

Biologic species give rise to other species through genetic changes and natural selection. Over the last 500 million years or so, crustaceans have diversified into perhaps several million species.

overwhelming amount of evidence. Because there exists no significant, verifiable evidence from the natural world that is contrary to either scientific theory, both are widely accepted by scientists. It is true that scientists argue about some details, such as the positions of certain tectonic plates at various points in the geologic past, or the rate at which species evolve, but disagreements over particulars will not invalidate an entire scientific theory unless it can be shown that the central tenets lack support from scientific tests. In popular culture, supposed challenges to evolution or other scientific theories are usually based on appeal to supernatural causes for natural phenomena rather than on shortcomings of the theories themselves. It is important to understand that the natural sciences do not rely on supernatural explanations for what occurs in nature, and they do not assume infallibility of ideas.

CONCEPT CHECK STOP

What is Earth history?

What is science, and what distinguishes it from a belief system?

How does a scientific theory differ from a theory in the popular sense?

How would it be possible to correctly interpret Earth history without an understanding of plate tectonics and biologic evolution?

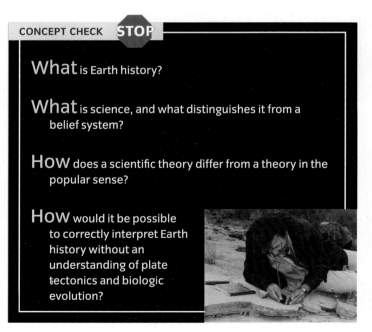

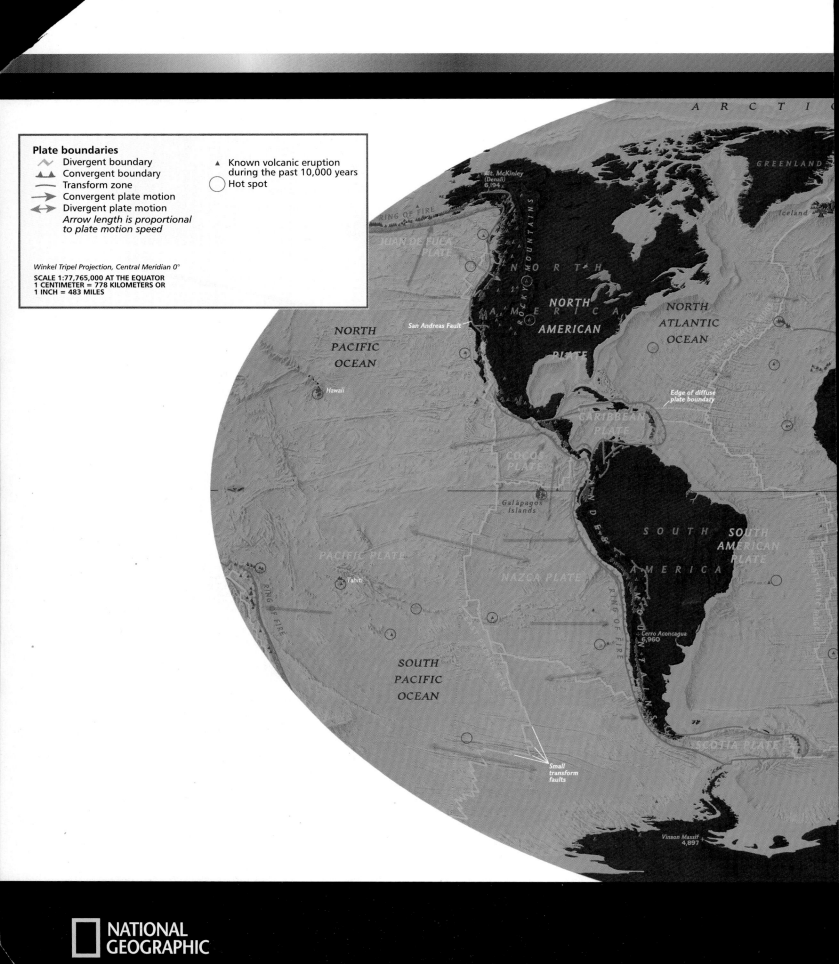

Plate boundaries
~ Divergent boundary
▲▲ Convergent boundary
— Transform zone
→ Convergent plate motion
↔ Divergent plate motion
*Arrow length is proportional
to plate motion speed*

▲ Known volcanic eruption
   during the past 10,000 years
◯ Hot spot

Winkel Tripel Projection, Central Meridian 0°
SCALE 1:77,765,000 AT THE EQUATOR
1 CENTIMETER = 778 KILOMETERS OR
1 INCH = 483 MILES

ARCTIC

GREENLAND

Iceland

Mt. McKinley
(Denali)
6,194

RING OF FIRE

JUAN DE FUCA
PLATE

ROCKY MOUNTAINS

NORTH
AMERICA

NORTH
AMERICAN

NORTH
ATLANTIC
OCEAN

San Andreas Fault

NORTH
PACIFIC
OCEAN

Hawaii

Edge of diffuse
plate boundary

CARIBBEAN
PLATE

COCOS
PLATE

Galápagos
Islands

SOUTH   SOUTH
AMERICAN
PLATE

AMERICA

NAZCA PLATE

ANDES

RING OF FIRE

PACIFIC PLATE

Tahiti

Cerro Aconcagua
6,960

RING OF FIRE

SOUTH
PACIFIC
OCEAN

MOUNTAINS

Small
transform
faults

SCOTIA PLATE

Vinson Massif
4,897

The Earth's surface is a mosaic of lithospheric plates whose movement is controlled by convection of hot material in the underlying asthenosphere.

OCEAN

EURASIAN PLATE

ASIA

URAL MOUNTAINS

Location uncertain

RING OF FIRE

EUROPE

ALPS

El'brus
5,642

Caspian Sea

IRANIAN
PLATE

Tibetan Plateau

ARABIAN
PLATE

HIMALAYA

Mt. Everest
8,850

NORTH
PACIFIC
OCEAN

PHILIPPINE PLATE

PACIFIC

Red Sea

AFRICA

INDIAN
PLATE

PLATE

RING OF FIRE

AFRICAN
PLATE

INDIAN
OCEAN

Kilimanjaro
5,895

GREAT RIFT VALLEY

SOMALI
PLATE

Edge of diffuse
plate boundary

CAPRICORN
PLATE

Madagascar

AUSTRALIAN

RING OF FIRE

TH
TIC
N

Location uncertain

PLATE

AUSTRALIA

Mt. Kosciuszko
2,228

ANTARCTIC PLATE

NTARCTICA

Mt. Erebus
3,794

Overview of Earth Systems

Earth and its inhabitants can be thought of as parts of a complexly intertwined set of systems collectively referred to as the **Earth system**. The systems approach to studying Earth is a way of breaking down large, complex problems into smaller components while remaining mindful of the connections between the components. Systems that make up

Earth system The sum of the physical, chemical, and biological processes operating on and within Earth.

the Earth system are the **geosphere**, the **biosphere**, the **atmosphere**, and the **hydrosphere** (FIGURE 1.4A). A change in one component commonly has effects in other components (FIGURE 1.4B). For instance, extraction of oil from Earth's strata, followed by refinement of the oil into gasoline and burning of that gasoline in an automobile engine, adds carbon dioxide, water vapor, and other gaseous

The Earth system FIGURE 1.4

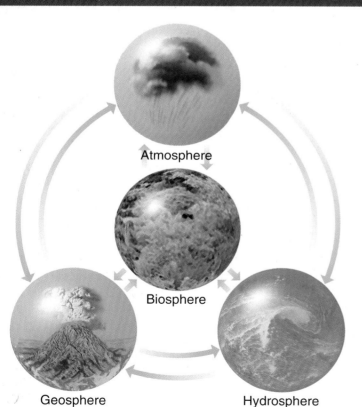

Atmosphere

Biosphere

Geosphere

Hydrosphere

A The Earth system has four principal components—the geosphere, hydrosphere, atmosphere, and biosphere—all of which interact and undergo change through time.

Global Locator

Mt. Fuji, Japan

B The geosphere, biosphere, atmosphere, and hydrosphere interact to create and modify Earth's landforms. Mt. Fuji, an active volcano, is being formed by magma rising from remelting of the subducting Pacific Plate. After the magma makes its way to the surface and hardens, the igneous rock is then modified by other Earth processes.

The water cycle FIGURE 1.5

The water cycle describes the passage of water from the atmosphere to Earth's surface, through the ground, and through living organisms.

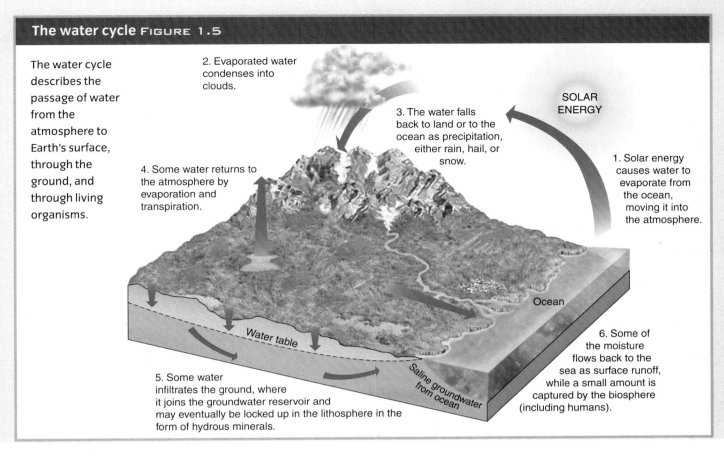

2. Evaporated water condenses into clouds.

3. The water falls back to land or to the ocean as precipitation, either rain, hail, or snow.

SOLAR ENERGY

1. Solar energy causes water to evaporate from the ocean, moving it into the atmosphere.

4. Some water returns to the atmosphere by evaporation and transpiration.

Water table

Ocean

Saline groundwater from ocean

5. Some water infiltrates the ground, where it joins the groundwater reservoir and may eventually be locked up in the lithosphere in the form of hydrous minerals.

6. Some of the moisture flows back to the sea as surface runoff, while a small amount is captured by the biosphere (including humans).

chemicals to the atmosphere. Addition of those gases limits heat loss from Earth's thin atmosphere and promotes an increase in temperature. Global temperature rise, in turn, tends to disrupt weather patterns and causes water locked in glacial ice to melt, forcing sea level to rise. Changing temperature regimes alter the distribution of animals and plants living on land and in the oceans.

Earth is a planet that differs markedly from others in the Solar System. Three important processes have substantially modified Earth's surface over time: (1) tectonics, which involves the movement of lithospheric plates across the planet's surface and which is responsible for the recycling of rocks; (2) the action of water in various forms (liquid, ice, and water vapor), which plays a major role in weathering and erosion, and in the formation of rocks; and (3) biologic processes. Living organisms are an underappreciated, but profound, source of change on Earth through time, and their effects involve surface and subsurface areas, water systems, and the atmosphere. Among other important consequences of biotic processes are the breakdown of rocks at and below the surface, precipitation

of minerals that form new rocks, production of oxygen, production of carbon compounds, and expulsion of gases that are involved in climate change. We are well aware that humans have dramatically altered this planet's surface, its water systems, and its atmosphere over recent millennia, but other organisms—even bacteria, plants, and fungi—have been working to modify Earth's surface, water, and atmosphere for billions of years. Although tectonic processes and water may have existed elsewhere in the Solar System, and the possibility of life elsewhere cannot be ruled out at present, the combination of its extensive tectonic, aqueous, glacial, and biologic influences sets Earth's long-term history apart from the histories of all other planets and moons orbiting our Sun.

Key elements of the Earth system are the cycles of essential ingredients for life and some of the forces of change: water, carbon, oxygen, and other chemicals. The water cycle (FIGURE 1.5) is used to describe the endless exchange of water among the atmosphere, oceans, lakes, and streams; through living organisms; and through the ground. Water precipitated from the atmosphere will fall to Earth's surface, become part of the surface water (oceans, lakes, or streams), or infiltrate the ground. As part of the surface or near-surface reservoir, water may be cycled through animals, plants, or other organisms and eventually be returned to the atmosphere through transpiration in plants or respiration in animals. Alternatively, it may be returned to surface water reservoirs or to the groundwater.

In addition to water, carbon, oxygen, and other nutrients are cycled through atmospheric, biologic, and geologic systems. Carbon and oxygen from the atmosphere may be used by living organisms. Animals use oxygen for respiration, and plants and some bacteria use carbon dioxide for photosynthesis. Carbon and oxygen can cycle back to the atmosphere, pass from one organism to another as food, become incorporated in the skeletons of organisms, become buried in sediment, or return to the atmosphere. Carbon can become buried in sediment such as peat or organic-rich shale, and oxygen can combine with other elements such as iron and likewise become buried. Eventually, the sediments or derivative rocks may be exposed to weathering and erosion, and the elements may be released to continue cycling through the Earth systems.

Earth is not entirely a closed set of systems. Biologic processes depend on both the presence of liquid water from Earth and radiation from the Sun. The surfaces of most planets and moons in the Solar System are modified to some extent by meteorite impacts, space dust, and other sources of particulate matter. Meteorites and other bodies from outer space have played a role in Earth's surface evolution, but apart from occasional perturbations to Earth systems, they have been significantly less important than tectonic and biologic processes. Aside from solar radiation, the combined Earth systems are influenced mostly by processes operating on, close to, or within the planet.

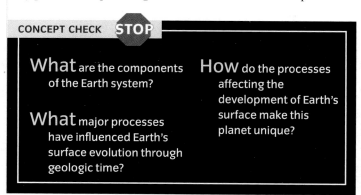

CONCEPT CHECK STOP

What are the components of the Earth system?

What major processes have influenced Earth's surface evolution through geologic time?

How do the processes affecting the development of Earth's surface make this planet unique?

Principles of Earth History

LEARNING OBJECTIVES

Explain Hutton's principle of uniformitarianism.

Describe Steno's three principles of stratigraphy.

Explain Lyell's two principles of stratigraphy.

Understand Smith's principle of biotic succession.

Seven important principles guide our interpretation of Earth history and provide a way of deciphering historical events in their correct relative time sequence: uniformitarianism, superposition, original horizontality, original lateral continuity, cross-cutting relationships, included fragments, and biotic succession.

UNIFORMITARIANISM

Fundamental to the interpretation of Earth processes is the principle of uniformitarianism, a concept that was first articulated by James Hutton (1726–1797), a Scottish physician, farmer, and geologist. **Uniformitarianism** states that processes operating on Earth today also operated in the geologic past. Hutton observed ripples produced by wave action along a shoreline and reasoned that rocks showing similar marks were once unconsolidated sediments rippled by currents along an ancient shoreline (see *What a Geologist Sees*). Hutton's views on the origin of rocks countered the prevailing view of the time—**neptunism**, the idea, attributed

uniformitarianism
The principle that processes acting on Earth today have also operated in the geologic past.

largely to Abraham Gottlob Werner (1750–1817) of Germany, that rocks crystallized out of an early universal ocean. Neptunism, which derives its name from Neptune, the Roman god of the sea, had its philosophic roots in a mythical creation story. Mountains were assumed to be irregularities developed in Earth's **crust** during its creation. Other rocks were thought to have formed as the sea gradually diminished in size and as land appeared.

Hutton argued that Earth was much older than most of his contemporary scholars in Europe thought it was. He also argued that Earth has had a dynamic history: he viewed change on Earth as operating cyclically. Hutton recognized that rocks exposed at the surface break down to form sand, gravel, and soil, and he inferred that rock debris was later incorporated in sedimentary strata. In his view, Earth's surface was constantly being rejuvenated, and the destruction of some rocks was balanced by the formation of new ones. Likewise, mountains rose up only to be eroded away, followed by another rise. Hutton originated the view that granitic rocks and mountains were formed from molten **magma** forcibly intruded upward into the crust because of subterranean heat, a concept that came to be known as **plutonism** (after Pluto, the Roman god of the underworld).

The Swiss naturalist Louis Agassiz (1807–1873) provided an important early illustration of the application of uniformitarianism, one that at the time revolutionized the interpretation of the age of Earth and the origin of part of its sedimentary record. Agassiz observed that glaciers leave behind **drift** (unconsolidated rock debris, or **regolith**) at their fronts and sides. The debris is material carved from bedrock as the glacier passes over it. The

Uniformitarianism

Ripples form in sand when currents of water or wind cause the sand to move (A). By applying the principle of uniformitarianism, geologists infer that ripple marks in ancient sandstone (B) also formed by current action.

A Ripple marks formed by wind on the surface of a dune in Idaho.

B Ripple marks formed by either water or wind in a Cambrian sandstone deposit of South Australia. ▶

What a Geologist Sees

material gets transported downstream through the ice sheet until it is dropped at a terminus, which is called a **moraine**. Large areas of North America and northern Europe are mantled with unconsolidated regolith ranging up to boulder size, sometimes thick and laterally extensive enough to be mined for sand and gravel. By applying uniformitarianism, Agassiz reasoned that the source of these sand and gravel deposits was glacial ice rather than a great flood, as thought by people at the time who were influenced by scriptural explanations for the natural world. Agassiz championed the idea of a "Great Ice Age," a time in the distant past when continental glaciers covered much more land area than they do now. Glaciers must have been responsible for transporting sediment from distant sources before depositing it, and then they ultimately receded to positions close to their present locations.

The concept of uniformitarianism, which is frequently expressed using the phrase "the present is the key to the past," is threaded through this entire book, often implicitly. You will be presented with ancient examples and asked to interpret them based on what you can observe today. One thing you must keep in mind, however, is that the scale of a process that operated in the past may have been different from that at which it operates today. In Agassiz's concept of a Great Ice Age, enormous continental glaciers would have covered vast areas of North America and Europe. Certainly large glaciers cover Antarctica and Greenland today, but we simply can't observe the processes of **continental glaciation** reaching to mid-latitudes as they must have during the Pleistocene Epoch.

Another demonstration of uniformitarianism, but at a scale different from that observable today, is in the numerous meteorite craters that are scattered over Earth's surface. The Chicxulub impact crater off the Yucatán Peninsula, Mexico, records the fall of a meteorite, an asteroid, or a comet that may have weighed as much as 10^{15} kg, ranged from 6 to 16 km in diameter, and left a crater, which is now buried under Cenozoic sediments, that is between 170 and 300 km across. The force of its impact was so great that it may have triggered large earthquakes and tsunamis, and it possibly brought about cataclysmic ecosystem collapse by setting off acid rain conditions and regional wildfires, and by reducing the amount of sunlight reaching Earth's surface because large volumes of dust were ejected into the atmosphere.

Those changes, in turn, may have resulted in mass extinctions of species living at the end of the Cretaceous Period (65.5 million years). All these phenomena—meteorite impacts, earthquakes, tsunamis, acid rain, wildfires, atmospheric dust clouds, and mass extinction of living animals and plants—can be observed occurring today, but not on the same scale or at the same rate as at the end of the Cretaceous Period. Also, humans have yet to observe most of these phenomena associated with meteorite impacts.

In the course of human history, even the largest observed meteorite falls, like the fall of the Nantan meteorite, which occurred in southern China during the Ming Dynasty (in May 1516), have not rivaled the Chicxulub impact when they finally struck Earth's surface. The iron-nickel–rich Nantan meteorite originally weighed more than 9,500 kg, but it apparently exploded in the upper atmosphere, scattering ore-bearing rocks across croplands in a belt 28 km long and 8 km wide.

Uniformitarianism is a tool that provides us with insight not only into the geologic past but also into the future. We can invoke uniformitarianism to help predict what changes the world will probably undergo, given a certain well-defined set of parameters, based on what we can observe occurring presently, and based on what we can infer has happened in the geologic past. In this way, we can use the past and the present as keys to the future. One of the most important scientific challenges of our time is grappling with issues related to global climatic change. The ancient stratigraphic record provides detailed information about ecosystem responses to changing climate regimes, and we can apply that information to help develop predictive models of future climatic change and ecosystem responses.

Stratigraphy, the science of layered rocks, developed from three principles first stated by Niels Stenson (1638–1686), a Danish physician (and later, Catholic priest) who moved to Florence, Italy, for part of his life. Stenson, who latinized his name to Nicolaus Steno, was the first to formally recognize the importance of stratification, or horizontal layering. Stratification of sediments occurs as the result of differences in particle size and density. When carried in a current (water or wind), heavier particles tend to settle first, and lighter ones tend to settle afterward. Changes in particle size cause layers to develop. Steno also recognized the importance of chemical dissolution and the precipitation of minerals from solution. Changes in the composition of layers through a sequence could thus be interpreted as reflecting changes in sedimentary conditions, primarily changes in current direction, current intensity, or temperature.

SUPERPOSITION

Steno's first principle, the **principle of superposition**, states that in an undisturbed succession of strata, the oldest layers are at the bottom of the sequence, and successively younger layers are above. This is the most basic concept for understanding the relative ages of strata and the fossils they contain. To correctly "read" the strata, it is important to begin at the bottom of the succession and work upward through the stratigraphic layers (see FIGURE 1.6, Amazing Places: The Colorado Plateau).

principle of superposition
In an undisturbed succession of strata, the oldest strata are at the base of the succession, and the youngest strata are at the top of the succession.

Amazing Places: The Colorado Plateau FIGURE 1.6

Spread across much of four U.S. states—Colorado, Utah, Arizona, and New Mexico—the Colorado Plateau exposes the thickest, most complete stratigraphic succession observable in any one region of the world.

Global Locator

▲ A Uplift of the Colorado Plateau during the Cenozoic Era triggered rapid downcutting by the Colorado River and its tributaries. At the top of the plateau, Mesozoic and Cenozoic strata, still in their original horizontal attitude, have been eroded into dramatic pinnacles, arches, and other structures. These erosional remnants are in Monument Valley, on the Utah–Arizona border.

▲ B The Grand Canyon of the Colorado River in Arizona offers a view through more than 1.5 km of Proterozoic, Paleozoic, Mesozoic, and Cenozoic strata.

C Near the bottom of the Colorado River valley, folded and metamorphosed Proterozoic rocks are overlain by horizontally bedded Phanerozoic strata. The layers become successively younger above this position at river level in Black Canyon, along the Arizona–Nevada border. ▼

An excellent place to illustrate the principle of superposition is in the Grand Canyon of Arizona (FIGURE 1.7). In FIGURE 1.7B, which is a diagrammatic representation of the lower part of the Grand Canyon succession, the stratigraphic interval encompassing the Cambrian through Carboniferous illustrates the principle of superposition. The younging direction of these strata is toward the top of the canyon.

ORIGINAL HORIZONTALITY

> **principle of original horizontality**
> Sedimentary strata were originally deposited nearly horizontally and parallel to Earth's surface.

Steno's **principle of original horizontality** states that sedimentary layers were deposited nearly horizontally and parallel to Earth's surface (Figure 1.7). Sedimentary particles tend to settle from water and air under the influence of gravity, and they do so in layers that are close to horizontal. Even rippled surfaces, which deviate from horizontal, are typically inclined only a small amount from horizontal. If strata are inclined steeply, some disturbance of the strata must have occurred following initial deposition.

In the Grand Canyon (Figure 1.7B), the Tonto Group (Cambrian) shows a nearly horizontal orientation over the eroded Proterozoic "basement." The Cambrian strata are in their original horizontal position. In contrast, strata of the Grand Canyon Supergroup (Proterozoic) are tilted. These layers were originally deposited horizontally. After deposition, they became tilted during a mountain-building episode.

ORIGINAL LATERAL CONTINUITY

Steno's **principle of original lateral continuity** states that at the time of deposition, strata extended continuously in all directions until they terminated by thinning at the edge of a basin, ended abruptly at a barrier to sedimentation, or graded laterally into a different sediment type. This principle allows us to reconstruct the original distribution of a sedimentary unit whose distributional area has been divided by erosion or faulting.

In the lower Grand Canyon succession (Figure 1.7B), the Cambrian formations continue laterally until reaching their depositional limits. In the case of the Tapeats Sandstone, the depositional limit is reached at the edge of an erosional remnant formed by the underlying Proterozoic succession. This erosional remnant projected out of the Cambrian sea as an island, and sediments of the Tapeats Sandstone filled in the beach and nearshore areas around the island.

The contact between the Tapeats Sandstone and the Bright Angel Shale is interfingering, as indicated by z-shaped "shazam lines" (Figure 1.7B). This gradational zone indicates that part of the Bright Angel Shale was deposited at the same time (but laterally adjacent to) part of the Tapeats Sandstone. The contact between the Bright Angel Shale and the Muav Limestone is also an interfingering relationship.

In 1830, an English geologist, Sir Charles Lyell (1797–1875), published the first edition of what was to become an enormously influential book, *Principles of Geology*, which provided important documentation for a great antiquity of Earth at a time when most scholars argued that Earth was no more than a few thousand years old. Lyell provided interesting illustrations of the concept of uniformitarianism and provided cogent explanations for geologic phenomena that countered the prevailing geologic philosophy of the time—**catastrophism**. Catastrophism was a paradigm that attempted to explain the development of erosional surfaces and the extinction of species by violent, rapid calamities, such as giant floods. Catastrophism was rooted in biblical scriptures, and extinctions were viewed by Georges Cuvier (1769–1832) of France, a leading proponent of catastrophism, as the plan of a supernatural power to bring life progressively closer to perfection.

> **principle of original lateral continuity**
> At the time of deposition, a sedimentary unit extended laterally and continuously in all directions until it thinned out or otherwise reached the limits of its depositional range.

Spectacularly exposed strata of the Grand Canyon in Arizona demonstrate the principles used to determine the relative ages of rock units.

A Overview of Proterozoic to Mesozoic stratigraphy visible at the west end of the Grand Canyon in Arizona. The oldest rocks, folded and metamorphosed Proterozoic strata, are exposed near the level of the Colorado River. Successively younger, horizontally bedded Paleozoic and Mesozoic strata overlie them. The exposed strata can easily be traced by eye along this part of the canyon, but the principle of original lateral continuity implies that these layers were once continuous over even greater distances.

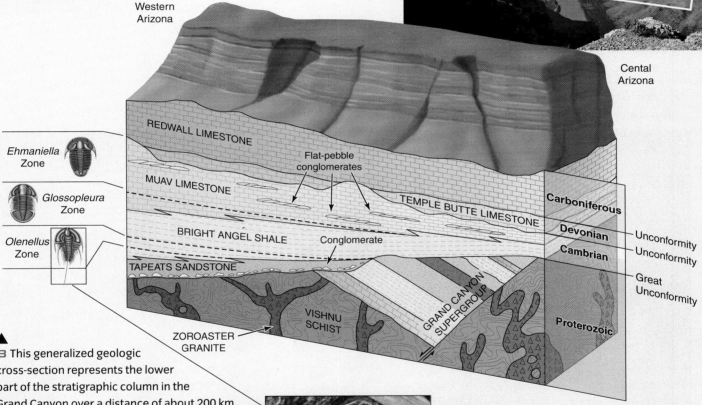

Western Arizona

Cental Arizona

Ehmaniella Zone

Glossopleura Zone

Olenellus Zone

REDWALL LIMESTONE

MUAV LIMESTONE

Flat-pebble conglomerates

TEMPLE BUTTE LIMESTONE

BRIGHT ANGEL SHALE

Conglomerate

TAPEATS SANDSTONE

VISHNU SCHIST

ZOROASTER GRANITE

GRAND CANYON SUPERGROUP

Carboniferous

Devonian — Unconformity

Cambrian — Unconformity

Great Unconformity

Proterozoic

B This generalized geologic cross-section represents the lower part of the stratigraphic column in the Grand Canyon over a distance of about 200 km. These units illustrate the stratigraphic principles of Steno (superposition, original horizontality, and original lateral continuity), Hutton (cross-cutting relationships and included fragments), and Smith (biotic succession). Fossils occur in strata in a regular, predictable sequence. By knowing the correct stacking order of fossils in the stratigraphy of one area, it is possible to determine what the pattern should be elsewhere unless the rocks have been disturbed.

C In the Grand Canyon and elsewhere around the ancestral North American continent, fossils of the common trilobite *Olenellus* define a narrow time interval about midway through the Cambrian Period. In the Grand Canyon, this time interval is represented by the lower part of the Bright Angel Shale. This specimen was collected from rocks of the same age in eastern Pennsylvania.

CROSS-CUTTING RELATIONSHIPS AND INCLUDED FRAGMENTS

Lyell not only provided key evidence for the antiquity of Earth and scientific tests that led to the rejection of catastrophism but also introduced two other general principles of geology. The **principle of cross-cutting relationships**, which Lyell first articulated, states that a rock unit or feature that cuts across a rock unit is younger than the rock that was cut (**FIGURE 1.8**).

During the Proterozoic in what is now the Grand Canyon (Figure 1.7B), lobes of igneous magma intruded (or cross-cut) the pre-existing (older) Vishnu Schist and cooled to form the Zoroaster Granite. A fault cross-cuts the Vishnu Schist, the Zoroaster Granite, and the Grand Canyon Supergroup (all of which are older than the fault), but does not affect the Cambrian layers (which are younger than the fault).

principle of cross-cutting relationships
A rock unit, sediment body, or fault that cuts another geologic unit is younger than the unit that was cut.

A cross-cutting relationship FIGURE 1.8

This outcrop in the Mohave Desert of Nevada shows an igneous dike (brown) of Cenozoic age that has cross-cut slightly folded Cambrian limestone layers (gray). The dike is at least 500 million years younger than the limestone beds.

> **principle of included fragments**
> Fragments of a rock or sediment body contained within another rock or sediment are from a preexisting (older) rock or sediment than the one in which they are contained.

A corollary to the principle of cross-cutting relationships is the **principle of included fragments**, which states that fragments of rock within a larger rock unit are older than the rock in which they are enclosed. Often, when an igneous intrusion cuts across other rocks, fragments of the earlier-formed rocks become incorporated in the intrusion. In Figure 1.7B, small pieces of Vishnu Schist are incorporated in the Zoroaster Granite indicating that the Vishnu Schist is the older rock type.

The principle of included fragments also applies to sedimentary rocks. In the Grand Canyon, conglomerate layers near the base of the Tapeats Sandstone (Cambrian) contain pebbles eroded from the underlying (and older) Proterozoic rocks. In the Muav Limestone (Cambrian), flat limestone pebbles form conglomerate layers in places. The pebbles are pieces of Muav Limestone ripped up from aready-hardened layers while new carbonate sediments were being deposited nearby.

The principles of cross-cutting relationships and included fragments are valuable tools for interpreting the relative order of events in rock successions in another type of circumstance.

Unconformities (erosional or non-depositional surfaces in sedimentary strata) are where truncation, or cross-cutting, of layers has occurred. This is nicely exemplified where a stream carves through sedimentary layers to form a valley and then leaves deposits in that valley. The valley deposits, especially in the sedimentary layer overlying the unconformity, are likely to contain sedimentary particles weathered and eroded from strata below.

In the Grand Canyon example, the contact between the Proterozoic rocks and the Cambrian layers is an erosional unconformity. The Cambrian layers cross-cut the Proterozoic rocks, and small pebbles eroded from Proterozoic rocks have been incorporated in the lower part of the Tapeats Sandstone. Both lines of evidence indicate that the Cambrian strata are younger than the Proterozoic rocks. This unconformity between the Proterozoic and the Cambrian is called the "Great Unconformity." Approximately 300 million years' worth of Proterozoic and Cambrian strata have "disappeared" at the Great Unconformity.

There are two other notable unconformities in the lower part of the Grand Canyon. One unconformity separates Cambrian strata from Devonian strata. The other unconformity separates Devonian strata from Carboniferous strata.

BIOTIC SUCCESSION

The final major principle used to interpret the relative ages of strata was proposed by the English engineer and surveyor William ("Strata") Smith (1769–1839). His concept, the **principle of biotic succession**, states that fossils appear in strata in a definite and determinable order. The order of appearance and disappearance of fossils in strata is non-repeating. As we understand it today (although unknown to Smith), the reason for the non-repeating succession of fossils is the evolution and extinction of species through geologic time. Because biologic evolution is a non-repeating process (that is, an individual species can evolve only once and can go extinct only once), rock layers can be organized in vertical fashion according to the fossils they contain. Those fossils are a proxy of "time's arrow," a unique record of Earth's changing life forms through time.

> **principle of biotic succession** Body fossils occur in strata in a definite, determinable order.

In the Grand Canyon succession (Figure 1.7B), trilobite fossils delineate three biozones through the marine part of the Bright Angel Shale and the Muav Limestone in western Arizona. The bases of the biozones define time lines. The same succession of trilobite zones (the *Olenellus* Zone followed by the *Glossopleura* Zone, followed by the *Ehmaniella* Zone) occurs around the margins of the ancestral North American continent, making it possible to correlate time-equivalent strata great distances (Figure 1.7C).

The principle of biotic succession is the basis for the science of **biostratigraphy**, which is stratigraphy based on the vertical, or geologic-time-based, ranges of fossils.

TESTING WHEN TO APPLY THE PRINCIPLES

Today the principles of uniformitarianism, superposition, original horizontality, original lateral continuity, cross-cutting relationships, included fragments, and biotic succession are routinely used to interpret the order of events in Earth history. However, that history is a complex one, and tectonic processes in particular can disrupt the original configuration of rock units. The principles are applied most easily to undisturbed strata. For this reason, when observing rocks in the field, it is critical to first test whether the strata are in undisturbed condition. So how can you determine whether sedimentary strata have been significantly disturbed?

One way of assessing stratigraphic order is to test using assumptions of one of the principles. For example, you should check to see whether the layers are horizontal or steeply inclined. If they are inclined, they were probably folded, although they may have been simply tilted and left in their correct order. If this is the case, you can apply superposition and other principles to their interpretation. Horizontality of layers does not alone guarantee that strata have been undisturbed because they could have been fully overturned. In that case, they are likely to contain fossils whose ranges are upside-down compared to their occurrences elsewhere. Also, strata juxtaposed at unconformities can be misleading. If included fragments are present, they should be in the layer overlying, rather than underlying, an erosion surface.

Another way to test for the original orientation of stratigraphic units is by using sedimentary features called **geopetal structures** to identify the top or bottom sides of beds (FIGURE 1.9). If an animal steps in mud

Geopetal structures FIGURE 1.9

Geopetal structures are used to determine the original orientation of stratigraphic layers.

A Footprints, like these left by a young woman walking along a Virginia beach, are concave skyward.

B Early hominids walked over a volcanic ash bed 3.6 million years ago, leaving these footprints near Laetoli, Tanzania. The bed is right-side-up, as indicated by the concave-up orientation of the footprints.

and leaves a footprint, that footprint will form a depression on the top surface of the mud layer (Figure 1.9A). After the mud becomes rock, the top of the bed can be identified as the side showing a concave footprint (Figure 1.9B). The bottom of the same bed can be identified by convexity of the footprint. Knowing this, you can determine which direction was stratigraphically up, and when you know that, the principle of superposition can then be applied to interpret the layered sequence. In addition to footprints, there are many other types of geopetal structures, including ripple marks, cross-beds, raindrop impressions left in damp mud, mudcracks caused by the drying of mud into polygons that curl up at their edges, and thinly layered sedimentary mounds called **stromatolites** that grow upward because they are formed by photosynthetic bacteria.

CONCEPT CHECK STOP

What is uniformitarianism, and how is it applied to the interpretation of Earth history?

What are the six major principles used to interpret the relative timing of events in Earth history?

What characteristic of fossils makes them useful for determining the relative ages of strata?

What techniques can be used to determine whether sedimentary strata at Earth's surface are in their correct order, with successively younger strata toward the top of the outcrop?

Ripple marks form upward-pointed crests. In cross-section, they form cross-bedding layers that thin toward the bottom of the sequence. Commonly, the tops of cross-bedded layers are cross-cut by overlying layers. All these patterns are clues to the direction that was stratigraphically up at the time of deposition.

C Ripple marks in Devonian sandstone of Pennsylvania. The top side of the layer is toward the observer.

D Cross-beds formed in an ancient dune field of Utah show that these beds are right-side-up. The cross-beds thin in the downward direction and are truncated by other sandstone layers on their upper surfaces.

SUMMARY

1 What Is Earth History?

1. **Earth history** (or **historical geology**) is the study of Earth's origin and its physical, chemical, and biologic changes through time.

2. A **scientific theory** is a unifying idea that incorporates multiple hypotheses, explains a large number of facts, has successfully withstood testing, and is regarded as the nearest a concept can come to being factual.

3. Correct interpretation of Earth history depends largely on an understanding of two scientific theories: **plate tectonics theory**, the concept that Earth's outer shell is composed of lithospheric pieces that interact with each other as they "float" on a hot, deformable asthenosphere; and **evolutionary theory**, the process by which living species give rise to other species by way of genetic changes and natural selection.

2 Overview of Earth Systems

1. **Earth systems** include all the physical, chemical, and biologic processes that operate on and within Earth. The principal components of the Earth system are the geosphere, biosphere, hydrosphere, and atmosphere.

2. Earth has a history that is unique among bodies in the Solar System because its surface has been modified through extensive tectonic processes, the action of water in its various states, and biologic processes.

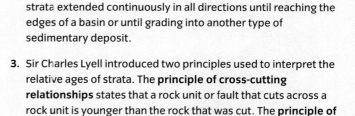

3 Principles of Earth History

1. James Hutton's concept of **uniformitarianism** is the idea that processes operating on Earth today also operated in the geologic past. Uniformitarianism has predictive value because processes operating today (and that operated in the past) will also continue to operate in the future.

2. **Stratigraphy**, the science of layered rocks, developed from three principles first articulated by Nicolaus Steno. The **principle of superposition** states that in an undisturbed succession of strata, the oldest layers are on the bottom of the sequence, and successively younger layers are above. The **principle of original horizontality** states that sedimentary layers were deposited nearly horizontally. The **principle of original lateral continuity** states that at the time of deposition, strata extended continuously in all directions until reaching the edges of a basin or until grading into another type of sedimentary deposit.

3. Sir Charles Lyell introduced two principles used to interpret the relative ages of strata. The **principle of cross-cutting relationships** states that a rock unit or fault that cuts across a rock unit is younger than the rock that was cut. The **principle of included fragments** states that fragments of rock within another rock unit are older than the rock in which they occur.

4. William ("Strata") Smith introduced the **principle of biotic succession**, which states that fossils appear in strata in a definite and determinable order.

KEY TERMS

CRITICAL AND CREATIVE THINKING QUESTIONS

1. Testing and rejecting ideas that do not agree with facts and observations about nature is a hallmark of scientific inquiry. In what ways would it be possible to test the now-rejected concept that Earth's rocks were crystallized out of a universal ocean?

2. How can you use the rock record as a tool for predicting future events, such as where earthquakes or tsunamis are likely to occur? What types of information would you need to collect from the rocks to assist in making predictions? Would you expect the types of clues left by present-day earthquakes and tsunamis to be the same as those recorded in rocks?

3. In this photograph of an unconformity (the thin line across the middle of the photograph), how can you determine whether the layers above the unconformity are younger or older than the ones below it? ▼

4. Are these strata undisturbed? How do you know? ▼

5. From the standpoint of "reading" Earth history, why is it important to determine the direction that was originally stratigraphically up?

Glaciers deposit unconsolidated particles of varying sizes called *drift*. *Till* is a form of unsorted drift often deposited in moraines at the fronts and sides of glaciers. Here, at the edge of the Shackleton Glacier in Antarctica, till accumulates in a moraine at the edge of the ice. Along the ridge, at least two other moraines are evident.

What do these other moraines imply about the thickness of glacial ice in this region in the past?

Global Locator

Shackleton Glacier, Antarctica

Pleistocene till deposit. Here, exposed along a cliff face on the Lake Erie shore of Pennsylvania are two Pleistocene sedimentary units, both of them unconsolidated. The unit below is a gray clay. Above it is a tan layer of drift that contains rocks of variable size.

What principle could you use to interpret the origin of this drift from the northeastern United States?

Applying that principle, what is the most likely origin of this drift?

Is the agent of sedimentary deposition still present? If not, what must have happened to it?

Which of the two stratigraphic units is younger? Assuming that the strata are undisturbed, what two principles allow you to make that determination?

SELF-TEST

1. How can an appreciation of Earth history equip you to more effectively tackle social and environmental issues facing the world today?

2. How does the popular use of the word "theory" differ from the scientific use of the word?

3. What scientific theory explains Earth's surface as a mosaic of moving plates, as depicted in the illustration? ▼

4. What are the components of the Earth system?

5. Which three principles, introduced by Steno, form the basis of stratigraphy?

6. The photograph shows a river valley cut through sedimentary strata. By what principle can you infer that rock layers on each side of the river valley were once continuous?

7. What geologic principle allows you to determine that the strata in the photo were deposited before the river valley existed? ▼

8. What principle allows you to interpret ancient geologic processes and their effects based on how they operate today?

9. What sedimentary features help you determine whether strata are in their original stratigraphic order?

10. When pieces of igneous rock are found in sedimentary layers, what stratigraphic principle can be used to infer that the igneous rocks are older than the sedimentary rocks?

11. What stratigraphic principle allows you to determine the relative ages of rocks based on the fossils they contain?

Earth Materials and Features

The study of Earth history begins with an understanding of rocks and the processes that form and modify them. Rocks are analogous to the books within which human history is recorded. They provide details about how Earth's surface formed and has been transformed.

Volcanic eruptions that carry hot magma to the surface to form new rocks are recorded as igneous rocks. The interaction of new igneous rocks with physical, chemical, and biologic processes and sediments formed by the breakdown and later deposition of those sediments leave their legacy in sediments that later become rock. The sediments accumulate in layers on beaches and in many other places. As more sediment layers form, become buried, and make the transition to solid rock, traces of the chemicals present in water are precipitated around the grains. In addition, marks are left by burrowing animals, and skeletons of organisms are added to the mix. Together, these features supply information about how the rocks were formed, how quickly, and in what chemical, physical, and biologic surroundings. The products of diverse Earth processes wait silently in the rocks until they can be deciphered using the right tools and training.

In this chapter, you will learn about the rocks and minerals that are most useful for interpreting Earth history. You will also review some of the geologic features that help us to understand Earth's history.

NATIONAL GEOGRAPHIC

Rocks and Minerals

Rocks and minerals are the basic building blocks of Earth. They contain key indicators of how Earth has developed through time. They provide testimony to their time of formation, the temperature of their surroundings, how they came to rest in the positions where they are now, and much more.

rock A mixture of minerals.

mineral A naturally occurring crystalline solid or a synthetic, chemically identical equivalent.

crystal A solid composed of atoms and molecules that have a regular internal structure and an external form defined by flat faces.

A **rock** is a mixture of minerals (**FIGURE 2.1**). Most **minerals** are crystalline solids that occur in nature. Most rocks are composed of more than one type of mineral, although a few are made of many crystals of one type. A **crystal** is a solid that has a regular internal structure of atoms and molecules and an outer form defined by symmetrically arranged flat faces. Synthetic (or laboratory-grown) substances that have the same chemical composition and atomic structure as minerals occurring in nature are normally included as minerals. For example, lab-produced sapphire (corundum), which is widely used for lasers and a variety of industrial and other purposes, is a synthetic mineral.

Often, noncrystalline solids (**mineraloids**) such as volcanic glass, opal, and amber are discussed in combination with minerals. Also, organic

A mixture of minerals FIGURE 2.1

A rock is a mixture of minerals. This rock is composed of a variety of minerals discernable by their different colors.

compounds such as **coal** and **petroleum** (oil) are referred to as "mineral resources," despite being noncrystalline.

Many minerals and rocks have inorganic origins, but a substantial part of Earth's rock record originates from the ability of organisms to form minerals as part of their normal growth activity (see *What a Geologist Sees*). Quartz sandstones are produced by the cementation of loose grains of quartz sand, an inorganic process.

Precipitation of Minerals and Mineraloids

Compounds that form Earth materials can develop through inorganic and organic processes.

▲ **A** Volcanic glass, a mineraloid, forms black rims around basalt "pillows" squeezed from a volcanic vent into an ancient Antarctic lake. Hot magma at the margins of the basalt pillows chilled so quickly that silica did not have time to organize into crystals. While the gray-black basalt pillows were cooling more slowly, reddish-brown iron-rich minerals also formed compounds from ions in solution. At the end of the process, white quartz crystallized out of solution in the cavities remaining between basalt masses.

B Many organisms have the ability to secrete minerals and potentially contribute to the sedimentary record. At least four types of organisms in this photograph have formed aragonite or calcite (calcium carbonate) skeletons or needles: the reef-forming corals and sponges, the starfish, and small green algae. The fish has formed bones and teeth of fluorapatite (calcium phosphate).

On the other hand, limestones are composed mostly of aggregates of calcium carbonate shells or tiny needles secreted by biologic organisms and then cemented together. The fossil record contained in layered sediments or rocks is largely a record of minerals secreted organically by biologic organisms. Organic secretion of minerals to form bones, teeth, and shells of animals; to form tiny needles coating the leaves and stems of marine algae; or to form other hard structures, is referred to as **biomineralization**. The rock record, which documents Earth's dynamic past, thus results from the combined effects of inorganic and organic processes.

> **biomineralization**
> Secretion of minerals as bones, teeth, shells, external coverings, or other structures by biologic organisms.

ELEMENTS, IONS, AND ATOMIC BONDS

The basic building blocks of minerals are atoms of chemical elements. Atoms bond to form crystal lattices. To understand minerals and rocks, we must first review a few essentials of the chemical elements from which they form. **Atoms** are the smallest individual particles that show all the distinctive properties of a chemical element (**FIGURE 2.2**). **Elements** are the most fundamental substances into which matter can be separated by normal chemical means. At the center of each atom is the nucleus, which contains most of the mass of the atom because of the presence of the **protons** (positively charged particles) and **neutrons** (neutral particles). Each atom has a specific number of protons, and that number determines the atomic number of an element. For example, the atomic number of carbon is 6, and the atomic number of oxygen is 8. Orbiting the nucleus within specific zones (or shells) are negatively charged particles called **electrons**. Overall, an atom must be electrically neutral. This means that the number of protons in the nucleus must be balanced by an equal number of electrons circling the nucleus. The number of electron shells in an atom varies according to the number of electrons present. As the number of electrons increases, so does the number of shells. The inner shell, close to the nucleus, contains no more than two electrons; the other shells each contain a maximum of eight electrons.

Chemical bonding, in which atoms join to form molecules, mostly results from the interaction of electrons among atoms (**FIGURE 2.3**). In an **ionic bond**, one atom loses an electron from its outer shell to another atom (**FIGURE 2.3A**). To form the mineral **halite** (rock salt), one sodium atom bonds with one chlorine atom. Sodium, with an atomic number of 11, usually has 1 electron in its outer shell. It loses that electron to an atom of chlorine (atomic number 17), which has 7 electrons in its outer shell. By exchanging an electron and filling the outer shells of both atoms, the two atoms form a stable molecule of sodium chloride (NaCl).

Model of an atom FIGURE 2.2

Atoms are the basic components of compounds, including minerals. As shown in this schematic diagram of carbon-12, six electrons orbit the nucleus in complex pathways, or energy-level shells, simplified for illustration as circles. The shells are most stable when they are filled with the full complement of electrons (two in the innermost shell and up to eight in each of the outer shells). Carbon-12 has only four electrons in its outermost shell.

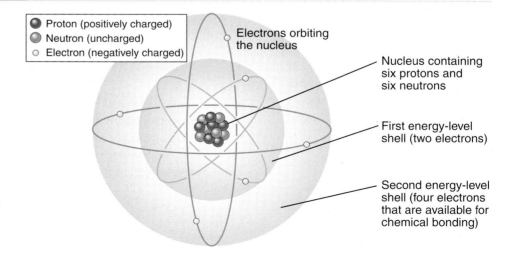

- Proton (positively charged)
- Neutron (uncharged)
- Electron (negatively charged)

Electrons orbiting the nucleus

Nucleus containing six protons and six neutrons

First energy-level shell (two electrons)

Second energy-level shell (four electrons that are available for chemical bonding)

A Ionic Bonds

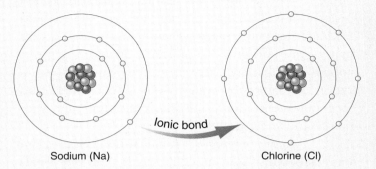

Sodium (Na)   Chlorine (Cl)

In ionic bonding, an electron is lost from one atom and gained by another.

Halite (rock salt), like most minerals held together by relatively weak ionic bonds, weathers (dissolves) easily.

B Covalent Bonds

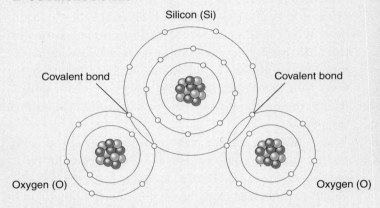

Silicon (Si)

Covalent bond   Covalent bond

Oxygen (O)   Oxygen (O)

In covalent bonding, electrons are shared between atoms.

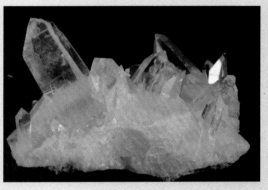

Quartz, like most minerals held together by covalent bonds, is relatively resistant to weathering. As a result, quartz particles often form sand and gravel deposits.

C Van der Waals Bonds

A Van der Waals bond is a weak secondary attraction between electrically neutral molecules, each of which has one positive end and one negative end.

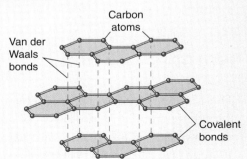

Carbon atoms

Van der Waals bonds

Covalent bonds

Graphite is composed of sheets of carbon. Strong covalent bonds hold individual sheets together, and weak Van der Waals bonds link the sheets. Graphite feels slippery when you rub it with your fingers because the Van der Waals bonds are easily broken.

D Metallic Bonds

Metallic bonds arise from the close packing of metal atoms. Electrons in the outer shells drift from one atom to another.

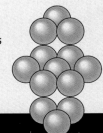

Metallic bonds make metals, such as gold, good conductors of heat and electricity.

The transfer of an electron between a sodium atom and a chlorine atom changes the charges of both atoms. Sodium, which loses an electron, ends up with 11 protons and 10 electrons, so its electrical charge becomes +1. Chlorine, which gains an electron, ends up with 17 protons and 18 electrons, so its electrical charge becomes −1. These charged atoms are called **ions**, and it is the attraction between a positively charged sodium ion and a negatively charged chlorine ion that bonds the molecule together.

Ionic bonds are easily broken through chemical weathering. An ionic bond links a calcium ion (Ca^{+2}) to the complex carbonate ion (CO_3^{-2}) to form the mineral **calcite** ($CaCO_3$), which makes up limestone. Water from rain, surface runoff, and water in the ground can rather quickly break the ionic bond and dissolve calcite. As a result, limestone deposits at and near Earth's surface are often full of cavities (or **karst**) formed through aqueous dissolution.

Some ionically bonded minerals can be quickly precipitated under the right sedimentary conditions. Such minerals as halite do not normally accumulate as loose sedimentary particles; instead, they precipitate in place. Ancient deposits of salt, like those along the coast of the Gulf of Mexico and under the Mediterranean Sea, precipitated quickly through evaporation in hot, arid regions.

Another common form of chemical bond in minerals is a **covalent bond**, in which electrons are shared, rather than exchanged, between atoms (**Figure 2.3B**). To form the mineral **quartz**, one silicon atom bonds with two oxygen atoms. Silicon, with an atomic number of 14, has four electrons in its outer shell, and oxygen, with an atomic number of 8, has six electrons in its outer shell. To reach the stable condition, silicon adds four electrons by sharing two electrons from each of two atoms of oxygen. Together, the silicon and oxygen atoms form a molecule (SiO_2) that has a stable structure.

Covalent bonds are strong. Minerals that have them offer considerable resistance to chemical weathering. Covalently bonded minerals, such as quartz, tend to accumulate in sand and gravel deposits. The minerals weather from the places where they originally crystallized, are eroded, and are subsequently deposited in stream beds and on shorelines. Some deposits of gem-grade, covalently bonded minerals formed through erosion of the minerals from rocks, followed by deposition and concentration in sedimentary deposits. Sapphire- and ruby-bearing gem gravels in Myanmar (formerly called Burma), Sri Lanka (formerly called Ceylon), and Australia represent ancient stream beds in which these minerals were concentrated along with quartz sand and gravel. Important diamond deposits along the west coast of Africa resulted from the concentration of diamonds eroded from igneous rocks and later deposited along the shoreline together with quartz sand and gravel.

Two other types of chemical bonds exist: Van der Waals bonds and metallic bonds. **Van der Waals bonds** are weak attractive forces that occur between electrically neutral molecules that have asymmetrical charge distributions. The positive end of one molecule is attracted to the negative end of another molecule (**Figure 2.3C**). Graphite, which forms the "lead" in pencils, is made of sheets of carbon atoms. Atoms in the individual sheets are held together by strong covalent bonds, but stacks of sheets are held together by weak Van der Waals bonds. When you press a pencil lead to paper, the slippery sheets of graphite slide onto the paper.

Metallic bonds, present in metals such as iron, nickel, copper, gold, and silver, result from the close packing of atoms (**Figure 2.3D**). In the outermost electron shell, electrons are held loosely enough that they can drift from one atom to another. This phenomenon provides metals with their special properties—good conduction of heat and electricity, malleability, and ductility.

ISOTOPES

The number of neutrons in an element may vary even though the number of protons does not change. Carbon may have 6, 7, or 8 neutrons, and oxygen may have 8 or 10 neutrons. The atomic weight of an atom is calculated by adding together the number of protons and neutrons in the nucleus. Each type of an element, as determined by its atomic weight, is called an **isotope** of that element. Carbon has three isotopes, carbon-12, carbon-13, and carbon-14, each distinguished by its atomic weight (**Fig-**

URE 2.4A). Oxygen has two isotopes, oxygen-16 and oxygen-18, which also have different atomic weights.

The isotopes of some elements are used to calculate the timing of events in Earth history. Radioactive, or unstable, isotopes decay to form other isotopes, which are referred to as **daughter products**. Carbon-14 (^{14}C) is a radioactive isotope formed from nitrogen-14 (^{14}N) in the atmosphere (FIGURE 2.4A). Carbon-14, along with other isotopes of carbon (^{12}C and ^{13}C), is incorporated in the tissues of living things. After it is taken up by organisms, ^{14}C undergoes radioactive decay, transforming it back to ^{14}N at a known rate. By calculating the ratio of ^{14}C to ^{14}N in wood or bone, geologists can calculate the amount of time elapsed since the tree or animal was alive. The ^{14}C dating method yields results extending back in time only about 50,000 years, but the ratios of other radioactive isotopes to their stable daughter products (such as uranium-235 to lead-207, potassium-40 to argon-40, and uranium-238 to lead-206) enable geologists to accurately calculate the ages of rocks that were formed millions to billions of years ago.

Isotopes of certain elements help scientists understand changes in seawater chemistry through time.

Isotopes FIGURE 2.4

Isotopes	Number of protons (p) in nucleus	Number of neutrons (n) in nucleus
Carbon-12 (^{12}C)	6	6
Carbon-13 (^{13}C)	6	7
Carbon-14 (^{14}C)	6	8

◀ A Carbon exists in nature in three forms, called isotopes. Carbon-12 (^{12}C), which has six protons and six neutrons in the nucleus, is the most common form. Carbon-13 (^{13}C) has six protons and seven neutrons. The radioactive form, carbon-14 (^{14}C), has six protons and eight neutrons.

B Carbon-14, used to accurately determine the ages of geologic materials, is formed when cosmic rays in the atmosphere bombard nitrogen-14 atoms, causing them to capture a neutron and expel a proton. The nucleus of a carbon-14 atom is radioactive. ▶

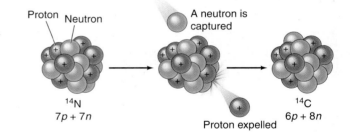

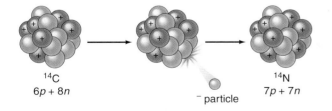

◀ C By beta (β) decay, carbon-14 reverts back to nitrogen-14.

D In carbon-14 dating, the ratio of radioactive carbon-14 is compared to the daughter product nitrogen-14 in organic material to determine how much time has elapsed since the organism was alive and incorporating carbon in its body. It takes about 5,730 years for one-half of the radioactive carbon in a substance to decay. ▶

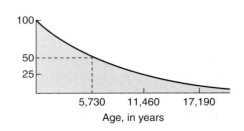

The ratio of oxygen isotopes in the oceans, for example, varies directly with the extent of glaciation and deglaciation globally, and this provides a proxy record of global temperature changes. During cold (or "icehouse") intervals of time, when continental glaciers were expansive, oxygen-16 (^{16}O) was preferentially locked up in glacial ice, and the amount of oxygen-18 (^{18}O) in oceanic sediments increased compared to the amount of ^{16}O. As glaciers melted during warmer (or "greenhouse") intervals, ^{16}O was released to the oceans, and the amount of ^{18}O in oceanic sediments decreased compared to the amount of ^{16}O.

Radioactive decay of isotopes yields heat that helps to drive tectonic processes. The amount of heat generated by each reaction may be small, but the total number of unstable atoms within Earth is enormous. The total amount of heat released through radioactive decay must be considerable, enough to drive convection within the asthenosphere. It is convection that propels the motion of tectonic plates at Earth's surface.

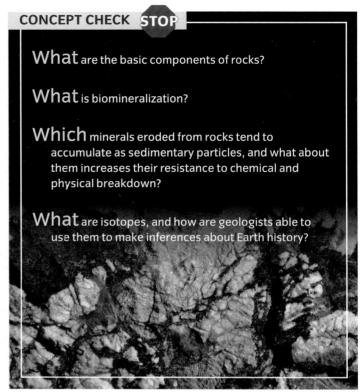

CONCEPT CHECK STOP

What are the basic components of rocks?

What is biomineralization?

Which minerals eroded from rocks tend to accumulate as sedimentary particles, and what about them increases their resistance to chemical and physical breakdown?

What are isotopes, and how are geologists able to use them to make inferences about Earth history?

Common Rock-Forming Minerals

Minerals that make up most rocks are divided into six groups (**FIGURE 2.5**), based on their chemical properties. About 20 rock-forming minerals are of primary importance for interpreting Earth history. The most common minerals in crustal rocks are silicates and carbonates.

Silicate minerals have silicate tetrahedra (SiO_4) as the basic chemical property. They are the dominant mineral group in igneous, sedimentary, and metamorphic rocks. Common silicate minerals are quartz, feldspar, olivine, mica, and clay minerals. Obsidian (volcanic glass) is an amorphous

silicate mineral
A mineral that has a silicate tetrahedron (SiO_4) as the basic chemical property.

carbonate mineral A mineral that contains a carbonate ion (CO_3^-).

(noncrystalline) silicate. Volumetrically, silicate minerals are the most important minerals in sediments that have undergone extensive weathering and erosion. Silicates have a wide variety of uses, including sandpaper, lasers, and current regulators in clocks and watches.

Carbonate minerals have calcium, magnesium, iron, or other ions attached to a carbonate ion (CO_3^-). They are important sedimentary rocks and can form the metamorphic rock marble. Common carbonate minerals are **calcite**, **aragonite**, and **dolomite**. Calcite has, in addition to its occurrences in sedimentary and metamorphic

Rock-forming minerals FIGURE 2.5

The rock-forming minerals are divided into six groups, based on their chemical properties. Silicates include quartz , feldspar, and mica. In the photo of mica, the screwdriver blade is separating muscovite mica into thin sheets along cleavage planes. Carbonates include calcite, aragonite, and dolomite. Gypsum is a sulfate; halite, or rock salt, is a halide; hematite is an oxide and an ore of iron; and pyrite is a sulfide. Other mineral groups, which are not as significant as rock formers, are native elements and phosphates.

SILICATES

| Quartz | Feldspar | Mica |

CARBONATES

| Calcite | Aragonite | Dolomite |

SULFATE   **HALIDE**   **OXIDE**   **SULFIDE**

| Gypsum | Halite | Hematite | Pyrite |

rocks, interesting occurrences in other geologic settings. For example, it often occurs associated with **hydrothermal deposits** and as vein fillings in rocks fractured through tectonic movement. Siderite forms some concretions, which are round nodules in sedimentary rocks that often contain fossils. Carbonate minerals have uses as food additives and antacids, among other things.

Sulfate minerals have calcium or other ions attached to a sulfate ion (SO_4^{-2}). Most rock-forming sulfate minerals, such as **gypsum** and **anhydrite**, occur in sedimentary rocks, particularly in coastal regions that have high evaporation rates. Gypsum and anhydrite are mined for wallboard and other uses.

Halide minerals have positive ions such as sodium and potassium attached to negative ions such as chlorine and bromine. Most rock-forming halides occur in sedimentary rocks, and the most important halide mineral is halite (rock salt). Deserts and other areas of high evaporation are where halide minerals usually precipitate. Halite is mined for table salt, road salt, and other uses.

Oxide minerals have metallic ions combined with oxygen. Oxides occur in igneous, sedimentary, and metamorphic rocks. The most important ones, **hematite** and magnetite, both of which are iron ores, are present in sedimentary rocks. Oxides are commonly mined as ores of metals, and metals have many applications.

Sulfide minerals have metallic ions combined with sulfur. They occur in igneous, sedimentary, and metamorphic rocks. The most important sulfide, **pyrite**, is common in sedimentary rocks. Some sulfides are metal ores.

CONCEPT CHECK STOP

What are the most important rock-forming minerals?

Why are rock-forming minerals important, both in contributing to the rock record and as raw materials of economic value?

The Rock Cycle

LEARNING OBJECTIVES

Summarize the rock cycle.

Define geosphere.

Explain the three main origins of rocks.

Central to an understanding of basic Earth materials (rocks, minerals, and sediments) and the evolution of Earth's **crust** is the concept of a **rock cycle** (FIGURE 2.6). Cycles operate continuously and have neither a beginning nor an end. The rock cycle, which describes the processes by which rocks are formed, decomposed, transported, modified, and formed again, is powered mostly by energy from Earth's internal heat and from the Sun. When rocks are exposed at Earth's surface, they react with the atmosphere, hydrosphere, and biosphere, all of which work to constantly modify Earth (or the

geosphere). The rock cycle is closely related to the tectonic cycle, which drives the movement of plates across Earth's surface, the formation of igneous rocks, and the metamorphism of rocks.

Rocks are grouped according to their origin as igneous, sedimentary (including biologic), or metamorphic. **Igneous rocks** (named from the Latin *ignis*, meaning "fire") begin as molten **magma** that cools and solidifies. Igneous rocks may be exposed at Earth's surface as tectonic processes cause mountain ranges to rise or when lava

rock cycle A conceptual model that describes the origin, alteration, and destruction of rocks through the action of Earth processes.

igneous rock Rock formed from the crystallization of magma.

magma Molten rock, including any suspended crystals (mineral grains) and dissolved gases.

The rock cycle FIGURE 2.6

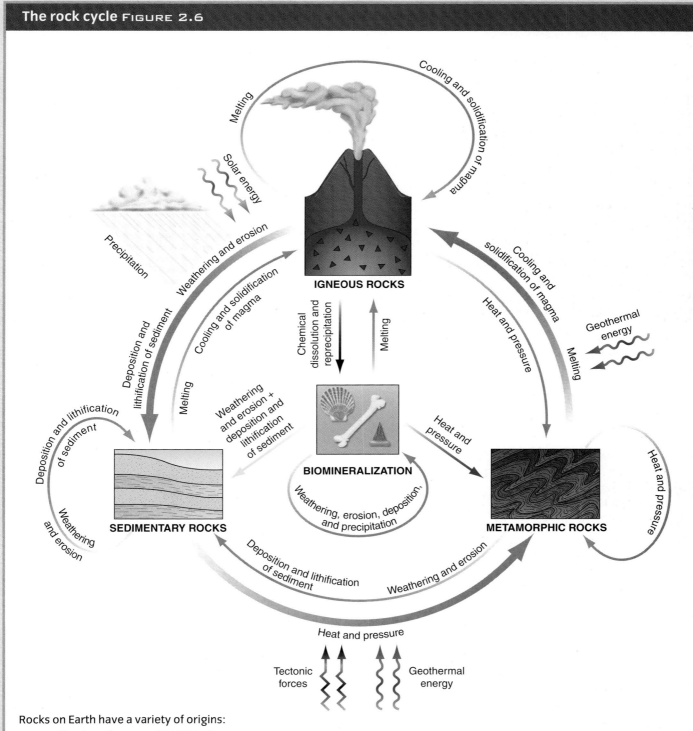

IGNEOUS ROCKS

Melting

Solar energy

Cooling and solidification of magma

Precipitation

Weathering and erosion

Deposition and lithification of sediment

Cooling and solidification of magma

Melting

Chemical dissolution and reprecipitation

Melting

Cooling and solidification of magma

Heat and pressure

Geothermal energy

Melting

Deposition and lithification of sediment

Weathering and erosion + deposition and lithification of sediment

BIOMINERALIZATION

Heat and pressure

SEDIMENTARY ROCKS

Weathering and erosion

Weathering, erosion, deposition, and precipitation

Deposition and lithification of sediment

Weathering and erosion

METAMORPHIC ROCKS

Heat and pressure

Heat and pressure

Tectonic forces

Geothermal energy

Rocks on Earth have a variety of origins:
- Crystallization of magma (**IGNEOUS**),
- Deposition of particles weathered and eroded from preexisting rocks (**SEDIMENTARY**),
- Biologic precipitation (**BIOMINERALIZATION**),
- Alteration of preexisting rocks (**METAMORPHISM**).

Biomineralization is commonly considered a sedimentary process.

Global Locator

Hawaii

Global Locator

South Victoria Land, Antarctica

A Mauna Loa, a volcano in Hawaii, erupts molten magma (lava). At the surface, the magma forms streams of lava that cool to form new basaltic rock. Simultaneous eruptions underwater form large rounded masses called pillow basalts.

B Pillow basalts of Jurassic age exposed in a mountainside in Antarctica. Lighter-colored, layered rocks flanking the pillows are sediments laid down in lake water between volcanic eruptions.

Formation of igneous rock FIGURE 2.7

sediment Unconsolidated particles of rock that have been transported by agents of erosion and unconsolidated particles formed as skeletal material through biomineralization.

sedimentary rock A rock, usually layered, formed from sediments and from minerals precipitated under aqueous conditions.

and volcanic ash are erupted (FIGURE 2.7). Quickly after exposure, these rocks begin weathering, or breaking down, through physical processes (including water and wind action, heating, and freeze-thawing), chemical processes (dissolution), and biologic processes (breakage and biochemical dissolution). Erosion, or transportation, of broken particles (**sediment**) by water, wind, glacial ice, or organisms is the next phase in the cycle. Sediments (FIGURE 2.8) derived from broken pieces of rocks, together with minerals precipitated by living organisms as

skeletal elements or in other ways, settle in layers to form sedimentary deposits that eventually become **sedimentary rock** (named from the Latin *sedimentum*, meaning "to settle"). Minerals precipitated by living organisms add to the sedimentary record through direct precipitation in sedimentary environments (for example, as reefs) or following breakdown of skeletal parts into sedimentary particles. **Metamorphic rocks** (named from the Greek *meta*, meaning "change," and *morphe*, meaning "form") result from the changing of rocks through heat and

metamorphic rock Rock whose original mineralogy or texture has been transformed through any combination of heat, pressure, chemical environment (including hydrothermal fluids), and shearing stress.

Formation of sedimentary rock FIGURE 2.8

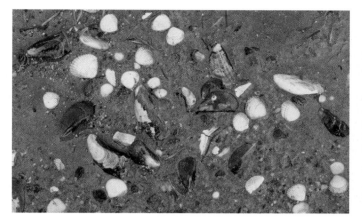

A Sedimentary particles accumulating on a beach ultimately come from two sources: weathering and erosion of pre-existing rocks and precipitation of shells and skeletons by organisms.

B After deposition in layers, compaction, dewatering, and precipitation of mineral cements, loose sediments are turned into solid stacks of layered sedimentary rocks.

Formation of metamorphic rock FIGURE 2.9

A Compression and heat associated with convergence of the Indian subcontinent (at left in the photo) against Asia to form the Himalaya Mountains (at right in the photo) is causing intense metamorphism and deformation of rocks.

B Metamorphism of granite through heating and compression forms a new rock called gneiss. This boulder of gneiss records an early mountain-building episode that contributed to the development of North America.

pressure, which is commonly associated with tectonic activity (FIGURE 2.9). Sediments, sedimentary rocks, and metamorphic rocks may undergo weathering and erosion, just like igneous rocks. Finally, tectonic processes may cause rocks to be heated until they melt to form new magma, and the cycle repeats.

CONCEPT CHECK STOP

What is the rock cycle, and why is the history of rocks explained in terms of a cycle?

What is the geosphere?

What are the three principal origins of rocks?

Types of Rocks

Rocks can be classified in two ways, and each way conveys a different type of information about Earth history. The first way is using a **descriptive classification** system. Normally in a descriptive classification, rocks are organized according to their texture or fabric (size of grains or crystals, their packing, etc.) and their composition (quartz, feldspar, mica, clay, calcite, dolomite, etc.). A rock containing sand grains composed of quartz, each of which is about 1 mm in diameter, could be classified as a quartz sandstone according to a descriptive classification system. A descriptive classification for a rock simply expresses the physical attributes of that rock, without any direct implications of its history or significance.

The second way of classifying a rock is using a **genetic classification** system, or a classification according to its origin. A quartz sandstone (using a descriptive classification term) might be classified as an eolian sandstone (meaning that the sand grains accumulated by wind action) using a genetic classification. The most common use of a genetic classification of rocks is the organization of rocks according to whether they were formed from the crystallization of molten rock (igneous rocks), from the compaction and cementation of unconsolidated sediment grains or biominerals (sedimentary rocks), or

intrusive rock
Igneous rock, usually coarsely crystalline, that resulted from the cooling and solidification of magma within Earth's crust. Also known as plutonic rock.

extrusive rock
Igneous rock, usually finely crystalline, that resulted from the cooling and solidification of magma erupted onto Earth's surface. Also known as volcanic rock.

from the alteration of preexisting rocks (metamorphic rocks).

IGNEOUS ROCKS AND PROCESSES

Igneous rocks, which formed through the cooling and solidification of magma, are classified into two broad groups, based on their place of origin (FIGURE 2.10). **Intrusive** (or **plutonic**) **rocks** are ones in which the magma solidified within Earth's crust, and **extrusive** (or **volcanic**) **rocks** are ones formed from magma that solidified at Earth's surface.

The sizes of crystals (or mineral grains) within an igneous rock provide key evidence of a rock's origin (FIGURE 2.11). Intrusive rocks usually have coarse crystals (**phaneritic** texture) because magma that cools within the crust does so slowly, probably over hundreds to thousands, or perhaps even millions, of years. This is enough time for large crystals to form. The largest ones, 2 mm or more in size, can easily be observed without magnification. **Granite** is a common type of coarsely crystalline, intrusive rock.

Extrusive rocks usually have fine crystals (**aphanitic** texture) because **lava** (magma that solidifies at the surface) cools quickly, commonly within days to months. This is not enough time for minerals to grow

Where igneous rocks form FIGURE 2.10

Igneous rocks crystallize from molten magma. Intrusive (plutonic) igneous rocks form within Earth, and extrusive (volcanic) rocks form at the surface. This diagram shows a variety of igneous plutons (a batholith, stocks, dikes, sills, and a laccolith), as well as volcanic cones. The plutons have intruded preexisting rock, as indicated by cross-cutting relationships and the presence of inclusions (**xenoliths**) in the batholith.

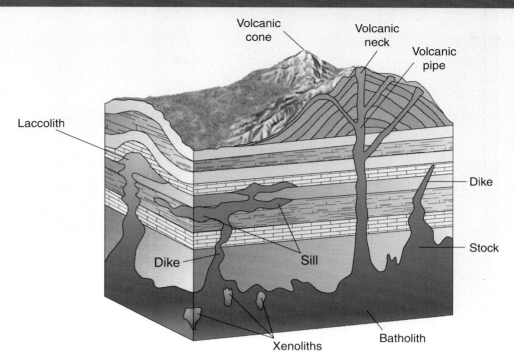

to large size. The largest ones are less than 2 mm in size and can be observed clearly only with magnification. **Basalt** and **rhyolite** are common types of finely crystalline, extrusive rocks. Often rhyolite is formed of **tuff**, which is fine volcanic ash mixed with larger particles called lapilli. Another term that may be used for some extrusive rocks is **pyroclastic**, which refers to rocks formed by the violent ejection of broken rock fragments from a volcano. Some rhyolites are a mix of tuff and pyroclastic debris.

Lava can cool and solidify rapidly at the planet's surface, sometimes so rapidly that it does not have time to organize into crystalline structures. When this happens,

Igneous rocks FIGURE 2.11

A Intrusive rocks usually have large crystals that result from slow cooling rates. They range from felsic (light-colored) granite through intermediate diorite and mafic (dark-colored) gabbro.

B Extrusive rocks usually have small crystals that result from fast cooling rates. They range from felsic (light-colored) rhyolite through intermediate andesite and mafic (dark-colored) basalt.

Granite

Diorite

Gabbro

Rhyolite

Andesite

Basalt

Volcanic glass FIGURE 2.12

Obsidian, or volcanic glass, is an amorphous solid that cools so quickly after eruption that it does not have time to form crystals.

obsidian (volcanic glass) forms (FIGURE 2.12). Obsidian is an amorphous or noncrystalline solid, also known as a mineraloid. Pumice, which consists of a mass of glassy bubbles, is another common type of glassy igneous rock. Finally, volcanic ash also consists largely of glassy fragments that cooled too quickly to crystallize when they were erupted.

Some igneous rocks show a mix of fine and coarse crystals. This textural type is referred to as **porphyritic**. Large crystals within a porphyritic rock are **phenocrysts**, and the finely crystalline material that encloses the phenocrysts is called groundmass. The history of a porphyritic igneous rock involves two phases: First, magma cooled slowly within the crust, allowing large crystals to grow; then the partly solidified magma moved quickly to Earth's surface, where it cooled rapidly, ensuring that the next crystals to form would be tiny.

A single magma will yield the same assemblage of minerals, regardless of whether it solidifies intrusively or extrusively. Only the textures of the resulting igneous rocks will differ according to the different environments where cooling occurred. The common igneous rocks consist mostly of six minerals or mineral groups: **quartz**, **feldspar**, **mica**, **amphibole**, **pyroxene**, and **olivine**. The relative percentages of these minerals, together with crystal size, help determine the name of the rock. Thus, the name of an igneous rock usually reflects not only its mineral composition but also its geologic history.

Light-colored (or **felsic**) igneous rocks are dominated by quartz, feldspar, and **muscovite** (potassium-rich) mica. Felsic rocks include granite and rhyolite. These granitic-type rocks form much of the continental crust.

At the other end of the spectrum, dark-colored (**mafic** and **ultramafic**) igneous rocks are dominated by **biotite** (a **ferromagnesian**, or iron- and magnesium-bearing) mica, amphibole, and pyroxene. Mafic rocks include **basalt** and **gabbro**. Basalt is the dominant component of oceanic crust and some oceanic islands, such as the Hawaiian Islands (**Figure 2.13**), which have formed over a magma plume originating in the mantle. Gabbro is essentially the intrusive equivalent of basalt. **Ultramafic** rocks, such as **peridotite**, are mostly composed of ferromagnesian minerals, especially **olivine** and pyroxene. Unlike mafic rocks, ultramafic rocks contain little or no feldspar. Gabbro and peridotite occur in both oceanic and continental crust. Their source magmas are generated in the mantle, and they usually make their way to the surface via igneous "pipes," or cracks in the rock. See *Amazing Places*, page 44 and 45.

Rocks that have mineral compositions midway between those of felsic and mafic rocks are called **intermediate**. They include **andesite** and **diorite** (or **diabase**), both of which occur in continental crust.

All intrusive igneous rock bodies, regardless of their shape or size, are called **plutons** (Figure 2.10). The word *pluton* is derived from Pluto, the Greek god of the underworld. Plutons include the complex of magma chambers and channels that feed volcanic vents. The largest pluton, a **batholith**, is an irregular body hundreds of square kilometers in diameter that cuts across pre-existing rocks. Intrusive bodies of irregular shape and smaller than 100 km² are **stocks**. Many batholiths are the solidified magma chambers of ancient volcanoes. They are usually composed of granite, diorite, or related coarsely crystalline felsic or intermediate rocks. Typically, they are composite masses resulting from solidification of

Basalt Figure 2.13

The Hawaiian Islands are broad volcanic cones composed of basalt. The gentle slopes of Mauna Kea (in the distance) were formed by fluid lava. In the foreground, solidified lava on one of the slopes of Mauna Loa illustrates a ropy texture.

Amazing Places: The Source of Diamonds FIGURE 2.14

Diamonds, treasured for millennia, were originally thought to be formed in rivers because they were first collected from alluvial deposits in India and Brazil. By 1859, diamonds had been found in river deposits of South Africa.

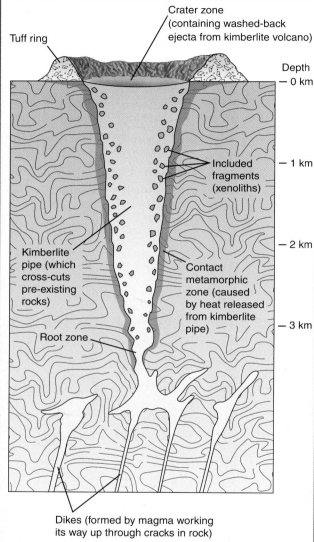

Crater zone (containing washed-back ejecta from kimberlite volcano)

Tuff ring

Depth
— 0 km

Included fragments (xenoliths)

— 1 km

Kimberlite pipe (which cross-cuts pre-existing rocks)

Contact metamorphic zone (caused by heat released from kimberlite pipe)

— 2 km

Root zone

— 3 km

Dikes (formed by magma working its way up through cracks in rock)

A In 1871, diamonds were discovered in an ultramafic igneous rock, peridotite, near Kimberley, South Africa. The rock, eventually called kimberlite, showed that the primary source of diamonds was in igneous rocks, not rivers.

Kimberley, South Africa

Global Locator

B Kimberlite makes its way to the surface by rapidly traveling through narrow igneous "pipes." The pipes have a conical shape as a result of being scored by a swirling mix of magma, crystals, rock fragments, and expanding gases during an eruption. The size and shape of an igneous pipe can be estimated from the hole in the photo on the right. This the "Big Hole," was left after mining at Kimberley, South Africa, from 1871 to 1914.

C Weathering and erosion cause the kimberlite pipes to release the crystals they contain. Covalently bonded diamonds, which are resistant to weathering and have a high relative weight, become concentrated in stream gravels, (sedimentary deposits). This photo shows diamond miners who work for a large company washing stream gravel in Angola. This Cenozoic river deposit illustrates the principle of included fragments: the diamonds, which crystallized in the Proterozoic Eon, are much older than the gravel in which they are found.

D These are rough diamonds collected from gravel deposits.

E After cutting and polishing, a diamond looks much different than it did in the rough.

The Sierra Nevada range, which includes the rocks forming Half Dome and other peaks of Yosemite National Park, California, is composed largely of eroded granitic batholiths.

separate magmatic intrusions. Batholiths of enormous size extend down the western side of North America and include the rocks that form the Sierra Nevada range (FIGURE 2.15) and the Coast Ranges.

Plutons sometimes contain zones of unusually large crystals referred to as **pegmatites**. Pegmatites are the last-cooled portions of igneous intrusive bodies, places where crystals could grow several centimeters or more in size (FIGURE 2.16). They are often enriched in highly desirable minerals that incorporate such rare elements as beryllium, boron, and lithium. Beryllium-rich pegmatites are among the primary sources of emerald and aquamarine crystals.

Minor igneous intrusions take on various shapes and can be of felsic, intermediate, or mafic composition. Dikes, sills, and laccoliths usually form when magma is squeezed into fractures in rock or along other lines of weakness. A **dike** is a thin, tabular, sheet-like intrusive

Pegmatite FIGURE 2.16

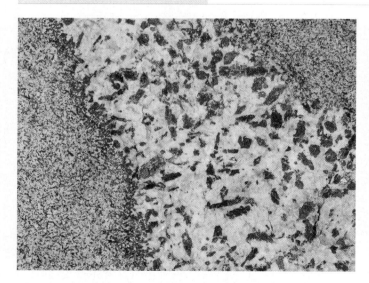

A Pegmatites are the last-crystallized parts of intrusive igneous melts. The largest crystals in a pegmatite exceed 2 cm in size.

B In pegmatites, crystals, like these emeralds (beryl) grow to large size partly because they have plenty of time and partly because the remaining fluids in the magma are often enriched in rare elements such as beryllium.

body that cuts across the layering or fabric of preexisting rocks. Dikes are probably the most familiar form of igneous intrusion because they occur in most places where intrusions have developed. A **sill** is a tabular, sheet-like intrusion that extends parallel to the layering or fabric of the rocks it intrudes. The Palisades, which line the Hudson River opposite New York City, are a magnificent example of an exposed sill. A **laccolith** extends parallel to the layering or fabric of preexisting rocks, but when it was emplaced, it caused the intruded rocks to bend upward into a dome. The Henry Mountains of Utah are domes formed by laccoliths. Another type of minor intrusion, a **volcanic neck**, is a roughly cylindrical structure resulting from the solidification of magma in the natural "pipe" that once fed a volcanic vent. Devil's Tower in South Dakota is a volcanic neck that has been exposed by erosion (**FIGURE 2.17**). It attests to an eruptive igneous history in a region where there are no longer any active volcanic peaks.

SEDIMENTARY ROCKS AND PROCESSES

Most sedimentary rocks originate as unconsolidated particles that undergo **lithification**, or a change to rock. Sedimentary particles derive from three

> **lithification** The processes involved in changing sediments to rock.

A volcanic neck FIGURE 2.17

Devil's Tower in South Dakota is an eroded volcanic neck of Paleogene age.

main sources: (1) fragments produced by the weathering and erosion of preexisting rocks; (2) skeletal debris produced by organisms; and (3) crystals precipitated from water and commonly mediated by the life activities of organisms. Sedimentary rocks are classified in various ways, but one practical technique is to categorize them based on their composition. Using this method, the principal categories of sedimentary rocks are siliciclastic rocks, carbonate rocks, and other rocks.

Siliciclastic sediments, which are the precursors of sedimentary rocks, are particles (or clasts) composed of silicate minerals (notably quartz, mica, and feldspar). Also called detrital sediments, siliciclastic sediments are ultimately derived from the weathering and erosion of rocks at Earth's surface. Particles are carried from their source areas by agents of erosion, mostly water, wind, and ice, but sometimes also animals, and deposited elsewhere. In general, sediments diminish in size as they are carried farther from their sources. They also diminish in size through multiple episodes of reworking by erosive agents. Erosion and sediment reworking (**FIGURE 2.18**) tend to break

Erosion and sediment reworking FIGURE 2.18

Weathering, erosion, and sediment reworking, especially by water, works to break down rocks and the minerals composing them into smaller and smaller pieces. On this cobble beach, wave action plays a significant role in particle weathering and sedimentation.

Descriptive classification of siliciclastic sediments and the sedimentary rocks lithified from them usually involves sediment grain size. Sediment size categories can also be used to describe carbonate sediments (for example, carbonate sand and carbonate mud).

◀ This chart shows sediment size categories and the names applied to both loose grains and sedimentary rocks.

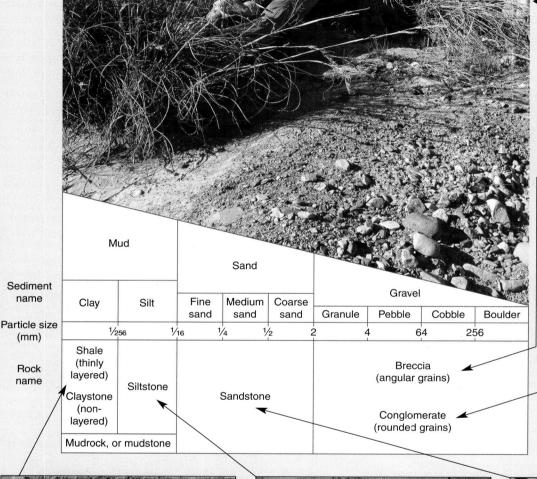

Sediment name	Mud		Sand			Gravel			
	Clay	Silt	Fine sand	Medium sand	Coarse sand	Granule	Pebble	Cobble	Boulder
Particle size (mm)	⅟256	⅟16	¼	½	2	4	64	256	
Rock name	Shale (thinly layered) / Claystone (non-layered)	Siltstone	Sandstone			Breccia (angular grains) / Conglomerate (rounded grains)			
	Mudrock, or mudstone								

E Breccia, like conglomerate, has grains of variable size and the largest ones are angular.

D Conglomerate has rounded grains of variable size, and the largest ones are greater than 2 mm in diameter.

A Shale is a thinly layered sedimentary rock composed of clay-size grains. Parting along the layers is characteristic of shale. If parting does not occur, the term claystone is usually used for sedimentary rocks having clay-size particles. Shales and claystones generally feel smooth to the touch.

B Siltstone is slightly more coarse than shale, and the grains are barely visible without magnification. It feels slightly gritty to the touch.

C Sandstone has grains large enough to be seen without magnification, but they are less than 2 mm in diameter. Sandstone feels like sandpaper to the touch.

NATIONAL GEOGRAPHIC

down minerals that are weakly resistant (like feldspar and mica) at a faster rate than other, more resistant minerals, such as quartz. As a result, quartz and clay minerals—the breakdown products of feldspar and mica—tend to be concentrated in more mature sediments.

Siliciclastic sediments and the sedimentary rocks they later form are classified mostly on the basis of grain size (FIGURE 2.19). The degree of sorting and roundness of grains (FIGURE 2.20) can also be used to help classify sediments composed of larger grains. The principal size categories of sediments are clay, silt, sand, and gravel. **Clay** is composed of sediment grains ranging up to 1/256 mm in size; the grains cannot be seen without high-powered magnification. Once it changes to rock, clay becomes **shale** if it shows fine layering (FIGURE 2.19A) or **claystone** if it lacks fine layering. **Silt** is composed of sediment grains ranging in size from 1/256 mm to 1/16 mm; at the larger end of the size range, the grains are barely visible without magnification. With a 10x hand lens or loupe, silt grains are commonly visible. Silt grains lithify to **siltstone** (FIGURE 2.19B). Clay- and silt-size grains are often grouped together as **mud**. Once lithified, mud becomes **mudrock**, or **mudstone**. **Sand** is composed of sediment grains ranging in size from 1/16 mm to 2 mm. Sand grains are clearly visible without magnification. Lithified sand is called **sandstone** (FIGURE 2.19C). Finally, grains larger than 2 mm are grouped as **gravel**. Gravel, once it is lithified, is called **conglomerate** if the grains are well rounded (FIGURE 2.19D) and **breccia** (FIGURE 2.19E) if the grains are mostly angular in shape.

Carbonate rocks are composed of minerals having carbonate ions, CO_3^{-2}, in their chemical formulas. The principal carbonate minerals are aragonite, calcite, and dolomite. Calcite and aragonite are different types

Sorting and shape of grains FIGURE 2.20

Sorting and shape of sand grains as they would appear in sandstones under a microscope.

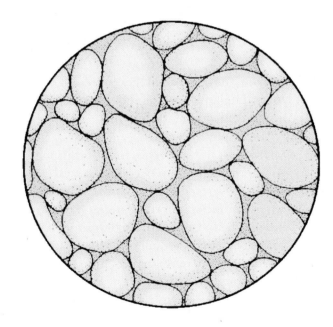

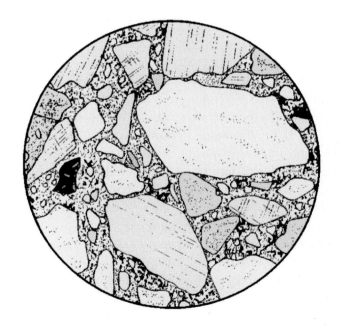

A Sand grains that are well sorted in terms of size. The edges of the particles are also well rounded. This is a sandstone composed almost exclusively of quartz grains held together by cement (pink).

B Sand grains that are poorly sorted in terms of size. The edges of the particles are more angular. This sand is composed of a mixture of minerals, including quartz (tan), feldspar (green), and rock fragments (orange). Spaces between the larger grains are filled with a matrix of clay and silt, which helps bond the rock together.

A Limestone is formed from calcium carbonate. Much of it is secreted by biologic organisms as tiny needles, shells, or other skeletal material.

B Cryptocrystalline quartz, called chert, or flint, is often associated with limestone deposited in warm, shallow, tropical seas. This is a limestone from the Devonian of Ohio that has relatively thin, blocky (and light brown to gray) layers of chert interbedded with it. To form the silica layers, skeletons of silica-secreting organisms such as glass sponges, and perhaps volcanic ash, were dissolved in the warm water and later reprecipitated as a silica gel that became lithified.

of calcium carbonate, and they both form **limestone** (FIGURE 2.21). Dolomite is magnesium-calcium carbonate and is the major constituent of **dolostone**.

Large limestone deposits usually form in warm, relatively shallow tropical or subtropical seas. Most sediment that forms limestone is secreted by living organisms. Photosynthetic algae and bacteria, which live within the photic zone, secrete tiny carbonate needles, and these needles make up most carbonate mud. Larger carbonate grains (sand and larger grain sizes) are mostly made of the broken shells or other skeletons of clams, snails, corals, and coralline sponges. However, sand-size spherical grains called **ooids** are formed by the concentric layering of calcium carbonate over a nucleus by the action of photosynthetic microbes. Ooids are rounded because they are washed back and forth in relatively high-energy, shallow water.

Chalk is a relatively soft type of limestone composed almost exclusively of tiny calcite plates secreted by marine microorganisms called **coccoliths**. The white cliffs of Dover, England, and neighboring areas of western Europe, are mostly made of thick chalk deposits (FIGURE 2.22).

Dolostone is formed in a variety of carbonate environments where magnesium becomes enriched in sedimentary layers. Restricted marine basins located in arid regions, where evaporation of seawater concentrates magnesium, may be locations where dolostone is found. Post-depositional changes to limestones (substitution of magnesium ions for calcium ions) seem to be responsible for forming some dolostones.

Other sedimentary rocks span a wide range of types that have diverse origins. Among them are iron ore deposits, evaporate deposits, and silica deposits.

Sedimentary iron ores are often rich in the iron-bearing minerals hematite or magnetite. Some sedimentary iron ores were formed in shallow marine areas where iron, weathered from surface rocks, was carried to the sea

Chalk FIGURE 2.22

Chalk, a soft form of limestone composed of the skeletons of microscopic Cretaceous coccoliths, makes up the white cliffs along the coast of Denmark.

by streams and then deposited. Water passing through the sediment, especially hydrothermal brines, further enriched the iron content of some sediments. Economically, **banded iron formations** (FIGURE 2.23) are the most important iron ore deposits. Banded iron formations are composed of thin hematite or magnetite layers interbedded with thin **chert** (cryptocrystalline silica) layers.

Evaporite minerals are precipitated when water evaporates. They normally form in arid regions such as restricted tropical seas or in small saline lakes in deserts. The most important evaporate deposits are salt (halite, or rock salt), gypsum, and anhydrite.

Silica deposits, which are usually called chert, or flint, are composed of cryptocrystalline quartz precipitated out of silica-rich water (FIGURE 2.21 B). Water becomes enriched in silica by the dissolution of silica from such sources as silica-secreting organisms and volcanic ash.

> **evaporite** A mineral deposited under evaporative (usually hot, dry) conditions.

Banded iron formation FIGURE 2.23

Banded iron formations are composed of numerous thin layers of iron oxide and silica. They are a major ore of iron.

Two main steps are involved in changing loose, unconsolidated sedimentary particles to solid rock: (1) deposition of sediments in layers followed by (2) lithification (the processes responsible for converting sediments to sedimentary rocks). Deposition of sediments, layer by layer, is a key feature of sedimentary strata—one of the primary distinguishing characteristics of sedimentary rocks. Sedimentary layering, which is also called **bedding**, or **lamination**, is visible at various scales ranging upward from millimeter-scale layering (FIGURE 2.24). In muds, fine layering can be destroyed by animals burrowing through the sediment while it is unconsolidated. However, even when this has

occurred, bedding may still be visible at the bottoms and tops of beds.

Lithification (Figure 2.24) involves compaction of sediments and cementation. When sediments are deposited, they are normally arranged with lots of pore spaces between grains. As more sediment is deposited, the weight of the overlying sediment causes layers below to become compacted, which results in a slight shifting of the grains and reduction of many pore spaces. If water is present in the pore spaces, it is squeezed out in a process called **dewatering**.

Cementation involves the precipitation of minerals out of water. In many cases, this mineral precipitation is related in part to the metabolic activities of microorganisms in the water. As cementation proceeds, thin mineral deposits grow on and between sediment grains, and

Sediments: loose and lithified FIGURE 2.24

Waves washing onto a barrier island along the coast of North Carolina rework unconsolidated sands. At the water's edge, wave action is also responsible for exposing earlier deposits of sand that have been lithified to sandstone. Distinct bedding planes are visible in the sandstone.

those minerals both glue grains together and further reduce the pore spaces between grains. Common cements in sedimentary rocks are calcite, aragonite, silica, and iron oxides such as limonite (or "rust") and hematite. During lithification, other changes, such as recrystallization and chemical alteration, can affect sediments. The collective term for all chemical, physical, and biologic changes that affect sediments between their deposition and their lithification is **diagenesis**.

METAMORPHIC ROCKS AND PROCESSES

Metamorphic rocks form through the alteration of other rocks at high temperatures and pressures. Metamorphism causes chemical (mineralogic) and textural changes in igneous, sedimentary, or other metamorphic rocks. Geologists describe metamorphism in terms of grades (low, intermediate, and high) that reflect temperature-pressure conditions during the time that rocks are altered. Low-grade metamorphism begins between 100°C and 200°C and at about 1000 atm (atmospheres) of pressure. At the other end of the spectrum, high-grade metamorphism usually occurs above 500°C and above 5000 atm of pressure. Metamorphism usually occurs beneath Earth's surface. Even low-grade metamorphism occurs at temperature–pressure conditions far exceeding those normally present at the surface, where temperatures rarely exceed 35°C, and pressures are usually close to 1 atm. Melting of rock can begin above 600°C and above 6000 atm. There is some overlap between the upper range of high-grade metamorphism and the point at which melting of rock to form magma occurs. Typically, rocks containing some water have lower melting temperatures than drier rocks.

A common distinguishing characteristic of metamorphic rocks is **foliation**. Foliation is due to an alignment of crystals that grow perpendicular to the direction of stress applied to the rock during metamorphism (**FIGURE 2.25**). Low-grade metamorphic rocks tend to be finely crystalline (fine grained), and individual crystals usually need to be magnified to be visible. The fo-

Foliated metamorphic rocks FIGURE 2.25

A Slate is formed from low-grade metamorphism of clay mineral-rich shale and sometimes siltstone. Further metamorphism of slate results in phyllite, schist, and sometimes gneiss.

B Phyllite is formed from intermediate-grade metamorphism of shale and siltstone. It has mica crystals that are just barely visible.

C Schist is a mica-rich, high-grade metamorphic rock.

D Gneiss, which has bands of mica and other minerals, is formed from high-grade metamorphism of shale, siltstone, or granite.

liation developed in these finely crystalline rocks, which mostly reflects the alignment of mica crystals and is characteristic of slate, is called **slaty cleavage**. High-grade metamorphic rocks tend to be coarsely crystalline, and individual crystals are readily visible without magnification. Foliation in coarsely crystalline rocks is often wavy or distorted, reflecting the presence and orientation of crystals of quartz, feldspar, and other minerals. This type of foliation is called **schistosity**, and it occurs in schist, gneiss, phyllite, greenschist, amphibolite, and granulite.

Some metamorphic rocks lack foliation and are called **nonfoliated** rocks (FIGURE 2.26). Two rocks metamorphosed from sedimentary rocks (metasedimentary rocks) that commonly lack foliation are **marble**, which is metamorphosed from either limestone or dolostone, and **quartzite**, which is metamorphosed from either sandstone or siltstone. These rocks lack foliation because the precursor sedimentary rocks were of only one mineral composition.

Metamorphic rocks are commonly classified according to their mineral compositions, which reflect their precursor rocks, and metamorphic grade. Metamorphism of shale and mudrock results in **slate**, **phyllite**, and **schist** or **gneiss** (in order of increasing grade of metamorphism). Original bedding planes are often still evident in slate, a low-grade metamorphic rock, but aligned mica crystals causing slaty cleavage are a clear indication that the rock has been changed by heat and pressure. Continued metamorphism of slate results in larger mica crystals and different mineral assemblages. Phyllite has mica crystals that are visible to the unaided eye, and it has a distinct foliation. Schist is a high-grade metamorphic rock that shows a distinct schistosity (foliation) due largely to coarse mica crystals. Gneiss, another high-grade rock, has coarse crystals, schistosity, and bands of mica crystals segregated from bands of quartz, feldspar, or other minerals. Metamorphism of granitic rocks (granite and rhyolite) also results in gneiss. Metamorphism of basaltic rocks (basalt and gabbro) results in greenschist (low-grade metamorphism), which contains the minerals chlorite, epidote, plagioclase feldspar, and calcite; amphibolite (intermediate grade metamorphism), in which amphibole replaces the chlorite; and granulite (high-grade metamorphism), in which pyroxene replaces the amphibole.

Metamorphism occurs in regional metamorphic zones associated with active tectonism, in contact metamorphic zones adjacent to igneous intrusions, in areas of hydrothermal activity, and in areas of deep burial (FIGURE 2.27).

Regional metamorphism occurs across large areas, such as in developing mountain chains or in subduction zones. During continental collision, sedimentary rocks along a continental margin undergo compressional stress that results in foliated slates, phyllites, schists, and gneisses. At subduction zones, regional metamorphism produces blueschist and eclogite under conditions of high pressure but low to moderate temperatures associated with subduction of a cold oceanic slab into the warm asthenosphere.

> **regional metamorphism**
> Metamorphism that affects large areas of the crust.

Contact metamorphism, which is localized in extent, occurs where the heat released from an igneous intrusion bakes the surrounding rock it has intruded. There is usually a strong gradient toward lower metamorphic grade away from the heat source.

In hydrothermal zones, hot fluids pass through rocks. This can occur along mid-ocean

Nonfoliated metamorphic rocks FIGURE 2.26

A Marble is recrystallized from limestone or dolostone. Marble is commonly white in color. Other colors (like the pink color of this piece) and wavy lines, if present, are due to impurities in the rock.

B Quartzite is formed from metamorphism of quartz sandstone or quartz siltstone. Pore spaces between the former grains of sand or silt are filled with silica, and the whole rock is recrystallized.

contact metamorphism
Localized metamorphism associated with the intrusion or extrusion of an igneous magma; heat and hydrothermal fluids may be involved in the metamorphic activity.

ridges where seawater circulates through hot, recently formed crust. In this setting, basalt often alters to greenschist. Hydrothermal alteration can also occur close to igneous intrusions. Many valuable ore occurrences, including gold- and silver-bearing veins, have a hydrothermal origin.

Burial metamorphism occurs when rocks are buried so deeply that high temperatures and pressures alter their chemical compositions. Coal is a low-grade metamorphic rock formed through burial compaction of peat deposits in which plant debris has been turned to nearly pure carbon. Peat, the precursor to coal, is commonly formed in swampy areas where large amounts of organic input (especially leaves and stems of plants and trees) quickly use up all the available oxygen in the water, allowing the organic remains to be preserved without much decomposition.

Where metamorphism occurs FIGURE 2.27

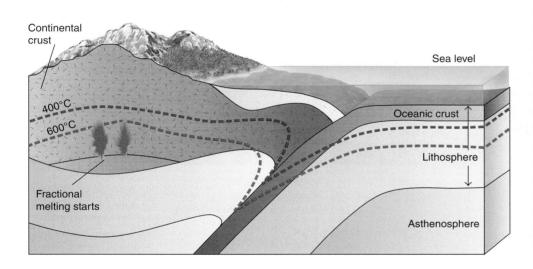

Regional metamorphism is commonly associated with the compressional stresses of mountain building. It also occurs at subduction zones where blueschist and eclogite form. Contact metamorphism occurs where hot granitic magma rises through preexisting rock (country rock) and releases heat to the rocks it intrudes. Burial metamorphism occurs in basins that subside under the great pressure of accumulating sedimentary layers. Hydrothermal metamorphism occurs where hot waters pass through cracks in rocks, and it may be associated with other areas where metamorphism of rock occurs.

CONCEPT CHECK STOP

What information enters into a descriptive classification of rocks?

What information enters into a genetic classification of rocks?

What are the key characteristics of igneous, sedimentary, and metamorphic rocks?

What are the felsic igneous rocks, what minerals are most common to them, and where do they form?

What are the mafic igneous rocks, what minerals are most common to them, and where do they form?

What are the primary size categories of siliciclastic sediment, and what are their sedimentary rock equivalents?

What are the major carbonate rocks, and how do they form?

What are the major metamorphic rocks, and how do they form?

SUMMARY

1 Rocks and Minerals

1. A **rock** is a mixture of minerals. A **mineral** is a crystalline solid that occurs in nature. A **crystal** is a solid that has a regular internal structure of atoms and molecules and an external form defined by symmetrical flat faces.

2. Many minerals and rocks have inorganic origins. However, a substantial part of the sedimentary rock record is composed of **biomineralized** material, which is secreted by living organisms.

3. Minerals are composed of chemical elements bonded together in crystal lattice structures. Elements are the most fundamental substances into which matter can be separated chemically. Minerals held together by **covalent bonds**, in which electrons are shared between atoms, are stronger than minerals held together by **ionic bonds**, in which one atom loses an electron to another atom. Minerals that have stronger bonds tend to be more resistant to weathering than minerals with weaker bonds.

4. Each type of an element, as determined by its atomic weight, is called an **isotope**. Atomic weight is calculated by adding the protons and neutrons in the nucleus of an atom. Radioactive isotopes can release heat that drives tectonic processes, and some can be used to determine the ages of rocks. Some stable isotopes provide information about global climate patterns.

2 Common Rock-Forming Minerals

1. Minerals that form rocks are divided into six groups based on their chemical properties. The most important rock-forming mineral groups are **silicates**, which include quartz, feldspar, mica, and clay minerals; and **carbonates**, which include calcite, aragonite, and dolomite.

3 The Rock Cycle

1. The **rock cycle** describes the origin, alteration, and destruction of rocks through the action of Earth processes.

2. The **geosphere** is the solid part of Earth, including its rocks, minerals, sediments, and Earth features.

3. Rocks are arranged in three groups according to their origin. **Igneous rocks** are formed from the cooling and crystallization of magma. **Sedimentary rocks** are normally deposited in layers and formed from sediments and minerals precipitated from water. **Sediments** are particles of preexisting rock or skeletal remains of organisms. **Metamorphic rocks** have mineralogies or textures that were transformed from their original minerals or textures through heat and pressure.

4 Types of Rocks

1. Rocks can be classified genetically (according to their origin, such as igneous, sedimentary, or metamorphic), or descriptively (according to their texture and composition).

2. Igneous rocks form from the cooling and solidification of **magma**, which is molten rock. Solidification of intrusive igneous rocks occurs within the crust, and solidification of extrusive igneous rocks occurs at Earth's surface. Magma at Earth' surface is called **lava**.

3. Sedimentary rocks form from unconsolidated particles that change to rock through the process of **lithification**, which involves compaction and cementation of sediment grains.

4. A key characteristic of sedimentary rocks is their tendency to form layering, which is called **bedding**, or **lamination**.

5. Sedimentary rocks undergo many chemical, physical, and biologic changes between the time of sediment deposition and the time of lithification. These changes are collectively referred to as **diagenesis**.

6. A key characteristic of many metamorphic rocks is **foliation**, which is caused by an alignment of crystals that grow perpendicular to the direction of stress applied to rock during metamorphism.

KEY TERMS

CRITICAL AND CREATIVE THINKING QUESTIONS

1. Why are sedimentary rocks composed mostly of silicate and carbonate minerals? Of the silicates, why are quartz and clay minerals predominant? What happens to the feldspar and other silicate minerals?

2. Do the terms *shale*, *siltstone*, *sandstone*, and *conglomerate* represent a genetic or descriptive classification system of sedimentary rocks?

3. Large igneous intrusions commonly form kilometers below Earth's surface. How would the batholiths that comprise the Sierra Nevada range of the western United States have made their way to the surface? What happened to the overlying rock?

4. Organisms are major contributors to the sedimentary rock record. Why do you think this is so? What places on Earth would you expect their contribution to sediment to be most important?

5. How would you distinguish sedimentary bedding from metamorphic foliation?

6. How would you distinguish a nonfoliated metamorphic rock such as marble or quartzite from its sedimentary derivative (limestone or quartz sandstone)?

7. If you were seeking to find the oldest rocks on Earth, where would be your best opportunity for success: on a continent or in a deep ocean basin? What type(s) of rock would you want to search for to gather evidence of the earliest-formed rocks on Earth— igneous, sedimentary, or metamorphic?

Weathering, erosion, and sedimentation in a mountainous desert.

This photo shows a craggy rock surface overhanging a cave. Below, a polished rock surface shows a thin veneer of minerals that have left vertical streaks.

What weathering and erosional processes led to removal and polishing of the rock?

Where have the removed rocks gone?

What is the source of the ions needed to reprecipitate the minerals forming the vertical streaks on the polished rock face?

This photo shows sediment stretching out across a plain in front of a mountain range.

What was the likely source of sediment deposited on the plain?

What processes created this sediment surface?

What might be the reason that the sediments are reddish in color?

1. What are the common rock-forming minerals?

2. What is the rock cycle?

3. Under what conditions do igneous rocks form?

4. What are the processes that form sediments?

5. What are siliciclastic sediments, and how do they form?

6. What are carbonate sediments, and how do they form?

7. What are the agents of weathering, erosion, and deposition of sediment?

8. How are unconsolidated sediments changed to sedimentary rock?

9. Under what conditions do metamorphic rocks form?
▼

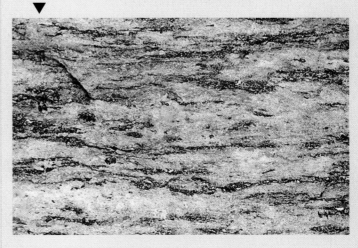

10. What is a descriptive classification of rocks?

11. What is a genetic classification of rocks?

Geologic Time

Earth has a turbulent, event-filled history that dates back 4.5 billion years or so. Our planet's deep past is chronicled in great detail in the layers of rock and sediment that are stacked like the pages of a book lying on its side, front cover down, with the first-written "pages" at the bottom of the stack, and more recently written pages appearing successively toward the top of the stack. The pages of our Earth history book are richly illustrated with remains of the life forms that successively appeared and disappeared through time. The layers (or pages) of the book are also ordered (or "numbered"), but the "page numbers" need to be unencrypted using the stratigraphic techniques first devised by Nicolaus Steno, Charles Lyell, and William Smith.

In this chapter, you will explore the means by which geologists unravel the story written in Earth's layers, how geologists use those layers to establish a timeline of events, and the reasons that geologists have separated deep time into the intervals that make up the geologic time scale.

In certain places, like the Grand Canyon of Arizona (shown here), reading the pages of Earth history and assembling a timeline is a fairly straightforward process. That is not always the case, however, because erosion and tectonic events have ripped, lost, or crumpled some of the pages of the book. To fully comprehend the rich story encoded in Earth's layers, you need to learn the methods geologists use to extrapolate the missing information or unravel the convoluted pages.

NATIONAL GEOGRAPHIC

Stratigraphic Correlation Techniques

Geologists describe the timing of events in Earth history using a "calendar" called the geologic time scale. That time scale has formal divisions that reflect major episodes in this planet's physical, chemical, and biologic development. Understanding the geologic time scale and how it was constructed is central to the study of Earth history because all events are linked to it. The development of both life forms and geologic features makes sense only when they are interpreted in their correct time context.

It has taken almost two centuries to build the geologic time scale to the point that we know it today, and fine-tuning of the time scale still continues. Part of the reason that assembly and adjustment of the time scale has been such a lengthy process is that there is no single place where we can go to see all of geologic time written in the stratigraphic record. Instead, we must piece together various segments of the stratigraphic record (FIGURE 3.1) from different parts of the world to build a complete chronology of Earth's changes and development.

Earth's layers FIGURE 3.1

The Badlands of South Dakota expose a thick succession of stratigraphic layers stacked like the pages of a book.

Badlands, South Dakota

Global Locator

FIGURE 3.2 Relative ages of strata

Using stratigraphic principles allows the geologist to determine the relative order of geologic events at this cliff high in the Cumulus Hills of Antarctica. A series of sedimentary strata, still in their original horizontal position, were deposited in a stream's floodplain. The strata were deposited in upward succession. After lithification, the rocks were uplifted, and agents of weathering and erosion exposed them along the side of a ridge. Debris eroded from this outcrop has fallen to the base of the cliff, where it is now forming a new sedimentary deposit. The contact between the erosional surface (the cliff face) and the recent unconsolidated deposit is an unconformity.

> **relative age dating** The technique of establishing a chronology of events arranged in relative sequential order.

Events that occurred in geologic time, or "**deep time**," are assessed in two general ways. The first way of unraveling the timeline is by using **relative age dating** techniques (FIGURE 3.2). Events such as the uplift of a mountain range, a rise in sea level, or the evolution of a new animal species, are placed in sequential order, beginning with those that occurred first and continuing with those that occurred subsequently. There is no direct implication of exactly when an event occurred in time, only whether a particular event occurred before or after some other event. The relative sequence of events can be determined through application of Nicolaus Steno's principles of superposition, original horizontality, and original lateral continuity; through application of Charles Lyell's principles of cross-cutting relationships and included fragments; and through application of William Smith's principle of biotic succession.

> **numerical age dating** The technique of establishing when events occurred according to how much time has elapsed since their occurrence. Geologic time that has elapsed is measured in thousands, millions, or billions of years.

The second way of unraveling a chronology of events is by using **numerical age dating** techniques (FIGURE 3.3 on pages 64–65), in which geologists calculate how many thousands, millions, or billions of years ago an event occurred. Numerical age dating techniques include counting tree rings, counting layers of ice drilled from glaciers, and counting sediment layers in ancient lakes. Especially useful for determining when events occurred in deep time, though, is radiometric dating, in which the time a sediment or rock unit formed is deciphered from information encoded in radioactive isotopes.

The geologic time scale was originally assembled using relative age dating methods. In practice, geologists today use relative and numerical age dating techniques side-by-side whenever possible to work out geologic problems.

> **stratigraphy** The study of layered rocks, including their compositions, origins, geometric relationships, and ages.

Stratigraphy is the study of Earth's rock and sediment bodies. Stratigraphy mostly deals with layered rocks and sediments (Figure 3.1, Figure 3.3), but igneous and metamorphic rocks are often included as well. One major goal of stratigraphic work is the matching of rock units from place to place based on evidence of their environment of formation or on the basis of when in time they were formed. Another major goal of stratigraphic work is determining the sedimentologic conditions under which rock units formed, whether sediment precursors were deposited in a stream bed, along a shoreline, in the deep ocean, or elsewhere.

> **correlation** Matching of strata from one location to another.

To piece together the segments of the stratigraphic record and place them in their correct time sequence, we rely on a variety of stratigraphic techniques: lithostratigraphy, biostratigraphy, chemostratigraphy, and sequence stratigraphy. Each of these techniques allows for the matching of strata, or **correlation**, from one locality to another (Figure 3.3). When enough overlapping segments of the stratigraphic record are matched together, we can eventually reconstruct the entire stratigraphic record.

Process Diagram

Rock units can be correlated using relative age dating techniques on the basis of their lithology (lithostratigraphy) or on the basis of criteria that are proxies for geologic time (chronostratigraphy). Fossils, variations in chemical isotopes, variations in physical properties (such as radioactivity levels), and sequences of unconformity-bounded strata are some of the variables that provide an understanding of the relative timing of events in geologic history and that can be used to correlate strata from one place to another. This hypothetical example shows how we can correlate among sections of relatively undisturbed rocks (**A**) using some commonly applied techniques (**B–D**).

A Stratigraphic sections and drill cores provide us with the basic stratigraphic information for correlating rocks from place to place. Here are four stratigraphic sections (W, X, Y, and Z), all from the same geographic region. In this example, study of the stratigraphic sections allows us to interpret lithostratigraphy (see **B**), and to determine depositional sequences (see **D**). They also provide fossils useful for biostratigraphic interpretation (see **C**).

The stratigraphic sections begin in granite "basement." A geochronologic method (the uranium-235 to lead-207 ratio in zircon crystals) yields a radiometric age of 2.1 Ga (2.1 billion years) for the granite at one site.

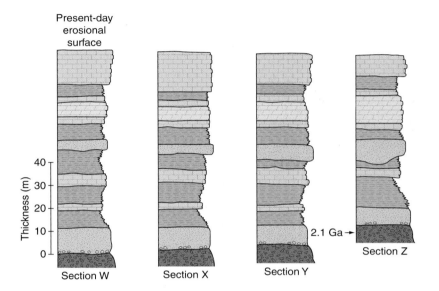

Lithostratigraphic Correlation

B In lithostratigraphic correlation, the boundaries (or contacts) between mappable rock types are correlated. Formations are the fundamental units of lithostratigraphy. Distinctive subunits of formations (members and beds) are also identified.

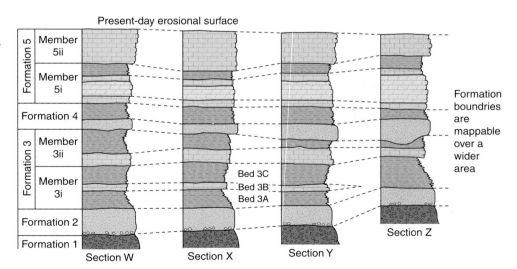

Biostratigraphic Correlation

⊏ In biostratigraphic correlation, the vertical stratigraphic ranges of guide fossils that define biozones (or zones) are used for correlation. One guide fossil is the trilobite *Olenellus*, shown below. The bases of biozones reflect evolutionary history, and define time-parallel lines. These lines do not necessarily coincide with lithologic (formation) boundaries.

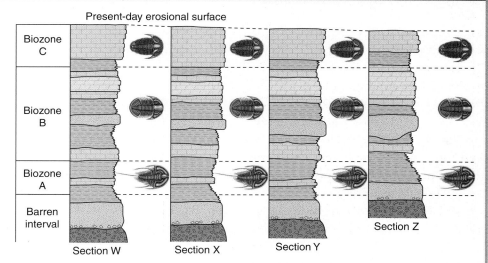

Present-day erosional surface

Biozone C

Biozone B

Biozone A

Barren interval

Section W Section X Section Y Section Z

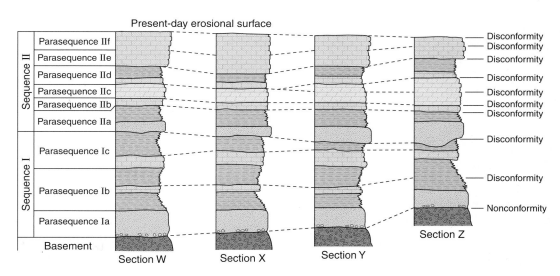

Present-day erosional surface

Sequence II	Parasequence IIf
	Parasequence IIe
	Parasequence IId
	Parasequence IIc
	Parasequence IIb
	Parasequence IIa
Sequence I	Parasequence Ic
	Parasequence Ib
	Parasequence Ia
	Basement

Section W Section X Section Y Section Z

Disconformity
Disconformity
Disconformity
Disconformity
Disconformity
Disconformity
Disconformity
Disconformity
Disconformity
Nonconformity

Sequence Stratigraphy

◻ In sequence stratigraphy, unconformities that bound genetically related strata are used for correlation. Sequence boundaries are generally indicated by erosion into underlying strata. Because all the strata deposited above an unconformity are younger than the unconformity, a sequence boundary parallels geologic time. A sharp but not deeply erosional lithologic change defines the base of a **parasequence**. The bases of sequence-stratigraphic units usually reflect sea level changes, and do not necessarily coincide with formation boundaries.

LITHOSTRATIGRAPHY

Lithostratigraphy involves the correlation of rocks on the basis of **lithology**, or rock type. The lithology of a rock usually involves some combination of the rock composition, grain or crystal size, and color. Lithostratigraphic units can be correlated on the basis of whether they are composed of sandstone, shale, conglomerate, limestone, or granite, for instance, but commonly additional qualifiers are needed to help distinguish one unit from another. Color of a rock is often used as a guide to correlation. Black or gray shale, green or red siltstone, white or red sandstone, and tan or black limestone units are commonly distinctive enough in certain regions as to be traceable across a wide geographic area. When illustrating formations, geologists often use standard symbols to distinguish various lithologies (**FIGURE 3.4**).

> ■ **lithostratigraphy** Stratigraphic correlation on the basis of rock type.

The fundamental unit of lithostratigraphy is called a **formation**. A formation is a clearly identifiable, mappable lithostratigraphic unit (**FIGURE 3.5**). Mappability refers to the ability to trace the outcrop pattern of a rock unit on a map, usually at the standard scale of mapping used in a region. The outcrop pattern of a rock unit is the total observed or inferred positions of that rock unit at Earth's surface, including that lying just below the soil layer. In most of the United States, the standard scale of mapping is 1:24,000, but in other countries and regions, the scale of mapping varies.

> ■ **formation** The fundamental unit of lithostratigraphy; it has a definable top and bottom and is mappable across geographic space.

Lithologic symbols FIGURE 3.4

A Formations are distinguished from each other by their lithologies, and their lithologies are illustrated using standard symbols. This hypothetical stratigraphic section illustrates standard lithologic symbols.

Dolostone	
Limestone	
Coal	
Black Shale	
Shale (gray shale)	
Siltstone	
Sandstone	
Breccia	
Conglomerate	
"Crystalline" Rocks — Igneous rocks	
"Crystalline" Rocks — Metamorphic rocks	

B Lithologic symbols are intended to reflect, in simplified form, rocks as they appear on outcrops. This is a limestone formation showing horizontal layers divided up by vertical blocks. The symbol for limestone (A) is similar, a pattern of stacked rectangles resembling bricks.

Mapping formations FIGURE 3.5

A A formation is a mappable lithostratigraphic unit that has a certain characteristic lithology, or rock type. In the Grand Canyon area of the Colorado Plateau, strata are predominantly flat-lying, and it is easy to trace out, or map, formation boundaries from a distance.

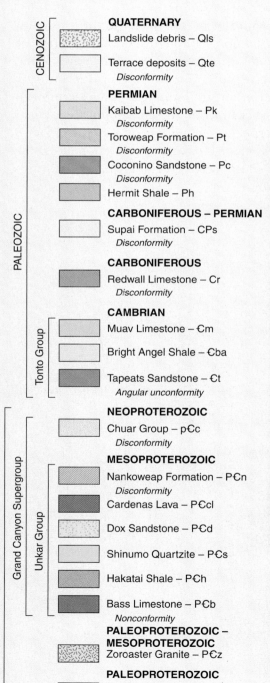

CENOZOIC

QUATERNARY
Landslide debris – Qls

Terrace deposits – Qte
Disconformity

PALEOZOIC

PERMIAN
Kaibab Limestone – Pk
Disconformity
Toroweap Formation – Pt
Disconformity
Coconino Sandstone – Pc
Disconformity
Hermit Shale – Ph

CARBONIFEROUS – PERMIAN
Supai Formation – CPs
Disconformity

CARBONIFEROUS
Redwall Limestone – Cr
Disconformity

CAMBRIAN
Tonto Group
Muav Limestone – Cm

Bright Angel Shale – Cba

Tapeats Sandstone – Ct
Angular unconformity

PROTEROZOIC

Grand Canyon Supergroup

NEOPROTEROZOIC
Chuar Group – pCc
Disconformity

MESOPROTEROZOIC
Unkar Group
Nankoweap Formation – PCn
Disconformity
Cardenas Lava – PCcl

Dox Sandstone – PCd

Shinumo Quartzite – PCs

Hakatai Shale – PCh

Bass Limestone – PCb
Nonconformity

PALEOPROTEROZOIC – MESOPROTEROZOIC
Zoroaster Granite – PCz

PALEOPROTEROZOIC
Vishnu Schist – PCv

B A geologic map is a two-dimensional representation of the formations making up the **bedrock**, or the rock at Earth's surface. Mappable formations of the Colorado Plateau, which are mostly flat-lying, similar to the example in Figure 3.2A, are shown here in different colors.

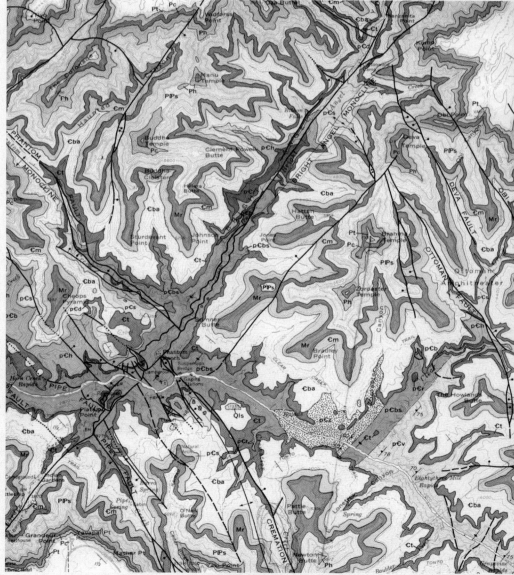

BIOSTRATIGRAPHY

biostratigraphy
Zoning of stratigraphic layers and arrangement of those layers according to relative time of deposition, using the ranges of fossils.

Biostratigraphy is the stratigraphic application of the principle of biotic succession. It involves the correlation of strata on the basis of the fossils they contain. Some fossil species, which are referred to as **guide fossils** (or **index fossils**), are useful for correlation purposes because they existed on Earth for relatively short spans of geologic time and have wide geographic distributions. Species are most useful for correlation if they have short stratigraphic ranges, occur in a variety of lithologies, are relatively common, and are readily identifiable.

The fundamental unit of biostratigraphy is called a **biozone** (or **zone**). A biozone (**FIGURE 3.6**) is a stratigraphic interval defined by its fossil content. It is usually given the name of a characteristic fossil present in that interval. Although there are several ways of defining a biozone, one common, precise way is by the first appearance of a species. With this method, the top of a zone is defined automatically by the base of the overlying zone. Other common, although less precise, ways of defining biozones are based on the total stratigraphic range of a species, the range through which two or more species overlap stratigraphically, an assemblage of species in a rock unit, and the maximum abundance (or acme) of a species. Depending on the type of guide fossil used to define a biozone, a zone can be used for correlation of strata locally, regionally, across a continent, or globally.

biozone A
stratigraphic interval defined by its fossil content and usually given the name of a characteristic fossil present in that interval. Also known as a zone.

Biostratigraphic zones FIGURE 3.6

Agnostoid trilobites, which are rarely more than 1 cm long, are excellent guide fossils in the Cambrian. They are widely distributed in marine strata, and they evolved quickly. Most species have short stratigraphic ranges that can be used to finely subdivide the vertical succession of strata into biozones (or zones). In the figure, dots indicate the stratigraphic levels where specimens of each species were found. Vertical lines indicate the stratigraphic ranges of the species.

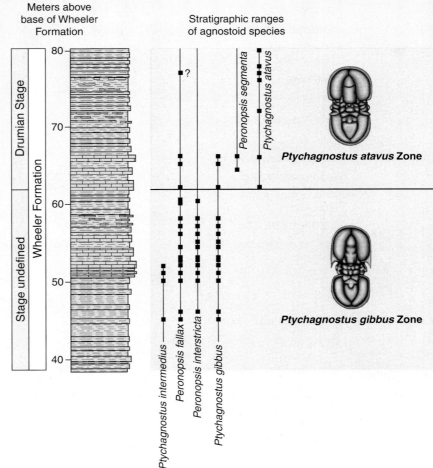

CHEMOSTRATIGRAPHY

Chemostratigraphy involves the correlation of strata using ratios of chemical isotopes. Excursions in isotopic ratios reflect global cycles and can usually be

■ **chemostratigraphy**
Correlation of strata using ratios of chemical isotopes.

correlated through marine sediments globally. After sampling numerous layers of limestone from a stratigraphic section and then analyzing the samples for the relative proportion of carbon-13 compared to the more common isotope carbon-12, a geologist can construct a curve of isotopic ratios through the stratigraphic layers (**FIGURE 3.7**). The curve may show some distinctive positive or negative shifts in carbon isotopic ratios, and those distinctive shifts, or **isotopic excursions**, can be

■ **isotopic excursion**
A positive or negative shift in the isotopic ratio of an element, as recorded through a succession of stratigraphic layers.

used to match isotopic curves from two or more sections. The excursions reflect chemical changes in the world ocean. In the case of carbon isotopic excursions, they are normally related to greater rates of either burial or exhumation of sediments containing organic matter.

Isotopes other than carbon may be used for correlation. One commonly used isotopic pair is the ratio of oxygen-18 to the more common isotope oxygen-16. Oxygen isotopic ratios are commonly inferred to be related to changes in temperature. The lighter isotope, oxygen-16, becomes preferentially locked up in glacial ice, leaving more oxygen-18 in the ocean. Higher proportions of oxygen-18 that have become incorporated in marine sediments thus reflect colder intervals.

Another isotopic pair that is being used with increasing frequency is the ratio of strontium-87 to strontium-86. The principal source of strontium that gets added to the world ocean is weathering of crustal rocks. Volcanic rocks tend to be richer in strontium-86 than they are in strontium-87. Development of a new chain of volcanic islands (a volcanic island arc), therefore, could be recorded by a trend toward a lower strontium-87/strontium-86 ratio as weathering acts on the newly

exposed mountains, releasing ions into solution and carrying them to the sea. The ionic composition of the ocean will change globally, so the record of weathering of a newly formed island arc will be traceable around the world.

There is no formal or fundamental unit of chemostratrigraphy. Some large excursions, or distinctive ones that can be constrained by other correlation techniques, however, are given names that help geologists quickly identify them.

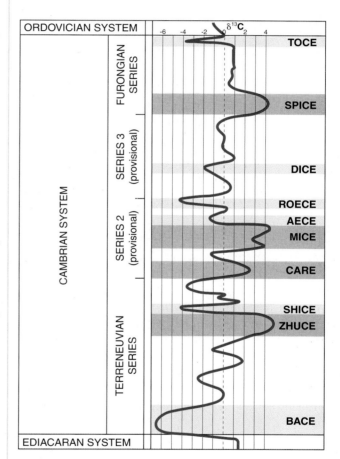

Isotopic curve FIGURE 3.7

Analyses of Cambrian strata yield some of the largest carbon-13/carbon-12 ($\delta^{13}C$) excursions preserved in the Phanerozoic stratigraphic record. These excursions are thought to relate mostly to rates of burial and exhumation of carbon-bearing sediments through time. Many of the excursions are useful in correlating strata worldwide and are known by such acronyms as BACE, SPICE, and DICE.

SEQUENCE STRATIGRAPHY

> **■ sequence stratigraphy**
> Correlation of strata on the basis of erosion surfaces that separate packages of sediments or sedimentary rocks called sequences (depositional sequences).
>
> **■ unconformity**
> A surface of erosion or nondeposition.

Sequence stratigraphy involves correlation on the basis of **unconformities**, or erosion surfaces, that separate packages of sediments or sedimentary rocks. Unconformities fall into three basic kinds: nonconformities, angular unconformities, and disconformities (see *What a Geologist Sees*).

Nonconformities and angular unconformities have little correlation value, at least on a fine scale, because the "gaps" in the stratigraphic record they represent often de-

veloped over intervals of many millions of years. Often disconformities can be used for precise correlation because they represent geologically short intervals of missing time.

Nonconformities occur where sedimentary strata have been deposited over top of crystalline (igneous or metamorphic) rock (*What a Geologist Sees* FIGURE A). Crystalline rocks are commonly formed deep within Earth, often several kilometers or more below the surface. Overlying rocks must be eroded so as to bring the crystalline rocks to the surface. Afterward, new, flat-lying sedimentary layers would be deposited. The process of eroding kilometers of rock to expose crystalline rocks at the surface may take tens to hundreds of millions of years to be completed.

A nonconformity should not be confused with an igneous intrusion. In a nonconformity, erosion has occurred, and sediments have been subsequently deposited in accordance with Steno's principle of super-

What a Geologist Sees

Unconformities

A Nonconformity

In a nonconformity, sedimentary strata overlie ``crystalline'' igneous or metamorphic rocks. The record of more than 500 million years is missing from this outcrop at the edge of the Adirondack Mountains in New York, where Cambrian sediments (now sandstone and conglomerate layers of the Potsdam Formation) have lapped onto slightly darker Proterozoic gneiss. The nonconformity surface is at lower left, at the level of the violet-colored flowers. The large white pebbles in the Potsdam Formation are included fragments derived from the underlying Proterozoic rocks.

position. In an intrusion, igneous rocks have been injected into preexisting rocks, and the principle of superposition cannot be applied. Instead, a geologist must rely on the principles of cross-cutting relations and included fragments to ascertain the relative order of formation of rock units.

Angular unconformities have sedimentary strata below and above the erosion surface, but the strata below are positioned at an angle compared to the beds above (*What a Geologist Sees* FIGURE B). An angular unconformity develops first through the deposition of flat-lying sedimentary rocks. Then, through tectonic processes, the strata are uplifted, commonly into mountains. Next, the uplifted strata are eroded, leaving angular layers at Earth's surface. Finally, new, flat-lying strata are deposited above the older, angular strata (FIGURE 3.8 on page 72). The time required to form the mountains, erode them, and deposit flat-lying strata above the erosion surface usually

ranges in the tens to hundreds of millions of years. Similar to nonconformities, angular unconformities have little correlation value because the stratigraphic gaps represent such large intervals of time.

Disconformities have parallel stratigraphic layers below and above the erosion surface (*What a Geologist Sees* FIGURE C). Marine sediment gets deposited on continental shelves when sea level is relatively high, and erosional disconformities develop when sea level falls. In many instances, these drops in sea level are **eustatic**, or global in scale, and triggered by the geologically rapid buildup of glacial ice in polar regions. Glacial buildup results in the removal of water from the world ocean.

> **disconformity** An unconformity in which strata below and above the erosion surface are parallel.

> **eustatic sea level** Global sea level.

B Angular unconformity

Sedimentary strata having an angular relationship to one another below and above an erosional surface define an angular unconformity. In Utah, flat-lying Cenozoic strata (red and gray) overlie older, reddish strata that were folded into mountains and then eroded. This unconformity cuts down into Jurassic strata and represents a loss more than 100 million years' worth of stratigraphic record.

C Disconformity

In a disconformity, sedimentary strata are relatively undisturbed below and above the erosional or nondepositional surface. The unconformity surface is often subtle, and the time gap represented in stratigraphy is often small. This disconformity between two Devonian limestone units in Ohio (at the level of the geologist's eyes) records the loss of less than 5 million years' worth of stratigraphic record.

A Along coastal South Australia, Ediacaran and Cambrian strata were folded through compressional tectonics early in the Paleozoic Era. On Kangaroo Island, exposed limbs of anticlines and synclines are now being weathered by ocean waters. ▶

B On the Pacific Ocean shore near Adelaide, sediments eroded from the former mountains are now being deposited over the upended strata as flat-lying stratigraphic layers, creating an angular unconformity. In time, lithification of the sediments will occur. The time gap represented by the developing unconformity is more than 400 million years.
▼

NATIONAL GEOGRAPHIC

Disconformities can form in only a few thousand years, which from a geologic perspective is a negligible amount of time. When sea level falls (FIGURE 3.9) across the world, erosion will occur simultaneously in many places. The resulting erosion surface, which approximates a time line, can therefore be used for correlation of strata globally.

The interval of strata bounded by a disconformity surface below and by another one above is called a **sequence**, or **depositional sequence** (FIGURE 3.9B,C). A sequence may include lithologies recording a variety of depositional environments, each of which could be formalized as a formation. A sequence is normally a large stratigraphic unit incorporating parts of several formations, whose boundaries are time-parallel.

> **sequence**
> A relatively conformable package of sedimentary strata that is bounded below and above by unconformities or their equivalent conformities. Also known as a depositional sequence.

Major Sea Level Changes Through Time

A Sea level falls triggered the development of widespread disconformities that differentiate depositional sequences. The times of significant erosion globally through the late Proterozoic (Neoproterozoic Era) and Phanerozic (Paleozoic, Mesozoic, and Cenozoic eras) are indicated on the sea level curve by horizontal (time-parallel) lines.

Seismic Reflection Profile

B Sequences (or depositional sequences) are genetically related strata, or strata all deposited between major changes in sea level and development of unconformities. Sequence boundaries and other large-scale stratigraphic relationships are easy to identify from seismic reflection profiles of subsurface layers, like this one taken in the Pacific Ocean basin off the west coast of New Zealand. To produce a reflection profile, scientists create sound waves that pass through some layers and reflect back to the surface off others. At the surface, receivers (called geophones) collect information about the reflected waves. Using the data on two-way travel time of the sound waves (indicated by the scale at left), computers can be used to calculate the distance of each reflective layer from the surface. The distance across this profile is approximately 150 km.

Eras	Periods	Relative changes of sea level ←Falling Rising→
Cenozoic	Quaternary	
	Paleogene	Neogene
Mesozoic	Cretaceous	
	Jurassic	Present sea level
	Triassic	
Paleozoic	Permian	
	Carboniferous	
	Devonian	
	Silurian	
	Ordovician	
	Cambrian	
Neoproterozoic	Ediacaran	

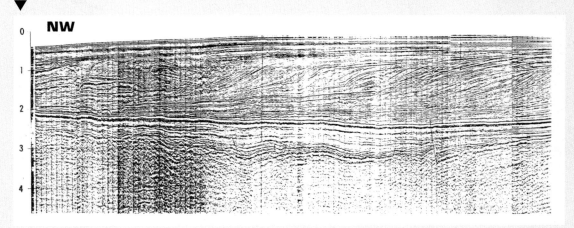

NW

Sequence Interpretation

C The same seismic stratigraphic section shown in B after the addition of geologic interpretation. Sequence boundaries are indicated by medium-weight black lines. The rocks indicated by a pink color are Paleozoic igneous and metamorphic units. Other colors are used to indicate the general depositional origins of strata within Cretaceous and Cenozoic sequences. The sequence boundaries are disconformities marking minor time gaps in strata; correlating these surfaces intercontinentally leads to the interpretation of global sea level curves. The near-vertical lines are faults.

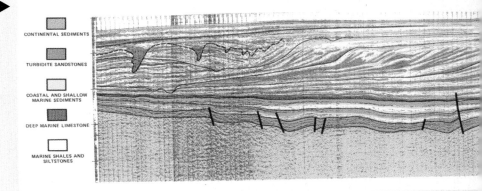

CONTINENTAL SEDIMENTS

TURBIDITE SANDSTONES

COASTAL AND SHALLOW MARINE SEDIMENTS

DEEP MARINE LIMESTONE

MARINE SHALES AND SILTSTONES

OTHER CORRELATION TECHNIQUES

Correlation of strata can be done using a variety of observable or measurable characteristics. In addition to the basic methods you have already read about, other techniques include measurement of the geophysical properties of rocks, most of which are indicators of lithology.

One geophysical technique used for correlating rocks according to a proxy for lithology is using a gamma-ray profile (FIGURE 3.10). Gamma rays are highly energetic electromagnetic rays that have short wavelengths. Emission of a gamma ray happens when an atomic nucleus undergoes radioactive decay and produces an isotope with a high energy state. By emitting a

Measuring stratigraphic radioactivity FIGURE 3.10

A Different rock types have distinctive gamma-ray signatures, reflecting the amount of radioactivity released from stratigraphic layers. Here is a profile of gamma-ray values measured from Devonian strata in Ohio.

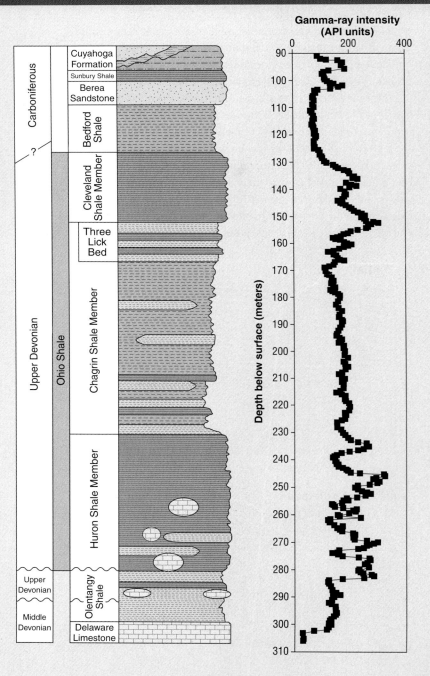

Key

- Black shale
- Dark to medium gray shale
- Gray shale or mudrock
- Carbonate concretion
- Siltstone
- Sandstone
- Limestone
- Disconformity

gamma ray, the newly formed isotope shifts to a more stable, lower-energy state.

Sedimentary rocks emit varying amounts of gamma rays, depending on the amount of radioactive uranium, thorium, and potassium they contain. Shales, for example, tend to be more enriched in these radioactive isotopes than limestones. Also, black shales tend to have higher radiation values than gray shales. By measuring the amount of gamma-ray emission from individual stratigraphic levels and obtaining a gamma-ray signature for each layer, it is possible to determine the type of rock in each of those layers. A gamma-ray profile can be used to correlate strata exposed at the surface or ones penetrated by drilling into the subsurface. This is especially helpful for correlating layers drilled in search of oil, natural gas, or other geologic reserves.

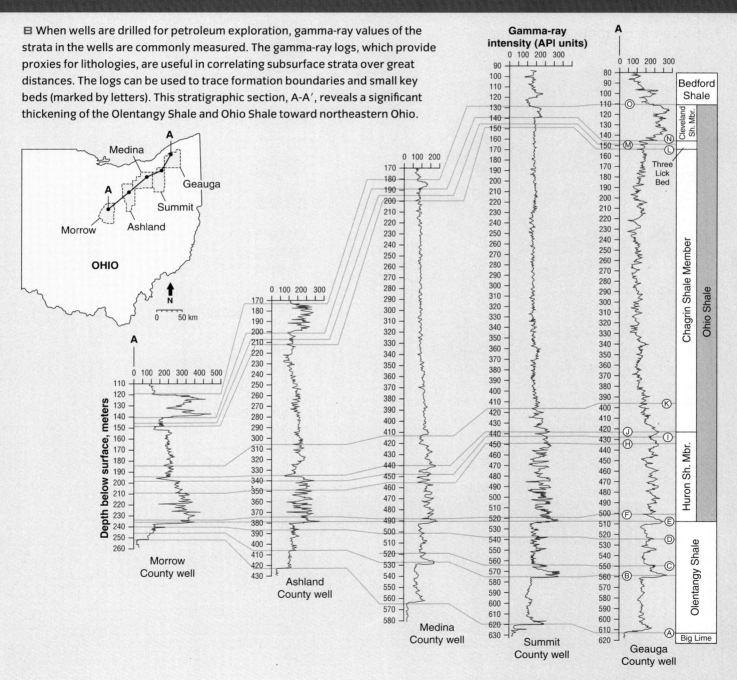

B When wells are drilled for petroleum exploration, gamma-ray values of the strata in the wells are commonly measured. The gamma-ray logs, which provide proxies for lithologies, are useful in correlating subsurface strata over great distances. The logs can be used to trace formation boundaries and small key beds (marked by letters). This stratigraphic section, A-A′, reveals a significant thickening of the Olentangy Shale and Ohio Shale toward northeastern Ohio.

Another important geophysical correlation technique involves measuring polarity directions recorded in rocks from the time the rocks were formed. Earth's polarity has switched many times during this planet's long history, and the sequence of normal and reversed polarity episodes provides a good basis for the correlation of strata (FIGURE 3.11).

Reversals in the direction of Earth's polarity through time are "frozen" in the magnetic minerals contained in rocks. Measuring the polarity direction recorded in oceanic basalts or other rocks allows the rocks to be correlated with others on the basis of magnetic signature. Figure 3.11B shows the four most recent "magnetic epochs," called chrons, each of which shows small polarity reversal episodes (or subchrons) but is dominated by either normal polarity (as at the present time) or reversed polarity.

Measuring magnetic reversals FIGURE 3.11

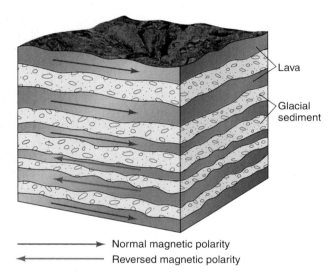

Normal magnetic polarity →
Reversed magnetic polarity ←

A When lavas cool, they instantly lock in a record of Earth's polarity field. Earth's polarity field has flipped back and forth from normal (the present field) to reversed through time, and the changes are recorded in lavas. This block diagram represents a stratigraphic section in Iceland, a place where mid-ocean ridge basalts were erupted intermittently between times of deposition of glacial sediment.

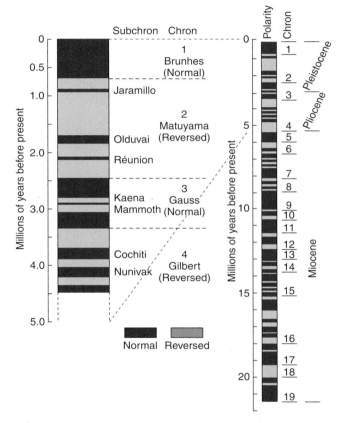

B Polarity reversals recorded in basalts date back to the mid-Jurassic Period. This diagram shows the record of polarity-defined time intervals, which are called magnetic chrons, through part of the Cenozoic Era.

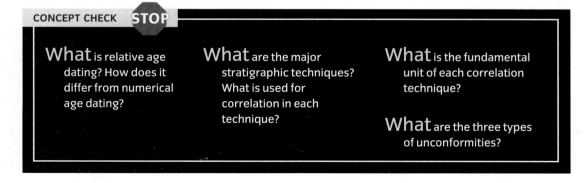

CONCEPT CHECK STOP

What is relative age dating? How does it differ from numerical age dating?

What are the major stratigraphic techniques? What is used for correlation in each technique?

What is the fundamental unit of each correlation technique?

What are the three types of unconformities?

The Geologic Time Scale

LEARNING OBJECTIVES

Understand the geologic time scale.

Distinguish geologic time units from chronostratigraphic units.

Summarize how geologic time is numerically calculated.

The **geologic time scale** (FIG-URE 3.12, on pages 78–79) is a chronology, or "calendar," of Earth history. Units of geologic time are not "natural" in the sense that there was a preset plan (or "chapter outline") that Earth adhered to. Instead, geologic time units have been defined and named in hindsight by geologists who have recognized that Earth has passed through a series of stages, each having its own distinct characteristics. The switchover from one stage to the next, or the boundary point between two time units, must be arbitrary but based on sound scientific reasoning.

By international agreement, each time-stratigraphic (or time-rock) boundary in the Phanerozoic Eonothem (plus one in the Proterozoic Eonothem) is marked by a physical point in a continuous stratigraphic succession, a unique horizon where time and stratigraphy meet. That position, referred to as a **GSSP (Global boundary Stratotype Section and Point)**, marks the point in time when that part of the stratigraphic succession began. The point, also sometimes called a "golden spike," is identifiable from an evolutionary event—usually the evolutionary first appearance of a species, but other correlation tools are also used to identify the equivalent position of the boundary horizon around the world.

The GSSP concept was introduced in the 1970s to provide precise, globally acceptable definitions of **chronostratigraphic units**, or time-rock units, and their equivalent geologic time units. Prior to the definition of GSSPs, the limits of time units were often interpreted differently in vari-ous areas of the world, sometimes with controversial results. Definition of GSSPs for each time-rock unit of Earth history now provides a standard chronostratigraphic scale with unambiguously defined time units. By implication, though, some concepts of geologic time units used in the past differ somewhat from their current definitions.

For all but the latest Proterozoic, the base of each chronostratigraphic unit in the "Precambrian" is defined by a numerical age, which is to say, it is defined by a geochronologic criterion. Where a chronostratigraphic criterion is applied, it is referred to as a **Global Standard Stratotype Age (GSSA)**.

CHRONOSTRATIGRAPHY

Eons, eras, periods, and epochs are formally defined units of geologic time. They are abstract entities, just like minutes and hours.

The stratigraphic record is a tangible proxy record of time. Its layers, formed during specific time intervals, record events that occurred during those geologic time intervals. The tangible stratigraphic representation of each time unit is referred to as a chrono-stratigraphic unit, or time-rock (time-stratigraphic) unit (FIGURE 3.12B). Correlation by integrating stratigraphic proxies for geologic time is called **chronostratigraphy**.

Each geologic time unit has an equivalent chronostratigraphic unit. An **eonothem** is the chronostratigraphic equivalent of an **eon**. An **erathem** is the chronostratigraphic equivalent of an **era**. The fundamental unit of chronostratigraphy is the

> **geologic time scale** A chronology of Earth history.

> **GSSP** Acronym for Global boundary Stratotype Section and Point, an internationally ratified point in strata marking the boundary between two time-rock (chronostratigraphic or time-stratigraphic) units and their equivalent time units.

> **chronostratigraphic unit** A time-rock or time-stratigraphic unit; the tangible representation of a geologic time (geochronologic) unit.

As of 2008, the International Stratigraphic Chart was still undergoing improvement. Some units (for example, series listed only by numerical designations) remained provisional and not formally defined. (Please see Appendix A for a detailed version of the Geologic Time Scale.)

▲ The International Stratigraphic Chart is the internationally recognized standard for the geologic time scale. Numerical geologic ages are indicated in millions of years ago (symbolized by Ma). Units in this chart are not illustrated to true scale.

Eonothem Eon	Erathem Era	System Period	Series Epoch		Geochronologic Age (Ma)
Phanerozoic	Cenozoic	Quaternary	Holocene		0.0118
			Pleistocene		2.588
		Neogene	Pliocene		5.332
			Miocene		23.03
		Paleogene	Oligocene		33.9
			Eocene		55.8
			Paleocene		65.5
	Mesozoic	Cretaceous	Upper		99.6
			Lower		145.5
		Jurassic	Upper		161.2
			Middle		175.6
			Lower		199.6
		Triassic	Upper		228.0
			Middle		245.0
			Lower		251.0

Eonothem Eon	Erathem Era	System Period	Series Epoch		Geochronologic Age (Ma)
Phanerozoic	Paleozoic	Permian	Lopingian		251.0
					260.4
			Guadalupian		270.6
			Cisuralian		299.0
		Carboniferous	Pennsylvanian	Upper	306.5
				Middle	311.7
				Lower	318.1
			Mississippian	Upper	326.4
				Middle	345.3
				Lower	359.2
		Devonian	Upper		385.3
			Middle		397.5
			Lower		416.0
		Silurian	Pridoli		418.7
			Ludlow		422.9
			Wenlock		428.2
			Llandovery		443.7
		Ordovician	Upper		460.9
			Middle		471.8
			Lower		488.3
		Cambrian	Furongian		~501.0
			Series 3		~510.0
			Series 2		~521.0
			Terreneuvian		542.0

Eonothem Eon	Erathem Era	System Period	Geochronologic Age (Ma)
Precambrian / Proterozoic	Neoproterozoic	Ediacaran	542
			~630
		Cryogenian	850
		Tonian	1000
	Mesoproterozoic	Stenian	1200
		Ectasian	1400
		Calymmian	1600
	Paleoproterozoic	Statherian	1800
		Orosirian	2050
		Rhyacian	2300
		Siderian	2500
Archean	Neoarchean		2800
	Mesoarchean		3200
	Paleoarchean		3600
	Eoarchean	Lower limit is not defined	

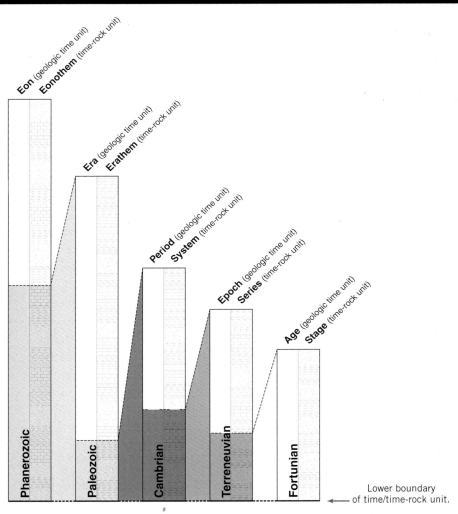

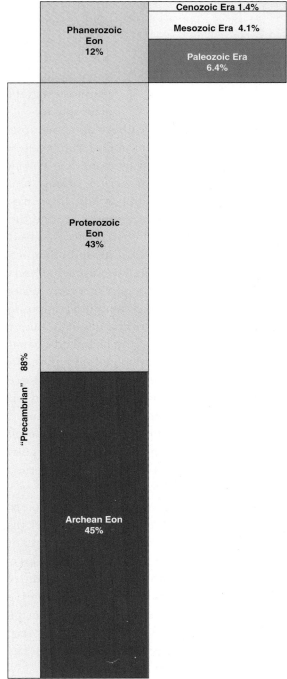

B Geologic time is organized into a hierarchy of formal units. The smallest unit recognized globally is an age, and successively larger units are epochs, periods, eras, and eons. Each unit of geologic time (which is an abstract concept) has a tangible representation in strata called a time-rock (or chronostratigraphic) unit. In other words, a time-rock unit represents all the strata deposited in a given interval of time. Geologic time and time-rock units are formally defined by their lower boundaries.

C The geologic time scale represented at true scale. The "Precambrian," comprising the Archean and Proterozoic Eons, is by far the longest interval of geologic time.

system The fundamental unit of chronostratigraphy.

period The unit of geologic time equivalent to a system.

system, and its time equivalent is the **period**. We speak of the Cretaceous Period when we are discussing time, and we speak of the Cretaceous System when we are discussing the strata deposited during Cretaceous time. Finally, a **series** is the chronostratigraphic equivalent of an **epoch**.

The geologic time scale shows abstract time units in their correct order and hierarchically arranged. The chart was developed by compiling chronostratigraphic information from numerous stratigraphic sections, including outcrops along roads, in quarries, and along mountain exposures, exposed around the world. Cores recovered from drilling into Earth's subsurface in search of petroleum reserves, coal, and other economically important raw materials have supplemented outcrop information. Stratigraphic sections from various areas must be spliced together to compile a composite record of the strata deposited during a geologic time unit. Separate stratigraphic sections can be linked when the tops or bottoms of sections overlap in time with the tops or bottoms of other sections. Biostratigraphic, chemostratigraphic, and sequence-stratigraphic correlation techniques allow us to match overlapping parts of sections exposed in various areas. Ultimately, geologists can develop a complete chronostratigraphic profile that represents an interval of geologic time by splicing together a series of smaller but overlapping profiles.

Correlation techniques used to establish chronostratigraphic units must be time-parallel; that is, they must define units representing the same "instant" (in geologic terms) everywhere in the world. The tops of the units are automatically defined by the bases of the overlying units. Biostratigraphic units (biozones, or zones) qualify as time-parallel units because biologic evolution and extinction is a non-repeating process. Each species has had only one time of appearance on Earth, and each has gone extinct, or eventually will do so, only once.

Chemostratigraphic profiles record variations in chemical isotopic ratios locked into sediments through geologic time. Details of the sine-wave pattern of positive and negative excursions provide a virtual "fingerprint" for time units. The peak of each positive and negative excursion was reached at the same time everywhere in the world.

Sequence-stratigraphic units, or sequences, are bounded by disconformities of short duration (geologically speaking). They are commonly the result of rapid global sea level changes, which means disconformities of equivalent age appear in stratigraphic sections around the world.

Lithostratigraphic units, especially formations, are, for the most part, not time-parallel. Formations usually represent a sedimentary **facies**, the type of sediment deposited in one area (say a beach, a reef, a lake, or a stream) and are not directly tied to time intervals. In the world today, many different sedimentary facies coexist in time, and this was also true in the geologic past.

A few types of lithologic units actually are time-parallel and can be used for chronostratigraphic correlation. These units were deposited rapidly in geologic terms and are usually so thin they are not normally separated as formations. A volcanic ash bed (**FIGURE 3.13**) is perhaps the best example of a time-parallel lithologic unit. When a volcano erupts, it spews ash, most of which will settle from the atmosphere within days or weeks of the event. The resulting ash bed is an excellent time marker within strata, but most ash beds are of relatively limited geographic distribution.

Volcanic ash bed FIGURE 3.13

A volcanic ash bed, about 2 cm thick, was deposited during the Devonian Period in Ohio. It represents a time-parallel lithologic unit that can be traced westward from northeastern Pennsylvania and upstate New York. The thin ash bed is at the top of the waterfall stream at left.

GEOCHRONOLOGY

Geochronology refers to the dating of rocks according to their numerical ages. Most geochronologic work today involves laboratory analyses of radiometric isotopes (TABLE 3.1). In radioactive decay, a radiogenic **parent** isotope changes through one or more steps, each referred to as a **daughter** product, until becoming a stable end product.

Radioactive decay occurs in a number of ways. Three processes useful in geochronology are called alpha decay, beta decay, and electron capture. **Alpha decay** is a type of nuclear fission, in which the parent splits into two daughter products. It tends to occur in isotopes that have large atomic numbers (that is, large numbers of protons in the atomic nucleus). Decay of uranium-235 (^{235}U) to radioactive thorium-231 (^{231}Th) occurs by this process. In this case, the atomic nucleus of the parent, ^{235}U, emits an **alpha particle**, α, which is composed of two protons and two neutrons (identical to the nucleus of a helium atom, ^{4}He), to form ^{231}Th and ^{4}He.

> **parent** In a radioactive decay series. An unstable isotope that decays, or transforms, into a daughter product.

> **daughter** An isotope formed from the radioactive decay of a parent isotope.

Beta decay takes two forms. In neither one does the atomic mass of a nucleus change. **Beta decay** can involve emission of a **beta particle** (an electron, $\beta-$), accompanied for reasons of symmetry by emission of an electron **antineutrino** from an atomic nucleus. (The term "beta particle" for an electron is a holdover from early studies on radioactivity.) Transformation of rubidium-87 (^{87}Rb) to strontium-87 (^{87}Sr), and conversion of carbon-14 (^{14}C) to nitrogen-14 (^{14}N) is by β-decay. Another means of beta decay is emission of an electron's antiparticle, which is called a **positron** ($\beta+$), accompanied for reasons of symmetry by emission of a **neutrino**. This process is alternately referred to as **positron emission** or **beta positive decay** ($\beta+$ decay).

In **electron capture**, a parent nucleus captures one of its own electrons and then emits a neutrino. The atomic mass of a nucleus does not change. Potassium-40 (^{40}K) transforms to argon-40 (^{40}Ar) by means of electron capture.

The amount of time it takes for one-half of the parent to decay to a daughter product is called a

Radiogenic isotopes used in dating events in Earth history TABLE 3.1

Isotopes			Half-Life of Parent (years)	Effective Dating Range (years)	Minerals and Other Materials That Can Be Dated
Parent	**Decay System**	**Daughter**			
^{238}U	α and β^- decay	^{206}Pb	4.5 billion	10 million–4.6 billion	Zircon and uraninite
^{235}U	α and β^- decay	^{207}Pb	710 million	10 million–4.6 billion	Zircon and uraninite
^{232}Th	α and β^- decay	^{208}Pb	14 billion	10 million–4.6 billion	Zircon and uraninite
^{40}K	Electron capture β^- decay	^{40}Ar ^{40}Ca	1.3 billion	50,000–4.6 billion	Muscovite mica Biotite mica Hornblende Whole volcanic rock
^{87}Rb	β^- decay	^{87}Sr	47 billion	10 million–4.6 billion	Muscovite mica Biotite mica Potassium feldspar Whole metamorphic or igneous rock
^{14}C	β^- decay	^{14}N	5730 ± 30	100–50,000	Wood, charcoal, peat, grain, and other plant material Bone, tissue, and other animal material Cloth Shell Stalactites Groundwater Oceanwater Glacial Ice

half-life. Radioactive decay diminishes the amount of parent in logarithmic fashion. Assuming a closed system, in which the total mass of material involved in a decay sequence does not change, daughter products accumulate at the rate at which the parent decays. After one half-life, 50% of the parent remains, and 50% of the atoms have become a daughter product. After two half-lives, 25% (or half of the 50% amount) of the parent remains, and 75% of the atoms have become a daughter product. After three half-lives, 12.5% (or half of the 25%) of the parent remains, and 87.5% of the atoms have become a daughter product. Further decay continues in the same way.

By knowing the half-life of a radiogenic parent isotope and obtaining the ratio of the unstable parent to its daughter product that has been locked in a rock (which allows you to calculate how many half-lives have elapsed), you can calculate how many millions of years ago the rock was formed. Today, the precision and accuracy of radiometric dating techniques is so good that geochronologic information is routinely used to supplement relative age dating techniques in the construction and testing of the geologic time scale.

Another type of radiometric dating is called **fission track dating**. This technique is based on the counting of tracks, or damage trails, left by the fission fragments of uranium-bearing minerals or glasses enclosed within crystals of zircon, mica, sphene, or apatite, or within volcanic glass. The fragments are emitted from uranium at a known rate, so the number of fission tracks in a crystal or glass correlates directly with the age of the sample and its uranium content. By examining polished sections of the crystals or glasses microscopically, an investigator can count the number of fission tracks. The uranium content of the sample is determined by annealing it through heating or some other method. When the uranium content has been calculated, it is then possible to equate the number of fission tracks in the sample to the time elapsed since it formed.

After its initial formation, a crystal or glass can undergo significant heating during metamorphism. If this happens, the sample can anneal naturally, and the process of fission track production will start over. This "resetting" of the sample's radiometric age then tells the time of most recent metamorphism.

CONCEPT CHECK **STOP**

What is geologic time?

What are the types of units of geologic time?

What is the relationship of a geologic time unit to a chronostratigraphic unit?

HOW are numerical ages assigned to geologic time units?

SUMMARY

1 Stratigraphic Correlation Techniques

1. Stratigraphic **correlation** involves the matching of strata from place to place. It is the basis for assembling the geologic time scale.

2. In **lithostratigraphy**, rocks are matched according to lithology. Lithostratigraphic formations reflect environments of deposition and are not generally useful as indicators of geologic time.

3. Correlation techniques that provide good indicators of geologic time include **biostratigraphy** (which is based on the ranges of fossils), **chemostratigraphy** (which is based on isotopic ratios in strata), **sequence stratigraphy** (which is based on evidence of sea level changes), and the polarity reversal record.

2 The Geologic Time Scale

1. The geologic time scale arranges Earth history into hierarchical units, similar to the chapters of a textbook. The fundamental unit of geologic time is the **period**, and its chronostratigraphic equivalent (all the strata deposited during that time) is the **system**.

	Cenozoic Era 1.4%
Phanerozoic Eon 12%	**Mesozoic Era 4.1%**
	Paleozoic Era 6.4%
Precambrian 88%	**Proterozoic Eon 43%**
	Archean Eon 45%

2. **GSSPs** are used to mark the bases of chronostratigraphic units in the Phanerozoic and uppermost Proterozoic eonothems. Through the Archean Eonothem and most of the Proterozoic Eonothem, numerical ages are used to define the boundaries of time units.

3. Estimation of the timing of events in geologic history, or **geochronology**, usually involves radiometric dating methods.

KEY TERMS

1. Why is it necessary to have a geologic time scale? In what ways can it serve as a platform for the study of Earth's physical, chemical, and biologic history? How might the geologic time scale assist people in the search for valuable natural materials such as oil, natural gas, gold, and diamonds?

2. The geologic time scale represents an arbitrary subdivision of time into units (periods, eras, etc.) according to recognizable characteristics preserved in strata, such as obvious changes in life forms. Why do you think changes in life forms provide a good basis for the establishment of periods and eras? ▶

Eonothem / Eon	Erathem / Era	System / Period	Series / Epoch	Geochronologic Age (Ma)
Phanerozoic	Cenozoic	Quaternary	Holocene	0.0118
			Pleistocene	
				2.588
		Neogene	Pliocene	5.332
			Miocene	
				23.03
		Paleogene	Oligocene	
				33.9
			Eocene	
				55.8
			Paleocene	
				65.5
Phanerozoic	Mesozoic	Cretaceous	Upper	
				99.6
			Lower	
				145.5
		Jurassic	Upper	
				161.2
			Middle	
				175.6
			Lower	
				199.6
		Triassic	Upper	
				228.0
			Middle	
				245.0
			Lower	
				251.0

Eonothem / Eon	Erathem / Era	System / Period	Series / Epoch	Geochronologic Age (Ma)
				251.0
Phanerozoic	Paleozoic	Permian	Lopingian	260.4
			Guadalupian	270.6
			Cisuralian	299.0
		Carboniferous	Pennsylvanian Upper	306.5
			Pennsylvanian Middle	311.7
			Pennsylvanian Lower	318.1
			Mississippian Upper	326.4
			Mississippian Middle	345.3
			Mississippian Lower	359.2
		Devonian	Upper	385.3
			Middle	397.5
			Lower	416.0
		Silurian	Pridoli	418.7
			Ludlow	422.9
			Wenlock	428.2
			Llandovery	443.7
		Ordovician	Upper	460.9
			Middle	471.8
			Lower	488.3
		Cambrian	Furongian	~501.0
			Series 3	~510.0
			Series 2	~521.0
			Terreneuvian	542.0

Eonothem / Eon	Erathem / Era	System / Period	Geochronologic Age (Ma)
			542
Precambrian	Proterozoic	Neoproterozoic — Ediacaran	~630
		Neoproterozoic — Cryogenian	850
		Neoproterozoic — Tonian	1000
		Mesoproterozoic — Stenian	1200
		Mesoproterozoic — Ectasian	1400
		Mesoproterozoic — Calymmian	1600
		Paleoproterozoic — Statherian	1800
		Paleoproterozoic — Orosirian	2050
		Paleoproterozoic — Rhyacian	2300
		Paleoproterozoic — Siderian	2500
	Archean	Neoarchean	2800
		Mesoarchean	3200
		Paleoarchean	3600
		Eoarchean	Lower limit is not defined

The Ediacaran GSSP

■ Why do you think the GSSP is not defined at the contact between the tillite and the overlying limestone?

■ Is the contact between the tillite and limestone continuous or unconformable?

A The base of the Ediacaran System is defined as a point in strata at a GSSP in Brachina Gorge, Australia.

B The horizontal line on the brass marker indicates the level of the GSSP.

C The layer beneath the limestone containing the GSSP is a Cryogenian-age tillite. The Cryogenian Period is named for its glacial episodes, and the Ediacaran Period is the time when animal fossils first appeared in the fossil record.

SELF-TEST

1. Do lithostratigraphic boundaries normally coincide with geologic time lines?

2. What is the basic unit of lithostratigraphy called? How is it defined?

3. What is stratigraphic correlation?

4. What is a chronostratigraphic unit?

5. What is a geologic time unit?

6. What is the chronostratigraphic equivalent of a period?

7. What type of unconformity is illustrated in the photograph? ▶

8. What kind of unconformity is most useful for fine-scale stratigraphic correlation?

9. How is radiometric dating used in assembling the geologic time scale?

10. What is a biozone? How is it used in stratigraphic correlation?

11. What is an isotopic excursion? How is it used in correlation?

Life on Earth and Its Fossil Record

4

Life on Earth has a rich history nearly 3.5 billion years in duration. For more than half that time, this planet was dominated by tiny, inconspicuous, single-celled archaebacteria and eubacteria. Close to 2 billion years ago, more complex multicellular organisms appeared, and in the last half billion years or so, their descendants proliferated into a bewildering array of creatures.

If we continue using the metaphor of the Earth's stratigraphic layers as "pages" in a book, then fossils, like this fossil fish from Cretaceous strata of Liaoning, China, are the "illustrations" that help us visualize this planet's biologic changes. Fossils not only provide us with direct insight into the evolution of life, they also provide hints about interactions among life forms and about the environments in which they lived. Animals, plants, and other organisms are sensitive indicators of changing habitat conditions. The distribution of life forms can tell us about climate history, sea level position, oxygen levels, consistency of the sediment, and other things.

To understand the fossil record, we first need to review what life forms dwell on Earth today, review their ecology, and then project backward in time. Some organisms have changed little since they first evolved, and others have undergone astonishing transformations.

In this chapter, you will be introduced to life forms, how they are classified, and how and where they live. Then you will discover how life forms have left a record of their existence as fossils and how a community of living organisms translates to an assemblage of fossils. You will learn what organisms have the highest odds of success in the fossilization process and what environments tend to favor the preservation of fossils.

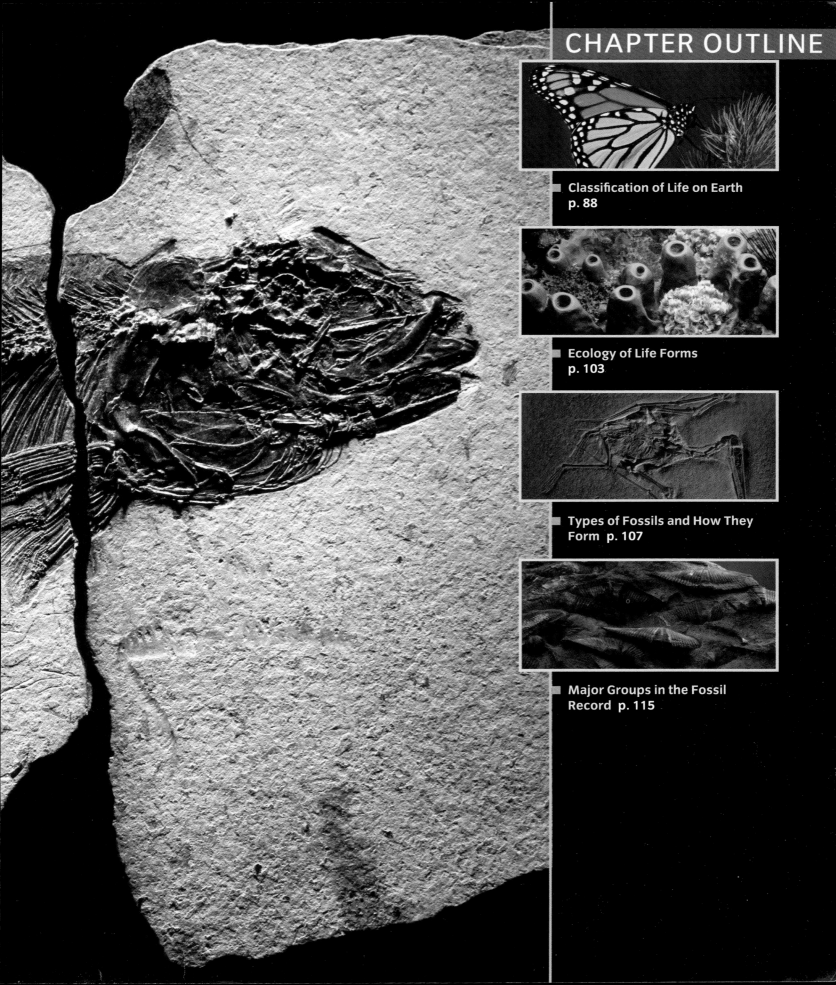

Classification of Life on Earth

LEARNING OBJECTIVES

Understand how living things are classified.

Differentiate between a biological species and a paleontological species.

Explain the differences between prokaryotes and eukaryotes.

Explain what characterizes the major life forms: archeans, bacteria, protoctists, fungi, plants, and animals.

Process Diagram

Classifying life on Earth FIGURE 4.1

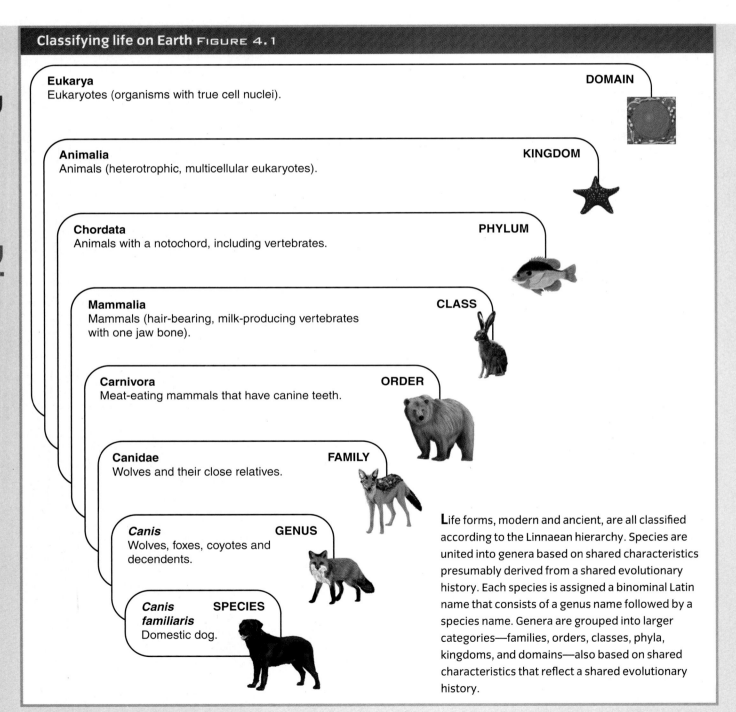

Eukarya
Eukaryotes (organisms with true cell nuclei).
DOMAIN

Animalia
Animals (heterotrophic, multicellular eukaryotes).
KINGDOM

Chordata
Animals with a notochord, including vertebrates.
PHYLUM

Mammalia
Mammals (hair-bearing, milk-producing vertebrates with one jaw bone).
CLASS

Carnivora
Meat-eating mammals that have canine teeth.
ORDER

Canidae
Wolves and their close relatives.
FAMILY

Canis
Wolves, foxes, coyotes and decendents.
GENUS

Canis familiaris
Domestic dog.
SPECIES

Life forms, modern and ancient, are all classified according to the Linnaean hierarchy. Species are united into genera based on shared characteristics presumably derived from a shared evolutionary history. Each species is assigned a binominal Latin name that consists of a genus name followed by a species name. Genera are grouped into larger categories—families, orders, classes, phyla, kingdoms, and domains—also based on shared characteristics that reflect a shared evolutionary history.

Since the time of the earliest scientific writing, humans have designed classification systems for natural things, including living organisms, the "elements" (Earth, air, fire, and water), and heavenly bodies such as stars and planets. In 1758, the Swedish biologist Carolus Linnaeus (who also published under the name Carl von Linné) introduced the system of biologic classification, which is also known as **taxonomy** or **systematics**, that we continue to follow today. This, the **Linnaean classification system**, is based on similarities and differences among organisms and recognizes the **species** as the fundamental unit of classification (**FIGURE 4.1**). Species are grouped in categories of ever-increasing size called genera, families, orders, classes, phyla (or divisions, for plants), kingdoms, and domains. Ideally, the modern Linnaean classification system would reflect inferred evolutionary history, but in practice, the classification of living and ancient organisms is a combination of artificial and evolutionarily based categories. Phylum (or division) boundaries, however, are mostly drawn between fundamentally different body plans, and domains are used to group organisms according to cellular organization and genetic information.

> **species** A group of organisms that can interbreed and produce fertile offspring.

Species are referred to using a two-part Latin name consisting of a genus name followed by a species name. This is called **binominal nomenclature**. The first part (the genus name) applies to a number of closely similar or related forms; with the addition of the species name, the full two-part name becomes unique to a species. The genus *Canis*, for example, refers to dogs and their close relatives. *Canis familiaris* is the domesticated dog, *Canis lupus* is the gray wolf, *Canis latrans* is the coyote, *Canis vulpes* is the fox, and *Canis dirus* is the extinct dire wolf. All these animals have a similar body construction resulting from a close, shared evolutionary history. By assigning these animals to different species (with a binominal identifier), we recognize that these animals cannot normally interbreed with each other.

> **binominal nomenclature** A technique of identifying organisms using a two-part Latin name, a genus name followed by a species name. The Latin names are set off from ordinary text using italics or underlining.

In some cases, domestic dogs and wild gray wolves *can* interbreed and produce fertile offspring. The reason is that these two species have been undergoing the trend toward reproductive isolation for only a short period of time, geologically speaking. They provide us with an excellent example of species-level evolution occurring right before our eyes.

Dogs are distinguished from cats by separation into a separate genus (*Felis*), represented by *Felis domestica*, the domestic cat, *Felis leo*, the lion, and others. Dogs and cats are grouped into larger groups in the Linnaean system, groups that reflect basic similarities between these and other animals. Dogs belong to the family Canidae (canids), cats belong to the family Felidae (felids), and members of both families belong to the order Carnivora (carnivorous mammals), along with bears, minks, seals, whales, and others. All these animals belong to the class Mammalia (mammals), along with anteaters, kangaroos, rodents, monkeys, apes, humans, and others; to the phylum Chordata (animals with notochords), along with tunicates, fishes, amphibians, reptiles, birds, and others; and to the kingdom Animalia (animals). Animals, plants, fungi, and protoctists, all of which have true cell nuclei enclosing their DNA, belong to the domain Eukarya.

Two rather different concepts of the species are in widespread use. From a biologic perspective, a species can be thought of as a group of organisms that can interbreed and produce fertile offspring. This is the **biological species concept**, and it can be tested using a reproductive criterion. For fossils, applying the reproductive criterion is impossible, so we use proxy evidence that individuals could have interbred and produced fertile offspring. Usually this means that similar-appearing fossils are placed in the same species. This is

> **biological species concept** The concept that members of a species can interbreed and produce fertile offspring, and that members of a single species are distinguished from other species by reproductive isolation.

Paleontological species concept FIGURE 4.2

These trilobites from the Devonian of New York show a certain amount of variation in their features but are similar enough to be considered one species. Reinforcing this interpretation, all the specimens are preserved in the same layer of strata, which indicates that they probably belonged to one population when they were alive.

paleontological species concept The concept that the limits of ancient species may be inferred from their preserved physical traits.

the **paleontological species concept**, and it is tested using patterns of morphologic similarities and differences (FIGURE 4.2).

CELLULAR ORGANIZATION

Living things fall into two fundamental categories based on the structure and biochemistry of their cell types: **prokaryotes** and **eukaryotes** (FIGURE 4.3). Prokaryotic organisms have simple cells. They are small (typically 1-10 μm) and lack nuclei and certain heritable organelles. An organelle is any visible structure within a cell. Prokaryotes have some organelles (carboxysomes and gas vacuoles, for example), but they cannot be inherited. Eukaryotic organisms have larger cells (typically 10-100 μm) that are more complex and incorporate true nuclei and heritable organelles. Some organelles (notably plastids, mitochondria, and Golgi apparati) have their own membranes separating them from the cellular cytoplasm. Most eukaryotes have mitochondria, the organelles that convert energy stored in simple organic compounds to a consumable form through enzymatic oxidation. Prokaryotes, which do not have mitochondria, release consumable energy from organic compounds using enzymes bound to their membranes.

Other characteristics differentiate eukaryotes from prokaryotes. Some eukaryotic cells have cilia and

prokaryote An organism that lacks a nucleated cell type.

eukaryote An organism that has a nucleated cell type.

Cell types FIGURE 4.3

Living organisms have one of two cell types: prokaryotic cells or eukaryotic cells. These cells have been stained to make their structure more visible under the microscope.

A A prokaryotic cell, typical of archaebacteria and eubacteria, lacks a true nucleus and most cell organelles. Genetic material is concentrated in a nuclear region.

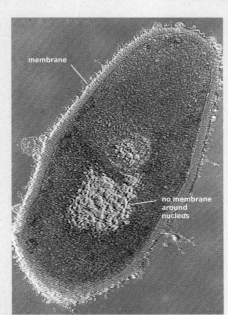

B A eukaryotic cell, typical of protoctists, fungi, plants, and animals, has a true nucleus that contains the genetic material and has heritable organelles.

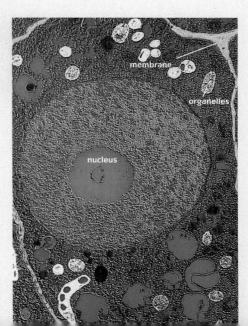

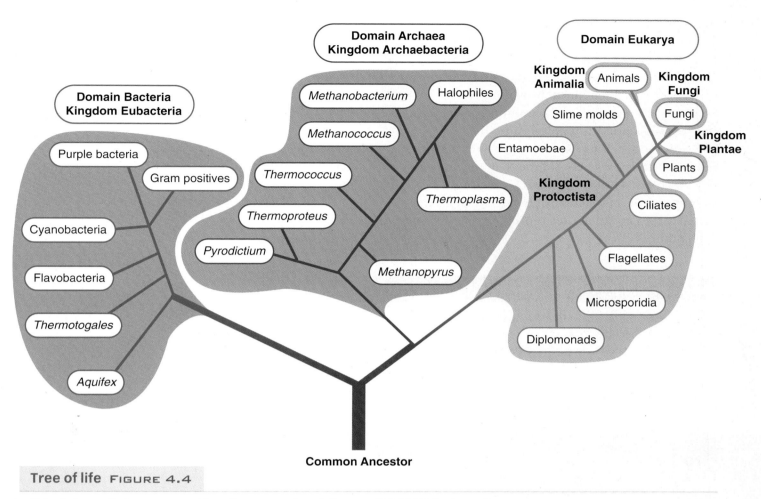

Tree of life FIGURE 4.4

This phylogenetic tree, prepared from the sequencing of ribosomal RNA (rRNA), shows the interpreted evolutionary relationships among living organisms, including domains and kingdoms. Archaebacteria and eubacteria are prokaryotes and include the most primitive of life forms. Protoctists, fungi, plants, and animals are all eukaryotes. The advanced eukaryotes (fungi, plants, and animals) evolved from eukaryotic protoctists.

undulipodia ("sperm tails") at some point in the life cycle, but these are lacking in prokaryotes. Chloroplasts and other plastids may be present in eukaryotes, but they are completely lacking in prokaryotes. A variety of prokaryotes and eukaryotes use photosynthesis to produce oxygen, but only prokaryotes can also produce sulfur and sulfate end products. All prokaryotes are single-celled microbes, and most live in colonies. Some eukaryotes are microbes, but most are large and multicellular, living either singly or in colonies.

Today, cellular organisms are usually organized into three domains and six kingdoms (FIGURE 4.4). Eukaryotes (domain Eukarya) comprise four kingdoms:

Protoctista, Fungi, Plantae, and Animalia. Prokaryotes comprise two main groups that are treated here as separate domains even though the best way of classifying them is not fully agreed upon at present. The domain Archaea comprises one kingdom, Archaebacteria. The domain Bacteria also comprises one kingdom, Eubacteria (also sometimes called Monera).

All prokaryotic and eukaryotic organisms form cells, but another group of arguably living things, the viruses, do not form cells. Viruses are composed of DNA (deoxyribonucleic acid) or RNA (ribonucleic acid) enclosed in a coat of protein. They replicate, which is an essential characteristic of a life form, but they cannot exist independently. Viruses must enter a cell and use the cell's biologic machinery to replicate. If they are outside a cell, viruses cannot feed, grow, or reproduce. Viruses apparently cannot sustain orderly internal chemical reactions on their own. Some can even crystallize like minerals, remaining in a state of "suspended animation" until they come in contact with the specific type of living tissue they need. Although viruses are of considerable importance in biology, little is known of their evolutionary history. Certain viruses may actually be more closely related to the cell types in which they reproduce than they are to each other.

ARCHAEA

Archaebacteria (or archeans) are the methanogenic, or methane-producing; halophilic, or salt-loving; and thermoacidophilic, or heat- and acid-loving prokaryotes (**FIGURE 4.5**). They are prokaryotes distinguished from eubacteria primarily by their gene sequences. A small subunit of the ribosome, called the 30S rRNA subunit, shows systematic differences from both eubacteria and eukaryotes. Other characteristics that set archaebacteria apart from other **microbes** (microscopic organisms) include the way lipids, or fats, are linked (using an **ether** rather than an **ester**, which

Amazing Places: Hydrothermal springs and vents FIGURE 4.5

In the spring, the varied colors of Yellowstone's hydrothermal pools reflect different temperature zones and their associated archaebacterial colonies. White crusts develop in places as archaebacteria mediate the precipitation of calcium carbonate.

In addition to hot springs, archaebacteria often make their homes surrounding hydrothermal vents along mid-ocean rifts. Water emitted from the vents is hot and rich in sulfur compounds.

B Archaebacteria are tiny prokaryotic cells that can survive under extreme conditions of temperature and other environmental variables. One group, the thermoacidophils, thrives in water above 80°C. If the water temperature dips below this mark, though, these archaebacteria "freeze."

A Steam rises from a hydrothermal spring in Yellowstone National Park, Wyoming, home to a variety of high-temperature-adapted archaebacteria.

is the organic compound that eubacteria and eukaryotes use to form links), the lack of a cell wall layer (the peptidoglycan layer) comparable to that in eubacteria, and the presence of only one DNA-dependent RNA polymerase (an enzyme).

Archaebacteria are among the most primitive life forms. They are perhaps best known from extreme environments: oxygen-depleted muds and soils, hot salt flats, salty sea shores, hot springs, boiling volcanic environments, and deep-sea hydrothermal vents. Archaebacteria metabolize by means of **chemosynthesis**, which involves con-

■ **chemosynthesis**
Synthesis of organic molecules using chemical energy released through oxidation of inorganic compounds.

version of chemical energy into biologically useful organic compounds. The ability of some archaebacteria to withstand, grow, and reproduce in hot, oxygen-deficient environments, including tectonically active areas, suggests that they may have been the first organisms to evolve on Earth. Our understanding of these fascinating microbes is still in early stages, though, and recent studies have shown that they are not restricted to extreme environments. They also live in seawater, lakes, soils, and other non-extreme settings. So far, the earliest fossil-bearing rocks have not yielded archaebacteria.

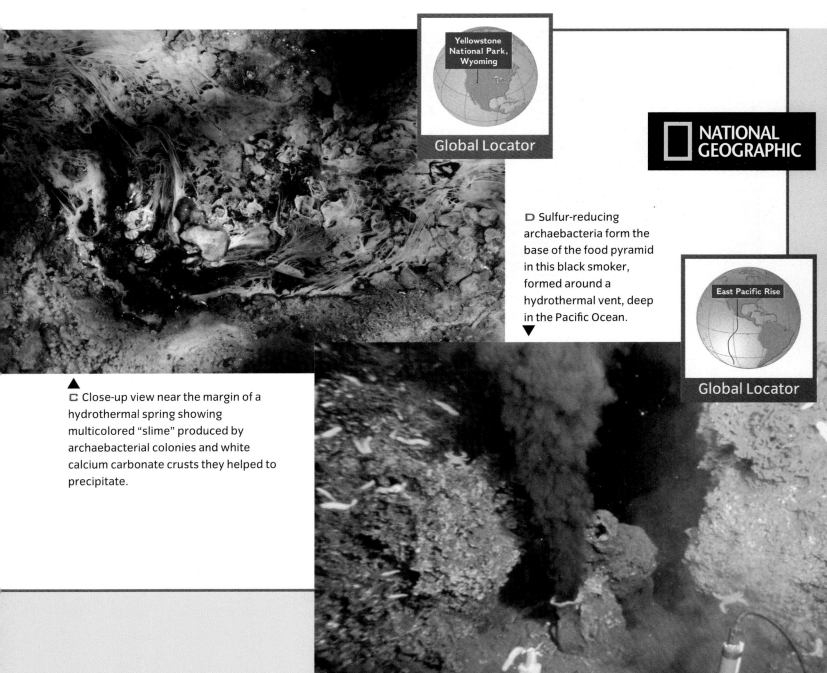

Yellowstone National Park, Wyoming

Global Locator

NATIONAL GEOGRAPHIC

D Sulfur-reducing archaebacteria form the base of the food pyramid in this black smoker, formed around a hydrothermal vent, deep in the Pacific Ocean. ▼

East Pacific Rise

Global Locator

▲ **C** Close-up view near the margin of a hydrothermal spring showing multicolored "slime" produced by archaebacterial colonies and white calcium carbonate crusts they helped to precipitate.

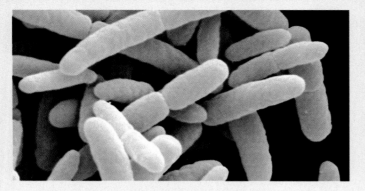

A The bacterium *Escherichia coli* (better known as *E. coli*) is harbored in large numbers in the digestive tracts of mammals, including humans, where it plays a vital role in the digestion of food. Some strains of *E. coli* are pathogenic.

B Bacteria play an indispensable role in the production of cheeses, adding greatly to the distinctive flavoring of varieties.

C Bacteria play a significant role in the decomposition of animal dung, which leads to nutrient recycling through ecosystems.

D A researcher exposes cyanobacteria to ultraviolet light, causing the photosynthetic chlorophyll to fluoresce red.

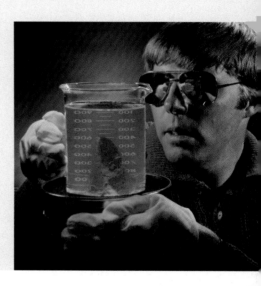

BACTERIA

Bacteria (or eubacteria) are all the prokaryotes except for those classified as archeans (archaebacteria). This group is exceedingly diverse in structure and metabolism. Eubacteria (FIGURE 4.6) range from simple, solitary unicells to more complex stalked, budding, and aggregated forms. Unicells are spherical (cocci), rod-shaped, or spiraled. Eubacteria are dominated by single-celled forms, but multicellularity abounds among the blue-green bacteria (cyanobacteria) and the actinobacteria. Some bacteria have flagella, whereas others form swimming colonies.

Eubacteria usually form organic compounds such as sugars, starches, phosphorous compounds, and proteins in one of two ways: through chemosynthesis (similar to archaebacteria) or through photosynthesis (similar to plants). **Photosynthesis** involves using sunlight to convert carbon dioxide in the atmosphere into organic compounds. In eubacteria, photosynthesis is essentially an anaerobic process occurring in dysoxic water. Conversion of light energy requires chloro-

photosynthesis The process by which plants, algae, and some bacteria create organic molecules from carbon dioxide and water using energy from sunlight.

phyll or some other pigment molecule such as rhodopsin. The chemical energy produced, called adenosine triphosphate (ATP), is a nucleotide used in the transformation reactions of all cells. Chemosynthetic and photosynthetic nutrition are referred to as **autotrophy**, which is the self-production of food and derivation of energy from inorganic sources. Eubacterial cycling of elements and compounds is essential for all life forms on Earth.

Respiration in eubacteria can be either anaerobic (in the case of photosynthetic forms) or aerobic. In many aerobic forms, when oxygen is present, O_2 is reduced to water (H_2O). If oxygen is absent, however, bacteria that are obligate aerobes stop growing. Others (facultative aerobes) do not stop growing but instead respire by reducing sulfate, nitrate, or nitrogen to sulfide, elemental sulfur, nitrite, and nitrous oxide. Many eubacteria are chemoheterotrophic and require reduced organic compounds for both energy and growth. A few oxygen-respiring eubacteria are predaceous, attacking other bacteria and reproducing in the cytoplasm of their prey.

All eubacteria reproduce asexually by binary fission. One cell divides, giving rise to two genetically identical offspring cells. Eubacteria can also donate DNA from one individual to another, but the processes by which this happens are not associated with reproduction. The processes (conjugation and transformation) are, nevertheless, regarded as sexuality.

The most numerous living things on Earth today are eubacteria. They are fast growing, indispensable to the health of animal digestive systems, essential components of soils, major producers of oxygen in the atmosphere, at the bases of microbial food webs, important decomposers, and essential for the production of some foods for humans. Approximately 10,000 different forms of bacteria have been described as "species," but this number underestimates the wide array of bacterial strains that exist.

The ubiquity and abundance of eubacteria is staggering. It has been estimated that a spoonful of garden soil contains roughly 10 billion eubacteria and that the number of individuals in your mouth exceeds the total number of people who have ever lived. Eubacterial cells account for some 10% of the dry weight of mammals. They cover our skin and especially inhabit warm, damp surfaces in the nose, ears, mouth, underarms, and between the toes. The digestive tract, especially the large intestine, is packed with eubacteria. One species, *Escherichia coli* (usually referred to as *E. coli*), is the subject of innumerable studies by biologists and may be the best studied organism on Earth (**FIGURE 4.6A**).

Some eubacteria are pathogenic (disease-causing), but most are not. Many antibiotics, which are compounds formed by one life form that inhibits the growth of another (usually a microbe) come from eubacteria. Streptomycin and erythromycin, for example, are antibiotics produced using eubacteria.

Eubacteria are the oldest-known fossils. Specimens have been dated to nearly 3.5 billion years. Aggregations of photosynthetic eubacteria, mostly filamentous cyanobacteria, form cohesive microbial mats. When microbial mats build mounds, they are called **stromatolites** (**FIGURE 4.7**) if they have fine internal layers and **thrombolites** if they have a clotted internal construction. Such microbially produced structures are known from modern and ancient strata.

Cyanobacteria and stromatolites FIGURE 4.7

Strands of photosynthetic multicellular cyanobacteria are important components of microbial mats that can stabilize sediment surfaces. Consortia of filamentous cyanobacteria trap fine sedimentary particles to produce thinly layered structures called stromatolites. At low tide in Shark Bay, Australia, these modern stromatolitic mounds are fully exposed.

Bacterial metabolic processes are partly responsible for concentrating certain metals in ore deposits. Bacteria promote the growth of manganese-iron nodules on the ocean floor. Gold in carbon-rich rocks of South Africa is associated with fossil eubacteria; the bacteria evidently mediated precipitation of gold during the late Archean. Other metals that may have been concentrated in ores through biochemical processes include silver, copper, lead, and zinc. Some sulfur deposits also may have a link to the biochemical processes of eubacteria.

PROTOCTISTS

Protoctists (**FIGURE 4.8**) are eukaryotic microorganisms and their descendants other than fungi, plants, and animals. They are thought to have evolved from the **symbiosis** of at least two kinds of eubacteria. More than 10,000 living species have

> ■ **symbiosis**
> A condition in which two or more dissimilar organisms live together in close association.

been described, but the total current diversity of the kingdom is estimated to range from 65,000 to more than 250,000 species. The best-known protoctists are algae, oomycotes (water molds, slime molds, and slime nets), and protozoans (the protists such as amoebas, ciliates, and diatoms). Protoctists have a good fossil record dating to the Proterozoic Eon.

Among the protoctists are unicellular and multicellular eukaryotes. Brown algae, oomycotes, and red algae are multicellular, and the rest (especially the protozoans) are single celled. All protoctists have nuclei and other cell organelles. Many also have plastids and mitochondria, and they are photosynthetic. Most protoctists are aerobes, and they have undulipodia at some stage in their life cycle.

Most protoctists live in saltwater, freshwater, or moist soil. Aquatic forms are a major component of plankton, and they include both heterotrophic forms, such as amoebas and photosynthetic algae. Together with microscopic animals, protoctists constitute the bases of many marine and freshwater food webs. Some species are parasitic or live symbiotically in the moist tissues of other organisms.

Protoctists FIGURE 4.8

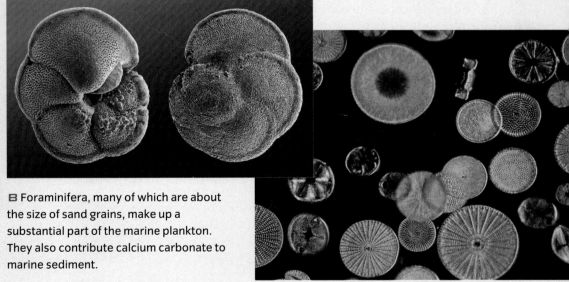

A Offshore of New Zealand, a fur seal peers between leaves of kelp, a brown alga.

B Foraminifera, many of which are about the size of sand grains, make up a substantial part of the marine plankton. They also contribute calcium carbonate to marine sediment.

C Diatoms are photosynthetic protoctists that secrete microscopic siliceous skeletons.

Fungi FIGURE 4.9

A Mushrooms play a vital role in nutrient recycling by breaking down organic matter. This is a photo of a "destroying angel" mushroom growing in a Massachusetts forest. ▶

B Yeasts are single-celled fungi. Here, yeast is added to buckets of grapes to ensure fermentation of wine. ◀

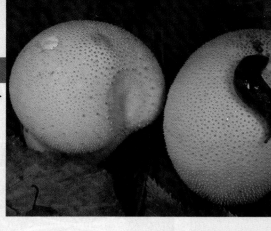

C Gem-studded puffballs expand from the leaf litter in Massachusetts. ▶

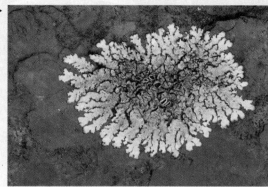

D This lichen, from Kangaroo Island, Australia, is a symbiotic association of a fungus and an alga. ▶

FUNGI

Fungi (FIGURE 4.9) are eukaryotic organisms that develop from chitinous fungal spores (propagules), have chitinous cell walls, and lack undulipodia at all stages of the life cycle. Most fungi are multicellular. They all acquire nutrients by digesting living or dead tissue through a process known as **absorptive heterotrophy**. Fungal cells can have more than one nucleus per cell. Reproduction occurs asexually, through mitosis, or sexually, in which meiosis produces haploid propagules.

Most fungi are terrestrial, but aquatic species also exist. Fungi range in size from tiny single-celled forms to enormous multicellular organisms, the largest known on Earth. A single fungus can extend its hyphal network over an area of tens of km². The diversity of living fungi is estimated to be about 1,500,000 species, of which only 60,000 have been described.

The known fossil record of fungi extends to the early Paleozoic. Evolution of the fungi may have resulted from conjugation of protoctist ancestors. Three phyla of fungi are recognized. The most primitive, Zygomycota, includes black bread mold and other forms, many of which live on decaying vegetation and produce tiny rootlike hyphae.

The phylum Basidiomycota includes the mushrooms, puffballs, jelly fungi, rusts, and smuts. These fungi have a microscopic club-shaped structure, the basidium, which usually bears four haploid spores (basidiospores) produced by meiosis. Basidiomycotes are heterotrophic, and many live symbiotically with plants, producing a sheath of fungal hyphae around plant roots that supplies plants with phosphorous, nitrogen, or other elements. In turn, the plants produce carbohydrates that the fungi can metabolize.

The phylum Ascomycota includes yeasts, blue-green molds, truffles, and morels. Most ascomycotes produce a microscopic reproductive structure called an ascus, in which hyphae of complementary mating types conjugate and produce haploid spores through meiosis. They absorb nutrients from living or dead plant material. Being able to break down cellulose, lignin, and collagen makes ascomycotes important for recycling nutrients through ecosystems. Many ascomycotes produce hyphae as they grow. Exceptions are the yeasts, which are single-celled forms that grow by mitosis.

Lichens (FIGURE 4.9D) are symbionts of ascomycotes and green algae (protoctists) or cyanobacteria. Lichens are photosynthetic.

Fungi are of huge economic importance. Baker's yeasts cause bread to rise by oxidizing sugars to produce bubbles of carbon dioxide. Various flavorings of cheeses result from the growth of fungi, and the holes in Swiss cheese are produced by bubbles of carbon dioxide released by yeasts. One of the best-known yeasts, *Saccharomyces cerevisiae*, ferments sugars into ethyl alcohol, which is necessary for making beer, wine, and spirits. Penicillin, erythromycin, and streptomycin are antibiotics produced from molds. Other pharmaceuticals produced from molds include vitamins, steroids (such as cortisone, hydrocortisone, and prednisone), and interferons (which prevent viral infections). Ergot fungi, which cause a disease of rye flowers, are poisonous to humans and farm animals, but drugs produced from ergot are used to treat migraines and uterine hemorrhages. Fungal diseases have decimated some desirable species of trees, such as the American elm and the American chestnut.

PLANTS

Plants (**FIGURE 4.10**) are multicellular organisms that have complementary sexes and develop from embryos. The life cycle of a plant is characterized by an alternation of generations beginning with an asexual

Plants FIGURE 4.10

▲ A Sporulating mosses populate a rock in the Utah desert.

B Ferns form the understory of a redwood forest in California.
▼

vegetative sporophyte (diploid) stage that produces spores through meiosis. Each spore germinates into a haploid gametophyte. Then the gametophyte reproduces sexually by producing gametes through mitosis. Fertilization of an egg by a sperm or pollen nucleus gives rise to a new diploid embryo and initiates a new sporophyte. Almost all plants are photosynthetic, transforming solar energy, water, and carbon dioxide into food, fiber, wood, and other forms of stored energy. Their metabolism yields oxygen as a by-product. Most plants are adapted to life on land, although some spend part of their lives in water.

About 500,000 species of living plants have been described, but the total diversity is probably at least twice that number. Plants range in size from small mosses to trees that are tens of meters tall. The fossil record of land plants dates to the early Paleozoic.

Plants can be divided into two major groups: nonvascular plants and vascular plants. **Nonvascular plants** include mosses, liverworts, and hornworts. Nonvascular plants often have "leafy" gametophytes with a stalk and rhizoids, but they lack the true stems, leaves, and roots that characterize vascular plants. **Vascular plants** include club mosses, psilophytes, horsetails, seed ferns, ferns, progymnosperms, gymnosperms (cycads, ginkgos, conifers, and gnetophytes), and the angiosperms (flowering plants).

D Rice, a crop plant important to millions of people, matures in a field in Guizhou, China.

C Conifers, like these from the Rocky Mountains of Colorado, are common in alpine areas.

E Flowering plants, like these daisies, have become the dominant form of land plants.

ANIMALS

Animals (FIGURE 4.11) are multicellular organisms, most of which develop from an embryo (blastula) that develops from a diploid zygote. A zygote is produced sexually through fertilization of a large haploid egg by a smaller haploid sperm. Gametes form by meiosis, and once a zygote forms, mitotic cell divisions result in a solid ball of cells that hollows out to become a blastula. In the embryonic development of many animals, the blastula develops an opening (the blastopore), which will eventually become one of the openings of the digestive system

Animals FIGURE 4.11

A Offshore of Vancouver Island, Canada, microscopic zooplankton, reflecting as tiny bright spots, are ensnared in the polyps lining the branches of a large frondlike sea pen (soft coral).

B Worms, which include representatives of several phyla, include forms important as sediment burrowers in the marine and terrestrial realms. This is a marine worm.

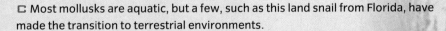

C Most mollusks are aquatic, but a few, such as this land snail from Florida, have made the transition to terrestrial environments.

D A diver off Belize encounters a starfish nestled in sand between corals, sponges, and algae.

(either the mouth or anus, depending on the type of animal). In animals showing spiral cleavage, however, the blastula remains a solid mass of cells. The cells of a blastula may migrate or differentiate to transform it into a gastrula, which is an embryo having a pocket that eventually becomes the digestive cavity.

Animals acquire nutrients through **heterotrophy**, and most ingest food through an oral opening (a mouth or an equivalent structure). Some animals are parasitic and lack a

heterotrophy A means of obtaining nutrients by ingesting or breaking down organic matter.

E Butterflies are pollinators of flowering plants. This monarch butterfly is alighting on a thistle flower.

F A red-margined wrasse fish is camouflaged among the branches of a pink soft coral in the Pacific Ocean offshore of Fiji.

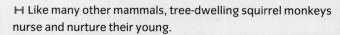

G A small lizard lies in wait at the opening to a termite nest (upper left) in the outback of South Australia.

H Like many other mammals, tree-dwelling squirrel monkeys nurse and nurture their young.

digestive system. Rarely, animals take in photosynthesizing symbionts that help them acquire essential nutrients. Reef-forming coral polyps harbor algal symbionts. Other animals that inhabit hydrothermal vents (deep-sea smokers) and cold seeps (where cold water rises from the sea floor) shelter chemosynthesizing archaebacterial symbionts. They either digest the archaebacterial cells directly or digest the organic molecules their symbiotic partners produce.

The multicellular condition, although present in all kingdoms except Archaebacteria, is most highly developed in animals. Animals normally have cells specialized for functions such as respiration, digestion, and protection, and these cells are grouped into tissues and organs. A wide range of body sizes, ranging from microscopic zooplankton to enormous whales, are known among animals.

Behavior of some sort is known from members of all kingdoms, but animals have the most complex and diverse behavioral patterns. In other organisms, behavior is usually limited to attraction to light, sensing of temperature and dissolved gases, and avoidance of harmful chemicals. Nervous systems that control behavior are unique to animals.

Most animal groups are dominated by marine forms. Exceptions are the terrestrial insects, non-insect hexapods, most chelicerate arthropods (spiders, scorpions, and their kin), and tetrapods (amphibians, reptiles, mammals, birds, and their kin). Some crustacean arthropods, such as crayfish, ostracodes, and branchiopods, are also terrestrial or freshwater adapted. Likewise, some mollusks, such as snails and clams, are terrestrial or freshwater adapted. Compared to the diversity of animal groups living in marine habitats, the number of animal life forms in freshwater is relatively small.

Animals are the only group to have successfully invaded the atmosphere. Although representatives of other kingdoms spend part of their lives airborne, usually as spores or seeds, some animals spend most or all their lives in the air. The capability for flight evolved several times in animals, but only in animals. Ones that evolved structures for true flight are insects and some vertebrates (pterosaurs, birds, and bats).

Animals can be divided in a couple different ways. One way is to distinguish animals that lack backbones (the **invertebrates**) from those that have back-bones (the **vertebrates**). Another way is to distinguish the animals that lack tissues organized into organs (the **parazoans**) from those that have tissues organized into organs and organ systems (the **metazoans**). Parazoans include placozoans and sponges. Metazoans comprise the cnidarians, ctenophores, flatworms, nematodes, priapulid worms, bryozoans, phoronids, brachiopods, annelid worms, sipunculid worms, echiurid worms, mollusks, echinoderms, arthropods, hemichordates, chordates, and other groups. The vertebrates make up only some of the chordates—the craniates (jawless fishes and gnathostomes). All the other groups can be referred to as invertebrates.

The fossil record of animals extends to the late Proterozoic. Some scientists have argued that they appeared perhaps 600 to 1,500 million years ago but did not leave fossils until much later in their history. Animals account for the largest proportion of macroscopic organisms preserved as fossils in Phanerozoic strata.

CONCEPT CHECK **STOP**

HOW are biologic organisms classified?

What is a species? What is a biological species? What is a paleontological species?

What are the main differences between prokaryotes and eukaryotes?

What are the major characteristics of archeans, bacteria, protoctists, fungi, plants, and animals?

Ecology of Life Forms

In order to survive, living organisms need food and water, tolerable physical and chemical conditions, and protection from enemies and extreme conditions. **Ecology** is the study of factors that influence the distribution and abundance of species in nature. The same factors also influenced the distribution and abundance of species in the geologic past, and we refer to the study of ecologic conditions affecting ancient organisms as **paleoecology**.

Habitats are settings on or close to Earth's surface inhabited by life forms. The distribution of organisms is largely controlled by access to water, so habitats can be broadly divided into **terrestrial** (land) and **aquatic** (water) settings. The type of water, whether **freshwater**, saltwater (**marine**), or **brackish** (slightly salty) water, also may be a limiting factor. In marine habitats (oceans and seas), organisms such as starfish

paleoecology
Study of the factors controlling the distribution and abundance of ancient species.

hypsometric curve A graph indicating the proportions of Earth's surface above and below sea level.

and corals depend on saltwater; in nonmarine habitats, organisms such as horses and trees depend on freshwater. Some creatures, such as oysters and fiddler crabs, can survive in brackish water environments where freshwater from rivers mixes with salty marine water. Terrestrial and freshwater habitats can also be grouped together as **nonmarine** environments.

The relative proportion of area above sea level and below it is expressed by the **hypsometric curve** (FIGURE 4.12). Today, about 70.8% of Earth's surface is covered by marine water, and about 29.2% is nonmarine (dry land and freshwater). Sea level position has not been constant through time, and the hypsometric curve shows that a relatively small increase in sea level will result in the flooding of a large area of the continents. In contrast, a small drop in sea level will not result in much gain of nonmarine area.

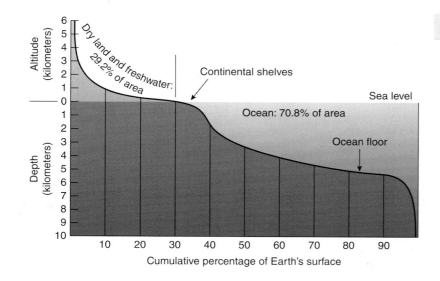

Land and sea habitats FIGURE 4.12

The hypsometric curve relates the relative position of sea level to elevation and bathymetry. It indicates the proportion of Earth's surface covered by marine water and can be used to predict how changes in sea level position will affect the relative proportion of marine and terrestrial habitats. Living organisms have adapted to nearly every conceivable terrestrial and aqueous habitat on Earth.

Aside from water and its salinity level, important physical limitations on the distribution of organisms are oxygen, sunlight, and temperature. Whereas large multicellular animals require access to oxygen (**oxic** conditions) to survive, many eubacteria and archaebacteria either do not require it or are poisoned by it. Oxygen-poor conditions are referred to as **dysoxic**, and conditions in which oxygen is lacking altogether are called **anoxic**. Access to radiant energy from the Sun determines whether plants, algae, diatoms, cyanobacteria, and other organ-

isms can photosynthesize and create the food necessary to sustain trophic webs. Extreme temperature conditions create inhospitable conditions that limit where organisms can live. Animals normally live in places where the temperature ranges from about −10°C to 40°C, but some archaebacteria that have adapted to hot water conditions "freeze" and die in water below 80°C.

Niches (or **ecologic niches**) are the ecologic positions of species in their environments, or the ways species relate to their environments. Niche spaces are usually de-

Food webs FIGURE 4.13

Energy is transferred within an ecosystem through a complex food web involving organisms that have varied life modes. A This diagram shows paths of energy transfer within a terrestrial ecosystem. Autotrophs (including photosynthesizers, I), produce energy by themselves.

Energy then passes through consumers (herbivores II and carnivores III, IV) and detritivores (or decomposers). When consumers release waste or die, they return nutrients to the system, where they are used again by biologic organisms. Detritivores have a crucial role in this nutrient recycling.

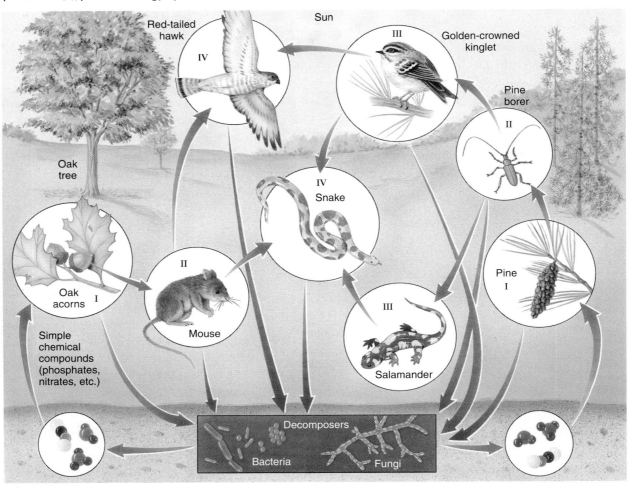

fined by nutrient requirements (chemosynthesis, photosynthesis, or heterotrophy), physical and chemical conditions, and interactions with other species, including predation deterrence or competition for the same food or dwelling space.

MODES OF LIFE

One way of categorizing organisms is according to the way they live within a niche. This is called **mode of life** (or **life habit**). It is a way of specifying how organisms acquire food or other nutrients, where they fit in the food web (FIGURE 4.13), how they reproduce, whether they are sessile or mobile, and where they physically live (on the land surface, on the sea floor, within the water column, or elsewhere).

Primary producers, or **autotrophs**, use raw materials in the environment to make their own food. Plants, algae, cyanobacteria, and diatoms usually use sunlight and chlorophyll to manufacture sugars. **Chemoautotrophs**, especially archaebacteria, derive their energy

B A marine food web operates similarly to a terrestrial food web. Autotrophs (I) produce the energy they need, and energy then passes through various levels of consumers (II–V) to detritivores (decomposers). Finally, the detritivores recycle nutrients to the system.

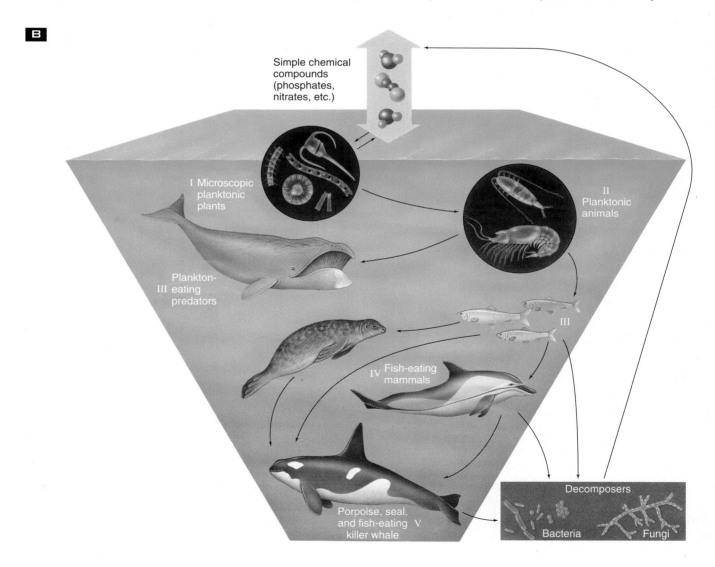

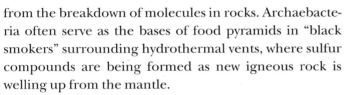

Benthic and pelagic animals FIGURE 4.14

A Benthic animals, like these fan worms and sponges, dwell on the floor of the Pacific Ocean.

B This white tip shark, a pelagic animal, swims in the waters of the Bahamas.

from the breakdown of molecules in rocks. Archaebacteria often serve as the bases of food pyramids in "black smokers" surrounding hydrothermal vents, where sulfur compounds are being formed as new igneous rock is welling up from the mantle.

Animals, fungi, and many eubacteria are **consumers**. They derive nutrients by either feeding on, or causing the decay of, organic tissues. **Herbivores**, such as cows, deer, and some snails, are consumers that feed on plant, algal, and fungal material. **Carnivores**, such as cats, dogs, and sharks, eat animals. **Omnivores**, such as humans and bears, subsist through a combination of herbivory and carnivory. Among carnivores, **predators** subdue live prey and eat it, and **scavengers** feed on dead animal carcasses. **Parasites**, such as tapeworms, live on or within other living organisms. Finally, **detritivores**, such as many fungi, obtain nutrients through the decay of tissues, usually after the organisms, or parts of them, are dead. **Deposit feeders**, such as earthworms, ingest mud and extract organic materials from it. **Suspension feeders**, such as sponges or crinoids, live in water and pass currents through a natural filter. Food particles, including plankton and other organics, are captured in the filter and then ingested.

Organisms that live on the floor of the sea, a lake, or a stream are referred to as **benthic** organisms (FIGURE 4.14A). If they live on the sediment, they are called **epibenthic** (**epifaunal** if they are animals or **epifloral** if they are other organisms). If they live in the sedi-

ment, they are called **infaunal** (if they are animals) or **infloral** (if they are other organisms). Corals are good examples of epibenthic animals, and many sea urchins and marine worms are examples of infaunal animals. Boring organisms, including some clams and fungi, make excavations into hard substrates such as shells or rocks.

Organisms that live in the water column are called **pelagic** (FIGURE 4.14B). Pelagic organisms can be further described as nektonic, planktonic, or pseudoplanktonic. **Nektonic** organisms, such as fish, are swimmers. Planktonic organisms, such as microscopic diatoms and kelp, are floaters. **Pseudoplanktonic** organisms live attached to other things that float or swim. Some barnacles, for example, live attached to floating logs.

In the terrestrial realm, organisms can be mobile (like animals), sessile (like plants), live on the surface, burrow or bore into the ground, or fly.

CONCEPT CHECK STOP

What are the major physical, chemical, and biologic factors that govern the distribution and abundance of life forms?

What is an ecologic niche?

What are the modes of life of marine and terrestrial organisms?

Types of Fossils and How They Form

LEARNING OBJECTIVES

Define the meaning of "fossil."

Distinguish the three types of fossils.

Explain what information can be inferred from body fossils, trace fossils, and biomarkers.

Explain the process of body fossil preservation.

Differentiate between unaltered and altered preservation of body fossils.

Explain what conditions favor fossilization.

Distinguish "normal" fossil preservation from "exceptional" fossil preservation.

fossil is any evidence of ancient life. To a geologist, the term "ancient" usually implies something that is pre-Holocene in age. We can use a figure of 11,800 years as a rough estimate for the Pleistocene–Holocene boundary, even though the position and age of the boundary have not yet been fixed. Evidence of past life that is younger than the Pleistocene Epoch is referred to as **subfossil**.

Fossils are of three basic forms: body fossils, trace fossils, and biomarkers. **Body fossils** (FIGURE 4.15) are direct or altered *remains* of ancient organisms. Any

fossil Any evidence of ancient life.

body fossil Direct or altered remains of an ancient organism.

portion of an organism's body, such as a bone, tooth, shell, test, leaf, or stem, is considered a body fossil. Even a mold, or an impression, of a body that was once buried in sediment is considered a body fossil.

Body fossils are valuable biologic records. They clearly indicate the size and shape of organisms, information that is useful for inferring the limits of variation within species, and, to varying extents, the mode of life and behavior of organisms. Body fossils give us the basis for understanding paleoecologic relationships. Perhaps most importantly, body fossils afford us with the best long-term record of the evolution of life on Earth.

Body fossils FIGURE 4.15

Body fossils are the direct or altered remains of ancient organisms.

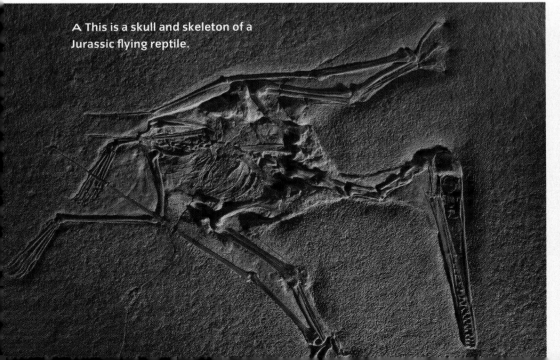

A This is a skull and skeleton of a Jurassic flying reptile.

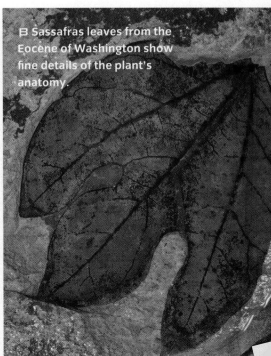

B Sassafras leaves from the Eocene of Washington show fine details of the plant's anatomy.

Body fossils are useful for solving a variety of geologic problems. They are one of our finest tools for stratigraphic correlation, useful especially for tracing time-equivalent strata across great geographic distances. The paleoecologic information they impart helps us recognize and interpret ancient sedimentary environments and even how those environments changed in terms of temperature, seasonality, aridity, oxygen availability, water salinity, or other conditions. They tell us about the long-term effects of invasive species. They also give us clues to large-scale patterns of land-sea configurations, water depth, and global climate.

Trace fossils, which are sometimes called **ichnofossils**, are evidence of the *activity* of ancient organisms, the prin-

■ **trace fossil**
Evidence of the activity of an ancient organism. Also called an ichnofossil.

cipal source of information about the behavior of ancient organisms (**FIGURE 4.16**). Behavior is most highly developed in animals, so most trace fossils are records of animal activity. They are especially informative for determining paleoecologic relationships. Tracks and trails document the walking, running, gliding, or other movement of animals. Burrows excavated by animals in sediment, and borings produced in hard substrates, such as solid rock, shell, or bone, provide insight into where creatures sought refuge, established homes, or nested. Bite marks, scavenging marks, scratches left in sediment as it was combed for organic nutrients, and other traces left as the result of food gathering, provide an understanding of the feeding behavior of

Trace fossils FIGURE 4.16

Trace fossils are evidence of the behavior of ancient organisms. They include indicators of walking, resting, burrowing, and feeding.

◀ **A** Three-toed footprints were left by a carnivorous dinosaur in wet sand that later lithified to sandstone in what is now Arizona.

B Tubular burrows in a Cambrian sandstone of Sweden are probably the work of infaunal worms.
▼

▲
C Subtle swirls in a Devonian siltstone bed in New York record the activity of benthic marine worms that extended from their vertical burrows and swept the sediment surface in search of food.

D A healed bite mark left ▶ on a Cambrian trilobite from Utah tells the story of an early predator-prey interaction and the prey that got away from its attacker. This trilobite is 3 cm long.

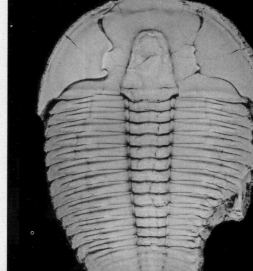

organisms. Coprolites, or fossilized fecal matter, tell us about the diets of ancient animals.

Trace fossils are usually classified a little differently from body fossils. The classification of forms is based on a combination of shape, ethology (or behavior) represented, and sometimes tracemaker. In many instances, formal Latinized names (**ichnogenera** and **ichnospecies**) are used for trace fossils, but the names are not exactly equivalent to names used for paleontological species.

Trace fossils provide a crucial type of paleoecologic information. Traces formed in sediments—footprints or crawling traces, for example—provide definitive evidence that an animal was in a particular location at some point in the geologic past. Traces in unconsolidated sediment cannot be moved by currents because the particles will disaggregate, and the traces will be destroyed. In contrast, bodily remains can be transported after death. In most environments, they are not transported far, but in a few instances, they can be moved to adjacent areas where the creatures may not have actually lived. When rivers enter flood stage, all sorts of objects are picked up and carried along by rushing water. Trees and animal carcasses may even be carried out to sea before being deposited. If they were transported, their remains can provide misleading clues about ancient environments. Similarly, hurricanes and typhoons can stir up and transport bodily remains great distances along marine coasts, again providing misleading paleoenvironmental clues.

Finally, trace fossils can assist in stratigraphic correlation. Their usefulness as proxies of geologic time is limited, however, because traces formed in sediment tend to be strongly controlled by sediment type.

biomarker An organic compound that serves as a "fingerprint" to a type of organism.

Biomarkers, which are sometimes referred to as **chemical fossils**, are organic compounds found in strata that serve as "fingerprints" to specific types of organisms. Some biomarkers are organic chemicals produced through the metabolic activities of organisms during their lifetimes, and others are chemical by-products of their decay.

Biomarkers provide evidence for the former presence of an organism in places where body and trace fossils may be lacking. In petroleum exploration, biomarkers are used to identify the organic sources of the oil, and this helps to determine ancient environments of deposition. It also helps to put limits on the timing of petroleum maturation and migration.

HOW BODY FOSSILS ARE FORMED

Understanding how organisms become preserved as fossils is fundamental to interpreting Earth history and the record of biologic evolution. Far fewer organisms leave documentation of their existence than those that do, and correct interpretation of the meaning of the fossil record hinges on knowledge of what, how, and why organisms get preserved as fossils. The study of the processes leading to fossilization is called **taphonomy**.

taphonomy Study of the processes of fossilization.

The transition from a living community of organisms to a fossil assemblage can be regarded as a series of taphonomic filters (**FIGURE 4.17** on pages 110–111). Taphonomic processes usually begin when an organism dies or releases body parts such as hair or teeth. Under most circumstances, organisms do not stay intact for long because of biologic recycling of organic tissues and biominerals through decay and disarticulation. Most of the breakdown that occurs in the first several days is caused by scavenging and microbial decay, and the first body parts to disappear are those prized as food or nutrient sources to heterotrophic organisms—internal organs, muscle, flesh, fruits, berries, and other edible parts.

Typically, after a few days or weeks, few remains of a carcass are left to the environment to potentially become a fossil. Bones, teeth, shells, wood, and other relatively durable and nearly inedible materials are the remains most likely to survive the initial taphonomic filtering.

Even relatively durable body parts may eventually break down through physical, chemical, and biologic means before they reach sufficient antiquity to be considered fossils. On beaches, pounding waves take their toll on shells. Wind abrasion, the Sun's radiation, and freeze–thaw cycles degrade even the toughest of

If an organism is to successfully make the transition from life to a fossil, it must pass through a series of taphonomic filters, and one of the most important of them is breakdown of body parts by consumers.

A Bodily parts most likely to vanish first are those used by other organisms for food. Here, a lioness in Botswana finishes a meal. When she is done with this recent kill, scavengers such as birds and hyenas will arrive for an opportunity at the scraps. Insect larvae and microbes will also play a big role in decomposing the soft parts of the carcass.

B Durable hard parts, like bones and teeth, often stand a good chance of surviving the effects of carnivores and decay agents, so they are the body parts most likely to become buried and preserved as fossils. This kangaroo skull in Australia is all that remains of a recently dead animal.

C Burial in sediment is generally a prerequisite for an organism to become fossilized. Burial takes place quickly in deltaic areas, near the mouths of rivers, where enormous quantities of sediment are delivered to marine shelves. In this photo, the Dungeness River is shown emptying sediment into the Strait of Juan de Fuca, Washington. The recently deposited sediment appears as pale areas in the ocean water. Bones, teeth, shells, and other hard skeletal parts that survive long enough to be buried in sediment stand good odds of becoming fossils.

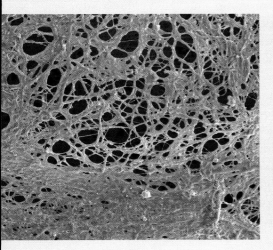

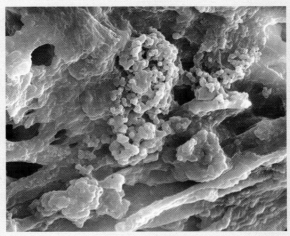

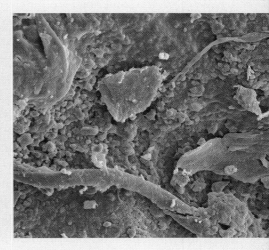

D Preserving "soft," or nonbiomineralized, tissues requires conditions somewhat different from simple burial of hard parts. Microbes play a vital role in this process. Shortly after death, an organism decaying in water becomes surrounded by a slimy halo of microbes. Through the scanning electron microscope (SEM) the decay halo comes "alive" as a network of stringy fungal hyphae and clusters of tiny, round eubacterial cells.

E Upon closer examination, the clusters of bacteria resolve into a great mass of tiny cells, each about 10 μm in diameter. The "string" is a fungal hypha.

F Under the right chemical conditions, many eubacteria can biomineralize. Once decay has proceeded far enough, microbes can initiate mineral precipitation on the "soft" tissues they were decaying, and that lithifies the tissues. These are fossilized microbes preserved in pyrite around the organic tissues of a Devonian cephalopod from New York.

skeletons exposed to the air. Bones may be crushed by scavengers or the feet of herding animals, and microscopic fungi may bore tiny cavities into shells that eventually break them into small pieces. Fungi and insects are likely to make short work of woody tissues once they bore into them.

To end the breakdown process, biologic remains need to be removed from the agents of decay and destruction (known as **biodegraders**) fairly quickly in order to survive long enough to become fossils. Burial in sediment and encasement in tree sap, tar, or glacial ice are some of the ways to remove biologic remains from scavengers and decay agents.

Even if organic remains are buried, there is no guarantee that they will become fossils. Animals digging in sediment may disrupt and destroy bones and shells, and slightly acidic water in the pore spaces between sedimentary grains can dissolve them.

At some stage, sediment usually lithifies, or turns to rock. Biologic remains still left in the sediment will become part of the rock. In the process of lithification, some remains are likely to change mineralogically, and others are not. Aragonite shells of mollusks often recrystallize to the more stable mineral calcite, but phosphatic bones and teeth rarely change in composition. The timing of lithification varies widely according to sediment type, environment of deposition, and geologic history of an area. In carbonate environments, sediments can turn to rock within a few years. In most siliciclastic environments, sediment lithification probably occurs within tens of millions of years. Many Cenozoic sediments of the deep sea, though, are still unconsolidated.

Enclosure of fossils in rock does not ensure that they have become indestructible. Rocks can become entangled in tectonic events where extreme metamorphism can destroy the fossils. When exposed at the Earth's surface, the rocks will be weathered and eroded. The fossils, after they are exposed, will inevitably be destroyed as physical breakdown and chemical dissolution turn them into smaller and smaller sediment grains unless they are collected and curated.

Now that you understand what reduces the chances of fossilization, it is easier to understand what increases the odds. Under ordinary circumstances, having relatively durable and inedible hard parts (bones, teeth, shells, tusks, or wood), is a prerequisite for fossilization. The fossil record is strongly biased in favor of biomineralized remains such as calcite or aragonite shells, and phosphatic teeth and bones. The fossil record is biased against nonbiomineralized insects, spiders, and plant leaves.

Another prerequisite for fossilization under most circumstances is relatively quick burial in sediment or encasement in some other sedimentary material such as resin, tar, or ice. The fossil record is biased in favor of organisms living in areas of high sedimentation rate (coastal marine areas, river deltas, and others; FIGURE 4.17C), and it is biased against those living in most terrestrial environments where erosion predominates.

The more body parts an organism produces in its lifetime, the greater its chances of leaving a record of its existence. A clam forms a shell of two valves, each of which can potentially become a fossil. A fish, however, could have 200 or more bones and teeth, plus a large number of scales, that could potentially become fossils. Arthropods, most of which molt their exoskeletons as they grow, can potentially leave a version of their outer covering for each growth stage they pass through. Trees that lose their leaves annually contribute many potential fossils to the environment.

Species that have high reproductive rates stand increased odds that some individuals from a population will fossilize. Tiny planktonic foraminifera, which reproduce quickly and in large numbers, leave many more fossils than the more slowly reproducing whales that depend on them for food.

HOW BODY FOSSILS ARE PRESERVED

Most body fossils are preserved in altered condition, but occasionally they are preserved in unaltered state (see *What a Geologist Sees*). Body parts that have not changed significantly in the process of fossilization are, except in rare instances, biomineralized. The ivory of Pleistocene mammoth tusks is almost always unaltered. Phosphatic bones and teeth of vertebrates, as well as phosphatic shells of some brachiopods, usually remain unaltered even for hundreds of millions of years. Some forms of calcite, and rarely other skeletal types, can be preserved in unaltered condition.

Types of Body Fossil Preservation

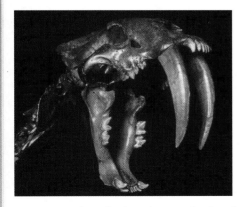

A Bones and teeth of a saber-toothed cat have been preserved in essentially unaltered state by encasement in the La Brea Tar Pits (Pleistocene) of Los Angeles, California.

B A cluster of corals from the Carboniferous of Texas showing recrystallization of the original aragonite skeleton to calcite, another form of calcium carbonate.

C Two matching sides of a concretion from the Devonian of Bolivia in which the exoskeleton of a trilobite has been dissolved away, leaving only internal and external molds.

D A Triassic crinoid from Germany whose original calcite skeleton has been replaced by another mineral—pyrite, or fool's gold.

E Colorful silica has permineralized this ancient wood. This fossil log is in Petrified Forest National Monument, Arizona.

F Fossil leaves preserved by carbonization in Eocene shale of Wyoming.

Forms of alteration in fossil preservation are recrystallization, moldic preservation, mineral replacement, permineralization, and carbonization (examples of which can be seen in *What a Geologist Sees* on page 113).

- **Recrystallization** involves incremental conversion of the original biomineral of a skeleton to another mineral (FIGURE B). Aragonite forming the skeletons of many mollusks and corals frequently recrystallizes to calcite, which tends to be more stable than aragonite over millions of years.

- **Moldic preservation** occurs when skeletal material in sedimentary rock dissolves away, leaving only a cavity and impressions in the matrix (FIGURE C). **Mineral replacement** usually occurs when the cavity formed by dissolution of a skeletal part is filled by a new mineral such as pyrite (FIGURE D).

- **Permineralization** results from the filling of pore spaces, usually by silica (chalcedony or opal), in a porous substance such as bone or wood (FIGURE E). The cellulose of wood also typically becomes replaced in the permineralization process.

- **Carbonization**, which usually affects plant or algal material, occurs once the material is buried deeply enough for heat and pressure to force volatile elements and compounds to escape, leaving behind black carbon (FIGURE F).

The fossil record is overwhelmingly dominated by biomineralizing organisms, although in unusual cases, nonbiomineralized parts, such as skin, organs, hair, and feathers, are preserved. **Exceptional preservation** is the term used to describe fossilization of these weakly resistant organic parts (FIGURE 4.18). In order for exceptional

Exceptionally preserved fossils FIGURE 4.18

Exceptionally preserved fossils are ones that, under ordinary circumstances, would not normally leave a body fossil record. The effects of predators, herbivores, and microbial breakdown agents must be mitigated for exceptional preservation to occur.

▲ B These Cambrian trilobites from the Burgess Shale of British Columbia, Canada, show a combination of normal preservation of the calcitic exoskeleton and exceptional preservation of the chitinous appendages.

▲ A An insect perfectly preserved by encasement in amber from the Neogene of the Dominican Republic.

C A fossil insectivore, related to the modern hedgehog, from the Neogene of Germany, showing preservation of skin, fur, spines on the back, and the animal's last meal in the stomach cavity.
▼

preservation to occur, the impact of predators, scavengers, and microbial decay must be minimized. Anoxia, salinity stress, desiccation, and rapid burial are some of the conditions under which the activity of biodegraders can be suspended. Exceptional preservation often involves precipitating fine mineral coatings over organic tissues, and the process apparently takes place within a few weeks of death. In other words, exceptionally preserved bodies become lithified long before the sediments do. Some of the microbes involved in decay can play a role in the precipitation of minerals that coat the nonbiomineralized tissues (Figure 4.17D–F on page 111).

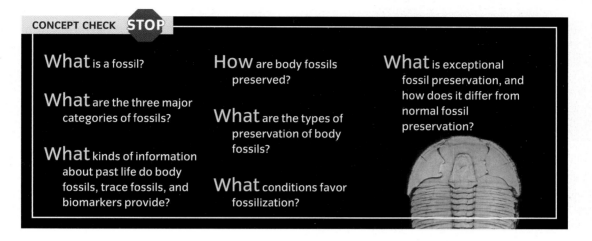

CONCEPT CHECK STOP

What is a fossil?

What are the three major categories of fossils?

What kinds of information about past life do body fossils, trace fossils, and biomarkers provide?

How are body fossils preserved?

What are the types of preservation of body fossils?

What conditions favor fossilization?

What is exceptional fossil preservation, and how does it differ from normal fossil preservation?

Major Groups in the Fossil Record

LEARNING OBJECTIVES

Explain what organisms are represented in the fossil record.

Explain what parts of organisms have been preferentially preserved as fossils.

R epresentatives of all six kingdoms are known from fossils. Fossilization processes, however, have led to uneven representation of major groups. In general, organisms that have a good fossil record are those with hard, biomineralized parts or those with resistant parts of other types (such as wood).

Archaebacteria may have the longest evolutionary history on this planet, but they have little recognized fossil record, partly because we have not necessarily looked for them in the right places or have not looked using the right technological tools. The handful of archaebacterial fossils that we know of today are mostly associated with volcanic vents (FIGURE 4.19). Such settings may seem to be unusual places for fossils, but they could be good places for us to enrich our knowledge of the evolutionary history of archaebacteria.

Ancient archaebacteria FIGURE 4.19

Fossil archaebacteria look remarkably similar to those from the modern world. Here, Jurassic archaebacteria are preserved as irregular organic masses in rock associated with a hydrothermal vent in Antarctica. The length across this SEM image is about 200 μm.

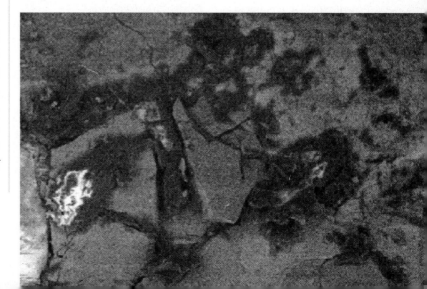

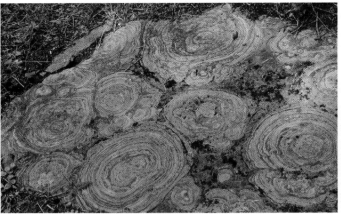

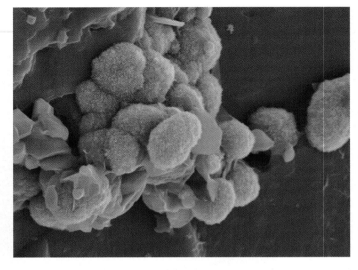

A Stromatolites are formed by cyanobacteria that cement numerous thin layers of sediment to form mounds. These are coalescing pillar-like examples from the Cambrian near Saratoga Springs, New York, that reveal the internal growth laminations. The columns were truncated by erosion shortly after they formed, showing that they were already solidifed before the spaces between the columns were filled in by loose sediments.

B Eubacteria seem to be widespread in the fossil record. Here is a coprolite (fossil excrement) from a Cambrian deposit of Nevada shows eubacterial cells that were probably responsible both for decay and fossilization. The fossil eubacterial cells are about 10 μm in diameter.

Eubacteria, whose remains have been reported from strata nearly 3.5 billion years old, are the oldest known body fossils. They may have a greater antiquity, however, because chemical breakdown products (biomarkers) attributed by some to eubacteria have been reported from even older rocks. Most of the known record of eubacteria comes from stromatolites (FIGURE 4.20A), thrombolites, and other structures they constructed.

These structures are the primary physical evidence of life through the Archean Eon and most of the Proterozoic Eon. Some tiny eubacterial cells have been found in association with exceptionally preserved organic tissues (FIGURE 4.20B), and in these instances, they were evidently agents of fossilization.

Protoctists are well represented as fossils (FIGURE 4.21). Some red algae make calcareous structures im-

A Foraminiferans are generally tiny marine protoctists that come in a variety of shapes, like those shown here, which were separated from deep sea sediments.

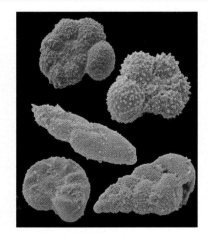

B Fusulinids are small rice-shaped foraminiferans common in Carboniferous and Permian rocks.

portant in the construction of Phanerozoic reefs, and green algae have been a major source of minute calcareous needles forming Phanerozoic carbonate muds. More than 40,000 species of **foraminifera**, a protozoan group, have been described from fossils dating back to the Cambrian Period. Most are small in size.

Other protoctists that have a rich fossil record are **acritarchs**, **coccolithophorids**, **diatoms**, and **radiolarians**. Tests of the organic-walled acritarchs extend back to the Proterozoic Eonothem. They are common enough in certain intervals of the Proterozoic and Phanerozoic to serve as guide fossils in stratigraphic correlation. Small, circular, calcitic plates that comprise the tests of coccolithophorids are the major components of Cretaceous chalk deposits. More than 70 genera of siliceous diatoms (Figure 4.8C on page 96) are known from fossils. They, and tiny, spinose radiolarians, are a major source of silica in sedimentary deposits.

Ancient fungi are known mostly from fossilized hyphae (Figure 4.17F on page 111), although their presence has also been inferred from some types of borings in wood and shell. Hyphal remains are known from both nonmarine and marine deposits.

Plants have left a good fossil record of leaves, stems, bark, and seeds or spores (FIGURE 4.22). Spores of putative terrestrial plants first appear in Cambrian strata. The first clear evidence of other tissues, mostly stalks and reproductive organs, comes from Silurian strata. Vascular tissues, stems, roots, and leaves are important components of many terrestrial sedimentary deposits.

Animals, principally those having hard skeletal parts, have perhaps the most diverse fossil record of any group (FIGURE 4.23). Animal fossils extend back more than 550 million years to the late Proterozoic. Some scientists have argued that animals appeared before that time, perhaps as much as 600 to 1,500 million years ago, but did not leave body fossils until much later in their history. Mollusks, brachiopods, bryozoans, corals, sponges, echinoderms, some arthropods, and vertebrates are well represented in the fossil record. In addition to a body fossil record, animals have also left an excellent trace fossil record.

Fossil plants FIGURE 4.22

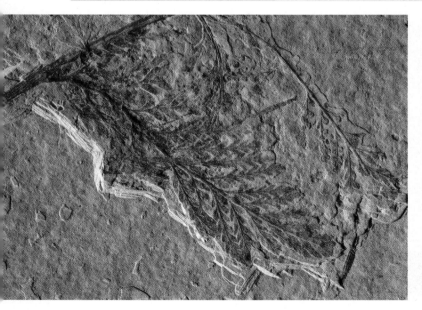

A A Cretaceous fern from the Liaoning deposit of China.

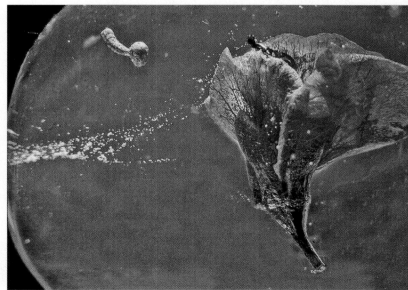

B A Miocene flower from the Dominican Republic was trapped in tree sap just as it was releasing pollen. The sap from a conifer tree later fossilized to amber.

A A colonial soft coral preserved in an Ediacaran sandstone layer of Australia.

B A cluster of Devonian brachiopods preserved in a limestone bed of Ontario, Canada.

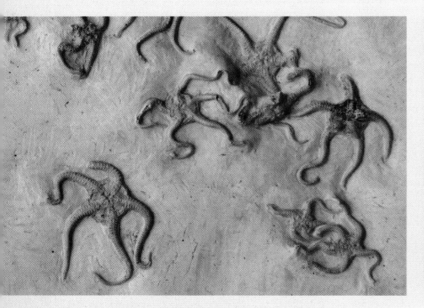

C Ophiuroids, or brittle stars, a group of echinoderms, are shown here on a slab collected from the Pliocene Series of Italy.

D A freshwater reptile from the Cretaceous of Lioaning, China, has been preserved with a small fish near its mouth.

CONCEPT CHECK **STOP**

Which kingdoms are represented in the fossil record?

What bodily structures have been preserved as fossils?

SUMMARY

1 Classification of Life on Earth

1. Life forms are classified taxonomically according to the Linnaean classification system. The **species** is the basic unit of classification. Increasingly larger groups of species are called genera, families, orders, classes, phyla (or divisions), kingdoms, and domains.

2. In the Linnaean classification system, species are referred to using **binominal nomenclature**. Genus and species names are written in Latin.

3. In the **biological species concept**, a group of individuals can interbreed and produce fertile offspring. In the **paleontological species concept**, species of ancient organisms are defined using morphologic similarities and differences.

4. Life forms are fundamentally categorized as **prokaryotes** or **eukaryotes**. Prokaryotic organisms have simple cells that are small, lack nuclei, and lack heritable organelles. Eukaryotic organisms have more complex cells that are relatively large, have nuclei, and have heritable organelles.

5. Archaebacteria (or archeans) are prokaryotes that are often methanogenic, halophilic, or thermoacidophilic. They have ribosomal RNA sequences and other characteristics that are distinctive from other organisms.

6. Eubacteria (or bacteria) are all the prokaryotes except for the archaebacteria. They include simple, solitary unicells to more complex stalked, budding, and aggregated forms.

7. Protoctists are the eukaryotic organisms other than fungi, plants, and animals. Most (such as amoebas, ciliates, and diatoms) are unicellular and microscopic in size, although the group also includes brown algae, oomycotes, and red algae, all of which are multicellular.

8. Fungi are eukaryotic organisms that develop from chitinous fungal spores, have chitinous cell walls, lack undulipodia, and are mostly multicellular. They acquire nutrients by digesting living or dead tissue through absorption.

9. Plants are multicellular, photosynthetic organisms that have complementary sexes, develop from an embryo, and have an alternation of generations.

10. Animals are multicellular, heterotrophic organisms, most of which develop from an embryo produced by fertilization of a large haploid egg by a smaller haploid sperm.

2 Ecology of Life Forms

1. Ecology is the study of factors that influence the distribution and abundance of species in nature. **Paleoecology** is the study of ecologic conditions that affect ancient organisms.

2. Habitats are settings inhabited by life forms. Habitats are broadly divided into terrestrial and aquatic settings. Aquatic habitats are marine (saltwater), freshwater, or brackish (slightly salty). Terrestrial and freshwater habitats are grouped together as **nonmarine** environments.

3. An ecologic niche is the way a species relates to its environment. A niche includes requirements for food, physical and chemical conditions, and interactions with other species.

4. Autotrophs, or primary producers, use raw materials in the environment to make their own food. Chemoautotrophs, including many archaebacteria, derive their energy from the breakdown of compounds in rocks. Consumers derive nutrients by feeding on or by decaying tissues of other organisms. Herbivores are consumers that feed on plant, algal, and fungal material. Carnivores eat other animals, and include **predators** and scavengers. Omnivores combine herbivory and carnivory. Parasites live on or within other living organisms. **Detritivores** obtain nutrients from decaying tissues. **Deposit feeders** ingest mud and extract organic materials from it. Suspension feeders live in water and pass currents through a natural filter.

5. Benthic organisms live on the floor of an aqueous environment. Pelagic organisms live in the water column and include swimming (nektonic) organisms, floating (planktonic) organisms, and pseudoplanktonic organisms (which live attached to other floating or swimming organisms).

3 Types of Fossils and How They Form

1. A **fossil** is any evidence of ancient life.

2. Evidence of ancient life is of three basic forms: (1) **body fossils**, or remains of ancient organisms; (2) **trace fossils**, or evidence of the activity of ancient organisms; and (3) **biomarkers**, or organic chemical signals left in sedimentary layers by organisms.

3. **Taphonomy** is the set of processes that lead to fossilization. Taphonomic processes begin either when an organism dies or releases bodily parts to the environment. Remains of most organisms normally do not stay intact for long because considerable breakdown by scavengers and microbes occurs.

4. Some body fossils are preserved in chemically unaltered state. However, most are preserved in altered states by recrystallization, as molds, by mineral replacement, by permineralization, or by carbonization.

4 Major Groups in the Fossil Record

1. Representatives of all six kingdoms are known from fossils. Eubacteria have left a good record primarily in stromatolites and thrombolites. The fossil records of archaebacteria and fungi are poorly known. Animals, plants, and protoctists have left good to excellent fossil records.

KEY TERMS

CRITICAL AND CREATIVE THINKING QUESTIONS

1. Why is it desirable to have a hierarchical system of biologic classification? What communication problems might result from simply referring to life forms as "kinds" or by colloquial names? How does a hierarchical organization of species promote advances in medicine, agriculture, genetics, and other sciences?

2. The limits of living species of organisms can be defined through a reproductive criterion, but this technique cannot be applied to ancient species. What types of proxy evidence could be used to identify living species? What techniques of identification would be most useful for ancient species?

3. The fossil record consists largely of the hard, biomineralized parts or lignin-impregnated parts of organisms. If you were to design an experiment to test the reasons for this, how would you do it? Is there someplace you could go to watch taphonomic processes in action that might lead to the outcome shown in the fossil record?

4. The organisms that bear hard parts comprise a small percentage of the life forms on Earth. How do you think this influences our perception of the history of life on Earth? If you wanted to fill in the details about life forms that do not normally fossilize, how would you do it?

▼

Carcasses of organisms can be transported by air or water currents, whereas footprints and other traces of activity left in unconsolidated sediment cannot be transported because they become destroyed by current action.

- Are body fossils or trace fossils more likely to faithfully record the presence of an organism at a particular site during the geologic past?

- Do you think most bodily remains are normally transported far from where the organisms actually lived?

- How might you be able to determine from fossil evidence whether a body was transported before being buried and fossilized?

- Under what circumstances might you expect bodily remains to have been transported? What clues would you look for in strata to indicate possible transport?

1. What is a species, and how does a biological species differ from a paleontological species?

2. What is binominal nomenclature?

3. What are the two fundamental types of cells?

4. What are the six kingdoms of life on Earth?

5. What are body fossils?

6. What are trace fossils?

7. What is an ecologic niche?

8. What is taphonomy?

9. What body parts are best represented in the fossil record?

10. What is exceptional preservation of fossils?

Biologic Evolution

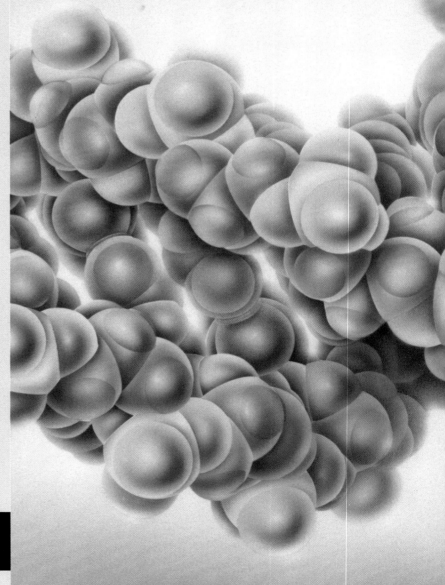

5

So far, you have learned that Earth's changes have been carefully recorded in stratigraphic layers and that we can use correlation techniques to decipher the "chapters" and "pages" of Earth's long saga. That narrative, however, is written not just in rock layers but also partly in biologic organisms.

Life forms are the source of two important types of information about Earth history. First, they are the most obvious outcome of biologic evolution over geologic time spans. Second, life forms carry a superb repository of evolutionary information—their genetics. The evolutionary history of every living organism is succinctly recorded in its genetic code. This code is contained in DNA molecules such as the one shown here. Researchers can use that information to determine species relationships and even to estimate when organisms diverged from each other.

In this chapter, you will learn how biologic evolution has left its mark in the great diversity of fossil organisms through geologic time. You will also learn about two factors that govern how evolution proceeds and that influence the morphologic outcome: natural selection and inheritance. Understanding these central concepts of evolutionary biology provides the metaphorical "character development" that is essential to understanding the story written in Earth's stratigraphic "pages."

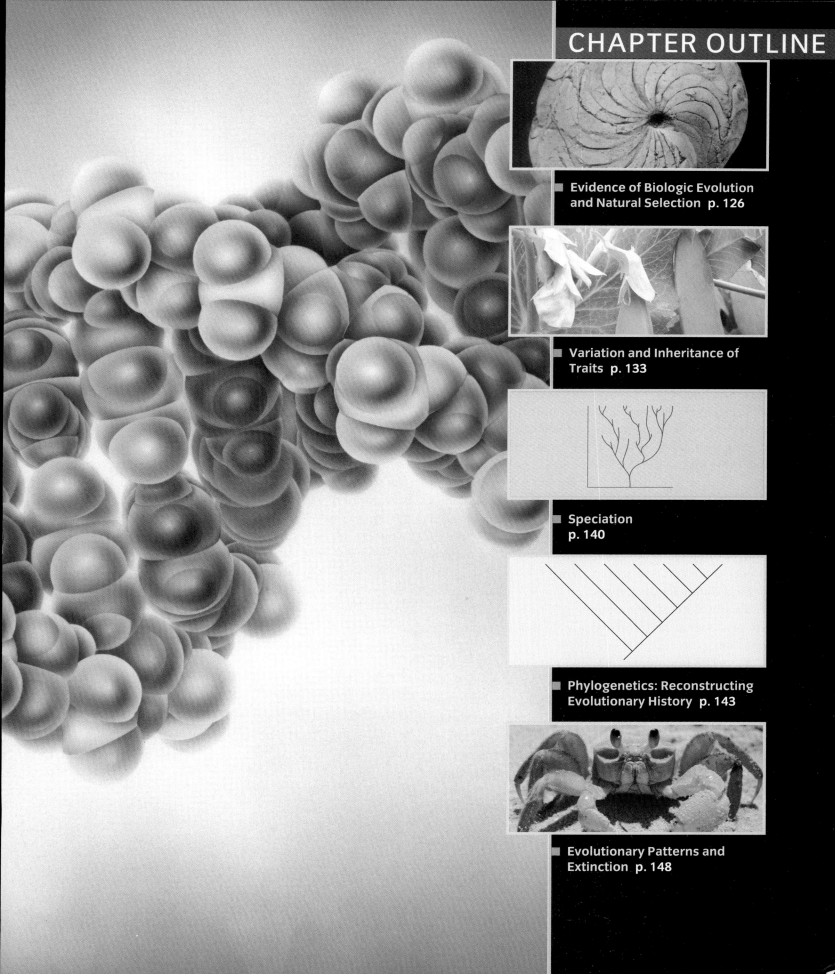

Evidence of Biologic Evolution and Natural Selection

LEARNING OBJECTIVES

Explain the meaning of biologic evolution.

Distinguish between microevolution and macroevolution.

Summarize the evidence for macroevolution.

Explain natural selection and how it operates.

Explain the role of adaptation in natural selection.

Evolution in organisms simply refers to change through time. That change can be thought of as occurring along a spectrum of scales. At the smallest scale, individual organisms go through changes during their lifetimes. Such changes are usually regarded as **ontogenetic** changes. Changes that occur within a species or population from one generation to the next, as parents pass their morphologic, behavioral, and other traits on to their offspring, represent **microevolution** (FIGURE 5.1). Microevolutionary processes are natural selection and genetics (mutation, genetic drift, and gene flow). Evolution at the species level, including changes that can be used to differentiate higher taxonomic groups, represent **macroevolution** (FIGURE 5.2). Macroevolutionary processes include speciation, changes in the tree of life (which comprises the genetic link among all life forms), and extinction. Microevolutionary and macroevolutionary processes are interrelated, and both are pertinent to unraveling the story of Earth's biotic history.

> **evolution** Change through time (usually applied to biologic organisms).

The notion of evolution in life forms has a long scientific and philosophical history. More than 2,000 years ago, philosophers in ancient Greece discussed biologic evolution in general terms. During the Middle Ages, though, strict adherence to theologic teachings suppressed scientific efforts to understand the natural world, and the concept of special creation of life forms held sway among Europeans for a long time.

In the 1700s and 1800s, the doctrine of **fixity of species** prevailed in scientific writing. According to this paradigm, plants and animals, which were the only major life forms recognized at the time, remained unchanged from the time of their first appearance on Earth. This doctrine was rooted in the religious belief of **special creation**, which was then the prevailing view about how species appeared on Earth. Even the Swedish physician and naturalist Carolus Linnaeus (also known as Carl von Linné), who is best known for originating the binominal classification of organisms in the mid-1700s, was an adherent of the concept.

During the late 1700s and early 1800s, when science was undergoing a period of reawakening in Europe, some of the framework concepts of biologic evolution were first articulated. Shortly after the French Revolution, the French naturalists Jean-Baptiste Lamarck and Georges-Louis Buffon demonstrated that species descended from other species and were not morphologically fixed. Lamarck, in his treatise *Philosophie Zoologique* (1801–1815), detailed the first general theory of evolution, which became known as the **doctrine of use and disuse**. In Lamarck's view, when individuals acquired or lost

Microevolution FIGURE 5.1

Changes evident from one generation to the next in a family demonstrate microevolution.

Macroevolution FIGURE 5.2

Changes in horses through time are a classic example of macroevolution.

A A simplified phylogeny of horses showing characteristic genera, regions of the world where lineages evolved, and the relationship of horse evolution to changing habitats in North America.

B Four key steps in the evolution of horses in North America. Horses increased in size but underwent a decrease in the number of toes, related in part to changes in running capability. Surfaces of the molar teeth increased in complexity, largely related to changes in dietary preferences associated with changing habitats through time.

Habitat in North America		South America	North America	Eurasia	Africa
Holocene 0.01 Ma				■ Wild asses	■ Zebras
Pleistocene 2.6 Ma	Prairie	■ *Hippidion*			
Pliocene 5.3 Ma	Savanna		■ *Equus* ■ *Hipparion* group		■ *Hipparion*
Miocene 23.0 Ma	Woodland	■ *Pliohippus* group	■ *Dinohippus* ■ *Merychippus* ■ *Hypohippus* ■ *Archeohippus*	■ *Anchitherium*	
Oligocene 33.9 Ma			■ *Parahippus* ■ *Miohippus* ■ *Mesohippus*		
Eocene 55.8 Ma	Forest		■ *Epihippus* ■ *Orohippus* ■ *Hyracotherium*		

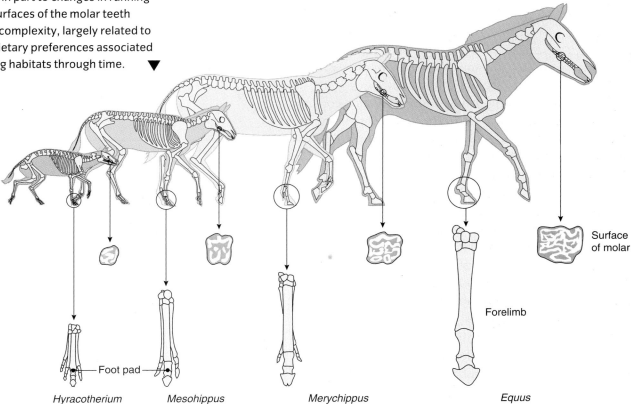

Surface of molar

Forelimb

Foot pad

Hyracotherium   *Mesohippus*   *Merychippus*   *Equus*

characteristics, the new condition became heritable. He championed the concept of **adaptation**, or modification of a living creature or its parts, so as to be better fitted to survival. Using Lamarck's reasoning, a modification of a body part, if it increased fitness for survival or reproduction, could get passed on into succeeding generations.

Lamarck and Buffon were correct in their general demonstration of biologic evolution. The concept of inheritance of acquired characteristics as a driving mechanism, however, is no longer accepted. The concept of adaptation, in a modified form, has survived as a component of modern evolutionary theory.

DARWINIAN SYNTHESIS

By the mid-1800s, biologic evolution had become widely accepted as a scientific theory. Although it was (and still is) accepted by practicing scientists with the highest degree of certainty, opinion differed on its driving force. The mechanism that drives biologic evolution—**natural selection**—was first put forward in separate 1859 publications by Charles Darwin and Alfred Russel Wallace. In natural selection, those individuals best suited to their environment are expected to survive through reproductive age and to contribute to the succeeding generation, whereas individuals that are less well suited tend to be eliminated. Darwin's book *On the Origin of Species by Means of Natural Selection* provided the starting point for much subsequent discussion of evolutionary theory.

> ### ■ natural selection
> The process by which individuals best suited to their environment survive and reproduce and less-well-suited individuals are eliminated from the population.

To provide supporting evidence for the concept of natural selection, Darwin addressed embryology, homologous structures, vestigial structures, breeding experiments, the fossil record, and adaptation.

Embryology is the study of unborn or unhatched offspring. Earlier work by the German anatomist Ernst Haeckel detailed the great morphologic similarities among embryonic vertebrate animals. Vertebrates such as fishes, amphibians, reptiles, birds, and mammals are nearly indistinguishable morphologically in their earliest embryologic stages (**FIGURE 5.3**). This similarity is the result of a shared common ancestry. As embryologic development progresses, species develop characteristics that more clearly differentiate them from other species. Progressive development of differential characters through embryology mirrors the macroevolutionary ancestry of each animal, in effect showing in microcosmic view how species have evolved in different ways from a basic vertebrate rootstock.

Homologous structures are organs, skeletal parts, or other features that have a similar position and evolutionary origin (**FIGURE 5.4A**). In separate species, they are not necessarily identical in structure, nor

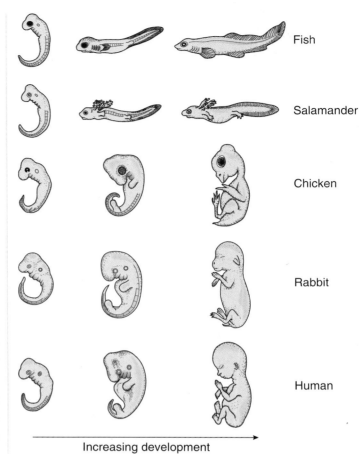

Fish

Salamander

Chicken

Rabbit

Human

Increasing development

Embryology FIGURE 5.3

Embryologic development of vertebrate animals shows that in their earliest stages they are nearly identical in structure, and this is the result of a shared common ancestry. Note particularly that all of these animals show gill slits and tails early in their embryologic history. Differences among groups become more evident in later development. Embryos of land-dwelling species lose their gill slits, and human embryos also lose their tails.

> ### ■ homologous structures
> Morphologic features in biologic organisms that have a similar position and evolutionary origin but not necessarily identical structure or the same function.

do they necessarily have the same function. Homologous structures in related species commonly have evolved into dissimilar forms as a result of adaptation to differing environmental influences or ways of life. Forelimbs of vertebrate animals all have the same evolutionary origins (as verified by the presence of bones such as the humerus, tibia, fibula, carpals, and metacarpals all in the same order), but the appendages have developed differently in response to differing uses for

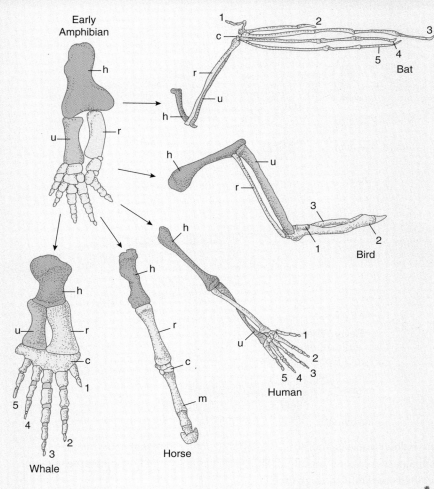

Homologous Structures

◀ **A** **Homologous structures** have a common evolutionary origin, meaning that homologous traits can be traced to a single ancestral species. The forelimbs of vertebrate animals all have a common origin in an early amphibian species. Although the living vertebrates have forelimbs that appear quite different as a result of adaptation to different life modes, the similarity of bone structure is compelling evidence of a deep, shared evolutionary history among all these species.

Key to the bones:

h = humerus
r = radius
u = ulna
c = carpal
m = metacarpal
1–5 = digits of the forelimb

NATIONAL GEOGRAPHIC

Analogous Structures

Analogous structures in organisms have essentially the same function but different evolutionary origins. Birds, bats, and insects have all evolved wings for flight, but they have done so by different evolutionary means. Analogous structures like these developed out of the need to adapt to similar ecologic conditions and are examples of convergent evolution.

B Birds have evolved wings through elongation of bones in the forelimbs and developing feathers. ▼

C Bats are flying mammals that evolved wings made of flaps of skin stretched between elongated finger bones. ▼

D Insects, which are invertebrate animals that have an external skeleton rather than internal bones, developed wings from flaps on the back surface of the chitinous skeleton. ▼

Vestigial structures FIGURE 5.5

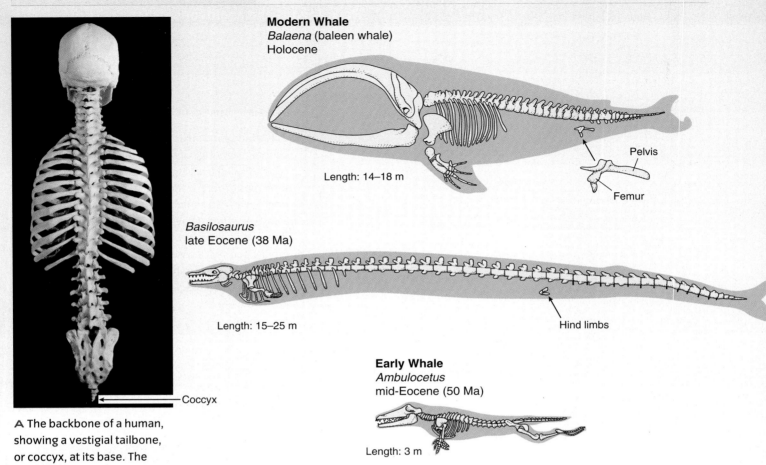

Modern Whale
Balaena (baleen whale)
Holocene

Length: 14–18 m

Pelvis

Femur

Basilosaurus
late Eocene (38 Ma)

Length: 15–25 m

Hind limbs

Early Whale
Ambulocetus
mid-Eocene (50 Ma)

Length: 3 m

Coccyx

A The backbone of a human, showing a vestigial tailbone, or coccyx, at its base. The coccyx is evidence that hominid ancestors bore tails.

B The pelvis of a modern baleen whale (*Balaena*) is a vestigial structure, one that had a function in its terrestrial and amphibious, four-legged ancestors. The pelvis has not yet fully disappeared, even though whales have lost their rear limbs through time.

the appendages. Amphibians have legs adapted to walking or crawling on the ground, a pattern that resembles the ancestral form of leg for a terrestrial vertebrate. The legs of most lizards are similar to those of amphibians. Wings of birds have been modified for flight, especially by the addition of feathers. Bats have elongated finger bones and a thin membrane of skin stretched between the fingers that form a wing. Whales have shortened but thickened bones in the forearm adapted for propulsion as they swim through water. The toes of a modern horse have been reduced to one enlarged toe. That, plus elongation of the lower extremities, provides the modern horse with the ability to run quickly across open plains. Humans have evolved rather gracile forearms and fingers, adapted for a variety of activities, including the use of tools.

Homologous structures are not the same as **analogous structures**, which are structures that have the same function in separate species but different evolutionary origins. Wings in birds, bats, and insects are all adapted for flight, but they clearly developed in different ways, and thus are analogous rather than homologous (**FIGURE 5.4B–D** on page 129).

vestigial structure

A structure in an organism, usually reduced in size or function compared to the homolog in earlier species of the evolutionary lineage, that is in the process of disappearing.

Vestigial structures are structures that are in the process of disappearing from a species or a lineage of species (FIGURE 5.5). When they are so reduced in size or function that they are effectively useless, their presence can be explained only as structures that were functional in an ancestral species. Modern humans have a coccyx, or rudimentary tailbone (FIGURE 5.5A), at the base of the vertebral column, even though we have no use for a tail. Modern whales still retain pelvic (hip) bones, even though they have lost their rear legs (FIGURE 5.5B). The main function of the pelvis in terrestrial mammals, including the ancient terrestrial ancestors of the whales, is to support the rear legs. That function has been lost in modern whales, yet a vestige of the ancestral condition remains.

Early whales show a progressive loss of the hind limbs and pelvis. The mid-Eocene whale *Ambulocetus*, descended from fully terrestrial ancestors, had robust rear legs, with which it could both swim and walk on land. By the late Eocene, the hind limbs were reduced to 60 cm-long structures in the giant *Basilosaurus*. The legs were not functional in swimming but may have been used as grasping structures during copulation.

Breeding experiments on farms, in greenhouses, and in laboratories are a means of concentrating desirable characteristics and accelerating change in a population. Human selection of certain animals or plants that will contribute offspring to the next generation allows us to observe microevolution as it is happening. Darwin analogized this artificial selection, which occurs on short time scales, to the natural world, where selection occurs on much longer time scales. In artificial selection, humans decide what traits are "desirable," but in natural selection, nature "decides" which ones are best. The many breeds of domestic dogs are an outcome of selective breeding. Dogs were originally domesticated from wild gray wolves, and today there are dozens of recognized breeding lines (FIGURE 5.6). Some of the many desirable traits that dogs have been bred for include large size, small size, protectiveness, herding ability, enlarged muscle mass, cold resistance, color, and ear length.

Finally, Darwin discussed the **fossil record** as long-term documentation of evolution (see *What a Geologist Sees* on page 132). In Darwin's view, evolution proceeded by the slow, gradual accumulation of characters or traits, which means that for a clear succession of species to be recorded in stratigraphic layers, Earth must have a deep prehistory. Earlier, the British geologist Sir Charles Lyell had cogently synthesized available geologic information indicating that Earth was at least several million years old. Darwin, acting with this interpretation in mind, figured that Earth was sufficiently old to permit species evolution in natural settings.

Darwin and Wallace separately proposed natural selection as the mechanism to explain how descent with modification (or evolution) happened. Natural selection is based on the observation that populations (or species) are usually composed of more individuals than the environment can support. Because more offspring

Selective breeding FIGURE 5.6

Selective breeding, in which humans decide which desirable traits should be concentrated in breeding, is exemplified in the many breeds of dogs, all of which have a single ancestor, the gray wolf.

Ammonoid evolution

Ammonoids lived from the Devonian Period to the end of the Cretaceous Period. Here are the ranges of three suborders, each showing a different suture type.

- Why do you think ammonoids with these three suture types are tied to distinct intervals of time rather than all appearing at the same time?
- Why do you think that ammonoids with goniatitic sutures are not found in Cretaceous strata?
- Why do you think that ammonoids with ammonitic sutures are not found in Devonian strata?

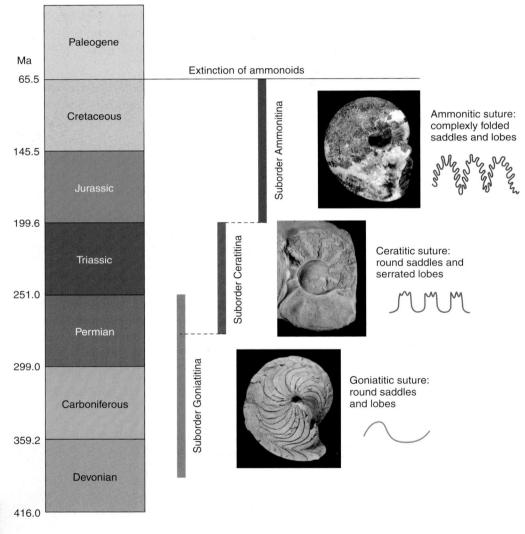

These shells of ammonoid cephalopods, a mollusk group, are arranged in stratigraphic order and show an unmistakable change through time in the saddles and lobes of the sutures, or the lines where internal walls (called septa) meet the external wall. Saddles point toward the outside of the spiral shell, and lobes point toward the inside.

are produced than can possibly survive to reproductive age, individuals in a population must engage in a **struggle for existence**. This struggle is a competition for food, living space, and mates; individuals must also avoid falling prey to other organisms. Individuals show variation in their traits, and those that are less successful at competing or that are less successful at escaping hazards are weeded out of the population. The remaining individuals—those with traits that make them competitively superior—will survive to reproductive age and contribute their traits to the next generation. Over time, the balance between production of offspring and the struggle for existence causes population sizes to remain relatively constant in many species. Although the differences that give individuals a competitive edge may be slight, over a long period of time, advantageous variations in traits will accumulate in the population. Reproductive isolation of a population can concentrate the favorable traits, leading to evolution of a new species.

> **adaptation**
>
> The process of modification of an organism or its parts to make it more fit for survival in an ecologic niche. The word can also refer to a trait that helps an organism survive in its ecologic niche.

The concept of natural selection depends on a modified definition of Lamarck's adaptation. Darwin viewed the process of **adaptation** as modification of an organism or its parts, making it more fit for survival in an ecologic niche. Organisms that have a selective advantage are the ones best adapted to their environment. Those that are poorly adapted tend to be selected out.

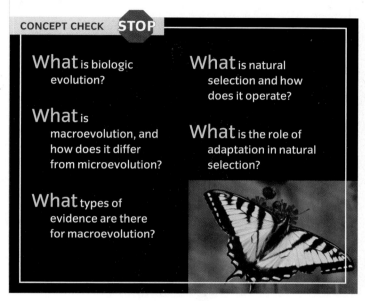

CONCEPT CHECK STOP

What is biologic evolution?

What is macroevolution, and how does it differ from microevolution?

What types of evidence are there for macroevolution?

What is natural selection and how does it operate?

What is the role of adaptation in natural selection?

Variation and Inheritance of Traits

LEARNING OBJECTIVES

Explain the origin of variation in populations and explain how changes can be inherited.

Explain what a mutation is and understand how it arises.

Understand the structure of genetic material.

Understand the concept of a molecular clock.

I n order for natural selection to take place, there must be variation in the traits among individuals of a species, and there must also be a means of passing those traits on to succeeding generations. The source of variation and the means of inheritance were unknown to Darwin and Wallace in 1859 and left the theory of natural selection open to criticism into the 1900s.

How do changes in organisms arise and accumulate in a species? This missing piece of the evolutionary puzzle was largely filled in by the work of Gregor Johann Mendel, a monk who worked in what is now the Czech Republic. Mendel, who carried out experiments on garden peas in the 1850s and 1860s, demonstrated distinct, replicable patterns underlying inheritance and developed a conceptual model for heritability that explained the patterns (FIGURE 5.7). His work was published obscurely in 1866 and went largely unnoticed until around 1900. The work laid the foundation for modern **genetics**, which is the branch of biology concerned with heredity and variation.

> **genetics** The branch of biology concerned with heredity and variation.

Why is genetics relevant to Earth history? First, genetic changes underlie the morphologic changes recorded in the fossil record. Genetic variation and inheritance of adaptive traits make natural selection possible. Second, genetic changes in modern organisms can be used to track evolutionary history and even to provide some sense of the timing of evolutionary events.

Mendel performed experiments involving contrasting traits of pea plants (white flowers vs. violet flowers, tall vs. dwarf, round seeds vs. wrinkled seeds, and other traits; FIGURE 5.7C). He began, in the parental generation (P_1), with **monohybrid** plants (in which each parent showed the same trait of a contrasting pair, such as violet flowers/violet flowers). When monohybrid plants were bred, the result was the same trait (for example, violet flowers) in the first offspring (or filial) generation (F_1), in the second generation (F_2), and in all subsequent generations.

In the next set of experiments, Mendel crossbred monohybrid plants. In the parental generation (P_1), for example, a violet-flowering parent was crossed with a white-flowering plant. The result in the first offspring generation (F_1) was only violet-flowering plants. Then the first-generation offspring were bred among themselves (self-fertilized, or selfed) and the result, in the second offspring generation (F_2), was a combination of violet and white flowers (roughly a 3:1 ratio between violet and white). In other words, the trait for white flowers, which was masked in the F_1 generation, reappeared in the F_2 generation.

Mendel performed similar crossbreedings of plants with other contrasting traits and arrived at similar results. In monohybrid crosses, contrasting flower color, size, pod shape, seed shape, and other traits seemed to disappear in the F_1 generation and then reappear in a 3:1 ratio in the F_2 generation.

To explain his experimental results, Mendel suggested the existence of **unit factors** for each trait. Unit factors are the basic components of heredity and are passed unchanged through a succession of generations.

Mendel put forward three principles of heredity. First, traits in individual organisms are controlled by a pair of unit factors. Second, if two contrasting unit factors for a character are present in an individual, one factor is dominant, and the other is recessive. In the outward expression of a trait, the dominant factor is always favored. Third, when gametes are formed, the paired unit factors separate, and each gamete has an equal likelihood of receiving each factor. If an individual has two of the same unit factors (for example, violet flower/violet flower), each gamete will receive that unit factor. We now call this the **homozygous** condition. If the individual has unlike factors (for example, violet flower/white flower), each gamete has a 50% chance of receiving each type of unit factor. This condition is called **heterozygous**. When gametes recombine to form offspring, a mix of traits in plants (for example, violet and white flowers) will result. This is the source of variation in heredity.

The reason for the outcomes observed in Mendel's peas can easily be visualized by constructing a Punnett square (FIGURE 5.7D), named for Reginald Punnett, who introduced the technique. Using the same crossbreeding example, a plant having violet/violet unit factors that is crossed with a plant having white/white unit factors in the P_1 generation will result in all plants of the F_1 generation having violet/white unit factors. Self-fertilization of the plants having violet/white unit factors will yield offspring having approximately 25% violet/violet unit factors, 50% violet/white unit factors, and 25% white/white unit factors. Because violet is dominant and white is recessive, the plants having violet/violet and violet/white unit factors will all appear violet. Only the plants having white/white unit factors will appear white.

Amazing Places: Gregor Johann Mendel's garden FIGURE 5.7

Johann Mendel was born in Heinzendorf, which is now part of the Czech Republic in 1822. In 1843, he entered the Augustinian Monastery of St. Thomas in Brno. There, he took the name Gregor. From 1851 to 1853, Mendel studied physics and botany at the University of Vienna. In 1854, he returned to Brno, where he taught for the next 14 years. Experimentation with common garden peas in the monastery garden led Gregor Johann Mendel to discover a primary source of heritable variation in living things. That variation is a fundamental element of the evolutionary process.

A Mendel began hybridization experiments on pea plants in his garden in 1856. The work continued until 1868, when he became abbot of the monastery. This is a photo of Mendel's garden as it appeared in the 1980s.

B The garden pea, *Pisum sativum*, was the focus of Mendel's experiments.

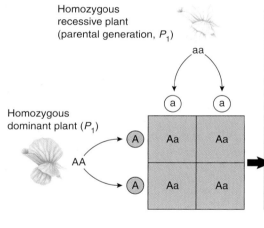

Homozygous recessive plant (parental generation, P_1)

aa

a   a

Homozygous dominant plant (P_1)

A   Aa   Aa

AA

A   Aa   Aa

First offspring generation (F_1) phenotypes:

Aa   Aa

Aa   Aa

C Results from one of Mendel's hybridization experiments, along with Punnett squares depicting the crosses. After crossing two homozygous pea plants, all the offspring in the F_1 generation bore violet flowers. In the F_2 generation, plants bore flowers in the ratio of three violet flowers for every white flower.

NATIONAL GEOGRAPHIC

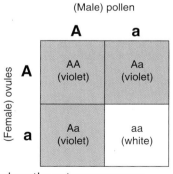

D A Punnett square is used to predict the probable outcome of a genetic cross. Uppercase letters represent dominant alleles, and lowercase letters represent recessive alleles. The different squares in this example show the outcomes among offspring of two heterozygous plants. Only the plant with the aa genotype will have white flowers. The others will have violet flowers.

(Female) ovules	(Male) pollen	
	A	**a**
A	AA (violet)	Aa (violet)
a	Aa (violet)	aa (white)

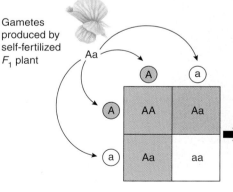

Gametes produced by self-fertilized F_1 plant

Aa

A   a

A   AA   Aa

a   Aa   aa

Second offspring generation (F_2) phenotypes:

AA   Aa

Aa   aa

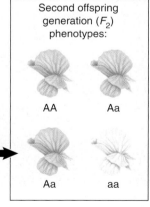

Variation and Inheritance of Traits   135

MODERN GENETICS: GENES, CHROMOSOMES, AND DNA

DNA (deoxyribonucleic acid) The basic storage vehicle for hereditary information.

gene A unit of chromosomal information about a heritable trait that is passed on from one generation to the next.

Cells store genetic information in **DNA (deoxyribonucleic acid)** molecules (FIGURE 5.8). In eukaryotes, the DNA resides in the cell nucleus, but prokaryotes lack a nucleus to contain the DNA.

The DNA molecule usually consists of two spiral strands, known as a double helix. It contains **genes**, which are the hereditary units that Mendel thought of as unit factors. Different versions of the same gene are called **alleles**. Genes are arranged in a line along a larger element called the **chromosome**. DNA stores information in a **genetic code**, which is a sequence of nucleotides in a DNA segment that constitutes a gene. Four different nucleotides (adenine, thymine, cytosine, and guanine) constitute the DNA "alphabet."

Each "word" in the genetic code is formed of a combination of three nucleotides. The code words specify the chemical makeup of the 20 or so **amino acids** that form the chemical building blocks of **proteins**.

chromosome A genetic structure by which hereditary information is physically transmitted from one generation to the next.

The genetic code FIGURE 5.8

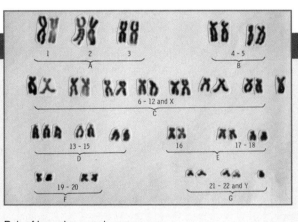

B Chromosomes are the elements of a DNA molecule that carry genes responsible for inheritance. These are human chromosomes.

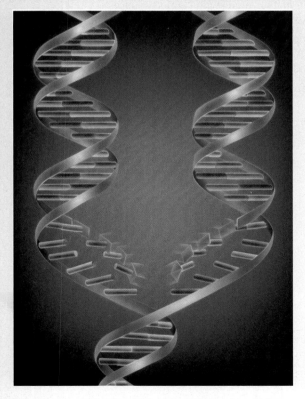

A The DNA molecule is a double helix made up of a number of tiny parts that carry genetic information. During replication, a strand of DNA separates along the chemical base pairs linking the two sides of the molecule.

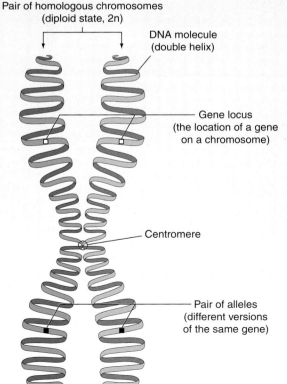

Pair of homologous chromosomes (diploid state, 2n)

DNA molecule (double helix)

Gene locus (the location of a gene on a chromosome)

Centromere

Pair of alleles (different versions of the same gene)

C A duplicated chromosome consists of two strands of DNA held together at a centromere. In meiosis, a diploid (2n) chromosome is duplicated, and then each splits at the centromere to form four haploid (n) chromosomal strands. Upon fertilization, a haploid chromosome is united with another haploid chromosome to form a new diploid individual.

In order to form protein molecules, information encoded in the DNA must be transferred into an **RNA** (**ribonucleic acid**) molecule during a process called **transcription**. Then, in the ribosome, nucleotides in RNA (adenine, cytosine, guanine, and uracil) allow **translation** into a protein.

There are two general types of proteins in living things, both of which are necessary for life functions and that ultimately contribute to an individual's survival potential. **Enzymes**, which are biologic catalysts, lower the activation energy needed for most biochemical reactions necessary for life activities. Nonenzymatic proteins have other roles, including oxygen transport (hemoglobin), immune response (immunoglobulins), and structural support and flexibility (collagen), as well as working as hormones (insulin, testosterone, estrogen, and others).

> **meiosis** Division of chromosomes to produce two haploid cells as gametes necessary for sexual reproduction are produced.

> **mitosis** Division of cell nuclei in which the parental chromosomal number is maintained; it is the basis for bodily growth and asexual reproduction.

Most species have a certain number of chromosomes, called the **diploid** number (2n), in each body cell. The number of different types of chromosomes is half the diploid number and is called the **haploid** number (n). In humans, for example, the diploid number is 46, and the haploid number is 23. Haploid gametes in animals, and haploid spores in most plants and protoctists, are produced through **meiosis**. In contrast, during **mitosis**, genetic material is duplicated and distributed in the process of cell division. Some organisms, such as plants, yeasts, and many cnidarians, are haploid during a significant part of the life cycle. Various plants evolved by acquisition of more than two sets of chromosomes; they are called **polyploid**.

MUTATIONS, ADAPTATIONS, AND RANDOM FACTORS

Variation within populations can lead to changes beneficial for survival, or adaptations. That kind of variation arises from

> **mutation** A heritable change in DNA or chromosomal structure that results in new versions of genes and, ultimately, in life's diversity.

mutations, or genetic changes. Mutations can be either **gene mutations**, caused by changes in the information stored in DNA when nucleotides are substituted, deleted, or duplicated, or **chromosomal mutations**, caused when chromosomal segments are duplicated, deleted, or rearranged.

How do mutations cause adaptations? The genetic makeup of an organism is called its **genotype**. Alleles, which are variants of the same gene, form through mutation. Individuals have different genotypes because they have different alleles contributing to their genetic code. Genetic variation can result in change of an organism's **phenotype**, or the way it appears and functions. Mendel was tracking phenotypic differences, due to different alleles, with his contrasting traits in peas (violet flower vs. white flower, tall vs. dwarf, and others). Most mutations are probably neutral or detrimental, but a few may be beneficial for survival. When beneficial changes become broadly spread through a population because of favorable selective pressure, they can improve the species' chances for longer-term evolutionary success.

Sometimes chance events cause random changes in gene frequency in a population. This is called **genetic drift**. Random factors include arrival of a small population on an isolated island, starvation, disease, overhunting, and various environmental calamities. If genetic drift causes a change in the genetics (and phenotypic expression) of a population, selection will act on the new population and, in time, possibly result in a new species.

One other relevant microevolutionary concept is **gene flow**. The frequencies of alleles change as individuals enter or leave a population, and this physical flow of alleles, called gene flow, helps neighboring populations stay genetically similar. Over time, differences between adjacent populations of a species are balanced out. If, however, gene flow between populations is disrupted, genetic divergence and speciation can result.

MOLECULAR EVOLUTION AND MOLECULAR CLOCKS

Using genetic changes to track evolutionary history is referred to as **molecular evolution**. All organisms have left a record of their changes in their DNA. With each evolutionary step, there is a substitution of a new allele for an older one. Mutations are responsible for the rise of new alleles, and favorable selection allows some of these new versions of genes to spread through populations. Closely related species share a large portion of their DNA sequences, and distantly related species have more differences in their genetic sequences.

In recent years, gene sequencing technology has led to an explosion of information about the genetic sequences of living species of organisms. Comparing the number of similarities and differences in chromosomal sequences of different organisms can be used to determine their degree of evolutionary relationship (FIGURE 5.9).

Comparing the times since species diverged and the number of differences in nucleotide sequences of chromosomes indicates that changes accumulate at a nearly constant rate through time. This leads to the concept of a **molecular clock**—the idea that the number of differences in the amino acid sequences of separate species can be used as a measure of the amount of time that has elapsed since they shared a common ancestor (FIGURE 5.10).

> **molecular clock**
> The technique of using the number of genetic differences between separate species to measure the amount of time that has elapsed since the species diverged.

Molecular evolution FIGURE 5.9

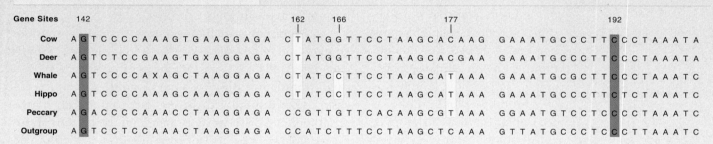

Gene Sites	142		162	166		177		192	
Cow	A G T C C C C A A A G T G A A G G A G A	C T A T G G T T C C T A A G C A C A A G	G A A A T G C C C T T C C C T A A A T A
Deer	A G T C T C C G A A A G T G X A G G A G A	C T A T G G T T C C T A A G C A C G A A	G A A A T G C C C T T C C C T A A A T A
Whale	A G T C C C C A X A G C T A A G G A G A	C T A T C C T T C C T A A G C A T A A A	G A A A T G C G C T T C C C T A A A T C
Hippo	A G T C C C C A A A G C A A A G G A G A	C T A T C C T T C C T A A G C A T A A A	G A A A T G C C C T T C T C T A A A T C
Peccary	A G A C C C C A A A C C T A A G G A G A	C C G T T G T T C A C A A G C G T A A A	G G A A T G T C C T C C C C T A A A T C
Outgroup	A G T C C T C C A A A C T A A G G A G A	C C A T C T T T C C T A A G C T C A A A	G T T A T G C C C T C C C T T A A A T C

▲ A One way to reconstruct the phylogenetic history of living organisms is to use genetic sequences. Here are nucleotides of the beta-casein gene, which encodes a milk protein in some mammals. Nucleotides are identified as A (adenosine), T (thymine), C (cytosine), and G (guanine). Yellow lines highlight shared traits that show a common evolutionary origin and help to cluster taxa. Red lines highlight traits that are shared widely among the studied species and that are uninformative for clustering taxa at this level.

B After calculating genetic distances between pairs of species, it is possible to construct a graphic evolutionary hypothesis by clustering according to genetic distance. Taxa with the least amount of genetic distance between them are clustered together.

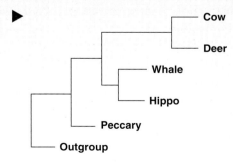

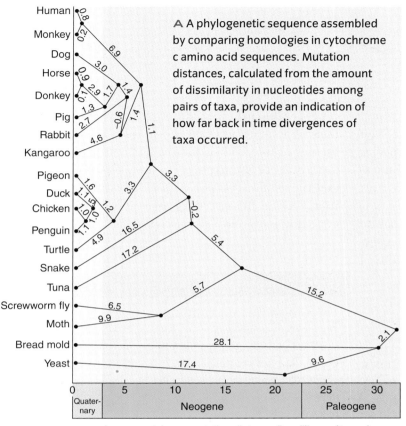

A A phylogenetic sequence assembled by comparing homologies in cytochrome c amino acid sequences. Mutation distances, calculated from the amount of dissimilarity in nucleotides among pairs of taxa, provide an indication of how far back in time divergences of taxa occurred.

Average minimum mutation distance (in millions of years)

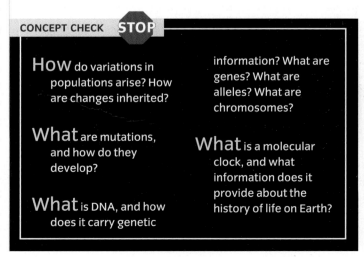

Time (Ma)

B Different proteins mutate and evolve at different rates. This leads to molecular clocks that "tick" at different rates. Cytochrome c (A) has a slow rate of change. Here are results of an analysis on primates using the protein carbonic anhydrase, which has a faster rate of change.

Various proteins can be used as molecular clocks, and there is evidence that they "tick" (or mutate and evolve) at different rates. To effectively use molecular clocks for interpreting divergence times of species in the distant geologic past, more than one method can be tried. The protein cytochrome c, for example, may be good for estimating divergence times among widely disparate life forms (FIGURE 5.10A), but to estimate divergence times among relatively closely related species of primates, another analytical technique may yield more reasonable results (FIGURE 5.10B). Further confidence in molecular clocks can come from calibration of results against the times of stratigraphic first appearances of fossils representing ancestors of the experimental species.

Some divergence times of life forms estimated from molecular clock evidence are in good agreement with divergence times suggested by the fossil record. In other cases, though, estimated divergence times are vastly different. This could mean that not all molecular clocks "tick" at a constant rate through all intervals of time, that the rate of "ticking" has been miscalculated, or that life forms had a much deeper ancestry than we have yet to verify from fossils.

CONCEPT CHECK STOP

How do variations in populations arise? How are changes inherited?

What are mutations, and how do they develop?

What is DNA, and how does it carry genetic information? What are genes? What are alleles? What are chromosomes?

What is a molecular clock, and what information does it provide about the history of life on Earth?

Speciation

The species is the basic level on which macroevolutionary processes operate. The defining characteristic of a biologic species is reproductive isolation (and, by implication, genetic divergence) from other species. For most sexually reproducing animal species, this concept holds well. Animal hybrids, or offspring of parents of separate but closely related species, are usually sterile. Plant hybrids, though, are commonly fertile, making the limits of species more difficult to determine.

SPECIATION PROCESSES

> **speciation** The evolution of a new species from an ancestral species.

Speciation, or the rise of a new species from an ancestral species, in sexually reproducing organisms, can occur according to one of three general models (**FIGURE 5.11**).

One speciation model, **allopatric speciation**, emphasizes disruption of gene flow between populations, usually by some sort of physical barrier (**FIGURE 5.11A**). Physical isolation can occur by geographic **dispersal**. A chain of islands colonized by a species may develop populations having limited gene flow between them. Alternatively, physical isolation can occur by **vicariance**, in which a species' range is split by imposition of a physical barrier. The splitting of a supercontinent into smaller continental fragments, or the change in course of a river, may divide the range of a species. Regardless of the means of isolation, geographically separated populations are likely to diverge genetically and become reproductively isolated. Even if the populations are later reunited, they may not fertily interbreed. If a geographically isolated population is small, it stands greater odds of experiencing significant change in allele frequencies because of genetic drift.

Allopatric speciation may be the most common means of speciation because most species do not span out continuously over their ranges. Instead, they occur as a series of populations separated by some distance, and they maintain limited gene flow with other populations. Under these circumstances, it is easy for barriers to gene flow to develop between populations.

Speciation processes FIGURE 5.11

A Allopatric Speciation

The most prevalent means by which species evolve is allopatric speciation. In this model, gene flow between adjacent populations is disrupted, and the populations then diverge genetically. Isolation can occur either by dispersal (geographic movement or displacement of individuals) or vicariance (division of a species' range into smaller areas).

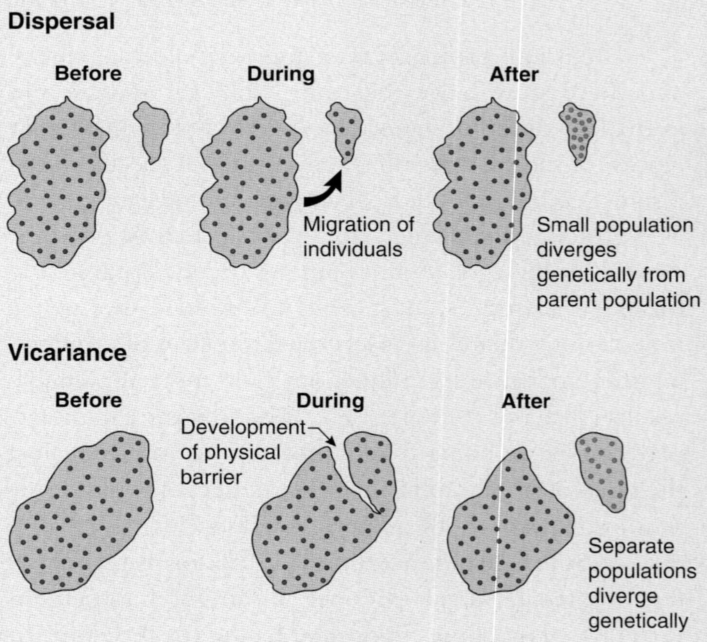

Dispersal

Before   During   After

Migration of individuals

Small population diverges genetically from parent population

Vicariance

Before   During   After

Development of physical barrier

Separate populations diverge genetically

Another speciation model, **sympatric speciation**, involves the rise of a daughter species from a group of individuals within the geographic range of an ancestral species (FIGURE 5.11B). Small differences in ecologic strategy, such as feeding habit, may account for the separation of sympatric species. Another way that sympatric speciation may occur is by polyploidy. Flowering plants are particularly susceptible to inheritance of additional chromosomes, some of which results from improper chromosomal separation during mitosis or meiosis. Polyploidy can also occur when a germ cell duplicates its DNA but does not divide and then functions as a gamete.

In the last speciation model, **parapatric speciation**, a daughter species might arise through hybridization of two populations. One way this could happen is when individuals along a common border between two populations having limited gene flow between them interbreed and hybridize (FIGURE 5.11C).

Cladogenesis and anagenesis FIGURE 5.12

New species can arise from ancestral ones in two ways, by cladogenesis and anagenesis.

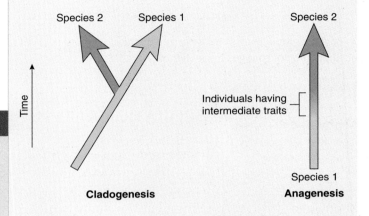

Cladogenesis

Anagenesis

A In cladogenesis, a population becomes reproductively isolated from the parent stock and then diverges genetically. The two species show few, if any, intermediate individuals.

B In anagenesis, traits grade from the parent species to the daughter species. Although endmember forms may quite obviously belong to separate species, the change will yield many intermediates that are not easily classified as one species or the other.

B Sympatric Speciation

In sympatric speciation, two diverging species have overlapping ranges. Genetic isolation follows a change in ecologic strategy, such as a change in feeding preference, among some individuals of the parent population.

Before　　　**During**　　　**After**

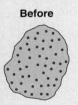 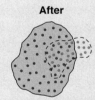

　　　　Change in ecologic　　After genetic divergence,
　　　　strategy among some　species have overlapping
　　　　individuals　　　　　ranges

C Parapatric Speciation

In the parapatric speciation model, two species may diverge from a single species following hydridization of populations that have limited gene flow. When partially isolated populations establish contact, a zone of hybridization can develop. Divergence of the nonhybridized stock would result in speciation.

Before　　　**During**　　　**After**

Two populations　Zone of hybridization　Speciation
with limited gene
flow

The fossil record shows two different speciation patterns: cladogenesis and anagenesis (FIGURE 5.12). **Cladogenesis** is a branching speciation pattern, and it applies to populations that become reproductively isolated and then diverge genetically from each other. **Anagenesis** refers to changes in allele frequencies that occur along a single evolutionary pathway. The parent species and daughter species are connected by intermediates, and the distinction between one species and its descendant may be ambiguous unless temporal endmembers are compared.

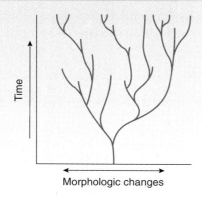

A Phyletic gradualism is characterized by the slow, gradual accumulation of changes through time.

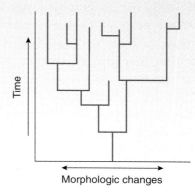

B Punctuated equilibrium is characterized by rapid speciation followed by a long interval of little morphologic change.

phylogeny The line or lines of descent in an evolutionary series.

phyletic gradualism Speciation rate characterized by a slow, gradual pace.

SPECIATION RATES

How fast do species evolve? In Darwin's view, evolution occurred slowly, as a result of a gradual accumulation of changes from one generation to the next and through **phylogeny**, or an evolutionary series. The two species linked by an ancestor-descendant relationship would be linked by numerous imperceptibly different intermediates. New species would arise anagenetically over time frames of millions of years in many cases. The notion of slow, gradual evolution is embodied in the concept of **phyletic gradualism** (FIGURE 5.13A). A classic example is the change through time in horses (Figure 5.2A on page 127).

As Darwin recognized, most species are distinguished from others by "gaps" in morphology. These apparent morphologic gaps were a weak point in Darwin's view of evolutionary process and were originally attributed to erosional gaps (unconformities) in the stratigraphic record that resulted in an incomplete set of intermediates.

In the early 1970s, Niles Eldredge and Stephen Jay Gould proposed an alternative interpretation for the morphologic gaps between species. They tracked the evolutionary histories of Devonian trilobites from North America and Pleistocene land snails from Bermuda

through minimally disrupted stratigraphic intervals representing a few million years. In their view, the morphologic gaps between species were real, and they resulted from the rapid evolution of new species. Populations separated at the periphery of the main geographic range of the species would be in a good position to undergo rapid evolution because of rapid gene flow through the isolated population. This is the essence of the allopatric speciation model. Once a new species had evolved, slow morphologic change (or morphologic stasis) within the species would ensue. Because new species could evolve quickly, perhaps on the order of thousands to hundreds of thousands of years, few intermediates would be left as fossils. This concept is known as **punctuated equilibrium** (FIGURE 5.13B). Ammonoid cephalopods (see *What a Geologist Sees* on page 132) commonly show a punctuational mode of evolution.

punctuated equilibrium Speciation characterized by rapid divergence of species followed by a long interval of little change.

CONCEPT CHECK STOP

What is speciation?

How does speciation happen?

How fast do new species evolve?

Phylogenetics: Reconstructing Evolutionary History

LEARNING OBJECTIVES

Explain the meaning of a phylogenetic tree.

Understand how an evolutionary hypothesis is represented on a cladogram.

Summarize how phylogenetic patterns are reconstructed using cladistics.

Biologic evolution is chronicled in the genetic material and morphology of organisms. Genetic profiles can be "read" from the DNA or RNA sequences of modern organisms, but it is rare to find useful genetic material in strata older than the Holocene. In both fossil and modern organisms, evolutionary history can be reconstructed from morphologic evidence. Prime evidence of evolutionary, or phylogenetic, affinity is afforded by homologous structures (Figure 5.4A on page 129). Homologous structures not only allow us to reconstruct ancestor-descendant relationships among ancient organisms but also allow us to link ancient organisms to ones living today.

Once the evolutionary relationships within a set of organisms are understood, they can be depicted in various ways. Traditionally, relationships are depicted as a "tree" showing the branching of one species from another (**FIGURE 5.14A**). This method has drawbacks, including the necessity to redraft the inferred evolutionary pattern to accommodate new species or higher taxa that are discovered and inexactitude in testability. Another method of depiction is the **cladogram**, which is an evolutionary hypothesis illustrated with a simple branching stick diagram (**FIGURE 5.14B**).

> **cladogram**
> A branching diagram that illustrates a phylogenetic hypothesis.

Phylogenetic trees FIGURE 5.14

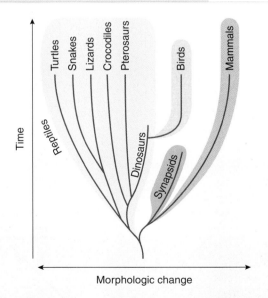

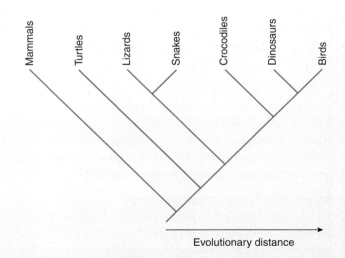

A One way of illustrating a phylogenetic sequence is by using a tree showing ancestor-descendant relationships. This tree shows inferred relationships among some terrestrial vertebrate animals.

B Another way of illustrating a phylogenetic hypothesis is by using a cladogram, which clusters taxa according to their degree of relatedness. Again, vertebrate animals serve as the example.

The core concept of a phylogenetic (or **cladistic**) analysis (FIGURE 5.15) is that species or other taxa are grouped based on evidence of common ancestry. In other words, two taxa are shown to be closely related by virtue of a shared common ancestry.

synapomorphy
A shared derived character.

A shared evolutionary history is evinced in derived characters, or modified versions of homologous characters. Derived characters, (or **apomorphies**, are of two types: **synapomorphies** (shared derived characters) and **autapomorphies** (unique derived

Constructing a cladogram FIGURE 5.15

A cladogram is an evolutionary hypothesis expressed in graphic form. Shared derived characters are used to build a diagram that reflects successively closer relationships among species or higher taxa.

To illustrate how phylogenetic reconstruction is done, let's use a group of vertebrate animals. A lamprey, which is a jawless vertebrate having few derived characters, will be used as the outgroup. The outgroup allows us to make inferences about the common ancestor of all the taxa being analyzed. Jawed vertebrates, such as the salmon, lungfish, eagle, cat, chimpanzee, and human, constitute the ingroup.

Step 1. Construct a list of characters.
To keep the analysis simple, we focus on the presence (+) or absence (−) of just seven characters, or traits.

Taxa	Characters						
	Jaws	Limbs	Lungs	Tail	Scales	Feathers	Hair
Lamprey (outgroup)	−	−	−	+	+	−	−
Eagle	+.	+	+	+	−	+	−
Chimpanzee	+	+	+	−	−	−	+
Cat	+	+	+	+	−	−	+
Lungfish	+	−	+	+	+	−	−
Salmon	+	−	−	+	+	−	−
Human	+	+	+	−	−	−	+

Step 2. Assign numeric values to character states.
A zero (0) conventionally represents the ancestral form of a trait, and a one (1) is used to designate a derived condition. A change from character state 0 to character state 1 is the polarity (or direction) of character evolution. Character states in the outgroup are primitive, or plesiomorphic, and changes in homologous characters (apomorphies, or derived characters) record the course of evolution.

Taxa	Characters						
	Jaws	Limbs	Lungs	Tail	Scales	Feathers	Hair
Lamprey (outgroup)	0	0	0	0	0	0	0
Eagle	1	1	1	0	−	1	0
Chimpanzee	1	1	1	1	−	0	1
Cat	1	1	1	0	−	0	1
Lungfish	1	0	1	0	+	0	0
Salmon	1	0	0	0	+	0	0
Human	1	1	1	1	−	0	1

Step 3. Search for shared derived characters.
Looking down the list of characters, you can see that chimpanzees, cats, and humans all have hair. Hair is a shared derived character, or synapomorphy. All the taxa except for the lamprey have jaws, a synapomorphy that shows a shared evolutionary ancestry among those species.

Step 4. Begin constructing a cladogram.
This simple cladogram shows how you would illustrate relationships based on the presence or absence of a jaw. This is the most basic level of differentiation—that between the outgroup and the ingroup. Among the analyzed taxa, the lamprey is the most closely related outgroup, so it is also known as a sister group. More distant outgroups that could have been chosen include other jawless fishes and chordates that lack backbones.

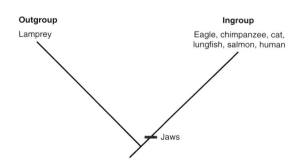

Outgroup
Lamprey

Ingroup
Eagle, chimpanzee, cat, lungfish, salmon, human

Jaws

characters). Shared derived characters are the key evidence of a shared evolutionary relationship between two or more species and are the basis for reconstructing evolutionary history using a cladogram. Unique derived characters only help distinguish one taxon from another, so they do not help us reconstruct common ancestry. Using a cladogram is an excellent way of depicting speciation that has occurred through cladogenesis, but it can mask cases in which anagenesis has occurred.

Outgroup

Lamprey Salmon   Eagle, chimpanzee, cat, lungfish, human

Lungs

Jaws

Step 5. Fill in the cladogram.
To complete the cladogram, you want to identify traits present in most, but not necessarily all, of the jawed animals. Lungs are present in all but the salmon, so you can add that character to the cladogram and see how things change.

All the vertebrates except the lamprey and the salmon have lungs. This means that the eagle, chimpanzee, cat, lungfish, and human together constitute a monophyletic group, or clade. This clade consists of an ancestral species that evolved lungs and all of its descendant species.

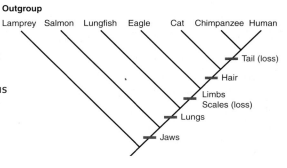

Outgroup

Lamprey Salmon Lungfish Eagle   Cat   Chimpanzee Human

Tail (loss)

Hair

Limbs
Scales (loss)

Lungs

Jaws

Step 6. Now you can successively add other characters to the cladogram. As you do, you will define smaller clades. Each clade must be supported by at least one synapomorphy.

Here is the cladogram that results from adding all of the synapomorphous characters.

Outgroup

Lamprey Salmon Lungfish Eagle   Cat Chimpanzee Human

Feathers
(autapo-
morphy)

Unique Derived Characters

Some traits that appear in the data tables do not appear in the final cladograms constructed from them. The presence of feathers, for example, does not appear because it is an autapomorphy, or unique derived character, and useful only for distinguishing the eagle from the other taxa.

Evolutionary Convergence

If we were to reconstruct the phylogeny of another set of animals—lamprey, salmon, lungfish, eagle, cat, bat, chimpanzee, and human, for instance—we would find that one character, the presence of wings, shows up in two separate places on the cladogram. Birds, such as the eagle, and bats developed wings independently. This is an example of convergent evolution, or homoplasy.

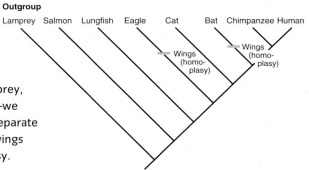

Outgroup

Lamprey Salmon Lungfish Eagle   Cat   Bat Chimpanzee Human

Wings
(homo-
plasy)

Wings
(homo-
plasy)

A shared relationship between two taxa is always expressed relative to a more distantly related third taxon (FIGURE 5.16). Humans and chimpanzees are more closely related to each other than either is to cats, for example. We can infer that lungfish, birds, and mammals share a common ancestry because all the taxa share a derived character, lungs. More distantly related are the jawed fishes that lack lungs.

A group of species that includes the ancestor and all of its descendants is called a **clade** (also known as a **monophyletic group** or **natural group**; FIGURE 5.17A). Ideally, the classification of species follows along the lines of clades, but in practice, this is not always the case. Instead, so-called artificial taxa are used in many circumstances. There are two basic types of artificial taxa: **paraphyletic groups**, in which one or more groups descended from the common ancestor are excluded from the group (FIGURE 5.17B), and **polyphyletic groups**, in which members of the group have separate ancestors (FIGURE 5.17C). If the birds are excluded from the group that includes dinosaurs, then the remaining taxa constitute a paraphyletic group. The sponges, which may have at least two evolutionary origins, are apparently a polyphyletic group. One type of polyphyletic group is a **grade**, or level of evolutionary organization achieved independently by organisms. The multicellular grade was achieved independently in animals and plants.

In a phylogenetic analysis, the group of interest (the group of taxa whose relationships you are trying to understand) is called the **ingroup**. A taxon that is not part of the ingroup is an **outgroup**, and the outgroup most closely related to the ingroup is called the **sister group**.

Reading a cladogram FIGURE 5.16

The cladogram expresses relative relatedness, not ancestor–descendant relationships per se. It also does not specifically express when in time evolutionary divergence occurred. In general, though, evolutionary distance is a function of time. According to the upper-left cladogram in the figure, humans and chimpanzees are more closely related to each other than either is to the cat. The cat, chimpanzee, and human are more closely related to each other than to the lungfish. They are members of a clade defined by the acquisition of hair, a shared derived character. The chimpanzee and human are members of a smaller clade defined by loss of a tail (and other characters).

The point where two branches meet on a cladogram is called a node. Because the cladogram expresses only relative relatedness, branches can be rotated at a node, providing different arrangements of names, even though the meaning of each cladogram is the same.

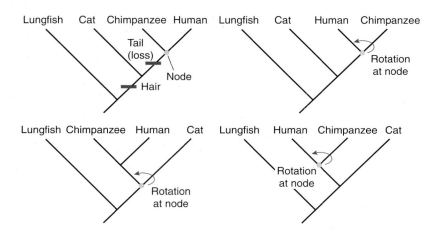

Monophyletic and artificial groups FIGURE 5.17

Taxonomic groups can be monophyletic (A), paraphyletic (B), or polyphyletic (C). Monophyletic groups, or clades, are natural groups that include an ancestral species and all of its descendants. In this cladogram, species A, B, C, D, E, and F comprise a monophyletic group. Smaller groups, such as D, E, and F or A, B, and C, could also be classified as clades.

Paraphyletic and polyphyletic groups are also called artificial groups. Paraphyletic groups contain some descendants of a common ancestor but exclude others. The group defined by species A, B, D, E, and F is paraphyletic.

Polyphyletic groups lack the most recent common ancestor of all the taxa included in the group. A group formed by species A, B, and F is polyphyletic because the common ancestor of all of these species is not part of the group.

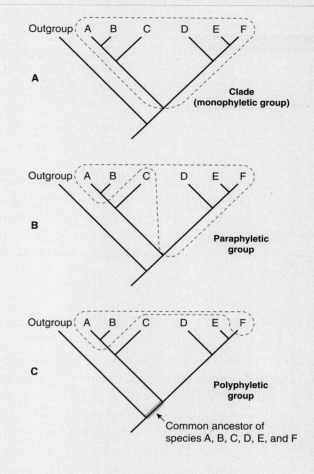

The direction, or **polarity**, of character evolution among a group of related taxa is determined using the character state in the outgroup. Character states in the outgroup used to polarize the analysis are assumed to be primitive (or **plesiomorphic**). Changes in homologous characters among members of the ingroup (derived character states, or apomorphies) register the evolutionary history within the ingroup. Synapomorphies (shared derived characters) provide documentation of shared common ancestry within the ingroup. Sometimes similar characters evolve in organisms by different means (such as wings in birds, bats, and insects). This condition is called **homoplasy** or **convergence**.

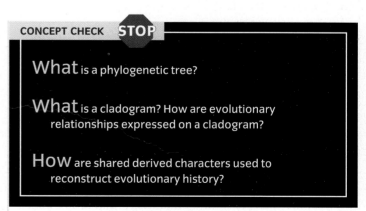

CONCEPT CHECK STOP

What is a phylogenetic tree?

What is a cladogram? How are evolutionary relationships expressed on a cladogram?

How are shared derived characters used to reconstruct evolutionary history?

Evolutionary Patterns and Extinction

LEARNING OBJECTIVES

Summarize the evolutionary patterns and trends commonly represented in the stratigraphic record.

Explain convergent evolution.

Understand circumstances leading to adaptive radiation.

Explain extinction and its causes.

Explain the role of mass extinctions in construction of the geologic time scale.

S tudy of evolutionary history shows that certain patterns or trends have occurred in group after group over a period of millions of years. In this section, you will learn about some of the most significant and pervasive evolutionary patterns.

IRREVERSIBILITY OF EVOLUTION

Late in the 19th century, the Belgian paleontologist Louis Dollo observed that evolution cannot produce exactly the same species more than once. This concept of the irreversibility of evolution has come to be known as Dollo's Law. Today, we understand why Dollo's Law is so: once a species has experienced genetic change, the change cannot be reversed. Also, once a species has gone extinct, it cannot come back into existence.

CONVERGENT EVOLUTION

Convergent evolution (or homoplasy, to use cladistic terminology) is the evolution of similar body forms in two or more biologic groups. Similar body forms apparently evolved in different taxa that used similar ecologic strategies, or as a response to similar ecologic pressures.

One excellent example of convergence is the acquisition of wings for flight. Animals that independently evolved wings are birds, pterosaurs (flying reptiles), insects, and some mammals (bats, flying squirrels, and flying phalangers). Flying phalangers are marsupial mammals that independently evolved a body form similar to that of flying squirrels, which are placental mammals. Flight mechanisms are different in each of the major

groups, which is cogent evidence that natural selection favors traits that are adaptive (characters that help organisms survive in their ecologic niches).

ADAPTIVE RADIATIONS

One recurrent pattern is the rapid evolution of groups when new ecologic opportunities develop. Numerous new species quickly evolving to fill open niche spaces is called an **adaptive radiation** (FIGURE 5.18).

> **adaptive radiation**
> Rapid evolution of organisms to fill new ecologic niches.

Adaptive radiations can occur under at least three circumstances. One scenario is the development of new body forms, or modifications of existing forms, that give organisms access to ecologic niches they did not have previously. This pattern can be referred to as an **adaptive breakthrough**. The evolution of many new, experimental body forms in the Cambrian Period is the result of adaptive breakthroughs by many marine animals almost synchronously. Other adaptive breakthroughs were the development of robust legs (in amphibians) from fins (in lobe-finned fishes), which allowed vertebrate animals to gain access to nonmarine habitats; the development of wings in insects, which allowed insects to fly; and the development of flowers in plants.

Another scenario under which adaptive radiation has occurred is when new ecologic habitats (new niche spaces) are opened or exploited. If new islands, lakes, oceans, or mountains develop, organisms are afforded opportunities to adapt to the new habitats. Speciation in Darwin's finches of the Galápagos Islands, off the coast of Ecuador, illustrates an adaptive radiation. The

Galápagos are volcanic islands formed within the last 10 million years. Migration of finches from mainland South America to the Galápagos Islands opened the way for their adaptation to a variety of niches. Tremendous variation in the size and shape of the beaks (**FIGURE 5.18**) is mostly related to their feeding preferences. Among seed eaters, those with larger beaks favor larger seeds, and those with smaller beaks favor smaller seeds. Some finches have beaks adapted for other ways of gathering food, such as catching insects.

Climatic change, such as global cooling, may open opportunities for animals, plants, or other organisms that can adapt to the changing conditions. In the Northern Hemisphere, a variety of new cool-weather-adapted mammal species (wooly mammoths, mastodons, and wooly rhinoceri) evolved as the world witnessed a cooling phase and expansion of glacial ice sheets during the Quaternary.

Thirdly, adaptive radiation can occur in one or more groups following extinction events. Mass extinctions, in particular, have opened up niche spaces that were previously occupied by other creatures. Extinction at the end of the Cretaceous Period, for example, left room for mammals to reoccupy niches previously held by dinosaurs.

INCREASE IN BODY SIZE

In the late 1800s, the American paleontologist Edward Drinker Cope observed a common tendency for animals to increase in body size through an evolutionary lineage. This trend is called Cope's Rule. The living horse *Equus caballus*, for example, is about four times the size of *Hyracotherium*, the earliest horse, of Eocene age.

Evolutionary size increase may help with survival or reproductive success. A larger predator may be more effective at capturing prey. On the other hand, potential prey may be better at avoiding predators if they are larger in size. In either case, longer survival leads to the possibility of producing more offspring. In species where males compete for potential mates, larger size is usually an advantage, so larger individuals are more likely to have more offspring than smaller ones.

Size increase cannot go on infinitely in animals. As the length of an animal increases, the size of tissues or organs that have surface-dependent functions (such as respiration and nutrient absorption) increase at a faster rate. This relationship creates an upper limit on the size that a creature can reach.

Adaptive radiation FIGURE 5.18

Adaptive radiation occurs when life forms rapidly diverge to fill a variety of niche spaces that have become open to them.

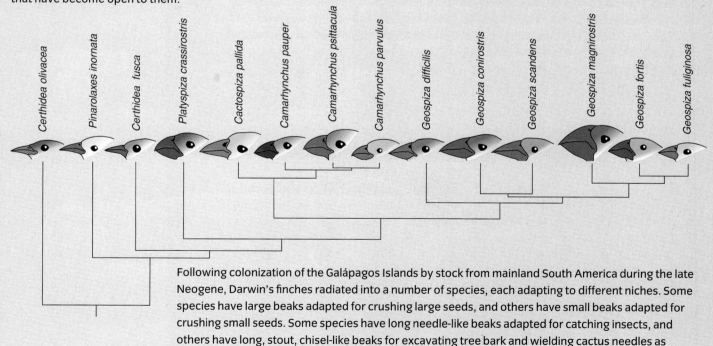

Following colonization of the Galápagos Islands by stock from mainland South America during the late Neogene, Darwin's finches radiated into a number of species, each adapting to different niches. Some species have large beaks adapted for crushing large seeds, and others have small beaks adapted for crushing small seeds. Some species have long needle-like beaks adapted for catching insects, and others have long, stout, chisel-like beaks for excavating tree bark and wielding cactus needles as tools in search of insects.

HETEROCHRONY

Heterochrony involves evolutionary changes in a lineage that result from changes in developmental (ontogenetic) timing (**FIGURE 5.19**). These changes cause the timing of onset of characters in the descendant species to be different from the timing of onset of homologous characters in the ancestor. There are two basic forms of heterochrony: paedomorphosis and peramorphosis.

In **paedomorphosis**, adults of the descendant species look much like juveniles of the ancestor. In other words, the descendant shows an earlier onset of sexual maturity relative to the ancestor. When this happens, the ancestor will retain "juvenile" characters of the ancestor into the adult phase of the descendant. If the descendant is the same size or larger than the ancestor, it is called **neoteny**; and if the descendant is smaller than the ancestor it is called **progenesis**. Neoteny seems to have played a big role in the evolution of humans (which look much like the juveniles of chimpanzees),

and neoteny probably played a role in the evolution of many tiny, smooth-shelled trilobites from larger, more ornamented forms.

In **peramorphosis**, the onset of sexual maturity in the descendant is delayed in comparison to the ancestor, meaning that the descendant continues to show development of characters that the ancestor did not show. The general tendency for size increase through evolutionary lineages (Cope's Rule) is largely due to peramorphosis.

In a lineage showing peramorphic development of descendant species, ancestral morphologic stages may be mirrored in part in ontogenetic development. This observation, first advanced by the German anatomist Ernst Haeckel in the 1800s, has been encapsulated in the phrase "ontogeny recapitulates phylogeny." Although this phrase is an oversimplification that does not necessarily apply to paedomorphically derived species, it can be usefully applied to determining shared relationships among some species in lineages (such as in the embryologic development of vertebrates).

Heterochrony FIGURE 5.19

Species in these trilobite lineages evolved by changing the timing of maturation in the descendant forms. Adults of later-appearing species that resemble the juveniles of their ancestors evolved through paedomorphosis. Large eyes, a small number of segments in the middle (thoracic) region, and large tail shields are all juvenile-type characters of the ancestor retained in the adults of the descendants. Adults of later-appearing species that developed more characters than those in their ancestors evolved through peramorphosis. An increased number of segments in the middle (thoracic) region and small tail shields are traits developed in the descendant species.

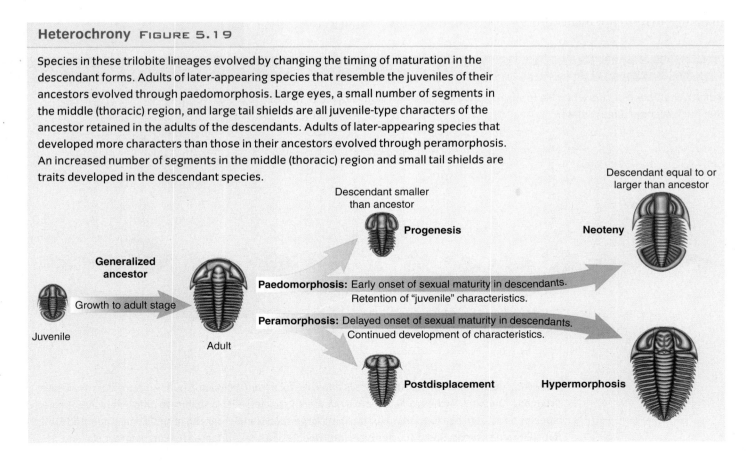

Descendant smaller than ancestor — Progenesis

Descendant equal to or larger than ancestor — Neoteny

Generalized ancestor

Growth to adult stage

Juvenile

Adult

Paedomorphosis: Early onset of sexual maturity in descendants. Retention of "juvenile" characteristics.

Peramorphosis: Delayed onset of sexual maturity in descendants. Continued development of characteristics.

Postdisplacement

Hypermorphosis

Asymmetry is a fundamental aspect of the biology of organisms. Selective pressure has reinforced patterns of asymmetry, some more obvious than others, through evolutionary history.

B A typical snail shell provides an excellent example of spiral morphology.

A The tusks of elephants, which are used in foraging for food, among other purposes, curl in a distinctly asymmetric fashion. This is a Pleistocene mammoth showing asymmetric tusks.

D Leaves and stems on plants are typically arranged in the form of a spiral. Here, leaves on a wild blueberry plant in Denali National Park, Alaska, show a loose helical pattern.

C Crab claws have distinct differences in size and function between the right and left sides, even though these animals have evolved a nearly bilateral symmetry.

ASYMMETRY

One pervasive pattern in biologic organisms is morphologic and functional asymmetry, and we can see this pattern expressed both in fossils and in living organisms (FIGURE 5.20). Although the right and left sides of many animals appear to be mirror images (meaning that the creatures are bilaterally symmetric), close examination reveals subtle differences between the right and left sides. Those differences may show up even better in behavioral differences between the two sides, a pattern known as **handedness** (from the right-left differences in the function of human hands). An even more obvious pattern of asymmetry is the spiral, which may be the most common shape in biologic organisms. Spiral organisms include various eubacteria, protoctists (such as foraminifera), plants (stem-and-leaf branching pattern), and animals (such as mollusks, brachiopods, and corals).

Some interesting patterns of asymmetry have been documented. In the evolution of the primates, it seems that the switch from a predominantly left-handed condition to right-handedness occurred near the base of the hominid lineage with the rise of australopithecines. Some planktonic foraminifera of the Cenozoic coiled preferentially in one direction during times when ocean waters were cold and in the opposite direction when waters were warm.

Asymmetry is not confined to the whole-organism level, but is the norm in organ systems, in the DNA molecule (right-handed double helix; Figure 5.8B on page 136), and in organic molecules, including amino acids, which are mostly left-handed molecules. The tendency toward asymmetry at the whole-organism level is possibly dictated by patterns imposed at these lower levels of organization, which suggests that it is "hardwired" into the biology of life forms. Morphologic asymmetry has been linked to the function of the nervous system and behavior of animals. In humans, asymmetric function of the hemispheres of the brain (so called "right brain/left brain" function) has been linked not only to functional differences between the right and left sides of the body but also to language development and cognitive skills.

Natural selection has had a long history of reinforcing right/left behavioral differences in species. Ambidexterity, or equal capability of the right and left sides, has tended to be selected against. It seems more adaptive for individuals to have a preferred lead side if they are to perform well in most functions, especially running or swimming from predators. Individual variation in the direction of handedness is characteristic of most species, and this, too, may be adaptive at the species level. If selective pressure reduced right-handed individuals of a species, left-handed individuals would still be available to repopulate the species.

Other adaptive reasons for asymmetry have to do with parsimoniously accommodating surface-dependent functions (such as digestion) and ensuring rapid reproductive isolation leading to speciation. A spiral shape allows more surface area to be confined in a smaller space than is possible with a nonspiral shape. A long digestive tract, for example, can be best accommodated within a confined space by coiling. In some snails, mutations causing switching of the direction of shell coiling have apparently led to the rapid divergence of species because reproductive isolation is immediately achieved when the direction of shell coiling changes.

EXTINCTION

Extinction is the annihilation of all individuals of a species. Every species that evolves eventually becomes extinct, so extinction can be viewed as a natural part of the evolutionary process. A certain percentage of the world's species can be expected to become extinct during any interval of geologic time, and the normal flux of extinction can be thought of as the background level of extinction.

Some intervals of geologic time, when many species were annihilated within a geologically short time frame, have attracted special attention. Such intervals, typically tens of thousands to a few million years in duration, are known as **mass extinctions**. Mass extinctions have punctuated the history of life on Earth and have had profound effects on its long-term development. Not only have such events eliminated many species, but they have served as bottlenecks (a form of genetic drift) for surviving species. Certain mass extinctions have stimulated short-term adaptive radiations. The coupling of a mass extinction with a subsequent adaptive radiation is a leading cause of some of the most dramatic biotic turnovers in Earth history (**FIGURE 5.21**), including those at the end of the Permian and Cretaceous periods. Some of the most important mass extinction intervals mark the ends of the Ordovician, Devonian, Permian, Triassic, and Cretaceous periods and have served as the historic basis for recognizing system (period) or era (erathem) boundaries.

Extinctions of species have multiple root causes, including overhunting, infectious disease, loss of habitat, and loss of a major food source. Presumably, mass extinctions are the result of major environmental perturbations, and they may have multiple causes. Rapid climatic changes, extraordinarily high levels of volcanic activity, meteorite impact and subsequent changes to the atmosphere, rapid spread of infectious diseases, falling of sea level and exposure of the continental shelves, and deforestation of the land all may result in ecosystem collapse. One of the great challenges in unraveling the origins of ancient mass extinctions is linking causes (evidence of the root causes of extinction) with effects (fossil evidence that biotic turnover has occurred). Interpretations of the processes leading to extinction from the geologic past are mostly based on circumstantial evidence, and in many cases, it is uncertain whether evidence of causes is coincidental or real.

> **mass extinction**
> A situation in which numerous species become extinct within a geologically short time interval.

Mass extinctions FIGURE 5.21

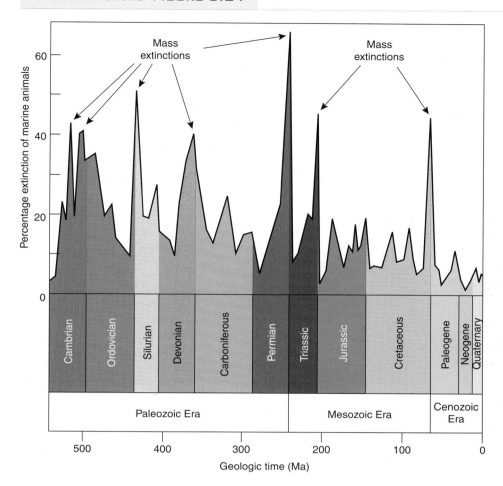

This graph illustrates the percentages of genera experiencing extinction through Phanerozoic time. Large spikes mark intervals of mass extinction. Such mass extinctions are a root cause for some of the biotic turnovers that distinguish the fossil records of successive units of geologic time.

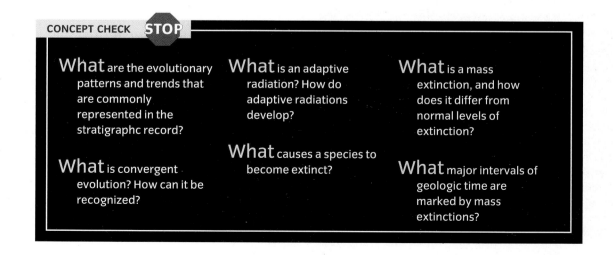

CONCEPT CHECK STOP

What are the evolutionary patterns and trends that are commonly represented in the stratigraphc record?

What is convergent evolution? How can it be recognized?

What is an adaptive radiation? How do adaptive radiations develop?

What causes a species to become extinct?

What is a mass extinction, and how does it differ from normal levels of extinction?

What major intervals of geologic time are marked by mass extinctions?

SUMMARY

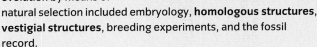

1 Evidence of Biologic Evolution and Natural Selection

1. **Evolution** in organisms refers to change through time. Changes occurring within a species or population from one generation to the next represent microevolution. **Natural selection** and **genetics** (including mutation, genetic drift, and gene flow) are microevolutionary processes. Evolution at the species level (and above the species level) constitutes macroevolution. Speciation, changes in the tree of life, and extinction are macroevolutionary processes.

2. **Natural selection** is the mechanism that drives biologic evolution. In this process, individuals best adapted to their environment are expected to survive through reproductive age, whereas less-well-adapted individuals tend to be eliminated. Darwin's evidence for evolution by means of natural selection included embryology, **homologous structures**, **vestigial structures**, breeding experiments, and the fossil record.

3. Natural selection depends on the process of **adaptation**, or modification of an organism or its parts, so that it is better fitted for survival in an ecologic niche.

2 Variation and Inheritance of Traits

1. Natural selection acts on variation within populations. As Mendel demonstrated, the source of variation and inheritance of traits are hereditary units that undergo independent assortment during reproduction.

2. **DNA** is a molecule that stores genetic information in both eukaryotes and prokaryotes. In viruses, either DNA or RNA stores genetic information. The DNA molecule, usually a double helix, contains **genes**, which are Mendel's hereditary units. Genes are arranged along a linear **chromosome**. The genetic code specifies how proteins are to be constructed.

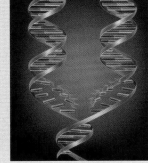

3. **Mutations** are the source of genetic variation. Mutations can develop in genes and in chromosomes. Alleles are different versions of the same gene that form through mutations. Few mutations are beneficial for survival, except for those that spread through a population because of favorable selective pressure.

4. Random events can contribute to changes in gene frequency in a population. This is called genetic drift.

5. The frequencies of alleles present in a population change as individuals enter or leave a population. This is called gene flow, and it helps balance out changes between populations that are introduced through mutation, genetic drift, and natural selection. Genetic divergence and speciation can occur if gene flow between adjacent populations is disrupted.

6. Molecular evolution refers to the use of genetic changes to track evolutionary history. Closely related species share a high percentage of gene types, and more distantly related species share fewer genes. If we know the rate of occurrence of genetic mutations, we can use the number of genetic differences between separate species as a measure of the amount of time that has elapsed since the species diverged. This concept is referred to as a **molecular clock**.

3 Speciation

1. **Speciation** is the rise of a new species from an ancestral species. Allopatric speciation involves disruption of gene flow between separate populations. Sympatric speciation involves the rise of a new species from individuals within the range of an ancestral species. **Parapatric speciation** involves the rise of a new species from individuals along a common border between two populations.

2. The fossil record shows two different speciation patterns. Cladogenesis is a branching speciation pattern that occurs as populations become reproductively isolated and then diverge genetically from each other. Anagenesis involves changes along a single evolutionary pathway, and the distinction between one species and its descendant may be ambiguous.

3. Speciation may happen slowly or relatively quickly. **Phyletic gradualism** is a speciation mode that involves slow, gradual evolution. **Punctuated equilibrium** is a speciation mode that involves rapid genetic divergence followed by a long interval of minor morphologic change.

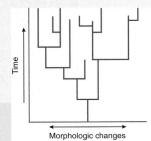

4 Phylogenetics: Reconstructing Evolutionary History

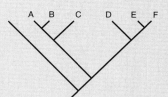

1. Cladistics, or phylogenetics, is the grouping of species based on evidence of shared common ancestry. A clade is a group of species that includes an ancestral species and all of its descendants. A **cladogram** is an evolutionary hypothesis that illustrates shared relationships inferred from a cladistic analysis.

2. A **synapomorphy** is a shared derived character and key evidence of a shared evolutionary relationship between two taxa.

5 Evolutionary Patterns and Extinction

1. Certain patterns or trends appear repeatedly in the fossil record. Irreversibility of evolution is a statement of the observation that a species can arise only once, and when it is extinct, it is gone forever.

2. Convergent evolution is the evolution of similar body forms in two or more biologic groups in response to similar ecologic strategies or similar ecologic pressures.

3. **Adaptive radiation** involves the rapid evolution of groups to quickly fill open niche spaces.

4. Animals commonly show a tendency to increase in body size through an evolutionary lineage.

5. Heterochrony involves evolutionary changes in a lineage that result from changes in developmental timing between the ancestral species and its descendant. In paedomorphosis, one form of heterochrony, adults of the descendant species look much like juveniles of the ancestor. In peramorphosis, another form of heterochrony, the descendant species continues to show development of characters beyond those of the ancestor.

6. Morphologic and functional asymmetry, including subtle deviations from bilateral symmetry and spiral asymmetry, is pervasive in fossil and modern organisms.

7. Extinction is the annihilation of all individuals of a species. Every species eventually becomes extinct, and thus extinction is a natural part of the evolutionary process.

8. In a **mass extinction**, many species representing multiple groups of organisms are annihilated within a geologically short time frame. Mass extinctions are thought to result from instances of major environmental collapse.

KEY TERMS

- **evolution** p. 126
- **natural selection** p. 128
- **homologous structures** p. 128
- **vestigial structure** p. 131
- **adaptation** p. 133
- **genetics** p. 134

- **DNA (deoxyribonucleic acid)** p. 136
- **gene** p. 136
- **chromosome** p. 136
- **meiosis** p. 137
- **mitosis** p. 137

- **mutation** p. 137
- **molecular clock** p. 138
- **speciation** p. 140
- **phylogeny** p. 142
- **phyletic gradualism** p. 142
- **punctuated equilibrium** p. 142

- **cladogram** p. 143
- **synapomorphy** p. 144
- **adaptive radiation** p. 148
- **mass extinction** p. 152

CRITICAL AND CREATIVE THINKING QUESTIONS

1. Why is genetic, morphologic, and functional variation important to life forms? If there were no source of genetic variation, what would life on Earth be like?

2. Why is understanding natural selection so important to evolutionary theory? How would you interpret the selective pressures that influenced ancient species from their fossils?

3. Do humans possess vestigial structures? If so, what are they, and what do they imply about the evolutionary history of humans? What other aspects of human biology give us clues to our evolutionary history?

4. Why are shared derived characters (synapomorphies) preferred for reconstructing evolutionary relationships among species? If you were to use all observable characters (including autapomorphies and homoplasies), might you arrive at other conclusions about relationships?

5. Why do you think evolutionary patterns and trends develop? Why do the same patterns and trends occur repeatedly in different groups of organisms?

Outgroup

Lamprey Salmon Lungfish Eagle Cat Chimpanzee Human

Tail (loss)

Hair

Limbs
Scales (loss)

Lungs

Jaws

What is happening in these pictures ?

Similar ecologic pressures can produce similar structures independently in animals. Dolphins, a group of mammals (above), and sharks, a group of fishes (below) evolved similar body forms as an adaptation to life as marine predators.

Are the fins and streamlined shapes constructed in the same way in both animals?

Is the fin morphology in these animals an example of homology or analogy?

1. What is macroevolution, and how does it differ from microevolution?

2. What is natural selection?

3. What are the principal lines of evidence of biologic evolution?

4. What is the most important contribution of the fossil record to our understanding of biologic evolution?

5. What is adaptation?

6. How has the science of genetics contributed to our understanding of evolution?

7. How does variation arise in natural populations?

8. What is a molecular clock?

9. Does evolution proceed slowly or quickly?

10. What is a cladogram?

11. How are shared evolutionary relationships between species identified using cladistic methodology?

12. What is an adaptive radiation?

13. What is extinction? What is a mass extinction?

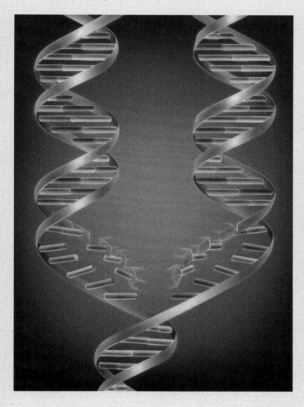

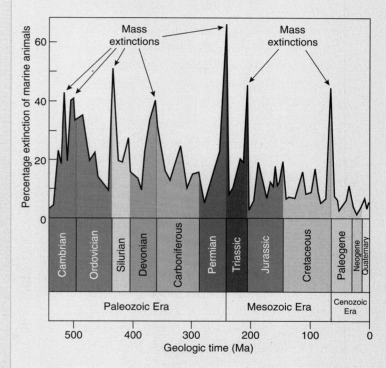

Interpreting Sedimentary Environments and Global Change

The "pages" of Earth's history book are its sedimentary layers, the places where information about biologic evolution, climate change, tectonic movement, and tectonic episodes become archived. The pages are not all stacked in one place, though. Instead, pieces of the text are separated and scattered, and in order to reconstruct the continuity of the information, we use correlation tools that allow us to determine the order of the pages. Once the pages are stacked in correct order, it becomes apparent that they record information from different parts of the ancient world, almost as though the text were written by multiple "authors" working from different perspectives. Separate environments, represented by differing types of sediments, provide multiple perspectives on each individual time frame of Earth history.

In this chapter, you will learn about sedimentary deposits and features, the circumstances under which they form, and the way they are recognized. This is useful for determining ancient sedimentary environments using clues left in sediments, from fossils, and from the relationships of neighboring sedimentary environments. You will also learn about techniques that help us understand how the ocean-atmosphere system has changed through time.

Knowing about ancient sedimentary environments and how they have changed through time provides the element of context for many of the subplots in Earth's multifaceted story. We need this information to develop a coherent picture of Earth's past.

Modern and Ancient Sedimentary Environments

Sediments are deposited widely across Earth—from mountain streams to broad plains, in deserts, along beaches, and in the ocean—anywhere that water or air currents can reach and release their loads when velocity decreases. Each environment in which sedimentary particles are deposited has a distinctive set of characteristics and a predictable relationship to other sedimentary environments. To correctly interpret ancient environments, we must apply a uniformitarianistic approach, comparing ancient sediment types, sedimentary structures, fossils, and stacking or lateral patterns of sediment types to those of the modern world.

Fundamental to determining ancient environments is the concept of facies. A **facies** is a sedimentary unit that has a set of characteristics particular to a local environment (**FIGURE 6.1**). Usually the word refers to a certain lithology (for example, sandstone, shale, or limestone), or **lithofacies**. Essentially, a lithofacies denotes the sediment of one depositional environment (such as a beach), as contrasted with the sediment in another depositional environment (such as a marine shelf). Lithofacies are often formalized as **formations**, which are the fundamental units of lithostratigraphy.

> **facies**
>
> A sedimentary unit that has a set of characteristics (such as lithology, color, or fossil assemblage) particular to a local environment.

Another common form of facies is a **biofacies**, which is based on a distinctive assemblage of fossils occurring within a sedimentary unit. In the modern world, living things are inextricably linked to particular environments, and we can expect that the same was true in the past. Fish are restricted to aqueous environments, and trees are exclusively terrestrial. All starfish and

Sedimentary facies FIGURE 6.1

The South Island of New Zealand provides a microcosm of the sedimentary environments and facies that are often spread out over the vast expanses of continents. In the distance, the great Southern Alps tower above a soil-mantled plain. A large braided stream winds its way across the plain toward the Pacific Ocean, where it empties its sediment load in a large delta at the mouth of Milford Sound. In the delta, white sands and gravels contrast with gray-green muds.

Grain size distribution of sediments FIGURE 6.2

In a delta along the shore of the Kamchatka Peninsula, Russia, sediment grains are distributed according to the amount of current energy. Wave agitation along beaches, including those on small islands, concentrates sands (identifiable as thin white strips). Finer-grained silts and clays are deposited in deeper, quieter waters just offshore (identified by brownish and bluish zones).

corals are limited to marine water, whereas different species of fish, clams, and snails live in marine and freshwater environments. Biologic organisms, which are sensitive indicators of environment, are an obvious guide to ancient environments. In some circumstances, the trace fossils they made are even more reliable because they cannot normally be transported from their point of origin as bodily remains can.

Sedimentary environments typically have distinctive characteristics that reflect whether they were aqueous or subaerial, freshwater or saltwater, high or low energy, oxygenated or anoxic, and inhabited by animals and plants or not. These physical, chemical, and biologic controls play a role in determining sediment type and sedimentary structures, both of which can be combined with other evidence to interpret ancient environments. Another widely used technique for determining ancient sedimentary environments is the regional pattern of coexisting facies. Finally, ancient environments can be interpreted in part using certain isotopes recorded in sediments or in the skeletons of organisms.

USING SEDIMENTS TO INTERPRET SEDIMENTARY ENVIRONMENTS

Studies of modern sedimentary environments show that grain size, sorting, and roundness are good indicators of the amount of current energy in an environment or the amount of transport from a sediment source.

Grain size is an indicator of the distance a sediment has traveled from its source or the amount of current energy available to move it (FIGURE 6.2). Coarse sediments tend to accumulate in environments

grain size The general dimensions (such as diameter or volume) of particles in a sediment or rock.

relatively close to the sediment source and in areas subject to high current energy. Fast-moving streams and rocky beaches are good places to find pebbles and cobbles. Sandy beaches and deserts are common places to find sands. Finer sediments (silts and clays) accumulate farther from the source and in areas of lower current energy, especially deep ocean basins, protected lagoons, and lakes.

Increased transportation results in greater size **sorting** of particles (FIGURE 6.3A). Sediments in the bed of a mountain stream close to the sediment source tend to be poorly sorted. The largest grains may be pebbles or cobbles, but sand and silt may also be present. As sediments are carried along the length of the stream, the range of sediment sizes diminishes as smaller grains become separated, or winnowed, from coarser ones and are carried farther along. Beach and desert sands, which are subject to intense current action, are often well sorted. Multiple episodes of transportation can increase the likelihood that a deposit will be well sorted. Beach sands that become buried, exhumed, and reworked as beach sands are generally very well size sorted.

Increased transportation in water tends to increase the **roundness** of grains (FIGURE 6.3B). A mountain stream close to a sediment source usually has lots of angular grains in the streambed. The farther the sediments are carried along by water, the fewer sharp edges will remain on the grains. Interestingly, sands

> **sorting** A measure of the range of grain sizes in a sedimentary deposit.
>
> **roundness** The degree to which a sedimentary particle's original edges and corners have been smoothed.

transported by wind are commonly quite angular. This is because blowing grains strike others and remove small chips at their edges.

Siliciclastic or detrital sediments accumulate in many environments. All that is required is a source of detrital grains (such as a mountain range) and a method of transporting the sediment (usually water, air, or glacial ice) from the source to the place of deposition.

Carbonate sediments, principally aragonite, calcite, and dolomite (the sources of limestone and dolostone) have restricted environmental occurrences. Carbonate sediments are produced largely through the activity of organisms. Mollusks, corals, and some sponges produce calcium carbonate skeletons, but an even larger share of carbonate sediment is produced through the activity of photosynthetic plants, algae, and bacteria. Conditions most conducive for the precipitation of calcium carbonate sediment (aragonite and calcite) are warm, shallow, well-lit marine waters (FIGURE 6.4). For the most part, carbonate environments are in or close to the tropics, and

Sorting and rounding of grains FIGURE 6.3

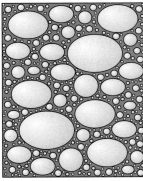
Poorly sorted

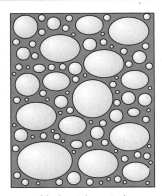
Moderately sorted

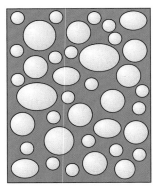
Well sorted

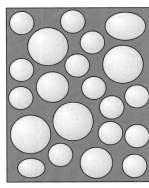
Very well sorted

A Sorting reflects the range of grain sizes in a sediment. Categories of grain size sorting range from poorly sorted to very well sorted.

Very angular

Angular

Subangular

Subrounded

Rounded

Well rounded

B A scale of rounding is a reflection of the amount a grain's original edges and corners have been smoothed. Categories of rounding range from very angular to well rounded.

Carbonate sediments form mostly in warm, shallow, well-lit tropical waters. Here, in the shallow waters surrounding Exuma Cays, Bahamas, photosynthetic algae, bacteria, and shell-secreting animals produce enormous quantities of light-colored carbonate mud and sand, which appear as plumes below the water line.

in water less than 30 m deep. Some carbonate sediments accumulate in deeper water, but they are mostly the result of the "raining" of skeletons of carbonate-secreting pelagic organisms to the sea floor.

SEDIMENTARY STRUCTURES

Sedimentary structures are features of sediment that reflect physical, chemical, or biologic controls on their accumulation.

> **bedding** Layering in sedimentary rocks.

The most basic feature of sediments is layering, or **bedding** (FIGURE 6.5). Layering occurs at scales ranging from millimeters to meters or tens of meters. Layering at fine scales is also referred to as **lamination**, and layering at larger scales is commonly referred to as **stratification**.

Layers, which are usually flat or nearly flat surfaces, result from short lapses of sedimentation or subtle changes in grain size or grain composition. Low current energy often leaves flat sedimentary layers. In a tidal setting, layering may develop when grains are deposited at high tide. At low tide, there is a lapse in sedimentation. The first grains deposited in a tidally deposited layer may be slightly larger than the last grains deposited in the previous tidal cycle.

In many settings, sediment is deposited in thin layers. However, the thin layers are not always preserved. Often, organisms living and burrowing in the sediment

Sedimentary stratification FIGURE 6.5

A Sedimentary layers shown here range from millimeters to several centimeters in thickness. This bedding was originally horizontal, but tectonic uplift has tilted the layers.

B Sedimentary layers can be followed for long distances along outcrops, like this one in western New York, if they are uninterrupted.

Bioturbation FIGURE 6.6

Sediments burrowed by benthic animals tend to be mottled and lacking the thin laminations they once had.

bioturbation

Reworking of sediment by organisms.

destroy the original laminations. If burrowing (or **bioturbation**) has been extensive, the sediment will appear mottled (**FIGURE 6.6**) and more thickly bedded. In general, bioturbated strata indicate areas inhabited by an array of bottom-dwelling animals, and these tend to be well-oxygenated, food-rich environments. Strata retaining their original thin laminations for the most part are ones limiting to bottom-dwelling animals because of low oxygen levels, salinity stress, or other reasons. Until the latest Cambrian or earliest Ordovician, relatively few animals had adapted to a burrowing lifestyle, so earlier-deposited strata tend to be well laminated, regardless of the setting.

Ripples (or ripple marks) are formed when air or water currents pass over sediment and shape the sediment surface

ripple

A sedimentary bed form that has a roughly triangular transverse cross-section and formed by the interaction of a moving air or water current with a mobile sediment.

into triangular peaks or ridges elongated perpendicular to the direction of the current (**FIGURE 6.7**). Most sediment accretion occurs in the downstream direction, on the lee side of a ripple, and sediment tends to be deposited at a slight angle. Ripples can range in size from centimeter-scale to dunes, which may be tens of meters across.

Ripples are of two basic shapes: symmetrical ripples and asymmetrical ripples. **Symmetrical ripples**, also called **oscillation ripples**, form where there is a back-and-forth (oscillatory) movement of water currents (**FIGURE 6.7A**). They usually form in relatively shallow, restricted areas, such as tidal flats, ponds, and swampy areas. **Asymmetrical ripples**, also called **current ripples**, form in a unidirectional current (**FIGURE 6.7B**). Consistent and predominantly one-way movement of water or air occurs in places such as rivers, on some beaches, and in deserts.

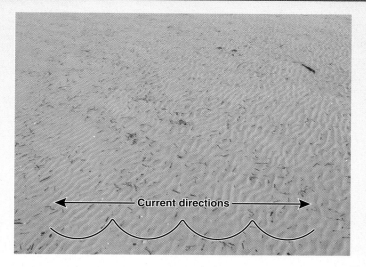

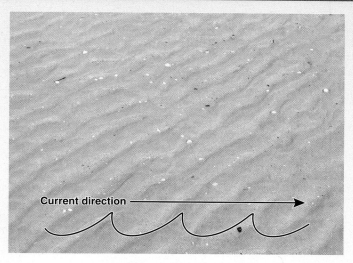

A In a shallow restricted bay, where gentle oscillating currents move the sand, and leaves of marine algae come to rest, small symmetrical ripples are forming.

B In a more open area of the same bay at low tide, asymmetrical ripples formed in unidirectional currents.

The bedding internal to a ripple is inclined, not perfectly horizontal. In fast-moving currents, the angle developed is usually greater than the angle developed in slow-moving currents by sediment of the same size and shape. Changes in current strength or changes in the direction of the current flow cause the bedding planes within the ripples to be inclined at different angles. This results in **cross-bedding** (FIGURE 6.8). Often, the tops of ripples are truncated by currents before the next set of ripples is deposited. However, the cross-beds record much of the internal geometry of the ripples.

cross-bedding
Strata inclined at different angles formed by the rippling of sediment.

Cross-bedding in strata
FIGURE 6.8

During the Jurassic Period, these giant cross-beds formed in the interiors of sand dunes in what is now Utah.

With **graded bedding**, sediments deposited during a single event show a vertical change in grain size (FIGURE 6.9). Usually, grain size diminishes upward. Graded beds are deposited in stream channels, especially on natural levees and on floodplains, when high-energy floodwaters carry a large sediment load and particles of variable sizes. As current energy decreases, the larger and heavier particles deposit first, and smaller and lighter ones remain suspended until current energy is reduced even further. Graded beds are also common in landslide deposits and in turbidites. **Turbidites** are deposits left by turbidity currents, the submarine equivalents of landslides, in which a turbid mix of water-laden sediment slides rapidly downslope in response to gravity.

> **graded bedding**
> A single-event sedimentary bed in which there is a progressive vertical change in grain size.

Graded bed deposits FIGURE 6.9

A Flooding of an ancient stream left this graded bed deposit. Pebbles up to 5 cm across are at the bottom of the layer. The sediments become finer toward the top. ▶

B In May 2001, the Lena River of Russia overflowed its banks, flooding the town of Lensk. When a river overflows its banks, it leaves behind a graded bed deposit. ▼

Mudcracks are polygonal structures formed by the drying and shrinking of mud (FIGURE 6.10A–C on page 168). They develop in damp or wet continental areas such as shallow and ephemeral ponds and lake and stream margins. Mudcracks form as water-laden clays and silts lose their moisture through evaporation. The mud shrinks and then splits into polygons that curl upward, causing them to appear dish-like.

Mudcracked sediment often has **raindrop impressions** (FIGURE 6.10A) associated with it. These sedimentary structures are small craters formed by raindrops while the mud was still wet. After the mud dries and cracks, the small craters become "frozen" on the mud polygons.

Mudcracks and raindrop impressions are excellent examples of a group of sedimentary features collectively referred to as **geopetal structures**. These are valuable tools for the field geologist because they show stratigraphic younging direction, or which way is stratigraphically up, even if the strata have been tilted or overturned. Mudcracks, because they curl toward the sky, are concave upward. Raindrop impressions, because they are little craters, are also concave upward. Once younging direction has been determined, a geologist can then apply the **principle of superposition**. Aside from mudcracks and raindrop impressions, some of the best geopetal structures are stromatolites and thrombolites, which grow upward toward sunlight; ripples, whose crests point upward; crossbeds, which thin downward and thicken upward; graded beds, which fine upward; flute casts, which are convex downward; and certain trace fossils, such as footprints, which are concave upward.

Flute casts are elongated, scoop-shaped structures on the bases of sedimentary beds (FIGURE 6.10D on page 168). They are caused by currents hollowing out the sediment, sometimes aided by sticks or stones carried in the current. Once the erosional current has passed, a sediment layer, typically a fine-grained

> ■ **mudcracks**
> Irregular, polygonal fractures formed by the drying and shrinkage of mud.

> ■ **geopetal structure**
> A sedimentary feature that shows the younging direction of strata.

layer, is deposited. The new layer covers the eroded surface and fills in the scooped-out areas, forming flute-shaped casts in sediment. Flute casts indicate the direction of current movement: the first-formed part of the scoop (the upcurrent side) is abrupt, and the "tail" of the scoop is elongated in the downcurrent direction.

Flute casts commonly occur in turbidite deposits, where they are arranged in unidirectionally oriented clusters on the bases of beds. Turbidites develop predominantly in deep water, especially on the continental slope and rise.

Flute casts also may occur at the bases of beds eroded during storms (hurricanes or typhoons) on shallow continental shelves. Storm-initiated deposits, called **tempestites**, generally show clusters of flute casts arranged in two or more directions.

Stromatolites are biogenically mediated sedimentary structures formed by microbes interacting with the environment. They are formed in shallow, aqueous carbonate environments and made of numerous thin layers (FIGURE 6.10E–G on page 169). During daylight hours, photosynthetic cyanobacteria (blue-green eubacteria) form a sticky upper surface on the stromatolites. At night, a thin layer of carbonate grains adheres to the cyanobacteria, and the organisms then lengthen strands between the sediment grains until they again reach the surface, where they can photosynthesize. Numerous thin layers accumulate after many such cycles. Stromatolites take on various forms, ranging from laminar through small bumps to tall domes or columns.

> ■ **stromatolite**
> A thinly layered biogenic-sedimentary structure that results from the trapping and binding of fine sediment in layers by photosynthetic cyanobacteria.

Thrombolites are similar to stromatolites and likewise form by cyanobacteria interacting with sediment grains. Instead of having distinct laminations throughout, they have a clotted texture internally.

Stromatolites form in shallow water because the photosynthetic cyanobacteria that form them are limited by the availability of sunlight. The height of a stromatolite is limited by water depth, and in intertidal areas, the height of a stromatolite indicates maximum water depth at high tide.

Sedimentary features are useful for interpreting ancient environments. A variety of them, distinguished by the term geopetal structures, also tell a geologist which way was stratigraphically up at the time of deposition. Even if the beds later become disrupted by tectonics, geopetal structures can be used to figure out the original upward direction of layers.

Features of dried mud

A Mudcracks and raindrop impressions form today in subaerial settings where wetting and drying occurs, such as along the margin of a pond. Here, small round craters were made in soft mud by raindrops before the mud dried out and formed polygonal mudcracks.

B In oblique view, mudcracks clearly have some relief. Polygons dry out and curl toward the sky. Knowing this helps geologists ascertain the younging direction of strata.

C Ancient mudcracks, like these, not only indicate a subaerial environment but also indicate which way is stratigraphically up. These ancient mudcracks curl slightly toward the observer, just like the mudcracks in A and B, meaning that you are looking at the top of the layer.

Features of wet mud

D Flute casts are formed when mud is scooped off the surface of a bed, and then a new layer is deposited over top, filling the hollowed-out scoops with sediment. These flute casts at the base of a turbidite deposit indicate both the flow direction of the turbidity current that formed the deposit (upper right to lower left) and which way was stratigraphically up. You are looking at the bottom side of the bed.

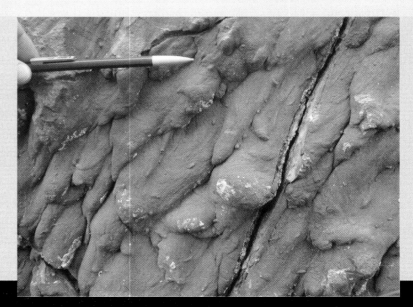

Interaction of photosynthetic bacteria and sediment

◄ **E** Stromatolites are thinly laminated structures that grow upward because consortia of photosynthetic eubacteria play a vital role in their development. These are modern stromatolites in Hamelin Pool, Shark Bay, Australia, as high tide is approaching. The stromatolites grow only as high as the level of water at high tide.

F Cyanobacterial mats that form stromatolites start out during daylight hours as sticky, uncovered surfaces. As sediment grains wash in with the tide or by other means, they adhere to the mat surface. Strands of cyanobacteria trap the grains by extending upward and around them (toward the sunlight), forming a sediment layer. At night, the mat surface is regrown, and the sediment layer below it is bound in place.

▼

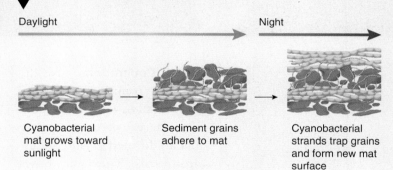

Daylight ──────────────────────────► Night ──────►

Cyanobacterial mat grows toward sunlight

Sediment grains adhere to mat

Cyanobacterial strands trap grains and form new mat surface

◄ **G** Ancient stromatolites, split longitudinally, show their internal structure. The thin, upward-arching layers result from upward growth toward the sunlight. As the modern stromatolites in E demonstrate, these structures also tell a geologist something about water depth: the rounded tops mark the high-tide level. These stromatolites are from the Cambrian of Utah.

Concretions, like this one in a Devonian shale along the Lake Erie shore in southwestern New York, result from precipitation of minerals such as calcite while the sediments surrounding them were still unconsolidated. Experiments suggest that mineral precipitation was initiated by chemical changes inside bacterial-fungal halos formed around decaying organisms. Because concretions form long before compaction and lithification of the sediment layers surrounding them, overlying layers bend around the structures.

> **concretion**
> A rounded body enclosed in sedimentary rock. Concretions are normally formed by microbially mediated precipitation of minerals during the decay of an organism.

Concretions are distinctive, rounded structures (FIGURE 6.11), usually calcite, siderite, pyrite, or quartz, that are formed in aqueous environments. They may range from a few centimeters to a few tens of centimeters in diameter. Rarely, they reach a meter or more in diameter. Concretions form around decaying organic matter, and sometimes remains of the partly decayed organisms are preserved inside them. In modern settings, concretions begin forming shortly after an organism dies. Microscopic fungi and bacteria responsible for decay form a balloon-like "halo" around the dead organism. The microorganisms then change the chemical microenvironment within the halo (causing dysoxia or an increase in pH) and initiate precipitation of minerals within a few weeks.

Paleosols are ancient soils (FIGURE 6.12), and they indicate terrestrial environments. Paleosols are commonly reddish or brownish-red in color, similar to the reddish soils and clays present today in many areas that were unglaciated in the recent past. The distinctive color is imparted by oxidation of iron-bearing minerals. In the stratigraphic record, deposits that have a reddish color are sometimes called "redbeds," and they are often cemented with iron oxides. Characteristics of some paleosols are mottling (due to burrowing animals), root casts, and caliche nodules (irregular calcium carbonate structures often formed around plant roots).

> **paleosol** An ancient soil horizon or profile.

WALTHER'S LAW OF FACIES

One job of a geologist is to reconstruct ancient environments (or facies) in relation to others that existed at the same time and in the same region. This is especially useful for predicting the occurrence of petroleum-bearing deposits, coal deposits, and aquifers.

▲ A Modern soils contain plant roots and burrows. Iron oxide, which imparts a reddish color, can serve as a cement as soil is lithified. This photograph shows two reddish soil profiles stacked on top of each other. Horizons rich in organic matter form the dark upper reaches of each soil horizon.

▲ B Paleosols show many of the same characteristics as modern soils, including a reddish color. These Mesozoic strata exposed in Arches National Park, Utah, were deposited subaerially, and they show a series of reddish soil profiles.

◄ C Caliche nodules, made of calcium carbonate, form around plant roots. These caliche nodules formed in a reddish, carbonate-rich soil in the Australian outback.

Walther's Law of Facies The concept that in an unbroken sequence, vertically superimposed lithofacies were laterally adjacent to each other at the time of deposition.

transgression A rise of sea level; marine facies shift landward.

How do geologists interpret the original lateral relationships of depositional environments? They do so using a method first articulated by the 19th-century Austrian geologist Johannes Walther. This method, known as **Walther's Law of Facies**, states that, in an unbroken sequence, vertically superimposed lithofacies were laterally adjacent to each other at the time of deposition (FIGURE 6.13 on page 172). The reason that facies become vertically stacked is that sea level changes cause the facies to shift and become positioned over the facies previously deposited in a location.

A **transgression** is a rise of sea level, when marine water floods the **continental shelf**. In a transgression, marine lithofacies shift landward, meaning that in any given location, they become stacked over shoreline facies. A **regression** is a drop of sea level, when marine water recedes from the land. In a regression, continental facies shift seaward, meaning that in any given location, they become stacked over marine facies.

Using Walther's Law of Facies, we can easily interpret whether sea level in an area was rising or falling. Most unbroken stratigraphic sequences consist of cyclic rises and falls, so midway through each cycle, there is a point above which the lithologies of strata are essentially a mirror image of those below. In a beach-to-offshore marine environment (FIGURE 6.13A on page 172), the cycle begins with a sandstone (representing sandy beach and nearshore environments). Finer-grained sediments are

continental shelf The part of the continental margin between the shoreline and the continental slope; characterized by a very low slope.

regression A drop of sea level; continental facies shift seaward.

Process Diagram

Walther's Law of Facies allows us to figure out what sedimentary environments were originally adjacent to each other from the way they are stacked in a stratigraphic column. In an unbroken succession, the vertical stacking pattern is a representation of the lateral facies after being rotated from horizontal to vertical.

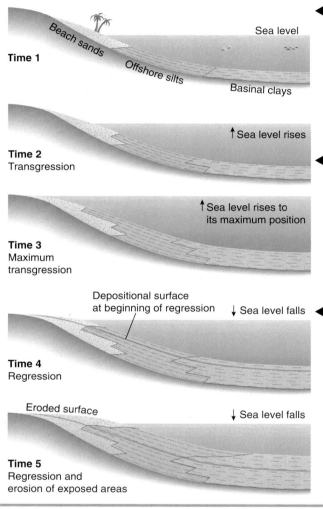

A **How do sedimentary facies become vertically stacked?** To visualize this, we will start with a generalized model of siliciclastic sedimentary facies. **Time 1** is our starting point. In the model, palm trees are growing in sand on the beach, close to the water's edge. Silts are being deposited just offshore, and clays are being deposited in the basin. As we progress through a series of time slices, pay particular attention to what happens at the point where the palm trees now stand.

B How does sea level change affect the sedimentary pattern? Rising sea level causes the position of sedimentary deposits to shift landward **(Time 2)**. This is called a *transgression*. The point on the beach where the palm trees once stood is now covered by marine water and receiving finer-grained sediments. As sea level continues to rise, even finer-grained sediments become stacked at that point. Eventually, sea level stops rising, and the transgression reaches its maximum extent **(Time 3)**.

C **How does the sea level cycle come to a close?** Falling sea level causes the position of sedimentary deposits to shift out to sea (**Time 4** to **Time 5**). This process stacks coarser-grained sediments over finer-grained sediments. In places that become subaerially exposed, erosion occurs. When sea level rises again, this erosional surface will be blanketed with sediment, creating a disconformity.

deposited farther offshore, so at some distance from the sandy deposits, silts are deposited. Still farther out to sea, clays are deposited in the quietest water, where current and wave energy is lowest. A rise in sea level over a relatively short period of time causes the marine facies to shift landward (FIGURE 6.13B), and silts are deposited over the beach and nearshore sands. The silts that became stacked on top of the sands will lithify to siltstones. A continued rise in sea level will stack clays over the silts. The clays will lithify to shales. Thus, a rise in sea level, or transgression, in this setting will be represented in a vertical stratigraphic column as a sandstone-siltstone-shale sequence (FIGURE 6.13D) reading the strata from bottom to top, as we always do. This pattern can also be thought of as a fining-upward sequence, with the coarsest deposits representing shoreline and nearshore deposits and the finer deposits representing increasingly deeper water deposits.

A sea level rise, which may be initiated by the melting of polar glaciers or other factors, can only pro-

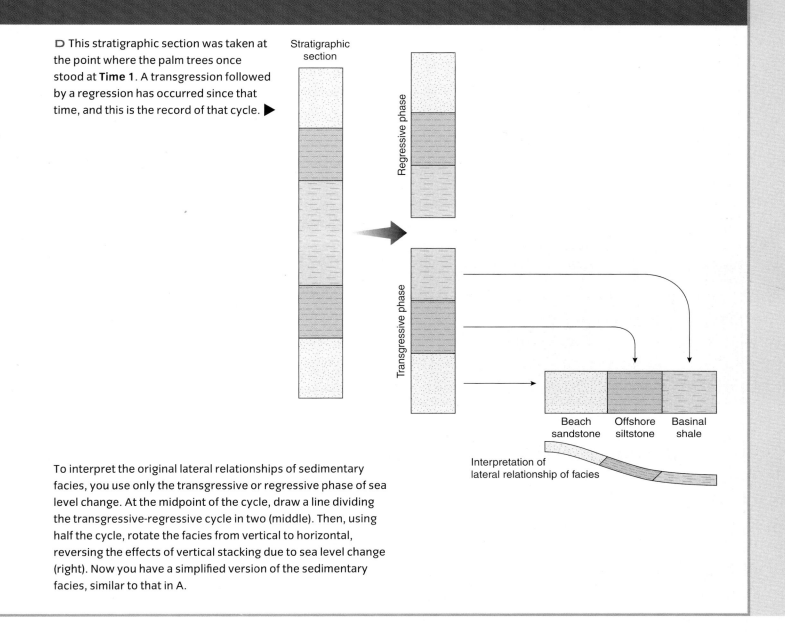

D This stratigraphic section was taken at the point where the palm trees once stood at **Time 1**. A transgression followed by a regression has occurred since that time, and this is the record of that cycle. ▶

Stratigraphic section

Regressive phase

Transgressive phase

Beach sandstone

Offshore siltstone

Basinal shale

Interpretation of lateral relationship of facies

To interpret the original lateral relationships of sedimentary facies, you use only the transgressive or regressive phase of sea level change. At the midpoint of the cycle, draw a line dividing the transgressive-regressive cycle in two (middle). Then, using half the cycle, rotate the facies from vertical to horizontal, reversing the effects of vertical stacking due to sea level change (right). Now you have a simplified version of the sedimentary facies, similar to that in A.

ceed so far. Eventually, the transgression will end. If a glacial cycle is the root cause of the sea level change, the renewed buildup of glaciers will spark a sea level fall. In our example, that point of reversal will be recorded in the clay layers (shales), which represent the maximum transgression (**FIGURE 6.13B**). If sea level falls, a regression will take place, and progressive recession of marine water should be recorded in a vertical stratigraphic column by shale, followed by siltstone and then by sandstone (**FIGURE 6.13C, D**). This is a coarsening-upward sequence, and it is the inverse, or mirror-image, pattern of the previous transgression.

The next transgressive-regressive cycle would begin with a sea level rise (recorded in our example with a sandstone). In other words, the sandstone marking the top of the regression also will serve as the base of the following transgression, and so the cycle repeats. The total time required for a transgressive-regressive cycle to form is trivial in geologic terms (normally in the 10,000-year to 100,000-year range), so we can largely ignore the effect of time on the stacking of sedimentary layers. The stacking of shales and siltstones over sandstones in a transgressive phase of a cycle can be thought of as almost "instantaneous" in geologic terms.

Now that you understand how to "read" a succession of layers from a stratigraphic column, you can lay out the depositional environments as they were at the time of deposition. Simply turning one transgressive phase (or one regressive phase) sideways yields a simple representation of laterally adjacent facies (FIGURE 6.13D on page 173).

In order to correctly apply Walther's Law, there can be no significant break in the sedimentary succession such as an unconformity or a fault. A fault will show offset beds, and an unconformity will show truncated strata overlain by other sediments.

One way to ensure that you have a complete succession represented is to examine the rocks for an "interfingering relationship," where the changeover from one lithology to the next is not perfectly sharp but involves a transition zone recorded by repeated switching from one lithology to the next before the new lithology becomes dominant. Repetition of the lithologies through a transition zone can result from sea level cycles occurring on a much smaller scale than the overall rise or fall recorded in the rocks.

CONCEPT CHECK STOP

What is a facies? How does a lithofacies differ from a biofacies?

What types of geologic evidence can be used to recognize ancient sedimentary environments?

What are the major forms of sedimentary structures? How can these sedimentary structures be used to interpret ancient environments?

What is Walther's Law of Facies, and how is it used to interpret ancient environments?

Interpreting Ancient Sedimentary Environments

LEARNING OBJECTIVES

Explain the major features of modern nonmarine, transitional marine/nonmarine, and marine environments.

Envision how nonmarine, transitional marine/nonmarine, and marine environments will leave a record in strata.

Explain the types of outcrop evidence used to interpret ancient sedimentary environments.

Explain the major types of sedimentary settings and the ways they can be recognized from stratigraphic successions.

What major depositional environments are represented in the sedimentary record? Just as in the modern world, ancient environments can be divided into three basic categories: nonmarine environments, transitional marine/nonmarine environments, and marine environments. In this section, you will learn the key characteristics of major depositional environments and how to identify these settings from outcrop evidence. The primary lines of evidence used for interpreting ancient sedimentary environments are (1) sediment types, (2) fossils, (3) sedimentary structures, and (4) Walther's Law of Facies. These information sources can be supplemented with other evidence, such as ratios of stable isotopes, which may help with interpreting water temperature or salinity. Also, computer modeling of strata can provide a geologist with an understanding of the three-dimensional geometry of a deposit. The overall shape and lateral relations of a deposit are important for reconstructing ancient sedimentary environments, and they constitute essential information for predicting the location and scope of economically valuable resources locked in the subsurface.

NONMARINE ENVIRONMENTS

Nonmarine environments in general are perhaps best identified by the presence of terrestrial or freshwater fossils and sometimes by reddish colors (Figure 6.12 on page 171) in the sediments.

One of the most significant nonmarine environments is the stream, or river, setting (see *What a Geologist Sees*). In addition to the stream channel, the stream environment includes the floodplain and sometimes swamps. Streams are actively erosive, cutting downward and laterally along the outsides of the bends (or meanders). They also deposit sediment within the channel, including on point bars, and on the floodplain.

In the stratigraphic record, sediment-filled cross-sections of stream channels are sometimes called **cut-and-fill structures**. The **principle of cross-cutting relationships** can be applied where the stream channel has cut into preexisting strata. Cut-and-fill structures are usually filled with coarse sediment (sandstone and conglomerate) deposited on point bars and at the bases of stream channels. Asymmetrical ripples and cross-beds are characteristic of stream channel deposits.

Floodplains and swamps receive mostly fine-grained sediments and are often the sites of rich organic deposits. Paleosols, coal beds (resulting from compression of organic matter), plant fossils, symmetrical ripples, and concretions are some of the features typical of

Stream Erosion and Sedimentation

A On the outsides of the Amazon's meanders, erosion is actively taking place. The white areas on the insides of the meanders are point bars, where sedimentary deposition is occurring. In the foreground is a large cutoff meander that is slowly being filled in by sediments. The floodplain of the Amazon is a vast, well-vegetated area.

This part of the Amazon River is far from the Andes Mountains, which are a sediment source. Because of the great distance of transport, coarse sediments the river carries tend to be rounded. The river carries a range of sediment sizes because frequent floods are able to carry particles of large size, in addition to the mud- and sand-size grains the river carries during times of normal flow.

The Amazon River of Brazil, like most streams, is a major transporter of sediment. Most of the sediment is ultimately carried to the ocean, but some remains in the stream's valley—in the channel, on the floodplain, in natural levees, in swamps, and in cutoff meanders.

B A stream channel emerging from the Andes Mountains of Chile is filling with sediment of differing sizes. The mountains are the source of the sediment, so the grains are poorly sorted and poorly rounded. The sediment fill in this channel will lithify to form a conglomerate.

Lake deposits FIGURE 6.14

A Loch Ness, a large freshwater lake in Scotland, has an elongate shape because it is formed along a fault. ▶

B Lake environments receive mostly fine silts and clays. Relatively few animals burrow the sediment of lakes, so thin sedimentary laminations are commonly preserved. This is a section through a Quaternary glacial lake in eastern Canada showing numerous light-dark sedimentary couplets called varves. The varves are arranged in larger bands grading from light to dark brown that also show cyclicity.

C Fish fossils are a feature of many Phanerozoic lake deposits. These specimens are from a Cretaceous lake deposit in Liaoning, China.

floodplains. Graded beds, resulting from deposition during flood stage, are normal on floodplains, and in some places, they build up to form levees.

Freshwater lake deposits (FIGURE 6.14) are generally quiet water areas in which fine sedimentation (clays and silts) predominate. Freshwater fossils, such as clams, snails, and fishes, may be present, but compared to many marine deposits, their species diversity is limited. Plant fossils, representing foliage from neighboring land, also may be present. Commonly, lake deposits retain thin sedimentary laminations because relatively few burrowing organisms live on the lake floor. **Varves**, or millimeter-scale couplets of a lighter layer deposited in summer months grading upward into a darker layer deposited in winter months (FIGURE 6.14B), are a common feature of lakes associated with glacial environments.

In arid regions where mountains have formed through fault blocking (such as the Great Basin of the United States), ephemeral mountain streams carry sediment from narrow canyons and deposit it at their mouths, which empty onto broad, flat areas between the mountain ranges. From the air, deposits that build up at the canyon mouths look much like hand-held fans extending from the openings. These deposits, called **alluvial fans**, thicken dramatically in the direction of the canyon and thin toward the other direction, where they merge with **playa lakes** that form in the flat basins (FIGURE 6.15). Sedimentation

> **alluvial fan**
> A relatively low, sloping mass of sediment, shaped like an open fan, deposited by a stream where it issues from a narrow mountain valley onto a broad plain.

Alluvial fans and a playa lake FIGURE 6.15

Alluvial fans form in deserts where ephemeral streams drop sediments at the mouths of narrow canyons. Playa lakes spread out in the broad basins formed between mountain ranges and in front of alluvial fans.

A An alluvial fan, photographed from a low-flying aircraft, in Death Valley, along the Nevada-California border. The thin line across the alluvial fan is a road.

B Death Valley is part of the Basin and Range Province, a place where large fault-block mountains rise above nearly flat valley floors. Broad, shallow playa lakes extend out in front of the mountains. From this vantage point, it is possible to see this relationship, as well as the narrow mountain canyons that fill with meltwater in the spring, and the alluvial fans that deposit sediment at the mouths of the canyons.

on the alluvial fans takes place mostly in spring, when snow melts and is then discharged as water into the basins. Playa lakes receive the water from spring melts, but because of high evaporation rates, they are commonly dry for most of the year.

Alluvial fan and playa lake deposits are relatively rare in the sedimentary record. However, where they occur, they may be indicated in part by graded bed and cross-bedded deposits, paleosols, few fossils, and evaporate minerals such as salt and borax.

Eolian environments are places where sediment accumulates through deposition by air currents (FIGURE 6.16). Dunes in deserts and along seashores are typical eolian settings in which sand-size grains predominate. Cross-bed sets, often ones with large, broadly rounded (or festoon) bases, are common, and fossils are rare.

Eolian deposits FIGURE 6.16

A Sand sculpted into large dunes, with few sedimentary features other than ripples and little evidence of life forms, is typical of many eolian (desert) environments. This is a view of the Sahara Desert of northern Africa.

B On the Colorado Plateau, the Navajo Sandstone, a very well-sorted quartz sandstone with large cross-beds, is what remains of a Jurassic dune field.

Glacially influenced landscapes include a great variety of subenvironments (FIGURE 6.17, FIGURE 6.18). Glaciation often leaves behind glacial grooves, which are elongate grooves carved by the scouring action of rocks lodged at the base of a glacial ice sheet as it slowly flowed over, sculpted, and polished the bedrock. Glacial grooves are parallel and elongated in the downstream direction.

Till, or unsorted sediment grains of markedly variable size, is another common feature of a glacial terrain. Tills are deposited in moraines at a glacial front, along the sides of a glacier, and as the glacier recedes. Sediment grains in the tills may have been carried great distances by glaciers, and when they are deposited in areas far from their source, they are referred to as **errat-**

ics. Tills may be reworked into other structures, such as **drumlins**, which are small hills of till elongated in the downstream direction of glacial flow.

Some other indicators of former glacial activity are outwash plain deposits, peat bog deposits, and dropstones. **Outwash plains**, where sediment is transported by streams away from the glacial margin, are indicated by size-sorted deposits with cross-bed sets deposited over wide areas. **Peat bog** deposits develop in outwash plains or in recessional moraines, often as icebergs become lodged in the sediment and slowly melt. Peat includes fossils of cool-weather-adapted terrestrial animals and plants, including perhaps conifer trees, mastodons, and mammoths. **Dropstones** (or **lonestones**) are rocks dropped into the muds of lakes from icebergs floating at the surface.

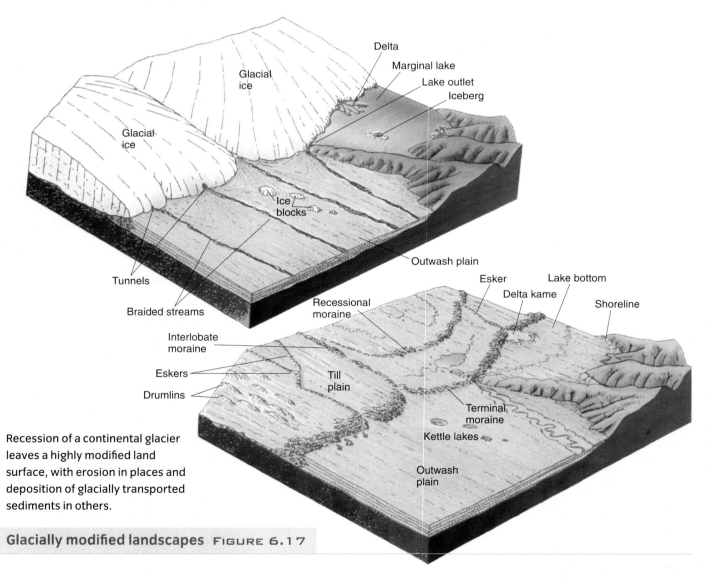

Recession of a continental glacier leaves a highly modified land surface, with erosion in places and deposition of glacially transported sediments in others.

Glacially modified landscapes FIGURE 6.17

Transantarctic Mountains, Antarctica

Global Locator

A Continental glaciers blanketing Antarctica are a modern analog for ancient glaciers that once covered North America, Europe, and parts of Asia and Australia. Along the shore of the Ross Sea, glaciers drape spectacular mountains, giving the landscape a subdued appearance.

B Glaciers carve and dramatically reshape the topography. This glacier is sculpting the sides of a mountain, or nunatak, that rises out of the ice.

C Large crevasse fields, or cracks in ice, form where a glacier passes over a large inflection in the underlying topography. These are crevasses in the Beardmore Glacier area.

D When glaciers pass over an area, they remove sediment and rocks, often carrying the particles great distances before depositing them in moraines. This moraine is forming at the edge of the McGregor Glacier in the central Transantarctic Mountains.

NATIONAL GEOGRAPHIC

TRANSITIONAL MARINE/ NONMARINE ENVIRONMENTS

Three transitional marine/nonmarine (or marginal-marine) environments have importance in the stratigraphic record: barrier islands and associated environments, salt flats, and deltas.

> **barrier island**
> A long, narrow, sandy coastal island above high tide level and parallel to the shoreline.

Barrier island environments include beaches, estuaries, **tidal flats**, and **lagoons** (Figure 6.19). Normally, on the seaward side, elongate, sandy barrier islands protect the continental shoreline, and a relatively shallow, semirestricted lagoon separates the barrier islands from the shoreline. Beaches on both the barrier islands and the continental shoreline are composed mostly of cross-bedded sand. Shells are often present, but if the sand is siliciclastic, the shells commonly dissolve before the sediment lithifies.

Much of the sediment arriving to the barrier island setting is through **estuaries**, or river valleys that are "drowned" by marine water during times of relatively high sea level. Some sediment also arrives through longshore drift. Estuaries frequently have brackish water, especially near their mouths, and they are home to few organisms that are readily preservable as fossils. Estuaries are subject to tidal influence and fluctuating salinity levels. These conditions limit the number of burrowing animals, and thin laminations, along with cross-beds typical of streams, may be preserved.

Lagoons are relatively shallow areas fed mostly by marine water through narrow access points to the ocean. Streams entering lagoons or high evaporation rates can cause significant changes in salinity, making the sediment inhospitable for most animals. Lagoons are generally quiet areas where silts and clays predominate, and thin laminations can be preserved because of the general absence of bottom-dwelling animals. During storms, coarser sediments, including shells, may wash in from a barrier island or another nearby area and be deposited in thin layers.

Tidal flats are mud- or sand-dominated zones that are covered by water at high tide and exposed at low

Barrier island complex FIGURE 6.19

Sediment is supplied to barrier island complexes from streams that empty their loads into the ocean at deltas. Ocean currents redistribute the sediment along the coastline, forming barrier islands and beaches, and isolating shallow lagoons.

A Barrier island complexes line the Atlantic coastline of the United States. This is a photograph of the barrier island complex at Cape Hatteras, North Carolina, taken from space aboard *Apollo 9*. The barrier islands form a thin white line in this view. The Atlantic Ocean is to the right (east). To the left (west) is a lagoon, an estuary, and a broad area of tidal flats and beaches.

B Beaches on barrier islands are harsh environments dominated by physical processes, and relatively few animals and plants live on them. This is Cape Hatteras as it appears from ground level, showing the impact of wave activity.

Salt flats FIGURE 6.20

A Great Salt Lake, Utah, is a broad, shallow desert lake. High evaporation rates help to maintain high salinity in the lake, and salt precipitates around the margins. In this photo, you can see white salt deposits rimming islands and the northern margin of the lake.

B The Bonneville Salt Flats extend across a wide area west of Great Salt Lake. The thickness of the salt that has accumulated here is about 2 m.

▼ C Salt flats accumulate around the hypersaline Persian Gulf. Here, in a remote area of Saudi Arabia, salt is deposited on mudcracks near the water's edge. In the distance is a sand dune.

tide. Symmetrical ripples and reddish-colored sediments are common. Relatively few animals live on tidal flats, so thin laminations are normally preserved. Mudcracks and raindrop impressions also may be preserved.

Salt flats (FIGURE 6.20), or sabkhas, occur at the margins of saltwater bodies, where evaporation rates are high. One place this occurs is in Great Salt Lake Desert, Utah. In springtime, snow melts from the mountains surrounding the Great Salt Lake, and water makes its way into the flat-bottom basin. Water levels rise in spring, but by summer, high evaporation rates cause shrinkage of the lake and precipitation of salt around the margins (FIGURE 6.20A). The vast Bonneville Salt Flats (FIGURE 6.20B), which spread out to the west of Great Salt Lake, are essentially dry all the time, having received no appreciable amount of water since the Pleistocene Epoch.

Another modern salt flat environment is the Persian Gulf (FIGURE 6.20C), a restricted basin. Freshwater combined from the Tigris and Euphrates rivers flows into the Persian Gulf from the north, but there is little outflow to the south across the Strait of Hormuz. Most water leaves the basin through evaporation, which leads to hypersalinity and deposition of salt.

Few organisms other than cyanobacteria can exist under the hypersaline and desiccating conditions that typify salt flats. Sedimentary features of salt flats include salt and other evaporate minerals, carbonate deposits (including perhaps dolostone), thin laminations, and stromatolites or thrombolites.

Deltas, where rivers and streams empty water and sediment into oceans or lakes, form a complex of interrelated

A Deltas form where rivers or streams empty their sediment loads as they enter less turbulent water of an ocean or a lake. Most of the sedimentation in deltas occurs below water level, as shown in this idealized cross-section.

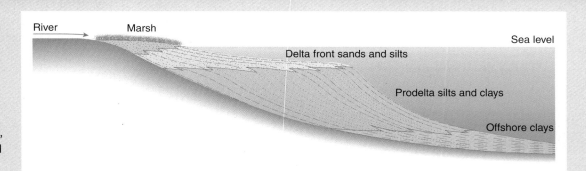

Deltas can have a variety of shapes, depending on the relative influence of various physical processes, including sediment load carried in by river currents and redistribution of sediment by waves or tides.

B This delta, more than 150 km wide, is forming at the mouth of the Lena River, which extends more than 4,400 km across Siberia, Russia, and empties into the Arctic Ocean's Laptev Sea. In this *Landsat 7* satellite image, taken in 2000, water appears in dark blue, and you can see many channels as the river meets the sea. Sediment deposited by the river forms new marshland, which appears in green. Wave activity at the ocean's edge has played a large part in shaping the rounded front edge of this delta.

C A satellite image of the Ganges Delta, formed at the edge of the Indian Ocean's Bay of Bengal. The Ganges River flows more than 2,400 km from the Himalaya Mountains, carrying fertile sediment to a vast delta floodplain covering more than 60,000 km² of India and Bangladesh. Strong tidal currents have redistributed sediments into a series of anastomosing ridges perpendicular to the coastline.

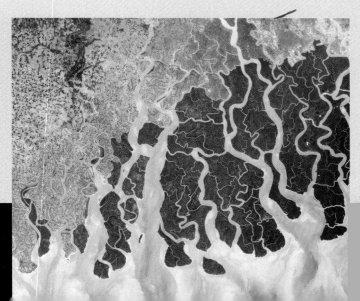

facies (FIGURE 6.21). Close to the river's mouth, coarse sediments are deposited, and progressively finer sediments are deposited farther from the mouth. Most of the volume of a delta is under water. In the delta plain, which is above water, stream channel deposits (representing distributary channels), swamps, and other typical terrestrial facies may be present. Delta front deposits, which are under water, consist of foreset (downwardly dipping) layers that thin in the offshore direction. Cross-beds, graded beds, and various fossils (including aqueous and terrestrial forms) are characteristic features of deltas.

Visualizing
Deltas FIGURE 6.21

D Part of the Mississippi Delta, Louisiana, where it empties into the Gulf of Mexico. From a low-flying aircraft it is possible to see how minimal the relief is in the marshy areas close to the sea. Sedimentation rate tends to dominate other physical processes in the Mississippi Delta, and this leads to a "birdsfoot" pattern of stream channels.

E The Yangtze Delta, with the city of Shanghai, China, in the lower part of the image, as it appeared in a 1989 *Landsat 5* image. The Yangtze River drains much of China over its 6,000 km course, and it empties great quantities of sediment into the East China Sea, where it is redistributed by longshore drift northward along the coast of Asia.

MARINE ENVIRONMENTS

At the present time, the marine realm occupies slightly more than 70% of Earth's surface. During the geologic past, however, that percentage has varied depending on sea level position. Much of the marine record in strata comes from relatively shallow water that covered the continental shelves. Other marine environments that have left a record are slope and rise settings, the deep ocean, and carbonate platforms.

Continental shelves are areas underlain by continental crust and covered by marine water (FIGURE 6.22), so the extent of shelf area during any interval of geologic time depends on sea level position. Because a considerable amount of area lies close to sea level today, a small transgression, or rise, will flood vast areas. A small regression, or fall, will have little effect on a global scale, but a large drop could expose wide areas of the continental shelves.

Shelf environments have supported a rich array of organisms since the early Paleozoic and are the source of most macroscopic body fossils and trace fossils. Fossils represented are of bottom-dwellers, swimmers, and floaters. In general, shelf sediments are coarser in nearshore areas and finer in more offshore areas. Crossbeds, ripples, concretions, flute casts, and other sedimentary structures are present in shelf environments. Thin laminations are variably present.

The continental shelf FIGURE 6.22

A An idealized bathymetric profile shows a nearly flat continental shelf, a steep continental slope, and a gradually sloping continental rise ending seaward at the abyssal plain. The shelf break is the point of inflection between the continental shelf and the continental slope.

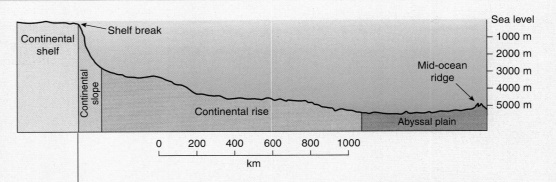

C Shelf areas are usually shallow, within the photic zone, and teeming with life. This photograph shows a small reef reef built just below the water line and not far from shore.

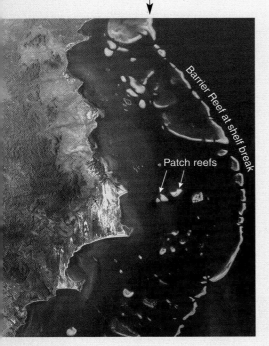

B The continental shelf is a broad, relatively shallow-water marine environment. In this view from space of the area around Cape Flattery, Australia, the Great Barrier Reef outlines the margin of the continental shelf, which is oceanward of the beach. Barrier reefs normally form at the shelf break, or the point where the continental shelf and slope meet. Smaller patch reefs form on the continental shelf.

Carbonate sedimentation FIGURE 6.23

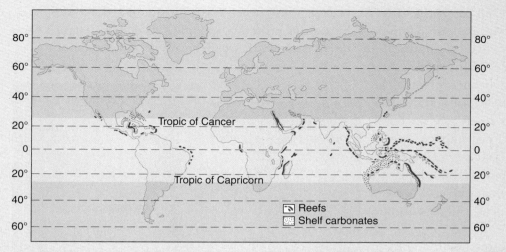

A Most carbonate sediments are deposited in warm, shallow seas in or near the tropics. This map shows the distribution of modern carbonates in shallow water.

Legend:
- Reefs
- Shelf carbonates

B Carbonate deposits, formed by living organisms, are the predominant sediment type in the Bahamian Platform. In this view from a satellite orbiting Earth, islands (such as Andros Island, at center) appear brown and dark green, carbonate platforms (rarely more than 10 m deep) appear as light blue areas, and deep oceanic areas (commonly 200 m or more in depth) appear in dark blue. North is to the right. Cuba is to the upper left, and the southern tip of Florida is mostly obscured under the satellite. Little Bahama Bank is to the lower right.

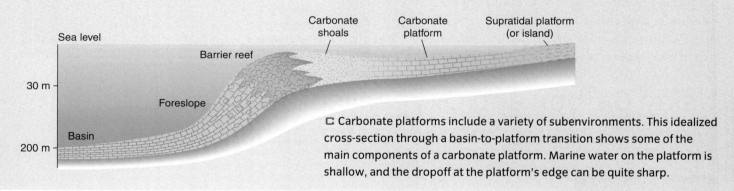

C Carbonate platforms include a variety of subenvironments. This idealized cross-section through a basin-to-platform transition shows some of the main components of a carbonate platform. Marine water on the platform is shallow, and the dropoff at the platform's edge can be quite sharp.

Carbonate sedimentation occurs preferentially in warm, shallow tropical shelves free of siliciclastic influence (FIGURE 6.23). Organisms that precipitate calcium carbonate, especially green algae and cyanobacteria, must live within the photic zone if they are to carry out photosynthesis. In addition, warm water and pH increase caused by the metabolic activity of organisms such as algae, corals, and mollusks tend to increase the rate at which bicarbonate ions HCO_3 come out of solution in seawater.

Carbonate platforms have a variety of subenvironments, but in general are characterized by limestone or perhaps dolostone deposits, a diverse marine fauna and flora, and cross-beds in places. Stromatolites and thrombolites occur in some carbonate shelf areas. Carbonate islands may show paleosols or karst topography (caves, sinkholes, or brecciated limestone).

One distinctive lithofacies associated with a carbonate platform is the reef facies. A **reef** is a wave-resistant structure constructed by organisms, often corals or sponges. Barrier reefs (FIGURE 6.22B) develop close to the shelf break on a carbonate platform, and patch reefs develop in scattered areas on the shelf.

> **reef** A wave-resistant structure constructed by organisms.

continental slope The part of the continental margin between the continental shelf and the continental rise, if there is one. The continental slope is characterized by a relatively steep slope.

The edge of the shelf, or **shelf break** (Figure 6.22A, Figure 6.22B on page 184), is a significant oceanographic boundary that separates the shallower continental shelf from the deeper ocean. The **continental slope** and rise environments receive mostly thinly laminated clay and silt, except when turbidity currents dislodge sediment and rapidly transport it downslope, depositing graded beds (turbidites), often with erosive bases and flute casts. Fossils are mostly of pelagic organisms, and trace fossils are rare.

In the deep ocean, fine mud is the predominant sediment type. Fine grains are often blown great distances from the continents by winds. Trace fossils are rare, and body fossils are mostly of microscopic planktonic (floating) organisms. Calcareous and siliceous oozes, which lithify to limestone and chert, respectively, are formed by the dissolution of shells of pelagic organisms under high pressure in deep water.

Although much of the marine realm today consists of abyssal plain, sedimentary rocks representing this environment are relatively uncommon. The main reason is that the abyssal plain is underlain by relatively dense oceanic crust, which is readily subducted and destroyed by tectonic processes operating at oceanic trenches. Continental shelves do not normally get subducted.

CONCEPT CHECK STOP

What types of outcrop evidence are used to interpret ancient sedimentary environments?

What features characterize modern nonmarine, marine, and transitional environments? How do these features translate into sedimentary features that can be used to interpret ancient sedimentary environments?

How can nonmarine settings be recognized from strata?

What are the most significant transitional marine/nonmarine settings, and how are they recognized from strata?

How can marine environments be recognizable from strata?

Using Carbon and Oxygen Isotopes to Interpret Global Climatic and Oceanographic Change

LEARNING OBJECTIVES

Explain how greenhouse gases cause increases in global temperature.

Differentiate greenhouse conditions from icehouse conditions.

Describe the two stable isotopes of carbon and how they move through Earth systems.

Describe the two stable isotopes of oxygen and how they move through Earth systems.

C hemical changes in the atmosphere and in the ocean can significantly affect vast areas of Earth. Today, rapidly increasing concentrations of **greenhouse gases**, or atmospheric gases that trap heat from the Sun, are making news because of the role they play in global climate change. In the atmosphere,

greenhouse gases Atmospheric gases that trap the heat from solar radiation near Earth's surface.

these gases act much like the glass windows of a greenhouse: solar energy enters and is absorbed, but the pane of glass limits reradiated heat from escaping. Greenhouse gases, including carbon dioxide (CO_2), water vapor (H_2O), and methane, among others, limit the escape of the Sun's heat from the lower atmosphere. When the atmospheric concentration of these gases is high, the atmosphere warms, but when the concentration is low, the atmosphere cools.

Carbon dioxide accumulation in the atmosphere has been a key contributor to the dramatic increase in atmospheric temperature in recent years. Some effects of rapid climate change are obvious: global warming has noticeably increased the rate of melting of continental glaciers, sea level has risen at a small but measurable rate, and instability of weather systems is becoming more commonplace. Longer-term effects of climate change, however, are more conjectural.

To understand how a continued increase in atmospheric temperature may play out in the future, we can look to sedimentary archives of global climatic change. Isotopes of oxygen bound up in limestone deposits, the skeletons of organisms, or glacial ice, for ex-

ample, reveal numerous cycles of warming (or **greenhouse conditions**) and cooling (or **icehouse conditions**) through time (**FIGURE 6.24**). Changes in the isotopic composition of sediments add to changes in the distribution of sedimentary lithofacies and biogeographic distributions of animals and plants, which also track changes in global climate.

In this section, you will learn how changes in the cycles of carbon and oxygen have influenced Earth's chemical, physical, and biologic history. Many of the shifts, or excursions, in carbon and oxygen isotopic values recorded in rocks are related to environmental changes on a global scale. In many cases, these changes occurred over geologically short time intervals, and the resulting isotopic shifts recorded in strata can be used for global correlation purposes.

Oxygen isotopic ratios in marine fossils FIGURE 6.24

A plot of $^{16}O/^{18}O$ ratios (or $\delta^{18}O$ values) from marine foraminifera shows fluctuations in oxygen isotopic values through the late Cenozoic Era. The cycles are related to changes in the ocean-climate system. Relatively heavy $\delta^{18}O$ values record colder climatic conditions when the lighter isotope of oxygen (^{16}O) was preferentially locked up in glacial ice and less readily available for making carbonate skeletons. Lighter $\delta^{18}O$ values record warmer conditions when glaciers melted and added more ^{16}O to the ocean where foraminifera could use it when making skeletons.

This plot shows records of oxygen isotopic values measured from benthic foraminifera in the Pacific Ocean, and in the Atlantic Ocean. Dashed lines between the curves show how time-equivalent positions in stratigraphic sections around the world can be correlated using oxygen isotopic records.

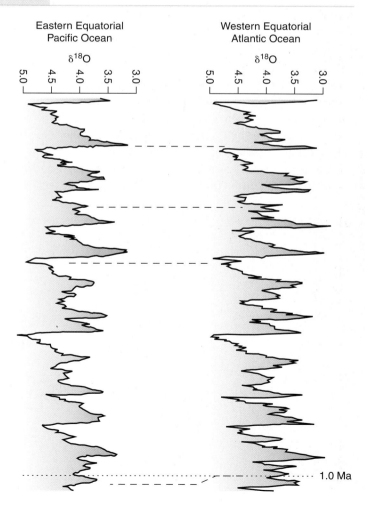

CARBON ISOTOPIC RATIOS THROUGH TIME

Two stable isotopes of carbon, carbon-12 (^{12}C) and carbon-13 (^{13}C), exist in nature, and they move through Earth systems, or chemical reservoirs, in subtly different ways. **Chemical reservoirs** are places in the Earth system, such as the ocean, atmosphere, crustal sediments, and biomass, where chemicals can reside for a while.

In the cycling of carbon and oxygen through reservoirs, life forms play a critical role. Carbon and oxygen flow back and forth between primary producers and consumers via photosynthesis and respiration. Photosynthesizing plants, cyanobacteria, and algae combine carbon dioxide (CO_2) and water to form sugars. Energy from sunlight, which fuels the reaction, gets stored in the sugars, and oxygen is a byproduct of the reaction. In respiration, consumers (animals) and decomposers (especially eubacteria and fungi) extract oxygen from air or water and combine it with sugars to drive the process.

Carbon dioxide and water are released as by-products of respiration.

Large volumes of carbon were buried in sediments (organic-rich shales, **peat** deposits, and coals) during certain time intervals. Because photosynthesizers preferentially extract the isotopically lighter form of carbon (^{12}C) from air and water, the large volumes of carbon locked up in the sedimentary reservoir were mostly ^{12}C. This left a greater proportion of ^{13}C in the air and water. As a result of an increasing ^{13}C/^{12}C ratio (or δ^{13}C value) in water, the ratio recorded in shallow-water limestones during times experiencing a high rate of carbon burial increased (**Figure 6.25**). Subsequent weathering of the carbon-enriched strata resulted in a return of more ^{12}C to the ocean and air, and it is recorded in limestones by a decreased ^{13}C/^{12}C ratio.

One of the most dramatic shifts in carbon isotopic values in the Phanerozoic occurred during the late part of the Cambrian Period, when δ^{13}C in limestones deposited on tropical continental shelves increased by about 5 parts

Carbon isotopic ratios in marine sediments FIGURE 6.25

Plots of ^{13}C/^{12}C ratios (or δ^{13}C values) in Cambrian limestones from what is now South China, Kazakhstan, Australia, and the western United States. A remarkable shift toward higher δ^{13}C values, called the SPICE excursion, reflects a time when the lighter isotope of carbon, ^{12}C, was preferentially buried in organic-rich shales, leaving an excess of ^{13}C to be deposited in limestones. This distinctive positive δ^{13}C excursion is used for stratigraphic correlation around the world. The dashed line indicates the base of the Furongian Series and the Paibian Stage. The base of the SPICE excursion is close to this horizon. A shift toward lower δ^{13}C values following the SPICE event may have been caused by weathering of carbon-rich shales in various parts of the world.

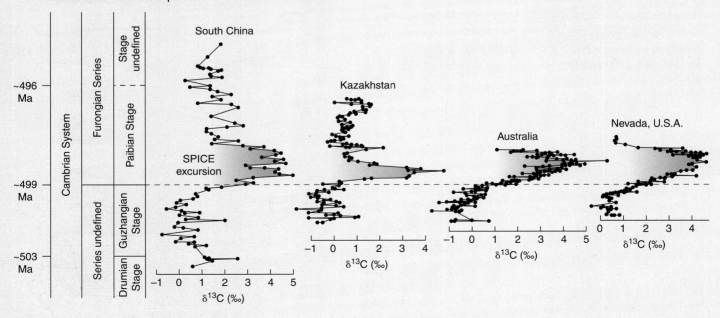

per thousand (or per mil, symbolized as ‰; Figure 6.25). This event occurred during a time of rapid transgression followed by a slower regression, when productivity in the oceans was high. Exposure of the continents led to weathering and erosion, which provided a source of fine-grained sediments in which to bury organic carbon. On the shelf of what is now Scandinavia, organic-rich black shales were deposited. Because photosynthesizers preferred to use isotopically light carbon, and because organic-rich shales locked up the lighter fraction of carbon rather quickly in geologic terms, high rates of organic-rich shale accumulation led to an excess of ^{13}C in the ocean-atmosphere system. When this happened, the carbon available to be precipitated in limestones became isotopically heavy.

Burial of organic carbon causes an increase in atmospheric O_2 but also generally causes a decrease in CO_2. Unlike the concentration of O_2, which depends largely on the rate of burial of carbon, though, the concentration of CO_2 depends more on chemical reactions occurring near Earth's surface. Carbon dioxide, when it combines with water to form carbonic acid (H_2CO_3), is important in the weathering of carbonate and silicate rocks. Weathering releases calcium (Ca^{+2}) and bicarbonate ions (HCO_3^{-}). The bicarbonate ions later recombine with calcium ions to form calcium carbonate ($CaCO_3$) skeletons and limestones. Weathering of shallow-water limestones or subduction of deep-sea limestones can later release CO_2 to the ocean-atmosphere system.

During the late Carboniferous, large volumes of organic carbon were buried in coal swamps, and atmospheric O_2 levels reached high concentrations. High weathering rates on continents at the time caused CO_2 levels to drop because the rocks being weathered were, for the most part, not rich in organic carbon. With low atmospheric concentrations of CO_2, global temperatures plunged, and extensive continental glaciers developed in the Southern Hemisphere. The buildup of glaciers is recorded by a trend toward positive $\delta^{13}C$ values in carbonate rocks.

OXYGEN ISOTOPIC RATIOS THROUGH TIME

Just like carbon, oxygen also has two stable isotopic forms in nature. Oxygen-16 (^{16}O) is much more common than oxygen-18 (^{18}O), and the ratio of the two is written $\delta^{18}O$. Isotopes of oxygen move through chemical reservoirs in slightly different ways (just like carbon isotopes), and study of $\delta^{18}O$ isotopic ratios preserved in carbonate fossils or sediments provides information about ancient temperatures and glacial ice volume.

Organisms that secrete calcium carbonate ($CaCO_3$) skeletons, such as mollusks, corals, and foraminifera, preferentially incorporate the lighter isotope of oxygen (^{16}O) in their skeletons. In addition, the proportion of ^{16}O compared to ^{18}O incorporated in skeletal material varies with temperature. At lower temperatures, ions containing ^{18}O react at slower rates than do ions containing ^{16}O, so ^{16}O more readily combines with other ions, and ^{18}O is left to be used by organisms in forming their skeletons.

Glacial ice volume has a large effect on the $^{16}O/^{18}O$ proportion available to organisms. When water evaporates from the ocean, water molecules that contain ^{16}O are extracted from the ocean at a faster rate than those that contain ^{18}O. Precipitation falling on the land becomes enriched in ^{16}O, and ocean water becomes enriched in ^{18}O. Continental glaciers, whose zones of accumulation are in continental interiors, tend to be highly enriched in the lighter isotope of oxygen. During times of glacial expansion (icehouse conditions), the ocean becomes enriched in ^{18}O, and that increased value is recorded in limestones and carbonate skeletons (see Figure 6.24 on page 187). When glaciers contract through melting during times of greenhouse conditions, large volumes of ^{16}O are returned to the ocean, and $\delta^{18}O$ values decrease both in seawater and in the skeletons of marine organisms.

CONCEPT CHECK **STOP**

How do greenhouse gases cause increases in global temperature?

What are greenhouse conditions? How do they differ from icehouse conditions?

What are the two stable isotopes of carbon? What causes shifts in the carbon isotopic composition of limestones through time?

What are the two stable isotopes of oxygen? What causes shifts in the oxygen isotopic composition of fossils and sedimentary strata through time?

1 Modern and Ancient Sedimentary Environments

1. A **facies** is a sedimentary unit that has a set of characteristics (such as lithology, color, or fossil assemblage) particular to a local environment. Usually, the word *facies* refers to **lithofacies**, which is a specific sediment or rock type. A **biofacies** is a facies identified by a distinctive assemblage of fossils.

2. Grain size, sorting, and rounding provide indicators of the current energy in a sedimentary environment or the amount of transport from a sediment source. Coarse grains tend to accumulate in environments relatively close to the sediment source and subject to high current energy, and finer sediments (silt and clay) accumulate farther from the source and in areas of low current energy. Increased transportation results in greater size sorting of particles, and transportation in water tends to increase roundness of grains.

3. **Siliciclastic sediments**, or detrital sediments, require a siliciclastic source from which the particles can be eroded. **Carbonate sediments** are produced largely through the activity of organisms that secrete calcium carbonate hard parts.

4. **Sedimentary structures** reflect a variety of physical, chemical, or biologic processes operating in sedimentary environments and provide one key to environmental interpretation. Examples of sedimentary structures are bedding (layering), ripples, cross-bedding, graded bedding, mudcracks, raindrop impressions, flute casts, paleosols, concretions, and stromatolites. **Geopetal structures** are sedimentary structures that indicate stratigraphic younging direction.

5. **Walther's Law of Facies** provides a means of interpreting the original lateral relationships among sedimentary lithofacies from a stacked succession of facies.

2 Interpreting Ancient Sedimentary Environments

1. Depositional environments can be characterized as **nonmarine**, **transitional marine/nonmarine**, and **marine**. Field evidence useful for distinguishing ancient environments includes sediment type, fossils, sedimentary structures, and application of Walther's Law of Facies.

2. Nonmarine environments, which are characterized by terrestrial or freshwater fossils and often reddish sediment, include stream systems, lakes, alluvial fans and playa lakes, and a variety of glacial-related settings.

3. Transitional marine/nonmarine environments include barrier island systems (incorporating beaches, barrier islands, lagoons, estuaries, and tidal flats), deltas, and salt flats.

4. Marine environments, which are characterized by marine fossils and a variety of sedimentary lithofacies, include marine shelves, carbonate platforms, continental slopes and rises, and abyssal plains.

3 Using Carbon and Oxygen Isotopes to Interpret Climatic and Oceanographic Change

1. **Greenhouse conditions** are times in Earth history of relatively warm global climate. Global warming can be influenced by **greenhouse gases**, which allow solar radiation to reach Earth's surface but limit the reradiation of heat to space. **Icehouse conditions** are times of relatively cool or cold global climate. Expansion of continental glaciers occurs during icehouse conditions.

2. Ratios of the two stable isotopes of carbon, ^{12}C and ^{13}C, can be used to assess the rate of organic carbon burial through time, and the atmospheric or oceanic levels of CO_2 and O_2.

3. Ratios of the two stable isotopes of oxygen, ^{16}O and ^{18}O, are closely tied to the volume of glacial ice on Earth, and this reflects global temperature conditions.

KEY TERMS

CRITICAL AND CREATIVE THINKING QUESTIONS

1. In the search for new petroleum reserves, it is helpful to model ancient depositional systems using modern analogs. Development of an economically viable deposit requires, at a minimum, an organic-rich source rock (such as dark shale) and a porous, permeable reservoir rock (such as sandstone or reef limestone) for matured oil or natural gas to flow into and become lodged. Where on Earth might you go to observe depositional environments that would serve as modern analogs of petroleum source rocks and reservoir rocks?

2. How can we use the stratigraphic record to determine changes in global CO_2 or temperature levels from the distant past? How can we assess the impact of past changes in CO_2 or temperature levels on other aspects of the Earth system such as biologic organisms?

3. Global temperature levels are currently rising, and the rise is evidently fueled in large part by the addition of greenhouse gases (principally CO_2) to the atmosphere. In what ways would the temperature rise be recorded in sediments that are being deposited currently?

What is happening in these pictures ?

Modern and Ancient Reefs

WHAT general similarities does the ancient reef (Capitan Reef) have to the modern example?

WHAT characteristics would you look for in the ancient reef to determine how it was formed?

WHAT does the presence of an ancient reef in the west Texas desert tell you about conditions there during the Permian?

These photographs show two different views of a modern reef in the Bahamas.

Capitan Reef, an enormous Permian reef in the Guadalupe Mountains of Texas, stands in relief above the surrounding desert.

SELF-TEST

1. What is a facies, and what are some of the types of facies?

2. What are the lines of evidence used to interpret ancient sedimentary environments?

3. How do modern environments help us interpret ancient environments?

4. What are the major carbonate subenvironments?

5. What features characterize nonmarine environments?

6. What is a delta?

7. What are the major features of continental shelf environments?

8. How is Walther's Law of Facies used to interpret the original lateral relationships of facies?

9. What are greenhouse gases?

10. What causes changes in the ratio of carbon isotopes through time?

11. How do changes in oxygen isotopes allow us to interpret global climatic changes through time?

Plate Tectonics in Earth History

O ver the span of geologic time, Earth's surface has been substantially reshaped by tectonic processes—volcanic eruptions (like this eruption of Mt. Etna, Sicily, Italy, in 2002), folding, faulting, mountain building, and crustal remelting. Reconfiguration of the continents and ocean basins has set in motion an array of physical and chemical changes. High spreading rates of oceanic plates, for example, have contributed to sea level rises. Some plate movements led to the collision of continents and the rise of mountain ranges. Weathering of mountains yielded sediments and chemical ions that were afterward delivered to basins. In the basins, the sediments accumulated and meticulously recorded "pages" full of details about Earth's development, while dissolved ions combined into sedimentary cements that lithified the loose layers. Without this combination of processes, the great history "book" that is the stratigraphic record would not have been assembled.

In this chapter, you will learn about plate tectonics and its influence on Earth's history. Plate tectonics theory explains Earth's surface as a mosaic of plates and summarizes the processes responsible for their movement. An ocean floor acts like a pair of giant conveyor belts, with new basaltic rock being formed at mid-ocean ridges, followed by splitting and spreading of that rock away from the ridges, where it cools and subsides. Oceanic slabs extend to deep trenches near the ocean margins, where they are subducted and remelted. Travel of the plates is driven by heat flow from within Earth. Volcanoes, mountain ranges, and deep-sea trenches owe their development to tectonic activity. The shifting of plates has also profoundly affected many species of life forms by altering their geographic distributions, stimulating their divergence, and contributing to their extinction.

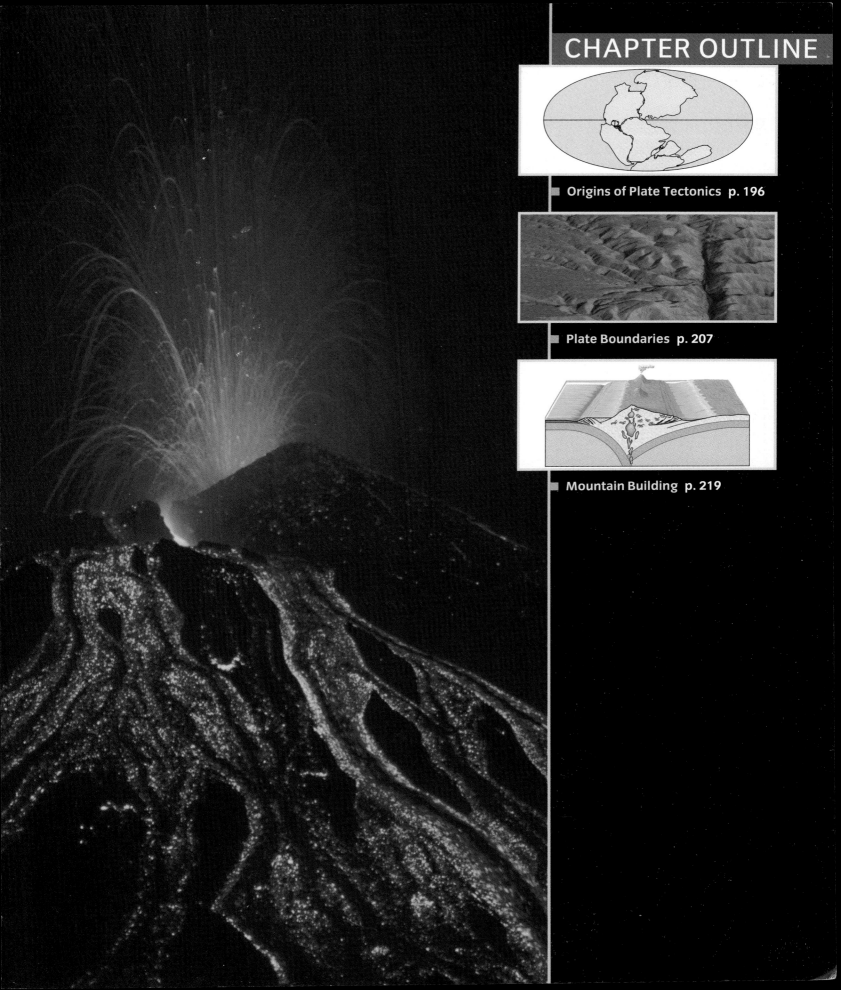

Origins of Plate Tectonics

Plate tectonics theory grew out of the fusion of two earlier hypotheses: continental drift and seafloor spreading. **Continental drift** is the idea that Earth's continents moved to their present positions after fragmentation of a larger landmass in the geologic past. **Seafloor spreading** explained certain oceanic features as resulting from the expansion of ocean **basins** by addition of new rock from spreading centers and destruction of older rock near the ocean margins. Seafloor spreading provided the driving mechanism for continental drift.

Prior to the widespread acceptance of plate tectonics theory in the 1960s, the inferred structure and behavior of Earth was considerably different from the way we now understand it to be. The crust was thought to have been rigid, with **continental crust** being permanently locked in place by adjacent **oceanic crust**. Much of the evidence used to displace the idea of a rigid Earth, or **permanence theory**, came from indirect sources—the apparent effects of plate movement on Earth's development, including the distribution of life forms whose geographic ranges were affected by continental movement. As evidence mounted, additional information, including rather direct observations of plate movement, came to light. Today, we recognize that it is impossible to adequately explain the varied aspects of Earth history except in the light of plate tectonics.

Understanding why the rigid Earth idea was rejected and why plate tectonics theory was finally accepted is useful not only from the standpoint of illustrating important examples of the historical development of Earth but also because it illustrates how science works. Science is a self-correcting process: explanations of natural phenomena are tested and, when found to be incorrect, rejected. They are then replaced by better explanations. Even long-held assumptions and interpretations can be rejected if evidence does not support them.

CONTINENTAL DRIFT

In 1915, a German meteorologist, Alfred Wegener, marshaled evidence that during the late Paleozoic, the continents were joined into a single landmass. That landmass, named **Pangea**,

plate tectonics theory A scientific theory that explains the lithosphere as cracked and composed of pieces that interact with each other as they float on a hot, deformable asthenosphere.

continental drift The hypothesis that the continents moved to their present positions after fragmentation of a larger landmass in the geologic past.

seafloor spreading The hypothesis that ocean basins expand through the addition of new rock from spreading centers and that older rock is destroyed near the basin margins.

basin A relatively depressed area of Earth's crust that receives sedimentary deposition.

continental crust The solid, outer part of the Earth underlying the continents and continental shelves, composed largely of granitic rocks.

oceanic crust The solid, outer part of Earth underlying the ocean basins, consisting largely of basaltic rocks.

Pangea The late Paleozoic to early Mesozoic supercontinent comprising most of the world's continental crust.

fragmented into pieces that formed the modern continents during the Mesozoic, and those pieces then moved, or drifted, toward their present positions (**FIGURE 7.1**).

Wegener published his revolutionary ideas more than 50 years before it was possible to directly measure the movement of continents from satellites circling the Earth, so his interpretations were based mostly on the secondary effects of continental movement recorded at the Earth's surface. What evidence supported the hypothesis of continental drift?

First, Wegener described the apparent jigsaw-puzzle-like fit of continental coastlines. The match between eastern South America and western Africa is particularly striking (**FIGURE 7.1 B, C**).

Pangea FIGURE 7.1

A In 1915, Alfred Wegener proposed that the world's continents were united like pieces of a jigsaw puzzle into a single landmass, called Pangea, in the late Paleozoic and early Mesozoic. This is a modern reconstruction of the supercontinent Pangea (with the names of major pieces labeled). It differs from Wegener's original version only in small details. In the middle and late Mesozoic, Pangea split into major pieces, Laurasia (the northern hemisphere continents) and Gondwana (the southern hemisphere continents), and ultimately into modern continental fragments.

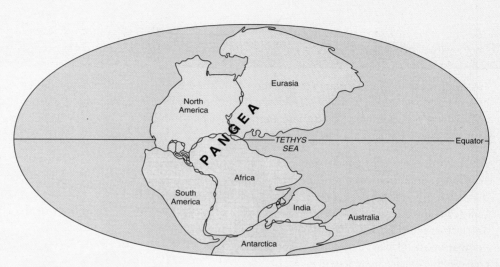

B Modern satellite imagery reveals a congruence in the coastlines of modern continents bordering the Atlantic Ocean, indicating that they were once joined but later split and are now spreading apart. The correspondence between eastern South America and western Africa is particularly striking.

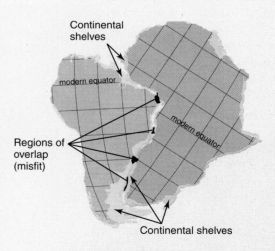

C When South America and Africa are fit together along their Atlantic shorelines, there is a close match. That fit is improved, though, when the continents are matched along the edges of the continental shelves (the true but submerged margins of the continents). The configurations of the shelf edges were unknown in Wegener's time.

Origins of Plate Tectonics   197

Another piece of evidence favoring continental drift was similarity in the geology of continental areas that Wegener thought were once joined (FIGURE 7.2). This was highlighted in a 1927 article by a South African geologist, Alexander du Toit, who presented a wealth of geologic information showing that areas of **Gondwana** (or **Gondwanaland**)—the southern hemisphere continents and islands—had a nearly identical geologic history from the early Paleozoic through the mid-Mesozoic. This reinforced the view that these areas were previously united in a single landmass. There are remarkable parallels in the strata of South America and South Africa. There are even parallels among Archean granites, Jurassic basaltic-type rocks, and diamond-bearing kimberlite pipes of the two regions.

One impressive stratigraphic similarity that du Toit discussed is a Carboniferous-Permian **tillite**, or lithified glacial **till**, in Brazil and a comparable tillite in South

Fossils distributed widely through the southern hemisphere help to link now-separated regions. Leaves of *Glossopteris* occur widely across the Southern Hemisphere. Permian lake deposits contain skeletal remains of the reptile *Mesosaurus*. Bones and teeth of synapsids (or "mammal-like reptiles"), including *Lystrosaurus*, occur in Permian and Triassic river deposits.

Finally, the Gondwanan stratigraphic sequence is capped by basalts that record the initial breakup of Pangea.

A Similar stratigraphic deposits and fossils across the southern hemisphere indicate that the entire region experienced the same climatic and geologic conditions at nearly the same time. In Carboniferous-Permian strata, glacial tillites (indicating a cold climatic phase) are followed by coal beds (indicating a warm climatic phase).

B Permian tillites, like this reddish-brown deposit with white pebbles, are widespread through Gondwanan regions. This is tillite of the Pagoda Formation, which overlies thinly bedded lake deposits, in Antarctica.

Period	South America (Brazil)	Africa (South Africa)	India	Antarctica (Transantarctic Mountains)
Jurassic	São Bento Basalt	Basalt / Stormberg Group	Rajmahal Basalt / Mahadeva series	Kirkpatrick Basalt / Ferrar Group / Prebble Formation / Hanson Formation
Triassic	Botucatu Sandstone / Santa Maria Formation	Beaufort Group	Panchet series	Falla Formation / Fremouw Formation
Permian	Estrada Nova Formation / Rio Bonito Formation / Irati Formation / Itararé Group	Ecca Group / Dwyka shale / Dwyka tillite (Dwyka Group)	Damuda series / Renigang beds / Barahar beds / Talchir Tillite	Buckley Formation / Fairchild Formation / Mackellar Formation / Tillite (Pagoda Formation)
Carboniferous		Dwyka shale		Tillite (Pagoda Formation)

Key
- *Lystrosaurus*
- *Glossopteris*
- *Mesosaurus*
- Coal
- Tillite
- Basaltic lava flow
- Sandstone
- Shale

Africa. Extensive tillites result from continental glaciations, and those that du Toit discussed imply that Brazil and South Africa were once connected.

Tillites are not only good stratigraphic markers but also excellent indicators of ancient climatic conditions. Wegener used paleoclimatic evidence to help demonstrate that present-day continents were in former times adjoining areas of a larger landmass. Some of the paleoclimate evidence is linked to paleolatitude because areas of the same latitude often have similar climates. Carboniferous-Permian tillites of the Southern Hemisphere, (FIGURE 7.2A–C), are indicators of cold climates and high paleolatitudes. Another climate-related lithofacies is coal, which usually indicates warm conditions. Gondwanan tillites are preceded stratigraphically by a coal succession (Figure 7.2A), which indicates that these areas simultaneously experienced a warm climatic phase followed by a cooler climatic phase.

Visualizing

Gondwanan stratigraphy FIGURE 7.2

C Tillite deposits and glacial striations left by late Paleozoic continental glaciers in the Southern Hemisphere have distributions inconsistent with the current latitudinal positions of landmasses.

D The glacial evidence makes more sense if the Southern Hemisphere continents were joined in the late Paleozoic. When this is done, glacial striations point to the radial spread of a continental glacial sheet from a point in southern Africa.

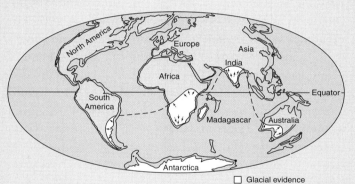

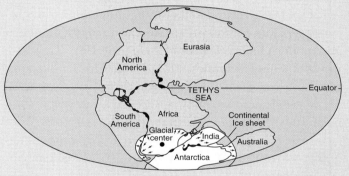

Glacial evidence
Glacial movement direction

E Tongue-like leaves of the land plant *Glossopteris* occur in Carboniferous and Permian strata of India, Australia, Madagascar, South Africa, and Antarctica. These fossils are from Australia.

F The distribution of *Glossopteris* fossils helps establish how Southern Hemisphere regions were juxtaposed in the Carboniferous and Permian.

G This skeleton of the freshwater reptile *Mesosaurus* is from a Permian lake deposit of Brazil. Similar specimens also occur in South Africa and help reinforce the interpretation that South America and Africa were joined during the late Paleozoic.

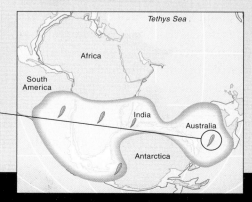

The present distribution of Gondwanan tillites extends from south polar regions northward to India, which lies north of the Equator. Wegener stressed the unlikelihood of continental-scale glaciation through the tropics, which are normally too warm to support continental glaciation.

As du Toit elaborated, Carboniferous-Permian glacial tills in Gondwanan areas are underlain by glacially polished and striated rock surfaces. **Glacial striations**, which are formed by rocks being dragged along the ice-bedrock interface, are elongate markings that thin out in the downstream direction. Continental glaciers flow outward from a central region, and glacial striations record that radiating flow pattern. Glacial movement in South America was from the southeast, in Australia it was from the west (present-day north), and in Africa it was outward from an apparent glacial center. The overall pattern of flow really makes sense only if the continents and major islands were joined in the way suggested by matches in their outlines (**FIGURE 7.2D** on page 199).

Closely related to the stratigraphic similarities between separate continents are similarities in their ancient and modern biotas (Figure 7.2A on page 199). There are two compelling examples of biotic similarity. Land plants, including the distinctive, elongate leaves of *Glossopteris*, occur widely across Gondwana (**FIGURE 7.2E, F** on page 199). In Wegener's time, the **Glossopteris flora**

> ■ **Glossopteris flora**
> An assemblage of plants, dominated by fossil leaves of *Glossopteris*, occurring in Carboniferous-Permian strata of the Southern Hemisphere.

was known from South America, South Africa, Madagascar, India, and Australia, but today it is also known from Antarctica. The other example is a small, carnivorous freshwater reptile called *Mesosaurus* (**FIGURE 7.2G** on page 199), that is known only from lake deposits of Brazil and South Africa. What makes these examples so compelling (but, ironically, not persuasive in Wegener's time) is that both *Glossopteris* and *Mesosaurus* are freshwater-dependent organisms. Their distribution across a deep, wide ocean separating South America and Africa needs explanation.

Some present-day animals are distributed in a way that suggests a former connection among present landmasses. A prime example is the distribution of terrestrial earthworms, which occur broadly across Europe, Asia, and North America. Their distribution across the Atlantic and Pacific oceans must be explained.

How were the distributions of organisms interpreted prior to acceptance of continental drift? The most common explanation was by invoking **land bridges** between continents, similar to the Isthmus of Panama. Before the Panama Canal was constructed, North America and South America were connected by a narrow strip of land that was a passageway for the migration of animals. Land bridges were thought to connect across the Atlantic Ocean, across the Indian Ocean, and across the Pacific Ocean. The lack of land bridges connecting those regions today was explained by sinking of the connectors into the ocean basins. With the advent of technology used to profile or photograph the ocean floor in the mid-1900s, it ultimately became clear that continental crust cannot disappear into the ocean.

Other explanations for the distributions of *Glossopteris*, *Mesosaurus*, earthworms, and other organisms were by means of island stepping-stones and by means of rafting (such as attachment to floating logs) or some other means of transportation. The concept of island stepping-stones failed because of lack of an adequate reason for the disappearance of the island chains. Rafting of organisms can occur, but only on a limited scale. Seeds of *Glossopteris* could have been transported by wind, but they are several millimeters in size and too large to have been carried across wide oceans by winds.

In Wegener's view, the process of continental drift was ongoing. This notion led him to seek contemporary geologic analogies. Many of Wegener's examples of continental drift were predicated on the splitting of Pangea and expansion of the Atlantic Ocean. Whereas the Atlantic Ocean represents an advanced stage of drift, an earlier stage of the same process is illustrated in Africa. The East African **rift system** comprises a series of valleys, lakes, and narrow oceans (**FIGURE 7.3**) that Wegener explained as an early step in the process of continental splitting.

Rifting in East Africa has developed along a three-armed system of faults, or a **triple junction**, meeting in the Afar Triangle area of Ethiopia and adjacent Eritrea. Crust at the triple junction is domed because of

Amazing Places: East African rift system FIGURE 7.3

Alfred Wegener explained the valleys, lakes, small ocean basins, and related topography in eastern Africa as an active rift system, the modern-day equivalent of an early phase in the breakup of Pangea.

A In this satellite view of Africa, the eastern one-third of the continent extending from the Red Sea and Gulf of Aden (upper right) southward to Mozambique appears as a series of valleys and lakes paralleled by highlands. This is the East African rift system.

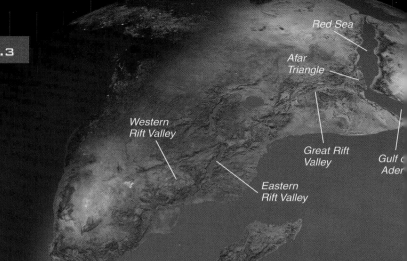

B Crustal splitting in the East African rift system has occurred along three major pathways, or arms, that meet at a triple junction. The three arms are the Red Sea, the Gulf of Aden, and the Great Rift Valley. Under the triple junction, which occurs in the Afar Triangle region, high heat flow has caused doming and extension of the crust.

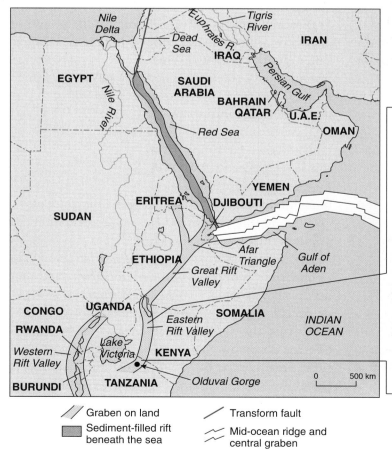

Graben on land
Sediment-filled rift beneath the sea
Transform fault
Mid-ocean ridge and central graben

C As viewed from a low-flying aircraft, this part of the Great Rift Valley in Kenya appears as a flat-bottom valley. The cliff face is an escarpment marking a fault plane along which a block of crust downdropped into a graben (valley).

D Faulting due to crustal extension in the Great Rift Valley has exposed fossiliferous Cenozoic strata to erosion. This is Olduvai Gorge, Tanzania, a locality famous for yielding remains of fossil mammals, including early hominids.

high heat flow underneath it. One arm extends to the northwest as the Red Sea, and another extends to the east-northeast as the Gulf of Aden. Both areas are actively expanding and now floored by basaltic ocean crust. The third arm of the triple junction, the Great Rift Valley, extends from the Afar Triangle southward through Kenya, Tanzania, and adjacent areas, to Mozambique. Various lakes such as Rudolf, Victoria, Tanganyika, and Nyasa, occupy **grabens**, or basins formed by the downdropping of fault blocks. Earthquakes, indicating active faulting, are a regular feature of the East African rift system. Although the Red Sea and Gulf of Aden are actively expanding, extension of the Great Rift Valley is now relatively slow, which means it probably will not open into an ocean basin. The Great Rift Valley is referred to as a failed rift, or **aulocogen**.

In 1937, du Toit recognized that Pangea did not become assembled until the late Paleozoic. Earlier in the Paleozoic, Gondwana existed as a supercontinent comprising most Southern Hemisphere continents and islands, and continents of the Northern Hemisphere were joined in a separate supercontinent, **Laurasia**.

Despite an abundance of supporting information, most geologists in Europe and North America viewed continental drift with skepticism for decades. That view began to change in the 1960s when the advent of the seafloor spreading hypothesis provided a reasonable mechanism for continental movement.

SEAFLOOR SPREADING

The primary objection to continental drift was the lack a known mechanism for moving continents over the Earth's surface, but that issue was solved by the American geologist Harry H. Hess in 1962. Hess postulated that continents did not plow through ocean crust, which was essentially Wegener's view of the process, but instead the entire crust was able to move (**FIGURE 7.4**). An American oceanographer, Robert S. Dietz, later dubbed this concept "seafloor spreading."

> **triple junction** A junction of three spreading edges of plates.
>
> **graben** An elongate basin formed through downdropping of a fault block and bounded on both sides by a normal fault.
>
> **aulacogen** A failed continental rift that has filled with sediment.

The seafloor spreading idea has its origins in the application of increasingly sophisticated and precise geophysical techniques used to study Earth's interior and surface features. Technological advances opened new research avenues and offered new approaches to testing large-scale geologic questions. Eventually, the combination of geophysical and other geologic information led to the eclipse of suppositions about the rigidity of Earth's surface.

Hess was a geologist and reserve naval officer called to active duty during World War II. When assigned to a transport vessel in the North Pacific Ocean, Hess surveyed the ocean floor using shipboard echo-sounding instruments. While crisscrossing the ocean, Hess obtained detailed profiles of the ocean floor, including its then recently discovered mid-ocean ridge crest and its deep ocean trenches. Trans-Pacific profiling led to the discovery of peculiar flat-topped undersea mountains, called **guyots**. Based on their size and shape, Hess reasoned that guyots were volcanic islands, or **seamounts**, that were first eroded by waves at sea level and then later sunk below the sea surface. In the 1970s, fossils of late Mesozoic shallow-water mollusks were collected from the tops of some guyots, and they reinforced Hess's interpretation of the origin of these peculiar structures.

> **guyot** A sunken seamount, or undersea volcano, that has a flat top due to erosion at sea level.

In a classic paper, *The History of Ocean Basins*, published in 1962, Hess outlined how crustal movement occurs: magma wells up from Earth's interior along mid-ocean ridges, creating new, warm seafloor that spreads away from the swollen ridge crest (**FIGURE 7.4A**). The rock cools and contracts as it gets farther from the ridge crest, and it eventually sinks into deep oceanic trenches. Sinking of the seafloor explains why seamounts would subside to sea level, where their tops get planed off by wave action, and then disappear below sea level (**FIGURE 7.4C**). It also explains why there are relatively few seamounts in the oceans (about 10,000). If they were continually forming and accumulating through time, the oceans should be crowded with them. Instead, they are destroyed when they reach the trenches.

Hess revived an older idea that Earth's mantle is capable of flow and rotation by means of giant thermal convection cells. Warm, low-density magma rises at a ridge, oozes out onto the seafloor, and then crystallizes to form basaltic rock. The crust then bends to form one flank of a ridge. The ridge's central furrow forms when the new crust splits and moves laterally.

Seafloor spreading provided a plausible mechanism for Wegener's continental drift. With seafloor spreading, continents were passively carried along as the

Ocean floor in motion FIGURE 7.4

A The mid-ocean ridge system is a continuous chain of undersea mountains. This is an artist's reconstruction of the Mid-Atlantic Ridge showing the ridge crest and transcurrent fractures (transform faults) that extend laterally away from the ridge crest. As newly formed igneous rock at the ridge crest cools and spreads away from the ridge crest, it contracts and sinks.

B The Hawaiian Island-Emperor Seamount chain formed by magma welling up from the mantle under the Pacific Plate. The ages of volcanic mountains increase away from the island of Hawaii, which is still actively forming over a hot spot. As the mountains increase in distance from where they are formed, they subside, along with the oceanic plate on which they ride, eventually sinking below sea level and become seamounts.

The Pacific Plate is currently moving to the north-northwest. From the Cretaceous to the Eocene (43 Ma), it moved to the northwest. Over the past 81 million years, the Pacific Plate has moved an average of about 10 cm per year.

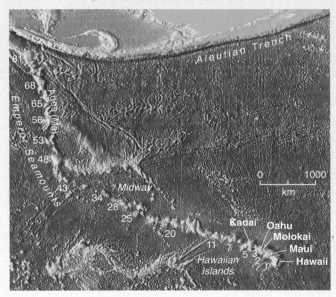

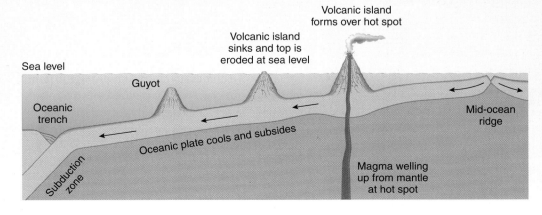

Volcanic island forms over hot spot

Volcanic island sinks and top is eroded at sea level

Sea level

Guyot

Oceanic trench

Mid-ocean ridge

Oceanic plate cools and subsides

Subduction zone

Magma welling up from mantle at hot spot

C Guyots are flat-topped seamounts that began as volcanic islands. As oceanic crust moves toward a trench, volcanic islands, like those in the Emperor Seamount chain, subside along with the cooling and contracting oceanic crust. The tops of the islands are eroded when the islands are near sea level. Eventually, the flat-topped mountains sink below sea level.

ocean floor spread from the ridges, instead of plowing through the ocean floor (which was physically impossible). Continents, which are composed of relatively light granitic rock (or **sial**), could be viewed as floating in denser basaltic rock (or **sima**).

Seafloor spreading also provided an explanation for another long-standing puzzle: why rocks in the oceans are no older than the Mesozoic, but rocks on land date back to the Archean. If rocks created at the ridge crests were carried away and ultimately destroyed in the oceanic trenches, extremely old oceanic crust would not be preserved.

Within a short amount of time, a variety of retrodictive tests of Hess's mechanism for moving crustal pieces were carried out. Some tests involved **paleomagnetism**, or the study of Earth's magnetic field and magnetic history.

> ■ **paleomagnetism**
> The study of natural remnant magnetism in rocks to determine properties of Earth's magnetic field in the geologic past.

Earth's magnetic history has been recorded in the alignment and dip of magnetic, iron-bearing minerals in igneous and sedimentary rocks that align with the direction of Earth's magnetic field, pointing like compass needles toward magnetic north.

It has long been known that Earth's magnetic field has reversed polarity from time to time. Magnetic surveying of the ocean floors showed bands of normal and reversed magnetic anomalies, the result of polarity reversals recorded in magnetic minerals locked in basalt on the seafloor (**FIGURE 7.5A**). The linear magnetic bands parallel the axis of the mid-ocean ridge and show lateral offsets along large fractures extending roughly perpendicularly to the ridge axis. The magnetic "stripes" on one side of the mid-ocean ridge are essentially a mirror image of those on the opposite side of the ridge.

In 1963, the British geophysicists Frederick J. Vine and Drummond H. Matthews explained the pattern of magnetic sea stripes as the result of new crust forming at the mid-ocean ridge, where it was later split and carried away laterally in either direction from the spreading center (**FIGURE 7.5B**). Later, measurement of the relative widths of magnetic sea stripes showed that they are proportional to the time intervals represented by the magnetic anomalies. Long intervals are recorded as wider stripes, and shorter intervals are recorded as narrower stripes.

Another interesting observation was that magnetic minerals in rocks provide information about the positions of the poles at the time the rocks were magnetized. Earth's declination is the difference between the magnetic North Pole (to which a compass needle points)

Magnetic seastripes FIGURE 7.5

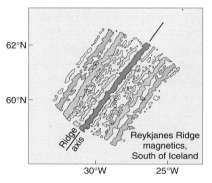

A Reversals in Earth's polarity field are recorded in mirror-image pairs of magnetic anomalies across the ocean floor. The colored bands indicate magnetically reversed intervals, and the intervening areas indicate intervals of normal polarity (or the same direction as in the present time).

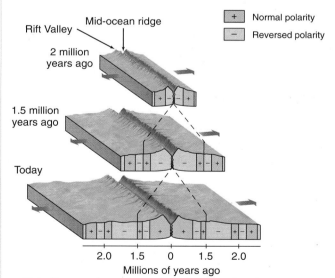

B This diagram shows how mirror-image pairs of magnetic sea stripes are formed. Lava extrudes along a mid-ocean ridge to form new crust, and the new rock becomes magnetized with the polarity of Earth's magnetic field. As rock along the ridge crest splits and separates, bands of oceanic crust having alternating normal and reversed polarity directions move away from the ridges.

A Earth behaves like a bar magnet, having north and south poles. The position of the magnetic north pole is about 15° from Earth's geographic North Pole, and this difference is called declination. At the time rocks are formed, magnetic minerals in them align according to Earth's north-south magnetic axis.

Curved lines in this illustration represent magnetic lines of force. Those lines have different angles compared to Earth's surface and are latitude dependent. At the time rocks are formed, magnetic minerals align with dip angles that match the angles of the lines of force.

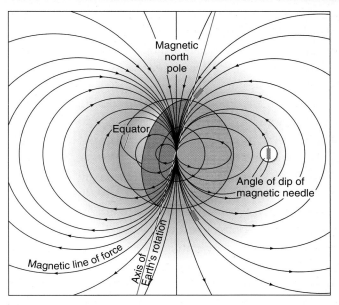

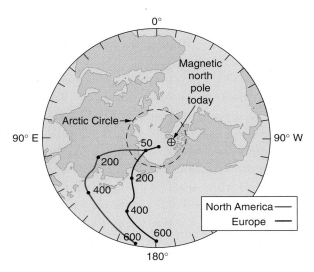

B Measuring the magnetic declination of ancient rocks shows that they have differences in declinations that increase with the age of the rocks. Plotting these changes on a globe reveals a path of apparent polar wander. In this illustration, the paths of apparent polar wander for North America and Europe are superimposed on a modern map of the globe. Ages along the polar-wander paths are given in millions of years. The polar-wander paths for the two continents are strongly divergent.

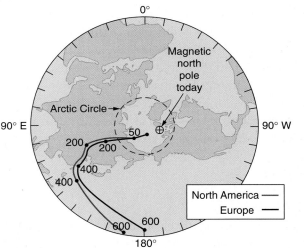

C If the Atlantic Ocean were closed, the apparent polar-wander paths for North America and Europe would nearly coincide for the Paleozoic Era (542 to 251 Ma), suggesting that North America and Europe were united during the Paleozoic.

and the geographic North Pole (**FIGURE 7.6A**). At the present time, the declination is about 15°.

Relatively young rocks were magnetized with a declination consistent with Earth's current magnetic field. Older rocks, however, showed disparities from the current magnetic field, and those differences increase with the age of the rocks (**FIGURE 7.6B**). The different pole positions indicated by older rocks came

to be known as **apparent polar wander** and indicated that either the magnetic north pole migrated great distances through time or the continents had moved great distances compared to the pole (which migrated only slightly). Even more intriguing, geophysicists discovered that pole positions recorded from North America were different from those recorded from Europe.

By the early 1970s, the riddle of apparent polar wander had been solved: by closing the gap across the Atlantic Ocean between North America and Europe, the polar-wander paths for the two continents were essentially identical for the Paleozoic, suggesting that North America and Europe were joined during that time (FIG-URE 7.6C on page 205) and that the position of the magnetic north pole was essentially unchanged. The paths diverged during the Mesozoic and Cenozoic, meaning that the two continents became increasingly separated during those eras.

Supporting the view that the continents had moved was information about magnetic dip preserved in rocks from the time of their magnetization. Dip is how much a compass needle, if allowed to rotate vertically, tilts following Earth's lines of magnetic force (Figure 7.6A). At the Equator, the needle is horizontal, but at higher latitudes, the needle dips at increasingly greater angles. If the continents had not moved at all, the remnant dip directions of magnetic particles in the rocks would have remained the same through time. Indicated dip angles varied, however, showing that the continents changed their latitudinal positions.

Confirmation of the production of new crust at mid-ocean ridges followed by lateral transfer across ocean basins came from study of **transcurrent fracture zones** (Figure 7.4A on page 203). These fractures are roughly perpendicular to a mid-ocean ridge at its crest but then extend in arcuate fashion across the ocean floor, gradually displacing crustal blocks to the left or right of one another. The fractures terminate near continental margins rather than continuing across continents. In 1965, the Canadian geologist J. Tuzo Wilson explained this paradox by assuming that the ridge crest is divided into segments, each of which produces new crust. Even though each block spreads laterally away along fractures, called **transform faults**, the ridge itself remains stationary. When a block reaches a deep-sea trench, near the margin of a continent, it sinks into the trench.

One of the final, and persuasive, supporting arguments for the rising tide of opinion favoring the developing plate tectonics model appeared with a fossil discovery in Antarctica. In 1969, Edwin H. Colbert and others reported remains of a primitive synapsid, a member of the phylogenetic line leading to mammals, from Triassic strata of the Transantarctic Mountains. This meter-long herbivorous animal, called *Lystrosaurus* (FIGURE 7.7), inhabited vegetation-rich areas near freshwater streams. *Lystrosaurus* had previously been known from South Africa and India (Figure 7.2A on page 198). The great distance and climatic conditions separating Africa and Asia from Antarctica minimized the likelihood that *Lystrosaurus*, and the plants on which it depended for food, had dispersed to the south polar region

Lystrosaurus FIGURE 7.7

Lystrosaurus was a squat, herbivorous synapsid (or "mammal-like reptile") that had beak-like jaws and a small pair of tusks. Its discovery in Triassic rocks of Antarctica in 1969 added to its known distribution in South Africa and India and reinforced the concept that Gondwanan areas were joined until the middle part of the Mesozoic Era. This reconstruction of *Lystrosaurus* is a sculpture by Margaret Colbert.

by way of ocean currents. By this time, hypotheses concerning sunken land bridges and island stepping-stones had been discounted because evidence for them failed to appear in geophysical profiles of the ocean floor.

CONCEPT CHECK STOP

What is the continental drift hypothesis, and what evidence was originally used to support it?

What is the seafloor spreading hypothesis, and what evidence supported it?

What was the mechanism for drifting of the continents provided by the seafloor spreading concept?

Plate Boundaries

LEARNING OBJECTIVES

Summarize the basic forms of tectonic boundaries: divergent plate boundaries, convergent plate boundaries, and transform plate boundaries.

Explain fault patterns associated with each type of plate boundary.

Explain active and passive margins and explain the types of stratigraphic information recorded in each setting.

In the modern concept of plate tectonics, which grew out of the aspects of continental drift and seafloor spreading that withstood testing, the Earth's surface is envisioned as a mosaic of mobile plates. At present, there are 7 large plates (carrying the continents and much of the Pacific Ocean) and about 20 smaller plates (FIGURE 7.8 on pages 208–209). Plate motion, typically 1 to 7 cm per year, causes plates to converge, diverge, or slide past one another, and this movement is the source of many earthquakes. Also, many volcanoes line up along or close to plate boundaries. Tracking patterns in the occurrence of frequent earthquakes and active volcanoes makes it possible to identify present-day plate margins.

Earthquakes also occur in some areas that are not along current plate margins. Occurrences such as those associated with volcanic activity in the Hawaiian Islands, the Galápagos Islands, Yellowstone National Park, and the Afar Triangle, are referred to as **hot spots**, or places where plumes of magma rise to the surface from deep within the mantle (Figures 7.4 on page 203 and 7.8 on page 206).

Other earthquake epicenters occur along ancient plate margins, ones that do not exactly line up with current margins, or along ancient spreading centers that failed to form wide ocean basins. In these places, **faults**, or cracks in crustal rocks along which movement has occurred, still permit infrequent tectonic adjustment. The valleys of many major rivers follow, in part, ancient fault lines. Earthquakes along the Mississippi River, centered near New Madrid, Missouri, during the winter of 1811–1812, are attributed to an ancient failed rift. Sporadic earthquakes near Lake Erie and Lake Ontario in Ohio, Pennsylvania, New York, and adjacent areas apparently result from continued readjustment along faults whose origin was separation of a Neoproterozoic supercontinent.

hot spot A volcanic center, often in the interior of a plate, caused by a plume of magma rising from the mantle.

A This map shows Earth's surface as a mosaic of mobile plates. Margins of the plates are indicated by the positions of numerous recent earthquakes and active volcanoes. Some ancient plate boundaries or failed spreading centers are indicated by the locations of infrequent earthquakes. Hot spots (denoted by red circles), such as active areas of the Hawaiian Islands, are indicated by earthquake epicenters and volcanoes but are unrelated to plate boundaries.

B Many plate boundaries are delineated by fault zones and lines of active volcanoes. Mt. Fuji in Japan is one of many volcanoes in the "Ring of Fire" rimming the Pacific Ocean. This volcano's last eruptive episode was in 1707–1708, and it is still considered active.

Each tectonic plate consists of rigid **lithosphere** (comprising the crust and uppermost part of the mantle) overlying a weak, partly molten region of the upper mantle called the **asthenosphere** (FIGURE 7.9). Below the

■ **lithosphere** The outer, relatively rigid layer of the Earth, comprising the crust plus the upper part of the mantle.

base of the lithosphere, which normally extends down 75 to 125 km, rocks are under conditions of such great pressure and temperature that they become capable of flow. Flow within the asthenosphere follows patterns of enormous

■ **asthenosphere** The layer within the upper mantle and below the lithosphere where rocks are relatively ductile and easily deformed.

Process Diagram

Tectonic plate motion FIGURE 7.9

The upper portion of the Earth can be thought of as a giant conveyor belt system in which new rock is created at spreading centers (mid-ocean ridges) and destroyed through subduction and remelting at oceanic trenches. Lithospheric plates are carried along by thermal convection cells within the underlying asthenosphere. Oceanic crust moves partly because of pushing from the mid-ocean ridges and partly because of pulling from the trenches at the ocean margins.

Divergent boundary
Divergence within continents yields grabens, or basins, flanked by steep-faced escarpments or fault-block mountains.

Convergent boundary
Where an oceanic plate meets a continental plate, the oceanic plate is subducted. Water-laden marine sediments reduce the threshold temperature required for melting rock so the descending slab can begin melting at a relatively shallow depth. Igneous melt rises and feeds a line of volcanoes, or a continental volcanic arc, on the overriding plate.

Convergent boundary
Continent-continent collision results in crustal thickening and rise of a mountain chain.

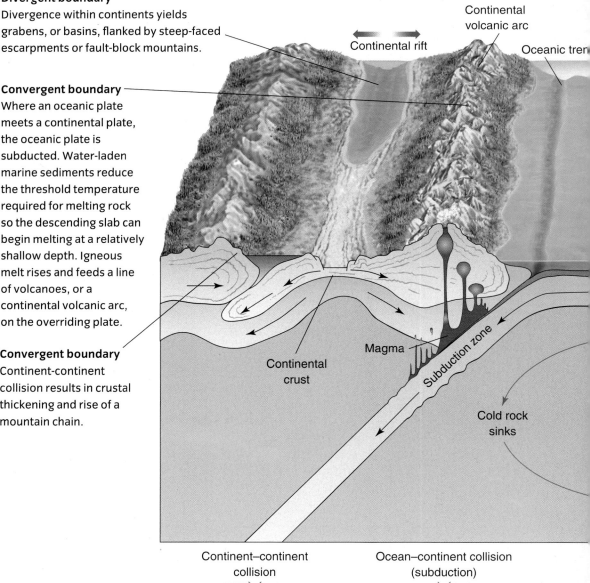

Continental rift

Continental volcanic arc

Oceanic tren

Continental crust

Magma

Subduction zone

Cold rock sinks

Continent–continent collision

Ocean–continent collision (subduction)

thermal convection cells, and this convection causes the overlying plates to move.

At **divergent plate boundaries**, plates pull apart under tensional forces (see *What a Geologist Sees*). Tension on the crust results in **normal faulting**, where grabens form as blocks drop downward in response to gravity. Blocks left in a relatively upthrown condition

are called horsts. Mountain ranges and their adjacent basins in the Basin and Range Province of western North America are outstanding examples of horsts and grabens developed during the Cenozoic Era. Other important divergent plate boundaries are along the mid-ocean ridges and in Africa's Great Rift Valley.

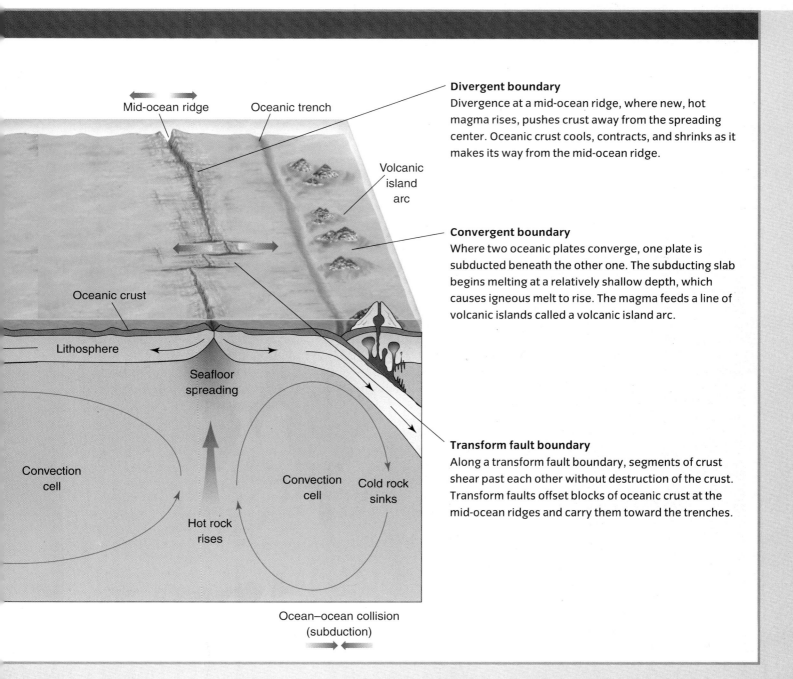

Mid-ocean ridge

Oceanic trench

Oceanic crust

Lithosphere

Seafloor spreading

Volcanic island arc

Convection cell

Convection cell

Cold rock sinks

Hot rock rises

Ocean–ocean collision (subduction)

Divergent boundary
Divergence at a mid-ocean ridge, where new, hot magma rises, pushes crust away from the spreading center. Oceanic crust cools, contracts, and shrinks as it makes its way from the mid-ocean ridge.

Convergent boundary
Where two oceanic plates converge, one plate is subducted beneath the other one. The subducting slab begins melting at a relatively shallow depth, which causes igneous melt to rise. The magma feeds a line of volcanic islands called a volcanic island arc.

Transform fault boundary
Along a transform fault boundary, segments of crust shear past each other without destruction of the crust. Transform faults offset blocks of oceanic crust at the mid-ocean ridges and carry them toward the trenches.

Signatures of Plate Motion

There are three basic forms of plate motion, each of which has a characteristic structural and stratigraphic signature.

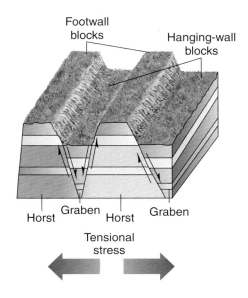

A Normal faulting, caused by extensional tectonics, results in downdropping of a graben (a basin), which then fills with sediments eroded from an adjacent horst (a mountain formed by block faulting). In a normal fault, the hanging wall (the block overhanging an inclined fault plane) moves down along a fault plane compared to the footwall.

B The Basin and Range Province of western North America shows numerous horsts and grabens. This is an aerial view showing horsts (mountains) and grabens (valleys) looking westward over the Oquirrh Mountains of Utah into the Tooele Valley.

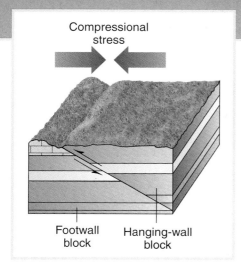

Compressional stress

Footwall block   Hanging-wall block

C Thrust faulting (or low-angle reverse faulting), caused by compressional tectonics, thickens the crust by stacking slices of strata on top of each other. In a thrust fault, the hanging wall moves up compared to the footwall.

D The Canadian Rockies in British Columbia are formed of stacked thrust slices. The mountain ranges are the upthrust edges of the thrust sheets.

Key

☐ Youngest Cambrian strata (Furongian Series)
▨ Mid-Cambrian strata (Series 3)
▨ Oldest Cambrian strata (Terreneuvian Series—Series 2)
→ Sense of movement on fault

0   10   20 km
Horizontal and vertical scale

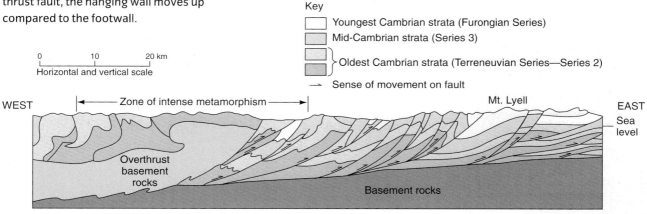

WEST
Zone of intense metamorphism
Mt. Lyell
EAST
Sea level
Overthrust basement rocks
Basement rocks

E This diagrammatic cross section through the Canadian Rockies illustrates the configuration of thrust slices. Stacked thrust slices have caused stratigraphic sequences to be repeated as older rocks were pushed on top of younger rocks.

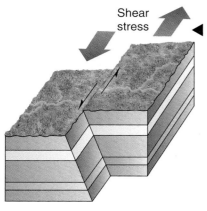

Shear stress

F Transform (strike-slip) faulting causes lateral shearing of rocks. In some places, such as western North America, crustal accretion has occurred; the juxtaposed blocks represent different tectonostratigraphic terranes.

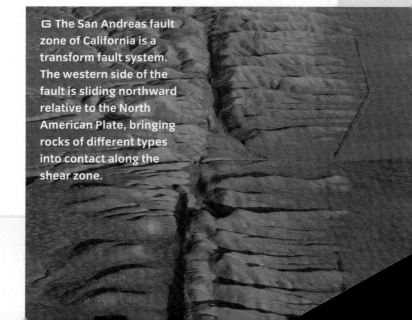

G The San Andreas fault zone of California is a transform fault system. The western side of the fault is sliding northward relative to the North American Plate, bringing rocks of different types into contact along the shear zone.

Development of a passive margin FIGURE 7.10

Passive margins receive much of the sediment eroded from the continents and deposited in the marine environment, along with evidence of life forms.

Diagrams A–C illustrate successive steps in the development of a passive margin.

A Rising magma heats and expands the crust, causing uplift and extension. The divergence causes crustal blocks to drop, forming sharp escarpments along fault-bounded horsts.

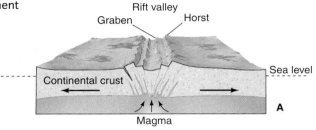

B If extension of the crust continues, basaltic magma welling up from the mantle will erupt and begin forming the floor of a narrow ocean basin. Erosion reduces the height of fault blocks on the diverging continental edges.

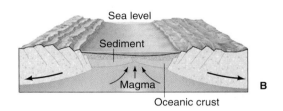

C Finally, the diverging edges of continental crust, now cooler, more contracted, and thinned by erosion, sink below sea level. These margins, which are no longer tectonically active, have become passive margins. They become subsiding basins, repositories for sediments carried by rivers to the ocean.

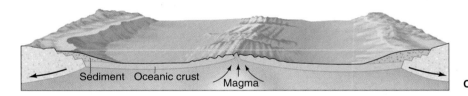

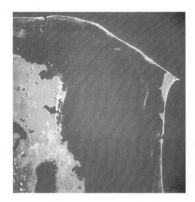

D The Atlantic coastal plain of the United States has remained a passive margin since the fragmentation of Pangea in the Mesozoic Era. This satellite view shows the Outer Banks area of North Carolina. The prominent point along the string of barrier islands is Cape Hatteras.

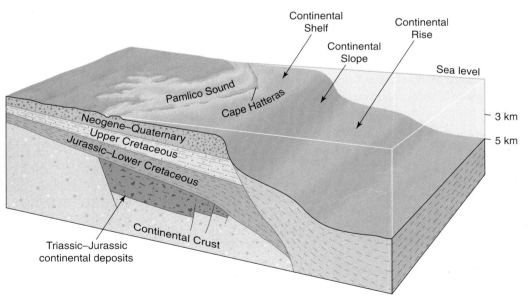

E A succession of Mesozoic-Cenozoic sediments, several kilometers thick, has accumulated in basins of the Atlantic coast. This is a simplified block diagram showing the succession off the coast of North Carolina from rift valley to passive margin.

Some divergent plate boundaries are **spreading centers**. As plates spread apart, new magma wells up from the mantle and solidifies in the fissure to form new rock. When this occurs at mid-ocean ridges, the new rock becomes new seafloor. The edge of a plate where new crust is added, and which is not involved in active tectonics, is its **trailing edge**, or **passive margin** (FIGURE 7.10). Passive margins are important from a stratigraphic standpoint because a substantial percentage of the sedimentary rocks used to "read" Earth history were deposited in basins on the passive margins of plates, including deltas and carbonate platforms. Just as importantly, passive margins can retain their stratigraphic records (or "pages" of Earth history) in good condition for millions of years if the margins remain tectonically inactive. The southern Atlantic–Gulf of Mexico coastline of the southeastern United States has been a passive margin continuously since the middle part of the Mesozoic Era. Recorded in its sediments is a rich, virtually undisturbed record of sea level history, climate change, and evolving biotas.

Passive margins of continents begin as bulges in the crust that split and form rift valleys (FIGURE 7.10A). High heat flow over a region, combined with tensional tectonic forces, causes continental crust to swell, bend, and break. New basaltic magma rises from the mantle and is extruded onto the floor of an incipient rift valley. Fissures that provide conduits for basaltic magma to reach the surface may solidify to form dikes, and some lava may squeeze out horizontally between rock layers to form sills. The Palisades along the Hudson River of New York and New Jersey (FIGURE 7.11) consist of spectacular sills formed by the injection of basaltic magma (which cooled to diabase) between sedimentary layers as Pangea was undergoing initial breakup during the Late Triassic and Early Jurassic. Early in the process of continental breakup, freshwater lakes, streams, and associated lake margin environments formed in the rift valleys. Triassic-Jurassic lake sediments of New Jersey, and dinosaur-track-bearing redbeds cropping out from Connecticut to New Jersey, are sedimentary deposits reflecting the early stages of sedimentary deposition in basins formed as Pangea was beginning to split.

If the rifting process continues over millions of years, a large ocean basin, with passive margins to either side, will develop (FIGURE 7.10C). In time, marine sediments will be deposited over the basaltic ocean floor. The modern Atlantic Ocean apparently formed from the splitting of Pangea in this way.

passive margin
The trailing edge of a tectonic plate, where active tectonic interaction with another plate is not occurring.

Evidence of an ancient rift FIGURE 7.11

A diabase sill called the Palisades is exposed along the Hudson River near New York City. The sill was formed through injection of magma along sedimentary layers as Pangea began to split during the Late Triassic and Early Jurassic.

Aulacogens and river valleys FIGURE 7.12

Some major river valleys follow in part the courses of ancient failed rifts, or aulacogens. When Pangea is reassembled, the positions of three-armed rifts formed as Pangea split along spreading centers become apparent. In many cases, two of the rift arms at spreading centers contributed to the fracture zone that eventually became the Mid-Atlantic Ridge. Failed arms filled with sediment or became river channels.

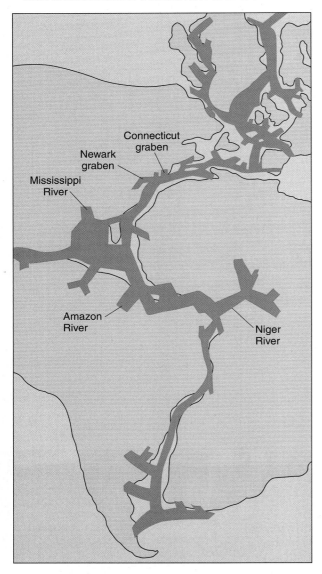

tem (Figure 7.3 on page 201), and one of the arms fails to continue rifting, the failed rift is called an aulacogen. The Great Rift Valley is just such an aulacogen. The lower Mississippi Valley region is part of a similar failed rift. The river's course follows a path of little resistance through a valley whose origin is far more ancient than the Mississippi. Other major rivers, such as the valleys of the Amazon and Niger, also seem to have origins as failed rifts (FIGURE 7.12).

Divergence at mid-ocean ridges is driven in part by the pushing effect of new magma ascending to the surface and partly by a pulling effect caused by subduction of oceanic crust at an oceanic trench (or **subduction zone**). A subduction zone is one form of **convergent plate boundary**, which is where plates move toward each other and collide. Because plates are involved in active tectonic interactions at convergent plate boundaries, they are also referred to as **active margins**. **Thrust faulting** (see *What a Geologist Sees* C, D on page 213) dominates along convergent plate boundaries.

▨ **subduction zone**
A long, narrow belt, usually including a deep-sea trench, along which subduction occurs.

▨ **convergent plate boundary**
A boundary between two plates that are moving toward each other.

At a subduction zone, a slab of cool, dense oceanic crust comes in contact with the margin of another plate, and it descends beneath that plate, eventually reaching depths where melting occurs under elevated temperature and pressure conditions. The rock will be remelted into magma and recycled. Eventually it will be carried by convection in the asthenosphere to a spreading center, where it will again rise to the surface and be extruded and solidified into new basaltic rock of the seafloor. The effect of this process is constant renewal of the ocean floor every 200 to perhaps 300 million years. Currently, the seafloor is a sedimentary archive that extends back in time no farther than the Jurassic Period (roughly 180 million years). For comparison, sedimentary archives on the continents provide Earth history information back to the early part of the Archean Eon (more than 4 billion years).

An oceanic slab that descends at a trench will have a relatively thin veneer of marine sediment deposited on it. As it is subducted into the asthenosphere, some of that sediment gets scraped off and piled up in an

Opening of a rift valley in a continent does not always guarantee that a new ocean basin will form. Rifting activity may be short-lived, and when it comes to a stop, a failed rift will be left in its wake. If a three-arm rift has developed, as we see today in the East African rift sys-

accretionary wedge (or accretionary prism) adjacent to the trench. Sediment that makes its way into this subduction complex is likely to become folded and regionally metamorphosed. One record of this activity is a chaotic, deformed mixture of rocks, or mélange. Metamorphosed and intensely folded rocks of the Franciscan Group exposed in the San Francisco Bay area of California represent deep-sea sediments that gathered in a now-vanished Cenozoic trench and its associated accretionary wedge (FIGURE 7.13).

Where a slab of oceanic crust collides with continental crust and the cool, dense oceanic slab is subducted, it melts, and new magma makes its way up to the surface to form volcanoes (Figure 7.9 on page 210). A line of volcanoes in the Andean region of South America, just east of the Peru-Chile Trench, has this origin. The Sierra Nevada range of the western United States represents a series of granitic plutons that were once the magma chambers of volcanoes formed in the same way but much earlier in the Cenozoic.

Subduction can also occur where one oceanic slab collides with another. As one slab descends, a series of volcanic islands (called a **volcanic arc**, **igneous arc**, or **magmatic arc**) normally forms in the overriding plate (Figure 7.9 on page 211). The Japanese islands are composed of a series of active volcanoes that form a volcanic arc.

The final type of plate boundary is a **transform fault boundary** (or **shear boundary**), where two plates slide past each other along transform faults without creating or destroying rocks (see *What a Geologist Sees F, G* on page 213). Transform faults offset crustal blocks at the mid-ocean ridges. Also, the San Andreas fault, which extends from Mexico to California, where the Pacific plate is sliding against the North American plate, is a transform fault.

> **volcanic arc** An arcuate line of active volcanoes and igneous plutons associated with a convergent plate margin where subduction is occurring.

> **transform fault boundary** A boundary between two crustal blocks that is characterized by a transform fault and where crust is neither created nor destroyed.

An ancient subduction zone FIGURE 7.13

A At a deep oceanic trench, a descending slab of cool oceanic crust is subducted below another crustal slab. Some deep-sea sediments that accumulated on the descending slab are scraped off to form an accretionary wedge, and others are dragged into the subduction zone, where they are mixed, folded, and metamorphosed. Continued descent into the subduction zone results in their remelting, along with the oceanic slab.

▼

B Metasedimentary and metaigneous rocks of the Franciscan Group were caught in a subduction zone and adjacent accretionary wedge in the area of present-day California during the Mesozoic and Cenozoic eras. Collisional processes, heating, and pressure associated with subduction folded and metamorphosed the former deep-sea sediments and rocks into a mélange. This exposure is in Marin County, California.

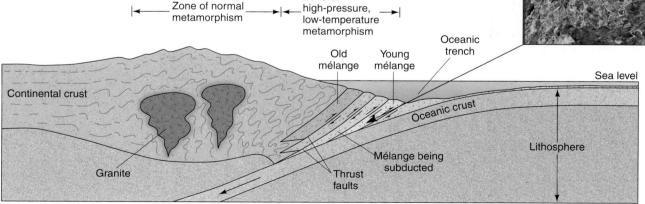

Tectonostratigraphic terranes are tectonic blocks that have different stratigraphic histories from the larger continental blocks to which they have docked.

This map of western North America shows the major tectonostratigraphic terranes added to the margin of the continent mainly through transform fault processes, during the Mesozoic and Cenozoic eras. Some terranes, such as the North Slope, may be displaced parts of the North American continent, whereas others, such as the Nixon Fork, probably originated as parts of other continents.

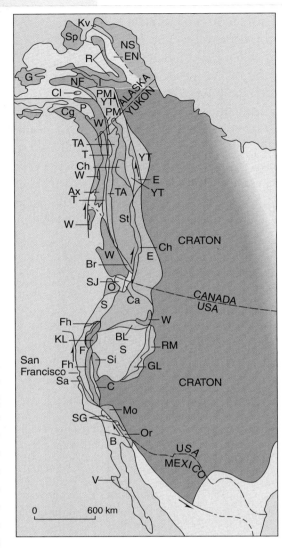

Symbol	Terrane Name
Ax	Alexander
B	Baja
BL	Blue Mountains
Br	Bridge River
C	Calaveras
Ca	Northern Cascades
Cg	Chugach
Ch	Cache Creek
Cl	Chulitna
E	Eastern assemblages
EN	Endicott
F	Franciscan and Great Valley
Fh	Foothills Belt
G	Goodness
GL	Golconda
I	Innoko
KL	Klamath Mountains
Kv	Kagvik
Mo	Mohave
NF	Nixon Fork
NS	North Slope
O	Olympic
Or	Orocopia
P	Peninsular
PM	Pingston and McKinley
R	Ruby
RM	Roberts Mountains
S	Siletzia
Sa	Salinian
SG	San Gabriel
Si	Northern Sierra
SJ	San Juan
Sp	Seward Peninsula
St	Stikine
T	Taku
TA	Tracy Arm
V	Vizcaino
W	Wrangellia
YT	Yukon-Tanana

Transform faults have had an important influence on the configurations of continents. By carrying oceanic rock away from ridges and toward the edges of continents (Figure 7.4A on page 203), transcurrent motion adds new blocks of crust to the continents, where they eventually become welded, thus reshaping continental margins. These blocks have tectonic and stratigraphic histories that are different from the continents to which they dock and are referred to as **tectonostratigraphic terranes**

tectonostratigraphic terrane

A rock body having an internally consistent geologic makeup and separated from a continental block or another terrane by bounding faults.

(also known as **exotic terranes**, or just **terranes**). Much of North America west of the Rocky Mountains is made up of exotic terranes welded to North America during the Mesozoic and Cenozoic (**FIGURE 7.14**). Substantial areas of eastern North America, stretching from Newfoundland to Florida, are composed of exotic terranes that joined North America in the middle to late Paleozoic.

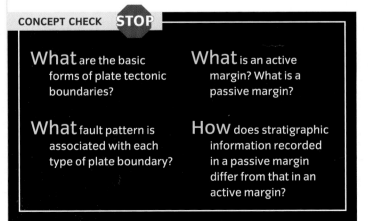

CONCEPT CHECK **STOP**

What are the basic forms of plate tectonic boundaries?

What fault pattern is associated with each type of plate boundary?

What is an active margin? What is a passive margin?

How does stratigraphic information recorded in a passive margin differ from that in an active margin?

Mountain Building

LEARNING OBJECTIVES

Describe orogenesis and in what tectonic settings it occurs.

Understand the origin and occurrence of igneous, sedimentary, and metamorphic rocks in an orogenic belt.

hains of mountains form in three principal tectonic settings: (1) where continental collision occurs, (2) in volcanic arcs adjacent to subduction zones, and (3) along mid-ocean rifts. The process of mountain building, at least where deformation of continental-type crust into mountain chains is involved, is called **orogenesis** (FIGURE 7.15). Block faulting and hot spot activity are not usually considered orogenic processes.

> **orogenesis** The process of building mountain chains and consequently deforming granitic-type crust.

Orogenesis FIGURE 7.15

When plates collide to form a mountain belt, a complex of environments develops. This is an idealized mountain chain formed by subduction of an oceanic plate beneath the margin of a continent.

Metamorphic belts are disposed to either side of the central volcanic arc. On the continent side, a fold-and-thrust belt develops. Loading of the crust by thrust slices causes adjacent crust to flex into a foreland basin. Sediments eroded from the mountains accumulate as a clastic wedge in the foreland basin.

On the oceanic side, a forearc basin is formed adjacent to the igneous arc. An oceanic trench, flanked by an accretionary wedge, marks the place where the oceanic slab is descending.

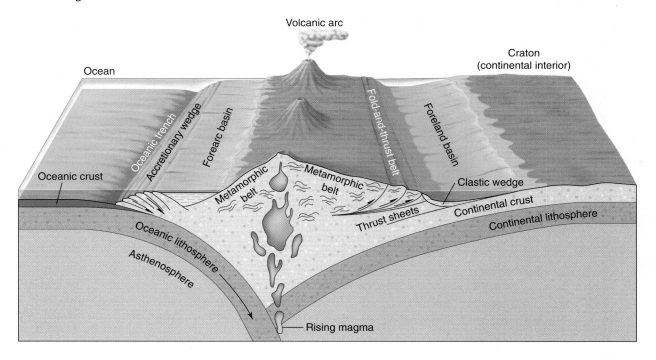

Himalayan orogenesis FIGURE 7.16

The Himalayan Mountains formed during the Cenozoic Era through the collision of two continental blocks. Subduction closed an ocean basin that originally separated India and the Tibetan (Xizang) region of Asia (**A**). As the continental crust of India approached Asia, it became wedged against the Tibetan margin (**B–C**). Thrusting and folding thickened the crust, and suturing of the two plates occurred. Today, India is fully sutured to the margin of Asia (**D**).

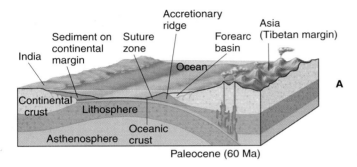

Paleocene (60 Ma)

A

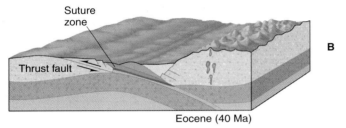

Eocene (40 Ma)

B

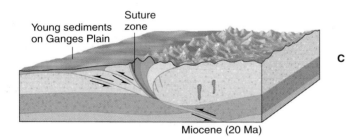

Miocene (20 Ma)

C

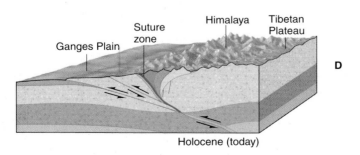

Holocene (today)

D

The Himalayan Mountains illustrate how an orogenic belt forms (**FIGURE 7.16**). An ocean basin once separated Asia and India, but during the Cenozoic Era, subduction of oceanic crust under Tibet (Xizang) closed the distance between the two areas until, finally, southern Asia and India were juxtaposed. Continental crust of the Indian plate, too buoyant to be subducted, was wedged under the Tibetan region of Asia, resulting in thrusting, folding, and thickening of both margins. As collision continued, the two continental blocks welded together along a **suture** zone. Collisional tectonics in this case has produced a mountain chain that extends down the middle of an assembled continent.

Sometimes in the process of continental collision, remnants of ocean floor are squeezed up along a suture zone. These remnants, called **ophiolites**, consist of pillow basalts sporadically extruded onto the ocean floor and ultramafic rocks of the upper mantle (**FIGURE 7.17**). Deep-sea sediments, including turbidites, black shales, and cherts, may also be associated with ophiolites.

> **ophiolite** An assemblage of ultramafic and mafic igneous rocks representing ocean crust.

Orogenesis in a volcanic arc setting, such as the Japanese islands or the Andes of South America, does not involve continental collision. Instead, as remelted magma rises toward the surface from a nearby deep ocean trench, **isostatic** adjustment (related to the addition of relatively low-density rock to the crust) and solidification of new rock cause crustal thickening. At the surface, volcanic peaks are formed.

A volcanic arc is often at the core of a major mountain chain (Figure 7.15 on page 219). Even mountains formed through continental collision normally incorporate volcanic arc-related rocks from early phases of their history, when subduction that caused shortening of the ocean floor was active, and continental collision had not yet been initiated.

Volcanic arcs ordinarily have an igneous core flanked by regional metamorphic belts. The metamorphism is a consequence of heating crustal rocks by the intrusion of hot magma emanating from the mountain core. Deformation also occurs by compression associated with subduction and by **gravity spreading**, or movement of thrust sheets near the bottom of the pile under the weight of rock stacked above.

Ophiolite FIGURE 7.17

Ophiolites are slivers of oceanic crust squeezed onto the edges of continents during plate convergence. This is an idealized ophiolite succession. Pillow basalts represent rocks of the ocean floor. They are commonly overlain by oceanic sediments. Beneath the basalt, sheeted dikes, gabbro, and ultramafic rocks represent mantle-derived material.

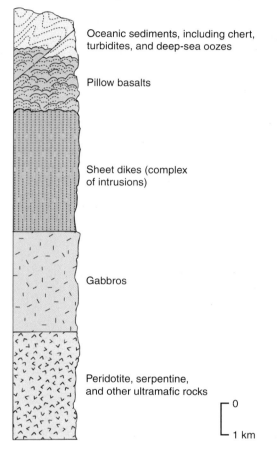

Oceanic sediments, including chert, turbidites, and deep-sea oozes

Pillow basalts

Sheet dikes (complex of intrusions)

Gabbros

Peridotite, serpentine, and other ultramafic rocks

0

1 km

The metamorphic belts are flanked by thrust sheets and sedimentary basins. Developed on the oceanward side of the igneous arc are a trench and an accretionary wedge. Behind these, a basin formed by compressionary folding and thrust faulting, called a **forearc basin**, fills largely with deep-sea sediments.

On the continent side of the volcanic arc, sedimentary rocks that are too far from the main zone of metamorphism to flow plastically undergo compression and break along thrust faults. Huge thrust sheets then slide away from the mountain core, stacking up, folding, and often becoming overturned, in a **fold-and-thrust belt**, where the crust is shortened and thickened.

Downwarping of the crust on the continental interior (or **craton**) beyond the fold-and-thrust belt produces a basin, called a **foreland basin**, that parallels the mountain chain. A foreland basin can develop quickly and become flooded with marine water entering through a passageway between or around mountains. The first sediments deposited in a foreland basin may be deep water muds and turbidites (historically referred to as **flysch**). As the mountain core rises, erosion sheds detrital sediment to the foreland basin. Eventually, deltaic sediments, and then nonmarine alluvial fan, stream, floodplain, and other deposits, historically referred to as **molasse**, fill the basin. Molasse deposits are thickest near the mountain chain and thin toward the craton. This wedge-shaped configuration of strata is called a **clastic wedge**.

A superb example of an ancient volcanic arc and associated environments occurs in eastern North America. During the Devonian, collision of North America with parts of Europe led to development of the Acadian Mountains. Volcanoes at the core of the mountain chain are represented today, after much erosion, by granites in eastern Maine. They were flanked by regional metamorphic belts represented today in part by the Green Mountains and Berkshire Mountains. Farther inboard, the fold-and-thrust belt, represented by the Piedmont region, opened into the great Appalachian Basin, an elongate foreland basin that extended roughly from New York to Kentucky. The Appalachian Basin contains a thick flysch-to-molasse succession known as the Catskill clastic wedge.

foreland basin A linear sedimentary basin that subsides in response to thrust loading of the crust.

clastic wedge A wedge-shaped deposit of sediments shed from an active thrust belt and filling a foreland basin.

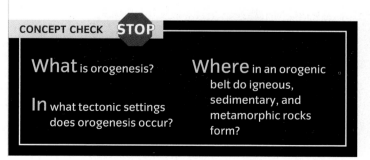

CONCEPT CHECK STOP

What is orogenesis?

In what tectonic settings does orogenesis occur?

Where in an orogenic belt do igneous, sedimentary, and metamorphic rocks form?

SUMMARY

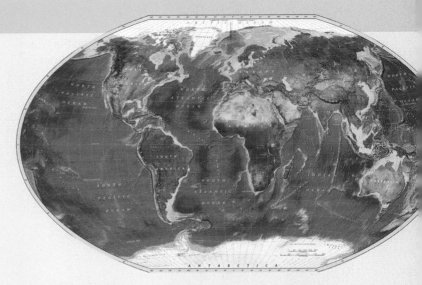

1 Origins of Plate Tectonics

1. **Plate tectonics theory** is a concept derived from melding the earlier ideas of **continental drift** and **seafloor spreading**. Continental drift postulates that Earth's continents have moved to their present positions after fragmentation of a larger landmass in the geologic past. Seafloor spreading provided the process by which continental drift could occur: the ocean **basins** expanded through the addition of new rock from spreading centers and the destruction of older rock near the ocean margins.

2. Alfred Wegener, Alexander du Toit, and others used various lines of evidence to support continental drift. These lines include the apparent jigsaw-puzzle-like fit of continental margins; stratigraphic and other similarities in the geology of continental areas thought to have been previously joined; paleoclimatic evidence; flow indicators from ancient glaciers; similarities in ancient and modern biotas; and contemporary evidence that continental drift is still ongoing.

3. Seafloor spreading was supported largely by geophysical evidence, including flat-topped seamounts **(guyots)**, the presence of mid-ocean ridges and their high heat flow patterns, paired magnetic "stripes" representing polarity-reversal episodes in **oceanic crust**, and paths of apparent polar wander.

2 Plate Boundaries

1. Plate tectonics theory states that Earth's surface consists of a mosaic of mobile plates. At present, there are about 27 plates, most of which are moving at an average rate of a few centimeters per year. Plate motion causes them to converge, diverge, or slide past one another, causing earthquakes and volcanic activity.

2. Tectonic plates consist of rigid **lithosphere**, which is made up of crust and the uppermost part of the mantle. The lithospheric plates overlie a region of the upper mantle called the **asthenosphere**, where weak, partly molten rocks are capable of flow by means of thermal convection. The flow along convection cells within the asthenosphere causes the lithospheric plates to move.

3. At **divergent plate boundaries**, plates spread apart by means of **normal faulting** as new magma wells up from the mantle and solidifies to form new rock. When this process occurs at mid-ocean ridges, the new rock becomes new seafloor.

4. The trailing edge of a plate, which is not actively involved in tectonic encounters with other plates, is a **passive margin**. Thick stratigraphic successions are recorded on passive margins of continents.

5. At a **convergent plate boundary**, plates move toward each other and collide. **Thrust faults** are formed. Convergence can occur at a subduction zone, where a cold, dense oceanic slab descends below another slab and becomes remelted. Convergence also occurs where continental collision takes place. Convergent plate boundaries, which are actively involved in tectonic activity, are **active margins**.

6. At a **transform plate boundary**, two plates slide past each other, and crust is neither created nor destroyed. Most **transform faults** offset rocks of the mid-ocean ridge systems. Transform faults are largely responsible for the docking of **tectonostratigraphic terranes** to continents.

3 Mountain Building

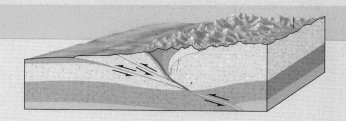

1. **Orogensis** is the process of forming mountain chains, primarily where continental collision occurs, and in igneous arcs. Mountains can also form on continents by block faulting and on continents or in ocean basins above hot spots.

2. Orogenesis resulting from collision causes welding of crustal blocks along a suture zone. Mountains formed in this way extend down the middle of an assembled continent.

3. Orogenesis in a volcanic arc setting involves subduction and remelting of a cold, dense oceanic slab. Remelted magma rises toward the surface and solidifies. Volcanic peaks may form at the surface.

4. A volcanic arc is usually at the core of a major mountain chain. The volcanic arc is flanked by regional metamorphic belts. They, in turn, are flanked by deformed thrust sheets and basins. On the oceanward side of the volcanic arc are a trench and an accretionary wedge. Behind these, a **forearc basin** forms. On the continent side of the volcanic arc, a fold-and-thrust belt develops where the crust is shortened and thickened. Beyond this, downwarping of the crust on the **craton** produces a **foreland basin**. Sediments filling the basin thin away from the volcanic arc, forming a **clastic wedge**.

KEY TERMS

- **plate tectonics theory** p. 196
- **continental drift** p. 196
- **seafloor spreading** p. 196
- **basin** p. 196
- **continental crust** p. 196
- **oceanic crust** p. 196
- **Pangea** p. 196
- *Glossopteris* **flora** p. 200

- **triple junction** p. 202
- **graben** p. 202
- **aulacogen** p. 202
- **guyot** p. 202
- **paleomagnetism** p. 204
- **hot spot** p. 207
- **lithosphere** p. 210
- **asthenosphere** p. 210

- **divergent plate boundary** p. 211
- **passive margin** p. 215
- **subduction zone** p. 216
- **convergent plate boundary** p. 216
- **volcanic arc** p. 217
- **transform fault boundary** p. 217

- **tectonostratigraphic terrane** p. 218
- **orogenesis** p. 219
- **ophiolite** p. 220
- **foreland basin** p. 221
- **clastic wedge** p. 221

CRITICAL AND CREATIVE THINKING QUESTIONS

1. The eastern margin of South America and the western margin of Africa look like matching pieces of a jigsaw puzzle. Other than separation of a once-single landmass, is there any other way that this matching pattern can be adequately explained? Why do you think it took so long for most scientists to accept the main ideas behind continental drift?

2. Thermal convection in the asthenosphere is invoked as the driving force behind tectonic plate movement. What do you think is the source of heat within the Earth? How could you test hypotheses about the source of heat within Earth?

3. Why do many mountain chains appear to be along the margins of continents?

4. Is there a limit to the altitude that mountain peaks can reach? What would limit the height of mountain peaks?

5. What makes a foreland basin a good place for recording the tectonic history of a volcanic arc?

6. How would you distinguish a thrust fault from a normal fault in the field?

7. If the present style of faulting along the San Andreas fault continues, what would you expect to ultimately happen to Los Angeles and adjacent areas of western California?

A Clastic Wedge

The Catskill delta complex is a Devonian clastic wedge that filled the Appalachian Basin, a foreland basin whose downwarping resulted from the loading of thrust sheets onto the cratonic crust behind the Acadian Mountains.

This is a generalized cross-section of the Catskill delta complex from the Catskill Mountains in eastern New York to Erie, Pennsylvania. Dashed lines indicate the extent of deltaic deposits removed by Cenozoic erosion. The Acadian Mountains were situated to the east of the present-day Catskill Mountains, and an epeiric sea spread out to the west.

Key:
- Limestone
- Shale
- Black shale
- Siltstone
- Sandstone
- Redbeds (reddish siltstones, shales, sandstones, and conglomerates)

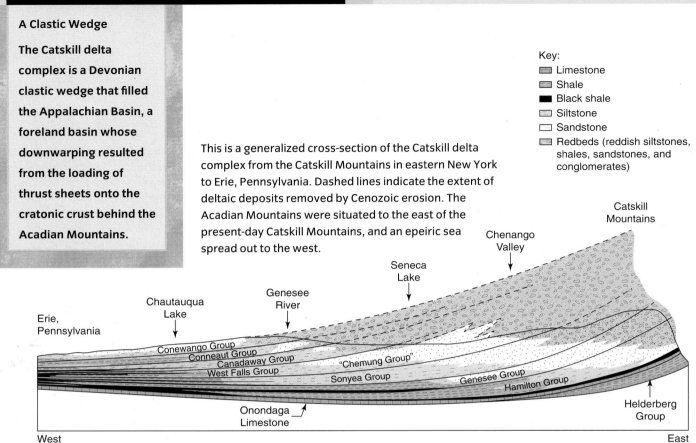

- Why does the clastic wedge thin toward the west?
- What was the sediment source for the clastic wedge?
- What direction was the sediment source?

This image shows reddish sandstones, siltstones, and shales from tidal environments of the Catskill delta complex near Altoona, Pennsylvania.

1. What two early concepts were incorporated in plate tectonics theory?

2. What was Pangea? When did it exist?

3. What types of evidence did Wegener use in support of continental drift?

4. What was Gondwana?

5. What types of geophysical evidence supported the concept of seafloor spreading?

6. What is a guyot, and how does one form?

7. What is the lithosphere? What is the asthenosphere?

8. What are the three types of plate boundaries?

9. How do mountains develop?

10. What is a foreland basin? Where does it form?

11. What are tectonostratigraphic terranes? How do they form?

12. What is a passive margin?

13. What is a subduction zone?

14. Where do aulacogens form?

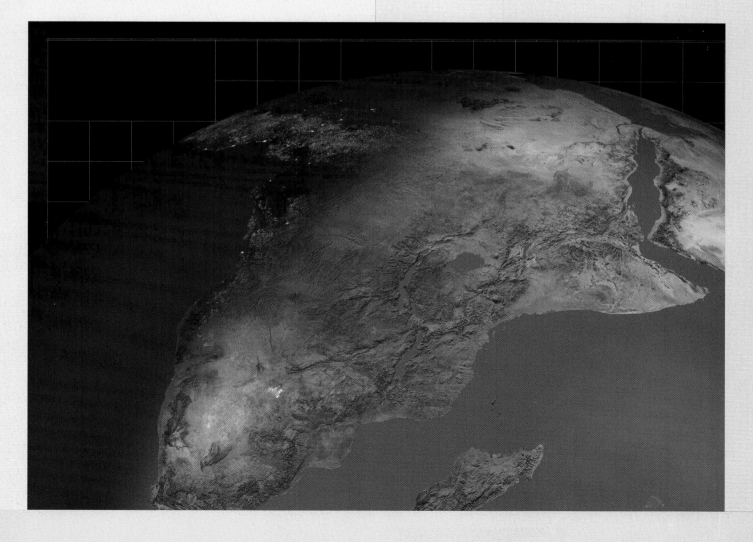

Archean World

So far, you have learned how to decipher Earth history from the record written in its stratigraphic layers (or "pages"). Now that you understand the clues to reading the history "book," we can begin to page through it. To interpret the elements of the story in their correct context, we'll start at the beginning and work toward the present. In interpreting the story, we'll often need to appeal, either implicitly or explicitly, to uniformitarianism and other principles. Using the present to help interpret the past may sound like skipping to the last page to determine how the story ends before reading through the entire book, but that's not quite true. In fact, the present is just another step in the process of Earth's evolution.

The dynamic events of the Archean Eon, the first of two eons informally called the Precambrian, comprise the first "chapter" of Earth history. It's a long chapter, roughly 45% of the "book," but many of the events in it are obscure because their record has been altered by subsequent metamorphism and erosion. Fossils in Archean rocks are all unicellular and difficult to classify, so they make poor biostratigraphic tools, and they limit our ability to correlate rocks in the way that we do for the Phanerozoic Eon. Another great contrast between the record of the Archean and that of the subsequent eons is that its rocks, like those shown here, are not arranged in a particularly orderly fashion because of repeated deformation by later mountain-building events. It's almost as if the pages of the Archean chapter were torn out, ripped in pieces, crumpled up, and then hastily replaced in the manuscript in disarray.

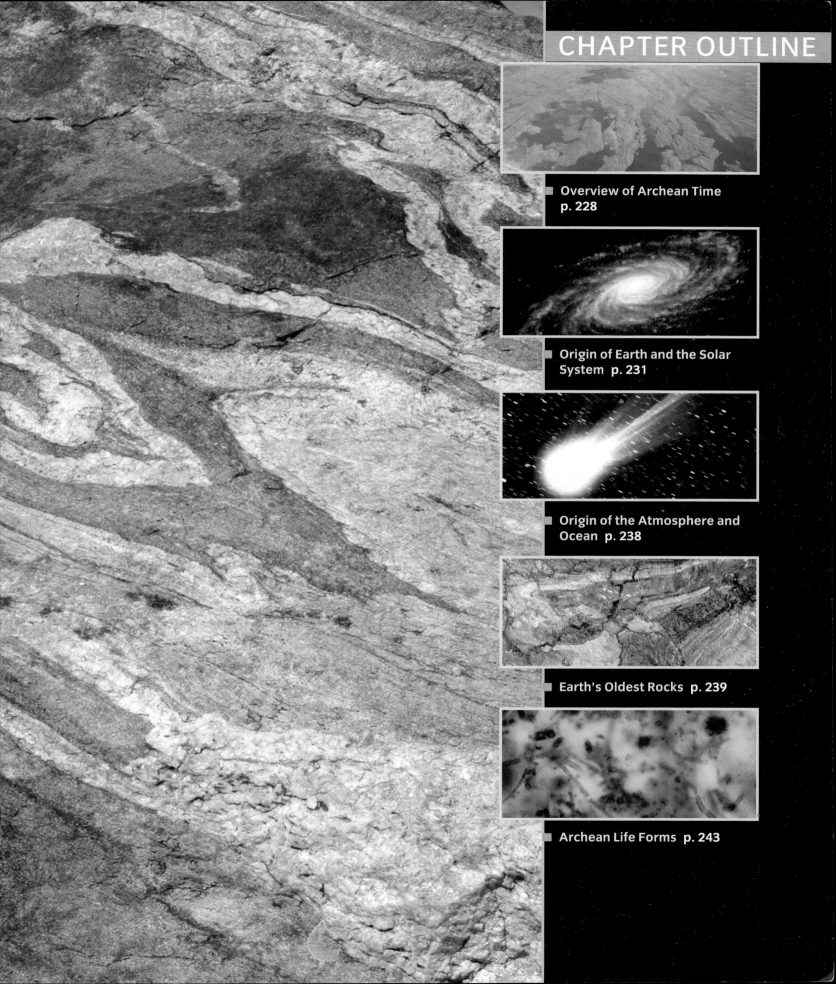

Overview of Archean Time

The "Precambrian," or the time of the Archean Eon plus the Proterozoic Eon, was a time in which many crucial events in Earth's physical, chemical, and biologic evolution took place. Among the major events of this time interval were initial formation of Earth, solidification of the crust, development of the cratons (cores of continents), initiation of plate tectonic activity, forma-tion of the oceans and atmosphere, the first prokaryotic and eukaryotic life, and early icehouse–greenhouse cycles, including glacial episodes.

The Precambrian rock record contrasts in some significant ways with that of the Phanerozoic Eon. Correlation of Precambrian rocks on the basis of fossils is really possible only close to the end of the Proterozoic Eon, and even then, the biostratigraphic precision is far inferior to

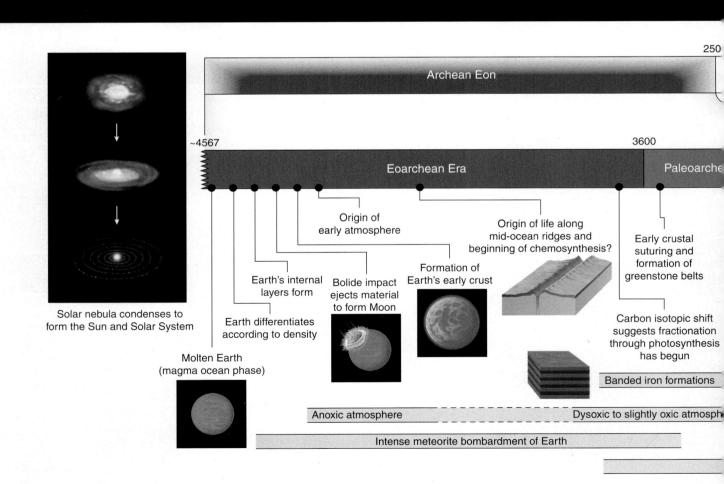

250

Archean Eon

~4567   3600

Eoarchean Era   Paleoarche

Origin of early atmosphere

Origin of life along mid-ocean ridges and beginning of chemosynthesis?

Early crustal suturing and formation of greenstone belts

Earth's internal layers form

Bolide impact ejects material to form Moon

Formation of Earth's early crust

Solar nebula condenses to form the Sun and Solar System

Earth differentiates according to density

Carbon isotopic shift suggests fractionation through photosynthesis has begun

Molten Earth (magma ocean phase)

Banded iron formations

Anoxic atmosphere   Dysoxic to slightly oxic atmosph

Intense meteorite bombardment of Earth

that of most of the Phanerozoic Eon. Also, many Precambrian rocks are either igneous or metamorphic, and thus they are fossil poor. These circumstances leave us little choice but to turn to other correlation techniques to figure out the age relationships of rocks. Isotopic dating, sometimes in combination with matching of distinctive stratigraphic sequences, can be an effective methodology. However, we must be careful about using isotopic methods because of the possibility that isotopic ratios have been reset in metamorphism, leading to anomalous isotopic ages.

The **Archean Eon** (FIGURE 8.1) begins with this planet's formation, approximately 4.567 billion years ago. The end of the Archean is automatically defined by the beginning of the next eon, the **Proterozoic Eon**, a point chosen as 2.5 billion years ago.

Earth had a violent early history. Its birth over the course of millions of years involved the powerful impacts of large orbiting asteroids, comets, and smaller particles, all melting into a sphere of molten magma. Crystallization of minerals to form the rocks of a thin outer skin of that magma ensued, but repeated impacts of incoming objects caused much of the surface to remelt into a cauldron of hot molten rock. Along with solidification of the crust came explosive volcanic activity and an outpouring of poisonous gases to the atmosphere. Amalgamation of protocontinents took place by the forceful collision of early-formed crustal blocks. Temperatures

Archean Eon The unit of geologic time beginning with Earth's formation, perhaps 4.567 billion years ago, and ending at the beginning of the Proterozoic Eon, 2.5 billion years ago.

Proterozoic Eon The unit of geologic time beginning 2500 million years ago (2.5 billion years ago) and ending at the beginning of the Phanerozoic Eon, 542 million years ago.

Visualizing
Archean timeline FIGURE 8.1

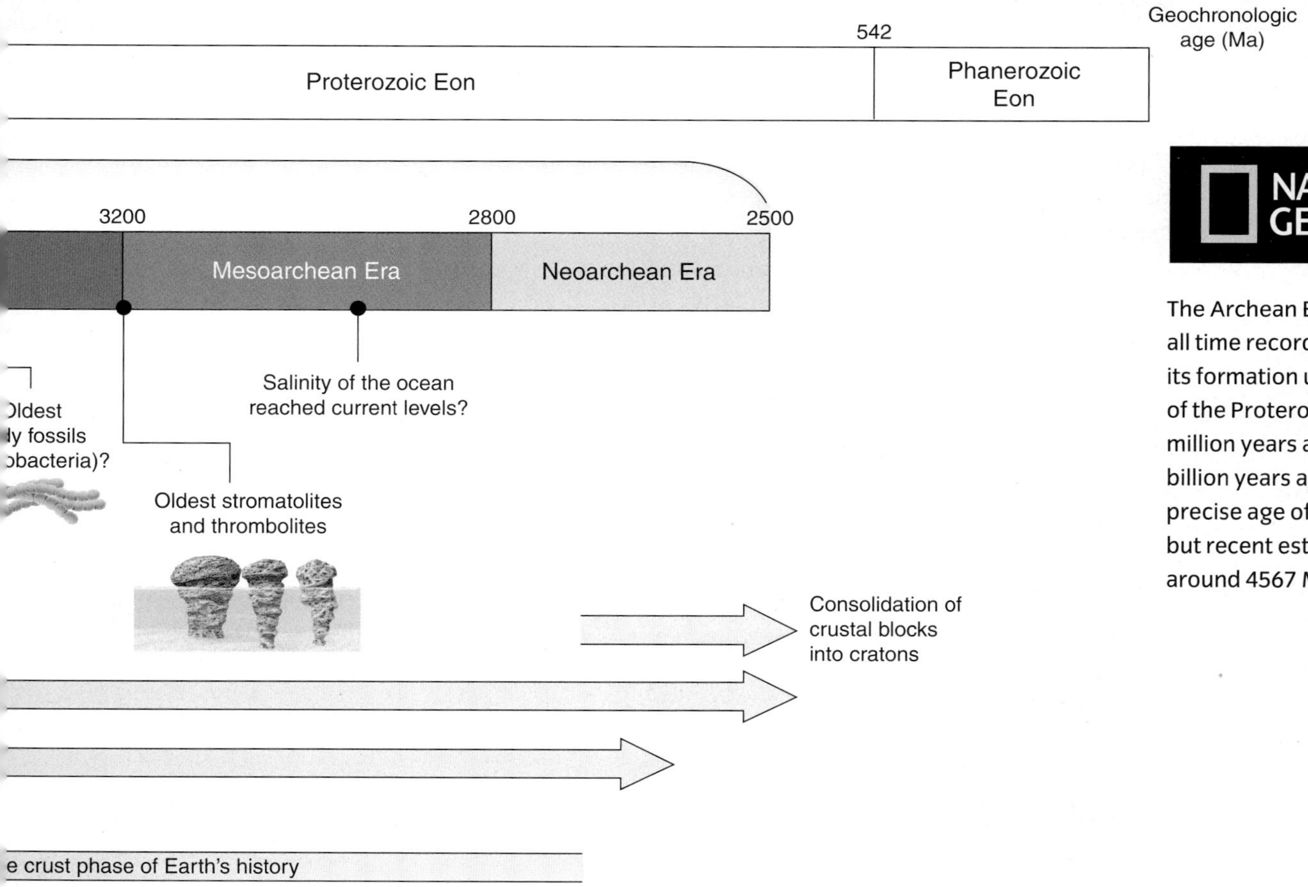

Geochronologic age (Ma)

542

Proterozoic Eon | Phanerozoic Eon

3200 | 2800 | 2500

Mesoarchean Era | Neoarchean Era

Salinity of the ocean reached current levels?

Oldest
ly fossils
obacteria)?

Oldest stromatolites and thrombolites

Consolidation of crustal blocks into cratons

e crust phase of Earth's history

NATIONAL GEOGRAPHIC

The Archean Eon encompasses all time recorded on Earth from its formation until the beginning of the Proterozoic Eon 2500 million years ago (Ma), or 2.5 billion years ago (Ga). The precise age of Earth is uncertain, but recent estimates place it around 4567 Ma (4.567 Ga).

In northern Canada, Archean rocks crop out at the Earth's surface in a vast area of low topographic relief. This, the Canadian Shield, is characteristic of the shield areas that form the foundations of all the continents.

---

craton The core of a continent—the part of Earth's continental crust that has attained relative stability and received little deformation for at least 1 billion years.

Canadian Shield A large area of Precambrian (Archean) basement rocks that make up the exposed core of North America and shows at the surface mostly across Canada and the northern United States.

and pressures reached in collisional events were so extreme as to cause separate rock bodies to weld together along metamorphic belts. Amid this chaos, a primitive ocean developed, and in it, the first life forms arose.

Our knowledge of the Archean comes mostly from rocks of the **cratons**, which are the large interiors of continents, places that have undergone little tectonic deformation since the late Proterozoic. On all the ice-free continents of the world today, there are vast exposed areas of low topographic relief, and the bedrock consists mostly of Archean rocks. These areas, which are

known as **shields**, are the major places where Archean rocks crop out at the surface. The **Canadian Shield** (FIGURE 8.2), which stretches across large expanses of Canada and parts of the northern United States, is a particularly good example of exposed Archean rocks. Archean rocks also crop out in a few other places, such as the Grand Canyon of Arizona, where deep incision of Earth's strata by the Colorado River has exposed early metasedimentary rocks. Other information about the distribution of Archean rocks comes from deep drill cores that have penetrated the Precambrian basement.

CONCEPT CHECK **STOP**

What were the major geologic and biologic events of the Precambrian?

Why is our understanding of the Phanerozoic Eon better than that of the Archean Eon?

Origin of Earth and the Solar System

When and how was Earth formed, and what is its place in the Solar System? Determining a precise age for the origin of Earth using terrestrial rocks poses some problems. Our planet is unique among those in the Solar System in having both active tectonism and a continuous supply of liquid water over a period of billions of years. Water is the principal agent of weathering and erosion. Tectonics, weathering, and erosion have substantially modified Earth's crust, making it unlikely that many (if any) igneous rocks at the surface date from the planet's inception. How then can we arrive at a reliable determination of Earth's age? To answer this question, we need to turn to extraterrestrial sources of rocks—rocks formed from the same initial processes and at the same time as the rest of the Solar System. If we presume that all the planets, their moons, and asteroids in the Solar System coalesced about the same time from a single mass of galactic dust, then sampling rocks from any of these bodies that has not been extensively reworked should yield radiometric ages close to the age of Earth.

USING METEORITES AND MOON ROCKS TO DECIPHER EARLY EARTH HISTORY

We have access to at least two extraterrestrial sources of information about the age of Earth: meteorites and Moon rocks. **Meteorites** are rocks that have fallen to Earth (or some other planet or moon)

meteorite
A relatively small rock that falls to a planetary surface from interplanetary space.

from space (**FIGURE 8.3**). They range in size from tiny particles to rocks 1 kilometer or more in diameter (see *What a Geologist Sees* on page 232).

Meteorites found on Earth can be divided into three basic groups, according to their composition: (1) stony meteorites (comprising chondrites, carbonaceous chondrites, and achondrites); (2) iron meteorites; and (3) stony-iron meteorites. **Stony meteorites** are composed of mafic or ultramafic igneous rocks. **Chondrites**, the most common meteorites, often contain small spherical bodies called **chondrules** that solidified from molten droplets ejected into space when large bodies swirling around in the solar nebula collided. **Carbonaceous chondrites** contain small amounts of organic molecules (up to 3%), including amino acids, chondrules, and sometimes diamonds. The organic compounds include the same ones that life forms on Earth use to build DNA and RNA, so it is possible that carbonaceous chondrites,

Meteorite FIGURE 8.3

Part of a carbonaceous chondrite meteorite that fell near Pueblo de Allende, Mexico, on February 8, 1969. This meteorite, estimated to be 4.567 billion years old, contains organic molecules and diamonds.

Meteorite Impact Craters

A The Moon's surface is scarred with craters of many sizes, reflecting the impact of thousands of meteorites. The obvious bombardment of the Moon over time by meteorites serves as an analog for meteorites that must have fallen to Earth. This photo of the lunar surface was taken in July 1969 from *Columbia*, the *Apollo 11* command module. ▶

B Meteor Crater, a large and unusually well-preserved crater, was produced by the impact of the Canyon Diablo meteorite in Arizona. Most meteorite craters on Earth have been significantly modified and obscured by weathering, erosion, and tectonic activity. Well-preserved craters, like this one, tend to be relatively young in age compared to the Moon's craters. ▼

NATIONAL GEOGRAPHIC

along with asteroids and comets, provided the basic compounds needed to construct life forms. **Achondrites** are stony meteorites made of angular rock fragments but lacking chondrules. **Iron meteorites** are formed of iron-nickel alloys. They have large crystal sizes, indicating a slow crystallization rate deep within the interior of a larger body, such as an asteroid. **Stony-iron meteorites**, which are rare, have a mixture of silicate minerals and iron-nickel alloys.

Most meteorites that have struck Earth are fragments of larger rock bodies that collided and shattered. When passing through the atmosphere, they also underwent partial burning, so they sometimes show fused crusts from the intense heat. Meteorites are basically of the same types as igneous rocks on Earth, probably because most represent asteroids, or essentially small planets, that have fragmented. Some, however, may be rocks that dislodged from other planets, such as Mars, or from the Moon, by the impact of asteroids or other meteorites.

asteroid One of many irregularly shaped planet-like rock bodies, including those orbiting the Sun between Mars and Jupiter.

Collectively, meteorite compositions suggest that asteroids, before fragmenting, were differentiated just like Earth into silicate-rich outer portions and metallic cores. Stony meteorites, which are of basaltic and ultramafic composition, are comparable to the silicate rocks of Earth's mantle and oceanic crust. Iron meteorites are comparable to the iron-nickel core of Earth. Finally, stony-iron meteorites probably originated near the boundary between the silicate mantle and the iron core of destroyed asteroids.

Rocks brought back from the lunar surface by U.S. astronauts during the Apollo missions of the 1960s and 1970s are silicate rocks of basaltic (mafic) and ultramafic composition. They are comparable in composition to stony meteorites (mostly ordinary chondrites) and to Earth's mantle. The Moon, just like Earth and the asteroids, was differentiated into a silicate outer region and a metallic deep interior. Native rocks on the Moon's surface are likely to reveal the age of the solidification of that body because there has been no liquid water to cause weathering and erosion, and tectonic activity has been slight, not enough to cause rejuvenation of the surface.

Radiometric analyses of stony meteorites and Moon rocks using uranium-lead, uranium-thorium, potassium-argon, and rubidium-strontium dating methods provide ages that cluster between 4.5 and 4.6 billion years. This is the probable age of the Solar System, an age that is supported by radiometric analysis of zircon crystals from Australia that were reworked from early igneous rocks and deposited billions of years later as sediment in detrital sandstones. Ages of the oldest detrital zircons found on Earth cluster between 4.1 and 4.4 billion years.

BIRTH OF THE UNIVERSE AND SOLAR SYSTEM

Astronomers refer to the formation of the universe as the **Big Bang**. Initially, all matter is assumed to have been concentrated at a single point. Upon explosion, matter shot out in all directions. Eventually, gravitational attraction caused its assembly into **galaxies**, which are disk-shaped clusters of stars (**Figure 8.4**). Our galaxy, composed of perhaps 250 billion stars that maintain constant distances among themselves, is called the Milky Way. The distance between the Milky Way and other galaxies is increasing, indicating that the universe is still expanding.

A spiral galaxy Figure 8.4
This galaxy consists of more than 200 billion stars spiraling from a bulge at the center.

The universe is thought to be 12 to 18 billion years old, based on calculations of the wavelengths of light radiating from distant stars. Light waves become stretched as they travel through space. Within the visible range of the electromagnetic spectrum, red light has the longest wavelength, so if light waves are being stretched, crossing the vast emptiness of space, they are shifting toward the red end of the spectrum, and this is called **redshift**. The farther that light waves pass through space, the more redshift will occur. After determining the amount of redshift in light emitted from distant galaxies to Earth, astronomers can calculate the distances to each of the galaxies and then reconstruct how far they have moved from their point of origin. Factoring in the present expansion rate of the universe using the Hubble constant (the ratio of velocity to distance in expansion of the universe) allows for estimation of the amount of time that has passed since all mass was concentrated at a single spot.

Gravitational collapse of dense gas clouds, mostly hydrogen, formed the stars. The Milky Way prob-

ably developed into a galaxy more than 10 billion years ago. Even after the initial burst of star formation has passed, additional stars may continue to be born from galactic matter in the spiral arms of the cluster. Our Sun, which was formed about the same time as the planets of our Solar System (less than 5 billion years ago), is a secondary star formed in the outer reaches of one spiral arm of the Milky Way.

Our Sun apparently formed from elements left over from the demise of another star. An earlier star collapsed violently, forming heavy elements. It then formed a **supernova**, or an exploding star, and ejected low-density matter, leaving behind a dense cloud of matter called a **solar nebula**. As the solar nebula condensed, it cooled and rotated. As the cloud shrank in size, its rotational speed increased.

During the rotation of the solar nebula, elements spun off from it were differentiated according to distance from the nascent Sun (**FIGURE 8.5**). Lighter elements were expelled the farthest, ultimately

The Solar System FIGURE 8.5

Process Diagram

A The Solar System began to develop after the explosion of an early-formed star. In the wake of this explosion (or supernova), a relatively dense cloud of matter, called a solar nebula, was left behind. The cloud condensed through the gravitational attraction of atoms.

B After millions of years, the condensation of matter created a dense cloud whose rotation helped differentiate matter within it. At the center of the cloud was a body that would eventually become the Sun. Condensation of matter in the outer part of the cloud formed the planets, moons, dwarf planets, and asteroids.

forming the outer planets of the Solar System (Jupiter, Saturn, Uranus, and Neptune), and heavier elements remained closer to the Sun, where they formed the inner planets (Mercury, Venus, Earth, and Mars). The dust cloud produced by the supernova flattened into a disk-like shape as it rotated and increased in density. Now the planets revolve around the Sun in essentially one plane.

The disk-shaped dust cloud of the solar nebula segregated into orbiting rings that later coalesced into planets and dwarf planets (such as Pluto), or asteroids. Some of the material between Mars and Jupiter remains in orbit around the Sun in an asteroid belt. From time to time, asteroids or pieces of shattered asteroids fell from their orbits, largely because the gravitational influence of Jupiter disrupted them. They then tumbled toward the inner planets, resulting in countless meteorite impacts on Earth and other planets, as well as on their moons.

EARLY EVOLUTION OF EARTH AND THE MOON

About 5 billion years ago, galactic particles orbiting the Sun were colliding and undergoing accretion to form what would eventually become Earth. Some of the impacts involved relatively small meteorites and asteroids, but others no doubt involved bodies of colossal size, maybe even the size of small planets. The energy released upon impact of these bodies helped ensure that early Earth was molten. Adding to that, the decay of short-lived radioactive isotopes and the force of gravitation would have generated still more heat. On a spinning, molten planet, heavier elements (such as iron and nickel) would tend to sink toward the center, and lighter elements (such as silicon, oxygen, and aluminum) would tend to make their way toward the outside of the mass. The molten outer surface, a so-called

◿ The inner planets (or terrestrial planets)—Mercury, Venus, Earth, and Mars—are relatively small and rocky in composition. Because of their general resemblance to Earth, they are referred to as the terrestrial planets. The outer planets (or jovian planets)—Jupiter, Saturn, Uranus, and Neptune—largely consist of gases. They are massive but have low densities. Saturn, for example, is 95 times more massive than Earth but is less dense than water. The rings of Saturn are made of rock fragments and ice. Pluto, which was once considered to be a planet, is now regarded as a dwarf planet.

magma ocean (FIGURE 8.6), cooled to form a silicate-rich crust, whereas the core became enriched in iron and other metals. When patches of the surface first cooled, they formed ultramafic rocks called **komatiites**, a group of igneous rocks that cool well above the 1100°C temperature needed for basalt to solidify. Komatiites may cool close to 1600°C. As surface temperatures continued to cool, molten areas between the komatiite patches could solidify to basaltic rocks.

Various hypotheses for the origin of the Moon exist, but the one that seems to best explain currently available data is that it formed in the wake of an impact by a body about the size of Mars. Soon after Earth had accreted, the **bolide** (or impacting body) struck Earth with enough force to cause separation of its core and explosion of its mantle. The dense material of the bolide's core penetrated Earth, sunk to its center, and joined Earth's core. Fragments of exploded mantle formed a cloud of material encircling Earth and held in orbit by its gravitational field. The force of the explosion caused **volatiles** (gases), including water vapor, to be expelled.

> **magma ocean** The condition of molten magma covering the outer surface of a planet or a moon.

Its intensity probably caused remelting of Earth's initial crust, forming a new magma ocean. In time, the orbiting material coalesced to form the Moon, which initially must have had a hot magma ocean at its surface (as suggested by komatiites in the lunar highlands). Because of the impact origin of the Moon, crystalline rocks comprising it are depleted in water as compared to terrestrial rocks. Also, the iron-rich core of the Moon is smaller than expected for a body of its size.

By about 4.5 billion years ago, the magma ocean of the Moon had solidified to form a crust, and Earth's second magma ocean had solidified to form its second basaltic crust. The density-based internal layering of Earth, which had been disrupted by a bolide impact leading to formation of the Moon, was redeveloped. Full differentiation of the planet resulted in formation of a solid iron-nickel-rich inner core, a liquid iron-nickel-rich outer core, a silicate mantle, and a silicate crust (FIGURE 8.7). Rotation of the planet on its axis, combined with the liquid iron-rich (and thus magnetic) outer core, established Earth's magnetic field.

Magma ocean FIGURE 8.6

Earth's early surface was probably molten silicate rock, called a magma ocean. This is an artist's rendition of an Archean magma ocean with blocks of komatiite, an ultramafic rock, solidifying in places. In the Archean, the Moon was much closer to Earth than it is today.

Internal differentiation of Earth FIGURE 8.7

Early differentiation of the molten Earth caused iron and other dense elements to sink toward the center and less dense elements to rise toward the surface. This cutaway view of the Earth shows layers of differing composition and zones of differing rock strength. The compositional layers are a silicate-rich crust, a silicate-rich mantle, and an iron-nickel-rich core. The Moho (or Mohorovičić discontinuity), marks the crust-mantle boundary. Zones of differing rock strength are the cool, rigid lithosphere (on the outside of the planet); the hot, plastic asthenosphere; and the hot but strong mesosphere.

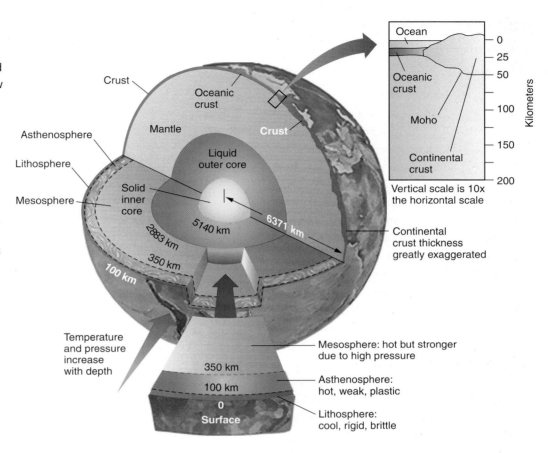

CONCEPT CHECK **STOP**

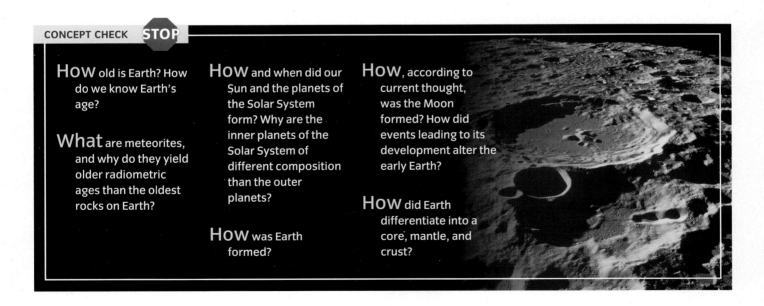

HOW old is Earth? How do we know Earth's age?

What are meteorites, and why do they yield older radiometric ages than the oldest rocks on Earth?

HOW and when did our Sun and the planets of the Solar System form? Why are the inner planets of the Solar System of different composition than the outer planets?

HOW was Earth formed?

HOW, according to current thought, was the Moon formed? How did events leading to its development alter the early Earth?

HOW did Earth differentiate into a core, mantle, and crust?

Origin of the Atmosphere and Ocean

LEARNING OBJECTIVES

Explain how Earth's atmosphere was initially developed.

Summarize the composition of Earth's early atmosphere.

Explain the sources of water for Earth's early ocean and explain the source of salts dissolved in it.

The early atmosphere must have formed after Earth had coalesced and was large enough to retain gases in its gravitational field. The gases comprising Earth's original atmosphere were lost during the time that the Solar System was forming. After this, and while Earth was in the molten state, volatiles easily escaped to the surface in a process called **outgassing** (or **degassing**). The outgassing process was responsible for Earth's second atmosphere. Magma is composed not just of liquefied rock but also of a large proportion of compounds and elements in the gaseous state. Most outgassing of Earth accompanied early differentiation of the surface and interior, but degassing, on a much smaller scale, still continues to the present day. Active volcanoes and deep-sea vents constantly emit atmospheric gases to the atmosphere.

By applying uniformitarianism, we can use the composition of gases from modern volcanic eruptions to determine what gases were released to the primitive atmosphere during the Archean Eon. Most of the gas released during a volcanic eruption is water vapor (H_2O). Smaller quantities of hydrogen (H_2), hydrogen chloride

> **outgassing** The process of releasing gases, including water vapor, from magma.

(HCl), carbon dioxide (CO_2), carbon monoxide (CO), nitrogen (N_2), and other gases are also emitted. The early atmosphere was probably anoxic, but it may not have been anoxic for long.

What is the origin of the world ocean? For many years, it has been assumed that water vapor released during the major phases of Earth's outgassing condensed to form the liquid water of the ocean. This seems reasonable, based on the fact that more water vapor is released from volcanoes than from any other source. This view is now changing, however. Some astronomers contend that the largest share of water in the early ocean was derived from comets passing through Earth's atmosphere. Comets, which come from outside the Solar System, consist of ice and cosmic dust (**FIGURE 8.8**). When their orbital paths bring them into the Solar System, heat from the Sun causes some of the ice to melt. Today, as many as 15 million comets, most of them small, pass through our planet's upper atmosphere per year. As they melt, they deliver a nearly continual supply of water to the surface.

Comets and volcanoes release water that is virtually salt-free, so the ocean's salt must have another

Comets and the ocean FIGURE 8.8

Comets like this one enter the Solar System and undergo partial melting because of heat from the Sun. The comet's tail is mostly composed of water and dust released by melting. Comets frequently enter Earth's upper atmosphere and are a nearly constant source of water raining to the surface. They may have been a major source of water for the world ocean.

source. The most likely source is chemical weathering of rocks, particularly on land and at the shoreline. Rivers began carrying dissolved ions to the oceans early in the Archean Eon, and it is hypothesized that the ocean reached its current salinity well before the end of the eon. Since the Archean, the salinity flux of the world ocean has remained relatively constant. Salts have been added to the ocean by rivers at about the same rate that they have been precipitated from seawater as sedimentary deposits.

CONCEPT CHECK STOP

How was Earth's early atmosphere developed?

What was the gaseous composition of Earth's early atmosphere?

What are the most likely sources of water for

Earth's early ocean? What is thought to be the most important source of water?

Why, if the likely sources of water for the ocean are essentially freshwater sources, is the ocean salty?

Earth's Oldest Rocks

LEARNING OBJECTIVES

Explain how Earth's continental crust became differentiated from oceanic crust.

Explain how continents evolved.

Summarize the evidence for Earth's oldest rocks.

As Earth's final magma ocean was cooling, crustal material could begin differentiating into mafic oceanic crust and felsic continental crust. High temperatures and heat flow within the mantle set up convection currents that helped drive early-formed crustal pieces across the planet's surface. This early tectonic activity caused slabs of dense komatiite to sink into the molten upper mantle, where **partial melting** and further compositional differentiation could occur. As the oceanic crust expanded and the mantle continued to cool, basaltic (mafic) rock replaced ultramafic rock at spreading centers.

There are at least two possible ways that felsic components could be differentiated from mafic rocks. In one hypothesis, subduction and partial remelting would occur, followed by the rise of felsic (granitic) magma toward the surface, where it could solidify as small patches of continental-type crust. One problem with this idea, however, is that in the modern world, volcanic rocks erupted above subducting slabs tend to have an intermediate (andesitic) composition. Usually, not enough felsic components are separated from mafic rocks to produce granitic crust.

Another model for crustal differentiation that could produce a protocontinent is that happening today in Iceland. The island is situated above a hot spot in oceanic crust along the Mid-Atlantic Ridge, and in the lower crust beneath it, felsic (granitic) igneous rock is segregating from the mantle. Because of the mid-ocean ridge passing through it, Iceland is broken by normal faults, and mafic (basaltic) magma flows along the faults. When the mafic magma comes in contact with felsic bodies, the felsic rocks melt and rise to the surface. Volcanoes producing basaltic rock and volcanoes ejecting rocks with a granitic composition are both present on Iceland, although the basaltic volcanoes greatly outnumber the granitic ones.

Another process operating in the Iceland hot spot also accounts for some crustal differentiation. As igneous rock continues to add to Iceland's surface, the island sinks isostatically, and its base undergoes partial melting. The melting removes some felsic components that later resolidify to form felsic rocks. Because Iceland is spreading at a relatively slow rate (about 2 cm per year, on average), it has enough time for crustal areas to undergo multiple episodes of remelting and concentration of felsic components of rocks before they spread beyond the active igneous zone.

EARLY CONTINENTAL CRUST

The cores of continents are amalgamated from large pod-like rock bodies welded along metamorphic zones called **greenstone belts**. The pod-like bodies are mostly high-grade metaigneous rocks rep-resenting the felsic crust of Archean protocontinents, and the greenstones connecting them are metavolcanic and metasedimentary rocks rich in chlorite, a green mineral formed under low-grade metamorphic conditions.

Greenstone belts formed through continental accretion processes during what we refer to as Earth's **mobile crust phase**, which lasted from the early Archean until the beginning of the Proterozoic Eon. The greenstones formed mostly along subduction zones adjacent to felsic protocontinents. Shortening of the

Archean shields FIGURE 8.9

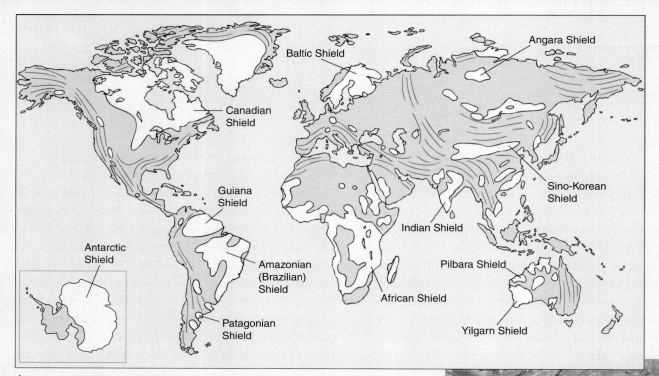

Baltic Shield
Angara Shield
Canadian Shield
Guiana Shield
Sino-Korean Shield
Indian Shield
Antarctic Shield
Amazonian (Brazilian) Shield
Pilbara Shield
African Shield
Patagonian Shield
Yilgarn Shield

▲
A Exposures of Precambrian basement rocks around the world, including the locations of Archean shields, are shown on this map. In areas such as Canada, Scandinavia, and Western Australia, cratonic areas are mostly exposed at the surface. In other areas, such as Brazil, Greenland, and Antarctica, sediments or glacial ice have covered large portions of the cratons.

B This satellite image shows the Pilbara Shield area of Western Australia. The shield consists of large pod-like metaigneous rock bodies welded along greenstone belts. Rocks exposed on this shield date to at least 3.5 billion years.
▶

ocean floor between protocontinental pieces caused collision of those pieces and the squeezing of oceanic sediments that would metamorphose to greenstones. Protocontinental pieces were numerous in the Archean Eon, and so were greenstones, meaning that subduction zones were much more numerous then than they are today.

The Archean cores of continents are present on all continents (**FIGURE 8.9A**). They are well exposed in such shield areas as the Hudson Bay region of Canada (part of the Canadian Shield), northern Sweden and Finland, and the outback of Australia. In most places, though, cratonic areas are covered by Proterozoic and Phanerozoic sedimentary strata.

The oldest-known block of continental crust is in the Acasta Formation, part of the Slave Craton, which is now part of northern Canada. Uranium-lead radiometric dating on zircon crystals provides ages ranging from 3.8 to 4.0 billion years for the Acasta Formation. Another block of continental crust exposed in Western Australia provides radiometric ages in excess of 3.5 billion years (**FIGURE 8.9B**).

Still older evidence of the former existence of continental crust comes from zircon crystals incorporated in sedimentary deposits. Zircons (**FIGURE 8.10**) are often hosted in micas. Micas form in igneous and metamorphic rocks. In order for the zircons to have survived weathering and erosion to become detrital sediment grains, they must have been large, and hosted in large micas. This implies that the zircons crystallized in intrusive igneous rocks or metamorphic rocks. The zircon crystals contain small amounts of uranium and the radioactive uranium allows them to be used for determining the ages of igneous or metamorphic rocks. Zircons are also extremely durable, resistant to both heat and weathering, and so they can survive relatively unscathed through the effects of weathering, erosion, incorporation in detrital sediments, and even some metamorphism. In an application of William ("Strata") Smith's principle of included fragments, zircons in Proterozoic and Phanerozoic sediments often yield radiometric ages that are considerably older than the materials from which they were collected.

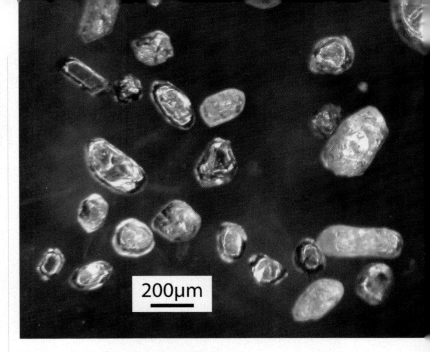

200µm

Zircon crystals FIGURE 8.10

The oldest direct evidence of solid crust comes from uranium-lead dating on detrital zircons. Crystals of zircon, like these, were weathered from igneous rocks and later incorporated in sedimentary deposits.

The oldest detrital zircons come from a slightly metamorphosed conglomerate in Western Australia that yields crystallization ages of 4.1 to 4.4 billion years. This can be considered a minimum age for the solidification of granitic continental crust. Although the conglomerate is much younger than the zircon crystals included within it, these grains unequivocally indicate that continental crust had begun to form within 150 million years of Earth's formation at about 4.56 billion years.

One group of sedimentary rocks, called **banded iron formations**, or BIFs, are almost completely restricted to the Archean and Proterozoic. These deposits consist of interlayered thin beds of chert (cryptocrystalline quartz) and iron minerals (commonly hematite or magnetite). Understanding the origin of BIFs by applying uniformitarianism is not easy because of the lack of good modern analogs. Some BIFs seem to have an origin with submarine volcanic zones, and

Amazing Places: Greenland Greenstone FIGURE 8.11

Southwest Greenland

Global Locator

A On the edge of the Inland Ice in Southwest Greenland sits the oldest-known contiguous block of Archean crust, the Itsaq gneiss complex–Isua greenstone belt. Uranium-lead dating of zircons indicates ages of 3.7 to 3.8 billion years for rocks in the Itsaq–Isua block.

B Greenstone at Qorqut Sound, Isua, Greenland, was formed through metamorphism of igneous and sedimentary rocks along the suture zone between early protocontinental masses.

C A banded iron formation at Isua is one of the oldest sedimentary rocks.

others have an origin with rift basins, both of which are places that would provide a source of dissolved iron and silica. One banded iron formation, from the Itsaq gneiss complex–Isua greenstone belt of Southwest Greenland, was deposited 3.7 to 3.8 billion years ago (FIGURE 8.11).

Radiometric ages recorded in the Itsaq gneiss complex–Isua greenstone belt range up to 3.9 billion years. Given what we know of events in the Solar System about this time, it is unlikely that we will ever find many large contiguous tracts of Archean crust older than 3.9 billion years. From impact craters on the Moon, it is clear that the Earth–Moon system underwent an episode of heavy meteorite bombardment peaking around 3.9 billion years ago. Some of the bolides may have initiated redevelopment of magma ocean conditions at Earth's surface, conditions that would have remelted much of the crust that had formed prior to 3.9 billion years ago.

Between 2.7 and 2.3 billion years ago, many continental areas experienced metamorphic episodes.

These episodes were not exactly simultaneous, and what triggered the metamorphism is unknown. However, this widespread metamorphism resulted in the consolidation of numerous small crustal bodies into larger cratons (continental cores) and reset their radiometric ages (Figure 8.9B).

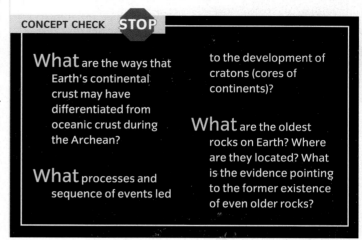

CONCEPT CHECK STOP

What are the ways that Earth's continental crust may have differentiated from oceanic crust during the Archean?

What processes and sequence of events led

to the development of cratons (cores of continents)?

What are the oldest rocks on Earth? Where are they located? What is the evidence pointing to the former existence of even older rocks?

Archean Life Forms

What is the origin of life on Earth? When did it first evolve? Is life restricted to Earth, or could it have arrived from elsewhere in the universe? Could life have developed independently in more than one part of the universe? These are probably the most vexing questions in any of the sciences. The fossil record provides us with a solid but incomplete set of answers, at least from a terrestrial perspective. Laboratory analyses and experiments provide additional insight, and, surprisingly, a little information comes from meteorites. Even still, our understanding of how and when life evolved is incomplete. Whether life exists or has existed elsewhere in the universe is still an open question.

THE EARLIEST EVIDENCE OF LIFE

The earliest fossils are of simple microbes preserved as molds with carbon in Archean chert. Chert, which is cryptocrystalline quartz, solidifies from a gel of silica. As the gel solidifies, it inhibits extensive decay of the organisms that get caught in it, much like the way tree sap inhibits the decay of organisms caught in it. The oldest fossils do not look like much—simple, tiny, and round or ovoid black cells or chains of such cells arranged like a string of pearls. The earliest fossils are presumably prokaryotic eubacteria, so there are no nuclei or other organelles preserved within the cells.

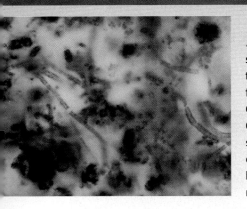

A This photomicrograph shows strands of cells, inferred to be cyanobacterial fossils, from the 2 billion-year-old Gunflint Chert of Ontario, Canada. These tiny Proterozoic specimens are among the oldest definitive body fossils, but similar material has been reported from Archean strata.

B A stromatolite of Archean age, constructed by consortia of cyanobacteria, from Western Australia, is shown above.

cyanobacteria
Blue-green eubacteria, most of which are photosynthetic.

photosynthesis
The process by which plants, algae, and some bacteria create organic molecules from carbon dioxide and water, using energy from sunlight.

autotrophy "Self-feeding" by means of either harvesting light energy from the Sun or from oxidation of inorganic compounds to make organic molecules.

heterotrophy
A means of obtaining nutrients by ingesting or breaking down organic matter.

stromatolite
A thinly layered biogenic-sedimentary structure resulting from the trapping and binding of fine sediment in layers by photosynthetic cyanobacteria.

How can we be certain that simple spheres or other shapes in Archean rocks are fossils? Our best evidence for life derives from one of the most important functions of a living organism—the capability to reproduce. Cells preserved either in the process of undergoing cell division or after having formed long chains of identical cells point to an organic origin. Carbonized material from 3.4 billion-year-old strata of South Africa and Western Australia shows what some scientists have interpreted as organic material—tiny cells with indications of asexual reproduction. Cells arranged in filamentous chains from 2 billion-year-old (Proterozoic) strata of Canada (FIGURE 8.12A) are even more convincing as genuine fossils because they closely resemble the filamentous strands of **cyanobacteria** (or blue-green bacteria).

Another indicator of the probable organic origin of small, simple Archean "fossils" is their chemistry. If their composition matches that of modern microbes, then it is possible that they are genuine fossils. There is controversy about this point, though, because some of the compounds found in simple Archean "fossils" can be synthesized abiotically.

The evidence for cyanobacteria is especially important because these microbes use **photosynthesis** to convert carbon dioxide and energy from the Sun into sugar and oxygen. Cyanobacteria, which are **autotrophic**, can thrive under low-oxygen conditions like those of Earth's early ocean–atmosphere system, but in producing oxygen as a by-product of their metabolic activities, they were paving the way for more advanced multicellular eukaryotes. Billions of years later, these **heterotrophic** organisms would respire in the water or open air.

Consortia of cyanobacteria are responsible for the buildup of **stromatolites** (FIGURE 8.12B) and thrombolites, especially on the seafloors of warm, shallow shelves where carbonate sediments accumulate. These thinly layered features are persistent in Archean and Proterozoic sedimentary environments. The oldest of them is from the Fig Tree Group of South Africa, where they have been dated at about 3.2 billion years. Stromatolites and thrombolites increase in abundance toward the upper part of the Archean and into the Proterozoic, reflecting an expansion of the shallow marine environments where cyanobacteria grew.

As we seek evidence of Earth's earliest life forms, it may be necessary to look beyond sedimentary strata, and into places where volcanic rocks formed. Small tubular, wormlike, granular, or coalesced spherical borings have been reported from 3.5 Ga pillow basalts of South Africa. These structures, however, now seem to be the work of modern contaminants, not Archean borers.

Indirect evidence that prokaryotic life forms existed on Earth even before the time of our earliest-known body fossils comes from carbon isotopic ratios in strata. Photosynthetic activity causes isotopically light carbon (carbon-12) to be preferentially removed from the air or water. If the photosynthesizers are buried in sediments, the strata should have a relatively high proportion of carbon-12. Metasedimentary rocks of the Isua greenstone belt of Southwest Greenland show just such a shift toward isotopically light values.

ABIOTIC SYNTHESIS OF AMINO ACIDS AND THE EARLIEST LIFE FORMS

Amino acids, composed of carbon, hydrogen, oxygen, and nitrogen, are the chemical building blocks of life. Twenty amino acids are used to build proteins, which are the organic compounds required for life forms to replicate by themselves and to control their internal chemical reactions. To understand how life first developed, we need to consider how amino acids form without biologic precursors. To gain insight into the possibility that life exists beyond Earth, it is also necessary to understand how amino acids can be present on meteorites that have fallen to Earth.

In 1953, Stanley Miller and Harold Urey published results of a simple laboratory experiment demonstrating that amino acids could be created under abiotic conditions without the availability of free oxygen (FIGURE 8.13). Their experiment was de-

Abiotic synthesis of amino acids FIGURE 8.13

A Stanley Miller and Harold Urey were the first to experimentally produce amino acids in a laboratory. Here, Miller stands next to the experimental apparatus used to abiotically synthesize amino acids.

B This is the setup of the Miller-Urey experiment. It is a closed system that links a primitive "ocean" to a primitive "atmosphere." The "ocean" water was boiled, causing evaporation of gases into the "atmosphere." Atmospheric gases consisted of hydrogen (H_2), water vapor (H_2O), ammonia (NH_3), and methane (CH_4). Electrical discharges, simulating lightning in the atmosphere, triggered chemical reactions and reorganization of the carbon-, hydrogen-, oxygen-, and nitrogen-bearing compounds. After cooling and condensation of the atmospheric gases, a variety of amino acids had accumulated in the simulated "ocean."

Geologists now think that Earth's primitive atmosphere was different from that simulated in the Miller-Urey experiment. Most notably, free oxygen (O_2) was probably available. A convincing process for synthesizing organic molecules under these circumstances has not yet been developed.

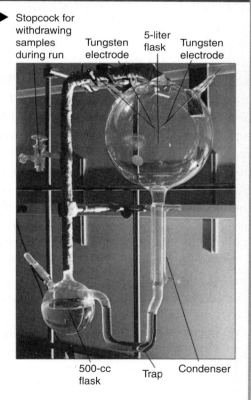

Stopcock for withdrawing samples during run

5-liter flask

Tungsten electrode

Tungsten electrode

500-cc flask

Trap

Condenser

signed to imitate the conditions under which life was thought, in the early 1950s, to have arisen on Earth more than 3 billion years ago.

Miller and Urey successfully demonstrated that the basic compounds needed to build life forms could be created abiotically from condensation of the types of gases expelled to the atmosphere during planetary outgassing. Although they assumed a complete lack of free oxygen in the atmosphere (a condition now thought unlikely), they did, nevertheless produce amino acids. The anoxic conditions they used for the experiment may be similar to those under which amino acids developed on asteroids (which fragmented into meteorites) or on comets. Since the early 1950s, various researchers have run experiments that differ in certain details from the original Miller-Urey experiment. Different energy sources and different "atmospheres" have been used. In most cases, the results are the same mix of prebiotic organic building blocks.

The step from amino acids to organized nucleic acids capable of coding for proteins has not been bridged in laboratory experiments. The prevailing view at present, however, is that in the earliest life forms, only RNA, rather than DNA, carried genetic information. This hypothesized early phase of life history has been referred to as an "**RNA world**." Even in modern organisms, RNA can replicate itself and can act as a catalyst for the production of certain proteins by helping to transmit genetic information. Eventually, though, DNA evolved (presumably through mutation of RNA) and superseded RNA as the molecule carrying the genetic code of organisms.

In hindsight, we now recognize a flaw of sorts in the Miller-Urey experimental setup—one that actually gives us another clue to how life may have first arisen on Earth. The experiment was run under anoxic conditions. Even a small amount of free oxygen would have oxidized and destroyed the organic compounds crucial for constructing life forms. Today, most free oxygen in Earth's atmosphere is produced through photosynthesis. However, a small proportion of it is produced by **photochemical dissociation** when ultraviolet light rays

■ **photochemical dissociation** The splitting of molecules into their components by means of energy from sunlight or other light sources.

■ **hydrothermal vent** An opening in Earth's crust, usually associated with magmatic activity, where hot water, often enriched in ions, is released.

■ **archaebacteria** Organisms belonging to the domain Archaea, including the methanogenic, halophilic, and thermoacidophilic prokaryotes.

from the Sun split molecules of water vapor, liberating oxygen from hydrogen. Even though the amount of free oxygen produced by this process may have been small, it is plausible that there was enough concentrated to oxidize fundamental organic compounds.

In order for life to have arisen abiotically, it must have first developed under anoxic, aqueous conditions. It is possible (although unlikely) that life arrived to this planet aboard a meteorite or a comet. The more likely option is that life arose on Earth. It is possible that life arose on Earth more than once, and it could have arisen in more than one type of environment. If life did in fact arise on Earth, it must have evolved in a place where contact with free oxygen of the atmosphere was limited. This tends to rule out small isolated ponds, lakes, and a range of marginal-marine settings, places that in earlier years were considered potential sites for the growth of a **primordial soup**, or a concentrated assemblage of organic compounds, including amino acids.

One anoxic environment where amino acids could organize into more complex genetic and organic materials is along mid-ocean ridges. Earth's early crust was probably rich in spreading centers and subduction zones. The spreading centers, especially **hydrothermal vents** along mid-ocean ridges, may have been the areas of origin for Earth's earliest life forms. Here, hot magma wells up from the mantle and spews out along faults and along the central valley of the ridge complex. Ocean water seeps into cracks in the rock and is heated where it makes contact with warm crust or magma. The hot water, often in excess of 100°C, readily dissolves ions while passing through the ridge system.

Remarkably, some **archaebacteria** (FIGURE 8.14) not only survive under the extreme conditions that exist along mid-ocean ridges but require them for their metabolism. Archaebacteria make their livings in various ways, but many require chemicals such as iron, phosphorus, zinc, and nickel dissolved from the rocks along the ridge system. They obtain energy through **chemosynthesis**, in which naturally occurring chemical

Life may have first evolved along mid-ocean ridges where anoxic conditions and newly formed igneous rock, rich in iron and other ions, provided an ideal environment for the growth of chemosynthetic archaebacteria. Modern hydrothermal vents along mid-ocean ridges provide an analog for the places where life may have first sprouted during the Archean.

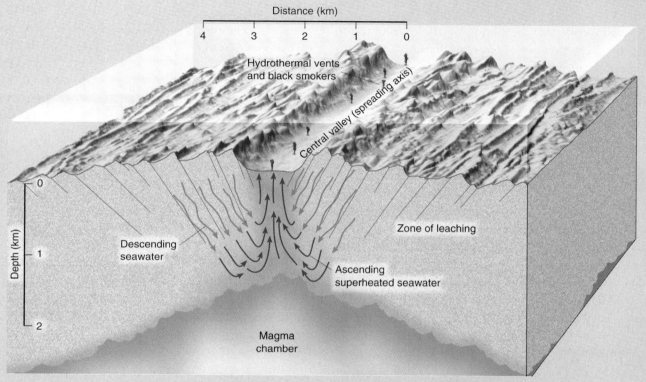

A The central valley of a mid-ocean ridge is a tectonically active setting where pillow basalts are erupted to form new rock of the ocean floor. Ocean water seeps into cracks along the ridge and helps to leach ions from minerals in the newly formed rocks. Ions of iron, magnesium, and other elements in solution will become nutrients for chemosynthetic archaebacteria that live around the margins of hydrothermal vents and black smokers.

B Archaebacteria thrive in hot waters rich in dissolved ions adjacent to hydrothermal vents. This vent is on the East Pacific Rise.

C These are thermoacidophile archaebacteria in highly magnified view. These modern chemosynthetic organisms may be little changed from their ancestors of the Archean Eon.

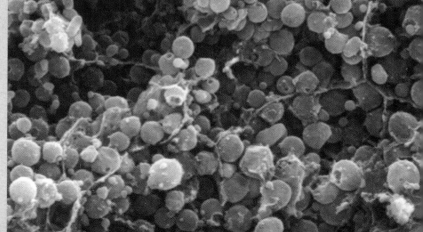

reactions are allowed to occur within their cells. Some archaebacteria (**thermoacidophiles**) combine hydrogen and sulfur to form hydrogen sulfide (H_2S), and others (**methanogens**) combine hydrogen and carbon dioxide to produce methane (CH_4).

So far, we have no direct fossil record of archaebacteria from early Archean spreading centers. Reasons for the lack of direct evidence include the low probability of preserving tiny prokaryotic body fossils at hydrothermal vents and problems of their recognition even if they were preserved. However, putative archeans have been identified from sedimentary strata associated with much younger hydrothermal vents, and it is conceivable that verifiable archeans will someday be discovered in Archean rocks.

DOES LIFE EXIST BEYOND EARTH?

Whether life exists elsewhere in the Solar System and whether the early Earth may have been "seeded" with life forms arriving from beyond our planet are questions that remain speculative. Reports of fossils in meteorites of Martian origin have so far failed under scrutiny. Although Mars shows topographic features that were, in the deep past, almost certainly produced under the influence of liquid water (a basic necessity for life as we know it), convincing evidence for present or ancient Martian life has yet to be discovered.

In the search and test for fossilized life forms from elsewhere in our Solar System or elsewhere in the universe, we can use the same criteria for meteorites or rocks collected directly from extraterrestrial bodies as we use in the study of Earth's earliest life forms. They include evidence of reproduction (especially asexual division of cells), carbonaceous chemistry, and ratios of carbon or other chemical isotopes that suggest biologic fractionation.

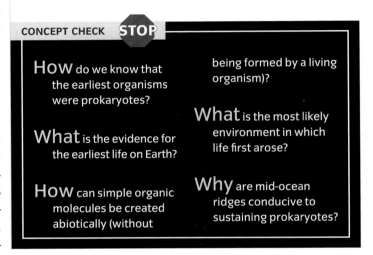

CONCEPT CHECK STOP

HOW do we know that the earliest organisms were prokaryotes?

What is the evidence for the earliest life on Earth?

HOW can simple organic molecules be created abiotically (without being formed by a living organism)?

What is the most likely environment in which life first arose?

Why are mid-ocean ridges conducive to sustaining prokaryotes?

SUMMARY

1 Overview of Archean Time

1. The **Archean Eon** began with Earth's formation approximately 4.56 billion years ago, and it ended 2.5 billion years ago.

2. Archean history is recorded mostly in rocks of the continental interiors, or **cratons**. **Archean shields**, cratonic areas of low topographic relief, are the major places where Archean rocks crop out at the surface.

2 Origin of Earth and the Solar System

1. The age of Earth is estimated to be greater than 4.5 billion years from radiometric dating of crystals in **meteorites** and Moon rocks. The oldest-known rocks on Earth are substantially younger than this because of tectonism and because of weathering and erosion of rocks.

2. The age of the universe is estimated at 12 to 18 billion years, and the Milky Way nebula is estimated at less than 10 billion years. Bodies of the Solar System coalesced around 5 billion years ago from materials captured in a solar nebula by the gravitational field of the Sun.

3. Early in its history, Earth was completely molten, and its rotation helped to differentiate it, causing denser metals to sink to the core and lighter silicates to rise to the mantle and crust.

4. The Moon formed after a bolide struck Earth, sending mantle fragments into orbit around the planet, where they could coalesce into the Moon.

3 Origin of the Atmosphere and Ocean

1. The early atmosphere formed through **outgassing** (the escape of gases from Earth's interior) after Earth had coalesced and could hold gases in its gravitational field.

2. The early atmosphere was different from the present atmosphere and may have been relatively rich in hydrogen, hydrogen chloride, carbon dioxide, carbon monoxide, and nitrogen.

3. The early ocean probably formed from the combination of water derived from the melting of comets entering Earth's atmosphere and from condensation of water vapor resulting from outgassing.

4. Weathering of rocks exposed at the surface is the source of most salt in the ocean.

4 Earth's Oldest Rocks

1. Cooling of a **magma ocean** at Earth's surface allowed crustal differentiation to begin. Early tectonic activity moved crustal pieces about the surface of the planet, and slabs of dense ultramafic rock sank into the molten upper mantle, where partial melting and compositional differentiation occurred. As the oceanic crust expanded, mafic rock replaced ultramafic rock at spreading centers. Differentiation of felsic rocks may have occurred in association with partial melting of descending slabs at subduction zones or associated with plumes of felsic composition rising through mafic rocks at hot spots.

2. During the Archean, early-formed blocks of felsic crust (protocontinents) were consolidated through suturing along **greenstone belts**.

3. The oldest-known rocks on Earth yield radiometric ages of 3.8 to 4.0 billion years. Detrital zircon crystals weathered from preexisting igneous or metamorphic rocks, however, indicate that crustal rock had formed by 4.1 to 4.4 billion years ago.

4. Consolidation of numerous small crustal bodies into larger continental cores (cratons) occurred mostly between 2.7 and 2.3 billion years ago.

5 Archean Life Forms

1. The earliest-known fossils are of **prokaryotes**. Cells of eubacteria dating to 3.4 billion years are tiny, and they lack nuclei and other organelles. Some cells of probable cyanobacteria are arranged in filamentous strands.

2. Stromatolites and thrombolites constructed by cyanobacteria date to 3.2 billion years.

3. Biomarker evidence (in the form of carbon isotopic ratios) suggests that **photosynthetic** organisms existed as early as 3.4 billion years ago.

4. Amino acids have been lab-synthesized abiotically under anoxic and low oxygen conditions from hydrogen, water, ammonia, and methane.

5. The earliest life forms may have arisen in association with **hydrothermal vents** along mid-ocean ridges. They probably metabolized using energy produced through chemosynthesis.

6. Earth's early fossil record provides a model for evaluating evidence of possible extraterrestrial life forms.

KEY TERMS

- **Archean Eon** p. 229
- **Proterozoic Eon** p. 229
- **craton** p. 230
- **Canadian Shield** p. 230
- **meteorite** p. 231
- **asteroid** p. 232
- **magma ocean** p. 236

- **outgassing** p. 238
- **greenstone belt** p. 240
- **mobile crust phase** p. 241
- **banded iron formation** p. 241
- **cyanobacteria** p. 244
- **photosynthesis** p. 244
- **autotrophy** p. 244

- **heterotrophy** p. 244
- **stromatolite** p. 244
- **photochemical dissociation** p. 246
- **hydrothermal vent** p. 246
- **archaebacteria** p. 246

CRITICAL AND CREATIVE THINKING QUESTIONS

1. How can the compositions of meteorites inform us about the composition and layering of Earth's interior?

2. Why does the surface of the Moon seem to have so many more impact craters than the surface of Earth? What does this say about the ages of most craters?

3. In one hypothesis, early life is thought to have arrived to Earth from a distant place aboard an asteroid or a comet. What evidence makes this idea seem plausible? What evidence makes this idea seem unlikely?

4. In early prokaryotes, what would have been the means of reproduction? Do you expect the diversity of early prokaryotes to have been great? How might new species have arisen from preexisting species?

5. The earliest terrestrial life forms may have lived along mid-ocean ridges. Where would you expect to find microbial life forms on other planets?

What is happening in this picture ?

Under parts of the frozen Antarctic landscape, tectonics are still active. In the distance, Mt. Erebus releases gases from a vent along the side of the volcano.

- What do you think is the composition of the emitted gases?

- Do you think the composition of gases emitted from a modern volcano is a reasonable analog for the gaseous emissions during Earth's early outgassing phase?

SELF-TEST

1. What two eons comprise the Precambrian?

2. How much of geologic time does the Archean Eon occupy?

3. What are the major challenges to correlation in Archean and Proterozoic rocks?

4. What are cratons, and how did they form?

5. Why do meteorites and Moon rocks yield older isotopic ages than the oldest-known terrestrial rocks?

6. Why does the Moon's surface seem to have more meteorite impact craters than Earth's surface?

7. By what process was Earth differentiated into a core, mantle, and crust?

8. What are the primary constituents of the core, mantle, and crust?

9. What are greenstone belts, and what do they tell us about Archean tectonic processes?

10. How was crustal rock able to differentiate into more mafic (oceanic-type) and more felsic (continental-type) endmembers?

11. What are the likely sources of water for the world ocean?

12. What is a banded iron formation?

13. How was Earth's early atmosphere different in composition from the way it is today?

14. What are the likely sources of free oxygen in the atmosphere?

15. What are the earliest-known life forms? Are they prokaryotes or eukaryotes?

16. Why is it more likely that life first evolved along mid-ocean ridges than in "primordial soup" environments along the margins of oceans, lakes, or ponds?

17. What are the prokaryotic life forms most likely to have evolved near hydrothermal vents?

Earth's surface became fully solidified within the first billion years of its history. By about 2.5 billion years ago, much of the framework for Earth as we know it today—a planet with distinct internal zonation, an outer skin fractured into a mosaic of mobile plates, well-defined continental and oceanic crustal areas, and mostly covered by water—had emerged from the chaotic events of the Archean. This turning point in Earth's history marks the beginning of a new eon, the Proterozoic.

Encompassing the time interval from 2.5 billion years ago to 542 million years ago, the Proterozoic Eon represents approximately 43% of geologic time. In contrast to the Archean, whose story written in the rocks is split up, jumbled, and in many places overwritten by later events, the Proterozoic shows a much more orderly pattern of stratigraphic "pages." Sprawling cratonic areas covered by shallow seas became sites of sedimentary deposition, and mountain-building events were much like those of the younger Phanerozoic. The Proterozoic witnessed Earth's first glacial episodes and the transition to an oxygen-rich atmosphere–ocean system. Microbes played a central role in oxygenating the atmosphere, a critical prerequisite for the evolution of animals.

The appearance of complex biologic organisms, such as the *Dickinsonia* fossils shown here, led to a fascinating new subplot in Earth's story, one that would ultimately dominate the story. In less than 2 billion years, life on Earth made the transition from minute cells—cryptic except for the stromatolitic structures some cells constructed—to some of the most obvious features of sedimentary strata.

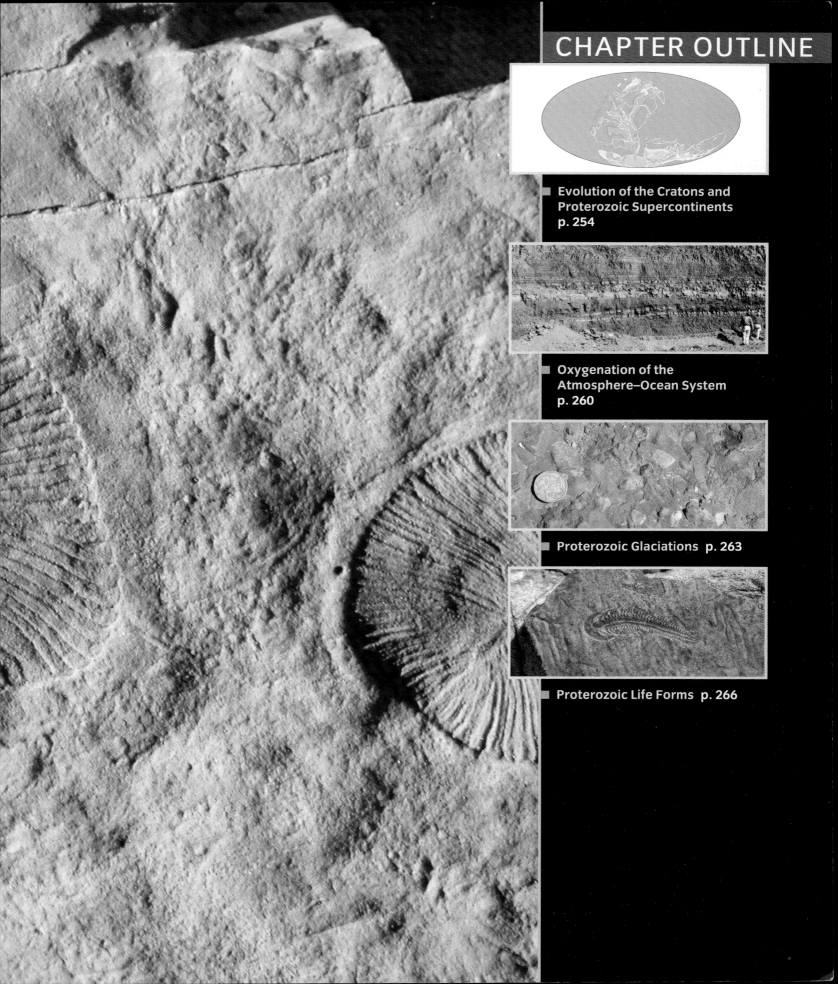

Evolution of the Cratons and Proterozoic Supercontinents

LEARNING OBJECTIVES

Contrast the nature of the crust in the Proterozoic Eon with that of the Archean Eon.

Explain what occurs in a supercontinent cycle.

Summarize how supercontinents were assembled and dispersed during the Proterozoic.

Explain the formation of Paleozoic continents from Neoproterozoic supercontinents.

A point 2.5 billion years ago was chosen as the beginning of the Proterozoic Eon because it approximates the time when Earth entered a new phase of its history. During the Archean, Earth experienced its mobile crust phase.

Between 2.7 and 2.3 billion years ago, most small crustal bodies were amalgamated through orogenesis into larger cratons, which are the continental cores (**FIGURE 9.1**). Beginning in the Paleoproterozoic Era, these areas became **platforms** for subsequent sedimentary

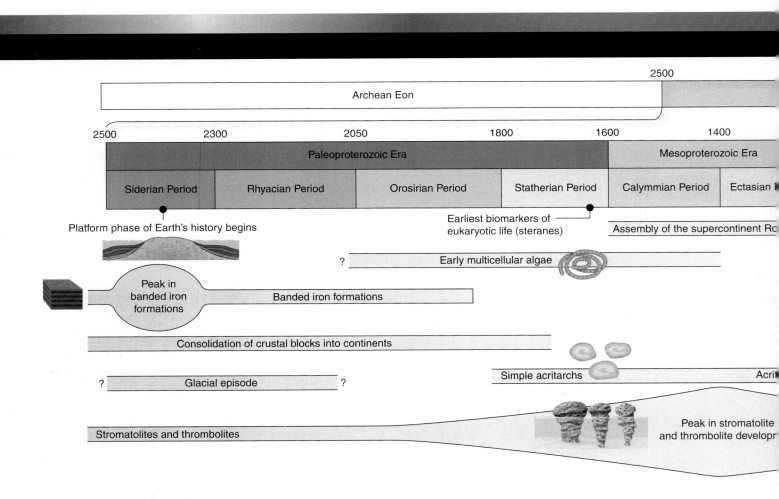

Archean Eon

2500

| 2500 | 2300 | 2050 | 1800 | 1600 | 1400 |

Paleoproterozoic Era | **Mesoproterozoic Era**

| Siderian Period | Rhyacian Period | Orosirian Period | Statherian Period | Calymmian Period | Ectasian |

Platform phase of Earth's history begins

Earliest biomarkers of eukaryotic life (steranes)

Assembly of the supercontinent Ro

? Early multicellular algae

Peak in banded iron formations

Banded iron formations

Consolidation of crustal blocks into continents

Simple acritarchs

Acri

? Glacial episode ?

Peak in stromatolite and thrombolite developr

Stromatolites and thrombolites

platform The part of a continent covered by flat-lying or gently tilted, mostly sedimentary, strata.

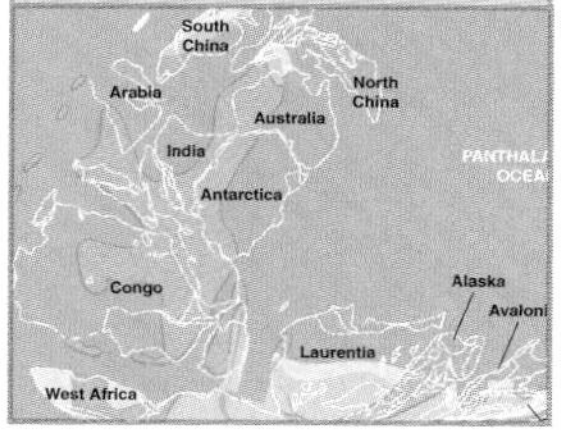

deposition, and Earth's surface entered a new stage, the **platform phase**. From this time forward, cratonic areas did not expand extensively. Instead, the assembled continental cores experienced docking of other crustal pieces and accretion at the margins, suturing during collisional events, separation along rifts, and, of course, sedimentary deposition. Initiation of the platform phase was possible only after the crust was clearly differentiated, and it marks the transition to operation of tectonic processes in a fully modern way. Stabilization of continental platforms provided a place for weathering, erosion, and the orderly accumulation of sedimentary layers (or the "pages" of Earth's history book) that are now so familiar to us.

platform phase An interval of Earth history, beginning with the Proterozoic Eon, characterized by relatively stable, amalgamated continental cores.

Visualizing

Proterozoic timeline FIGURE 9.1

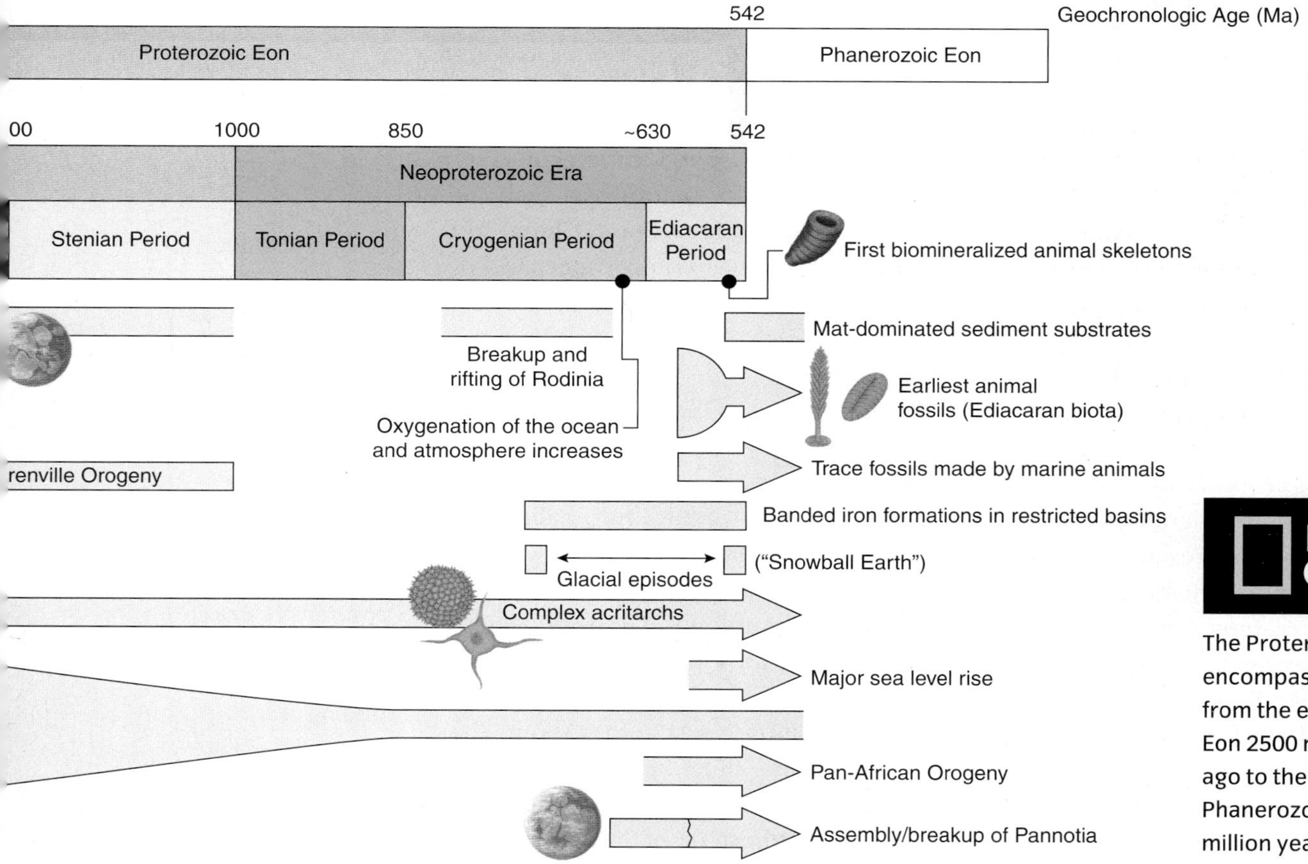

NATIONAL GEOGRAPHIC

The Proterozoic Eon encompasses all time on Earth from the end of the Archean Eon 2500 million years (Ma) ago to the beginning of the Phanerozoic Eon 542 million years ago.

The Canadian Shield, the craton of North America, illustrates how continents are formed of amalgamated protocontinental bodies. The individual protocontinental bodies solidified at different points during the Archean Eon and were later assembled along orogenic belts. Subsequent changes to the large North American, or Laurentian, craton include sedimentary deposition on the platform areas (beyond the shield) and rifting and accretion of various terranes around the margins.

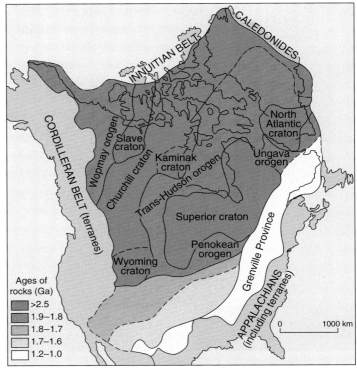

Ages of rocks (Ga)
- >2.5
- 1.9–1.8
- 1.8–1.7
- 1.7–1.6
- 1.2–1.0

0   1000 km

PROTEROZOIC SUPERCONTINENTS

Rodinia An early supercontinent, assembled in the Mesoproterozoic and separated in the Neoproterozoic.

Similarities in the style and timing of tectonic deformation across some of the world's major cratonic areas indicate that during the Mesoproterozoic, many of the cratonic platforms (FIGURE 9.2) had become sutured into a supercontinent. This supercontinent, called Rodinia (FIGURE 9.3), was considerably different in its configuration from Pangea, which would become assembled more than a billion years later.

Rodinia had at its center the North American–Greenland craton, which is called **Laurentia**. A central position is inferred by a series of faults, which are evidence of rifted margins developed during later breakup of Rodinia, ringing the craton. Attached to the Cordilleran margin (the present-day western margin) between 1.6 and 1.0 billion years ago were the **Australian** and **East Antarctic cratons**. The exact configuration is disputed, but similarities in Mesoproterozoic orogenic belts imply that Australia was connected to the western United States, and East Antarctica was connected along the southern Australian margin. **Baltica**, which is the Scandinavian craton, was linked to the eastern margin of Greenland. **Siberia**, another cratonic block, was at-

tached to the Innuitian margin (the northern margin of Laurentia). The locations of other cratonic platforms, such as **North China**, **South China**, and **Avalonia** (comprising parts of the British Isles and eastern North America), are less certain.

One important episode in the assembly of Rodinia was the development of an extensive **orogenic belt**, called the **Grenville orogenic belt**, through many parts of the supercontinent. Grenville tectonism took place from Labrador to Mexico between 1.3 and 1.0 billion years ago. About the same time, tectonism snaked through southern Australia, East Antarctica, and the southern tip of Scandinavia (Baltica). This distinctive oro-

orogenic belt
A linear or arcuate region subjected to folding and other deformation during a mountain-building cycle. Also known as an orogen.

Grenville orogenic belt An arcuate orogenic region that developed 1.3 to 1.0 billion years ago and that affected an extensive area of present-day North America and adjacent regions.

genic event is regarded as a **piercing point**, or a geologic feature used to link formerly adjacent parts of the same continental block. Grenville tectonism is one of the key lines of evidence for reconstructing the main crustal components of Rodinia.

The Grenville episode marked the end of a nearly 1-billion-year history of intrusive and extrusive magmatism, metamorphism, and volcanic arc accretion along the eastern and southern margins of Laurentia. One large within-plate magmatic body today forms the Adirondack Mountains of New York. Toward the end of the Grenville episode, two cratonic areas (**Amazonia**, or part of South America, and **Kalahari**, part of Africa) apparently collided along the southeastern margin of what is now the United States.

Proterozoic supercontinent cycles FIGURE 9.3

A supercontinent cycle begins with the collision and welding of tectonic plates to form an enormous mass of continental crust. It ends with the breakup and dispersal of fragments of the supercontinent. The Proterozoic Eon witnessed at least two supercontinent cycles.

A Rodinia was perhaps the first global supercontinent. Assembled in the Paleoproterozoic Era, between 1.6 and 1.0 billion years ago, it remained intact through the Mesoproterozoic Era and underwent breakup in the Neoproterozoic Era, around 800 million years ago, or a little later.

Final assembly of Rodinia was brought about by events related to the Grenville orogeny. Grenville tectonism affected an extensive band across Baltica, Laurentia, Australia, and East Antarctica.

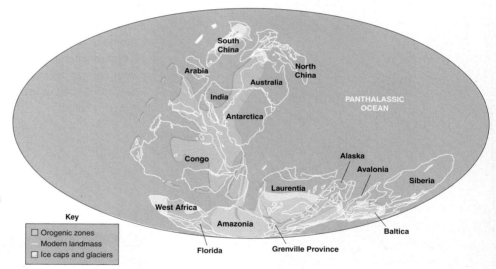

Key
- Orogenic zones
- Modern landmass
- Ice caps and glaciers

B Following the end of the Rodinian supercontinent cycle, another hypothesized supercontinent, Pannotia, is thought to have developed. Pannotia lasted from about 600 to 550 million years ago. Whether crustal blocks were fully assembled into a supercontinent or just lying in close geographic proximity during the Ediacaran Period is uncertain. Continental fragments dispersed from Pannotia became the continents of the early Paleozoic Era.

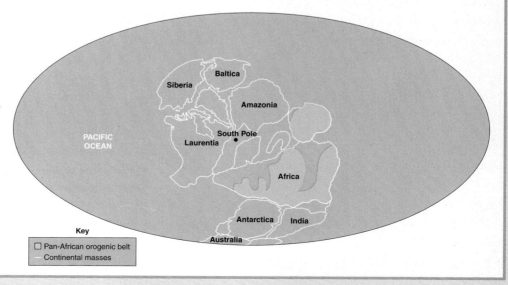

Key
- Pan-African orogenic belt
- Continental masses

Just a couple hundred million years before assembly of Rodinia was completed, the Laurentian craton was nearly ripped down through the present-day U.S. Midwest by the development of an extensive rift zone, the East Continent–Midcontinent rift system (**FIGURE 9.4**). Between 1.3 and 1.0 billion years ago, this rift system opened a large, arcuate gash in Laurentia extending from Kansas northeastward to the upper Great Lakes region, then turning to the south, passing through Michigan and western Ohio, to Kentucky. In the Lake Superior region, basaltic lavas surged into the rift basin, hardening to form rocks of the Keweenawan Group. The volcanic vents also released copper, and today the Ke-

weenawan Group in Michigan remains one of the world's most prolific sources of native copper. Similar basaltic eruptions occurred in other places along the rift system, but they are buried in the subsurface; they can be detected using magnetic anomalies (which record a contrast between the iron-bearing basalts of the rifts and iron-poor granitic rocks of the craton) and drill cores.

Associated with the East Continent–Midcontinent rift system are siliciclastic rocks, including conglomeratic redbeds. The conglomerates formed on alluvial fans as tensional forces ripped the craton apart. Where normal faulting initiated the formation of troughs, streams flowed from narrow canyons along the step-like rift margin and

A Proterozoic rift FIGURE 9.4

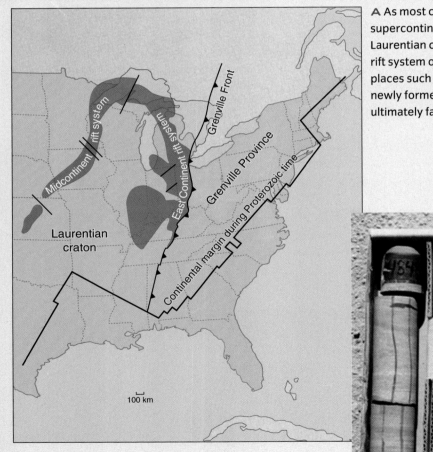

100 km

A As most cratonic pieces were converging into the supercontinent Rodinia, rifting threatened to tear apart the Laurentian craton. The extensive East Continent-Midcontinent rift system opened up between 1.3 and 1.0 billion years ago. In places such as the Lake Superior region, basaltic lavas filled newly formed rift valleys. For unknown reasons, the rifting ultimately failed.

B Part of the sedimentary fill of the East Continent rift system. Siliciclastic sediments that filled the rift valley are now buried deep in the subsurface. This is part of a core through siltstones and sandstones of the basin fill in what is now Ohio.

episodically deposited their loads in the basins.

For unknown reasons, rifting along the East Continent–Midcontinent system ended about 1.0 billion years ago. The faults, however, have continued to undergo readjustment even into modern times. Some of the most active earthquake zones in North America follow the pattern of the rift system.

Rodinia fragmented in the middle to late Neoproterozoic Era. Rift basins along the western margin of Laurentia and eastern Australia point to separation of this region in the Cryogenian Period (800 to 700 million years ago). This split between Laurentia and Australia is significant because it opened up the Pacific Ocean, a geographic feature that has remained to the present.

We can think of the assembly and later breakup of a supercontinent as a **supercontinent cycle**. The suturing of cratonic blocks to form Rodinia and eventual rifting of its pieces, 800 to 700 million years ago, represents one supercontinent cycle.

Conflicting ideas have been put forward about the fate of crustal pieces split from Rodinia, but in one view, fragmentation marked the transition from the first supercontinent cycle to a second Proterozoic supercontinent cycle. As Rodinia divided along the Laurentian–Australian rift margin, a block composed of Australia and East Antarctica rifted away. Also, between about 700 and 500 million years ago, many small crustal pieces were sutured together to form **Africa** in a series of events called the **Pan-African orogeny**.

During the Cryogenian–Ediacaran interval, the Pacific Ocean continued growing. At some point, the leading edges of Laurentia and eastern Gondwana may have collided with opposite sides of the newly assembled African craton, forming another Proterozoic supercontinent. This hypothesized supercontinent arrangement, which existed from about 600 to 550 million years ago, has been called **Pannotia** (FIGURE 9.3B on page 257). It is unclear whether Laurentia and Baltica had broken away from the cluster of cratonic fragments that were assembling prior to full suturing of the African craton. If Africa had not been fully assembled by the end of the Ediacaran Period, when

supercontinent cycle A tectonically driven cycle defined by the assembly of a supercontinent and later fragmentation and dispersal of its pieces.

Pannotia
A hypothesized late Neoproterozoic supercontinent.

Gondwana The Paleozoic to mid-Mesozoic landmass that included South America, the Falkland Islands, Africa, Madagascar, India, Australia, and Antarctica.

Laurentia and Baltica rifted away, Pannotia may have never really been a supercontinent but just a cluster of cratonic blocks lying in close proximity to one another.

As the Ediacaran Period (and the Proterozoic Eon) came to a close, Pannotia underwent separation. Laurentia, Baltica, and Avalonia rifted from the recently assembled African craton. Siberia had already started separating from the northern margin of Laurentia; that happened about 600 million years ago. Now, with Africa assembled, and Laurentia, Baltica, and Avalonia parting ways with Africa, about half of the world's continental landmass was assembled in one large continent, **Gondwana** (comprising most of the present-day Southern Hemisphere continents). Gondwana would remain united through the entire Paleozoic and into the early part of the Mesozoic. The separation of Pannotia represents the transition to the most recent supercontinent cycle, one that broadly corresponds in time to the Phanerozoic Eon.

Continental breakup stimulated a substantial rise in global sea level during the Neoproterozoic, as high rifting rates produced a large volume of warm, relatively expanded oceanic lithosphere. The eustatic rise lasted into the Cambrian Period.

CONCEPT CHECK STOP

How had the crust of Earth changed from the mobile crust phase of the Archean Eon to the platform phase of the Proterozoic Eon?

What is a supercontinent cycle?

What tectonic events led to the assembly of Rodinia?

What tectonic events led to the formation of Pannotia?

What continental blocks were formed following supercontinental breakup in the Neoproterozoic?

Oxygenation of the Atmosphere–Ocean System

LEARNING OBJECTIVES

Understand the ways that free oxygen can be released to the atmosphere.

Explain banded iron formations and what they signify about oxygen levels in the atmosphere and ocean.

Explain redbeds and what they tell us about atmospheric oxygen levels.

Since early in Earth's history, free oxygen (O_2) has been released in small amounts from the breakdown of water vapor in the upper atmosphere by the Sun's ultraviolet radiation. By about 3.5 billion years ago, photosynthetic prokaryotes (especially cyanobacteria) also began releasing oxygen to the atmosphere–ocean system. The production of oxygen by photosynthesizers greatly outweighed its production by chemical dissociation in the atmosphere. Stromatolites and thrombolites became much more numerous in the Proterozoic than they were in the Archean. This is an indication that oxygen reached increasingly higher levels in the atmosphere and ocean during this time.

Another indication that oxygen in the atmosphere increased from negligible levels (perhaps 1% to 2% of present levels) in the Archean to much higher levels in the Proterozoic is the occurrence of certain minerals in strata. Pyrite (or "fool's gold"), a common mineral composed of iron sulfide (FeS_2) in sedimentary rocks, forms under low-oxygen conditions, and when exposed to oxygen, it readily disintegrates (or "rusts") to iron oxide. Pyrite is rare in surface sediments younger than 2.3 billion years. Pyrite, rather than iron oxide, is present in shallow marine and nonmarine sediments of the Archean and earliest Proterozoic, however. This suggests that there was little free oxygen in the atmosphere or ocean before 2.3 billion years ago.

Banded iron formations (BIFs), which are composed of iron minerals interlayered with silica (**FIGURE 9.5**), offer further evidence of oceanic oxygen levels. Most banded irons were deposited between 3.6 and 1.9 billion years ago. A few Neoproterozoic and Phanerozoic examples exist, but they tend to be small in size compared to Archean and Paleoproterozoic examples. Banded iron formations serve as the world's major sources of iron ore.

Conditions leading to deposition of banded iron formations are controversial, and they probably did not all form in exactly the same type of environment. Some BIFs are associated with turbidites and apparently formed in deep water, whereas others are associated with glacial tillites and seem to have formed in shallow water. One common theme among BIFs is a relationship with igneous activity. In shallow deposits, vents associated with rift valleys emitted hot fluids that were the likely source of iron and silica, whereas in the deep sea, rift systems and magma plumes at hot spots spewed iron- and silica-rich water into surrounding areas.

Iron in BIF deposits is only weakly oxidized, which means that water injected with volcanic emissions had little oxygen in it. The rarity of BIFs younger than 1.9 billion years suggests that about that time, atmospheric oxygen levels rose sharply, primarily through photosynthetic activity. Oceanic mixing would have carried atmospheric oxygen to areas of deeper water, effectively ending the formation of banded irons except in restricted basins such as failed arms of triple junctions or in areas close to volcanic sources.

Buildup of atmospheric oxygen in the Proterozoic was probably enhanced by the filling of chemical sinks for oxygen such as reduced iron and organic carbon. Once large volumes of iron and carbon had been buried, more of the free oxygen released to the atmosphere through photosynthesis could remain there because of the availability of fewer reactive molecules for bonding. Deposition of iron in BIFs left more oxygen to accumulate in the atmosphere. That was followed up toward the end of the Proterozoic Eon by a negative shift in the carbon-13:carbon-12 ratio preserved in limestones

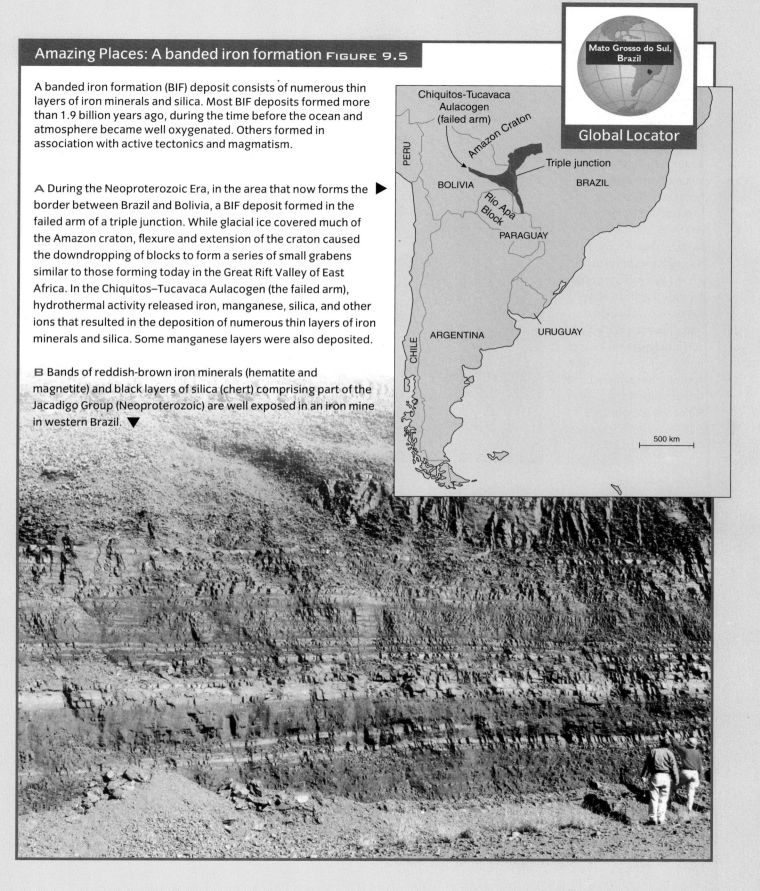

Amazing Places: A banded iron formation FIGURE 9.5

A banded iron formation (BIF) deposit consists of numerous thin layers of iron minerals and silica. Most BIF deposits formed more than 1.9 billion years ago, during the time before the ocean and atmosphere became well oxygenated. Others formed in association with active tectonics and magmatism.

A During the Neoproterozoic Era, in the area that now forms the border between Brazil and Bolivia, a BIF deposit formed in the failed arm of a triple junction. While glacial ice covered much of the Amazon craton, flexure and extension of the craton caused the downdropping of blocks to form a series of small grabens similar to those forming today in the Great Rift Valley of East Africa. In the Chiquitos–Tucavaca Aulacogen (the failed arm), hydrothermal activity released iron, manganese, silica, and other ions that resulted in the deposition of numerous thin layers of iron minerals and silica. Some manganese layers were also deposited.

B Bands of reddish-brown iron minerals (hematite and magnetite) and black layers of silica (chert) comprising part of the Jacadigo Group (Neoproterozoic) are well exposed in an iron mine in western Brazil. ▼

Redbeds, ancient and modern FIGURE 9.6

Redbeds, which owe their distinctive reddish color to oxidized iron-bearing minerals, first appeared in the stratigraphic record about 1.9 billion years ago, when the atmosphere became rich in free oxygen. In this scene from South Australia, cliffs of upturned Neoproterozoic redbeds rise above a spring that serves as a watering hole for desert animals. Clay minerals containing iron make up much of the soil surrounding the watering hole. Oxidation of the iron contributes to the reddish-brown color of the Holocene deposit.

(reflecting wholesale burial of isotopically light carbon-12). This indicates a greater availability of free oxygen in the atmosphere by 2.0 to 2.2 billion years ago.

Redbeds, which contain well-oxidized, iron-bearing sediments, show a clear relationship to atmospheric oxygenation. The characteristic reddish or reddish-brown color is imparted by oxidized forms of iron such as hematite and limonite (**FIGURE 9.6**). These minerals form either through oxidation of iron minerals at the time the sediments accumulate (red soils are a common feature of warm areas of the world) or through the secondary oxidation of iron minerals (such as magnetite or pyrite) in sediment. The occurrence of redbeds has an inverse relationship to banded iron formations: they are almost nonexistent in strata older than 2 billion years, and they are prevalent in strata younger than 1.9 billion years. The changeover

> **redbed** Reddish or reddish-brown sediments containing oxidized iron-bearing minerals.

from BIFs to redbeds marks the time when oxygen levels of the atmosphere and ocean approached levels of the early Phanerozoic. By the Proterozoic–Paleozoic transition, oxygen levels were high enough to support multicellular animals that depended on oxygen for their respiration.

CONCEPT CHECK STOP

In what ways can free oxygen be released to the atmosphere? How was free oxygen produced during the Archean and Proterozoic eons?

What are banded iron formations? What do they tell us about atmospheric–oceanic oxygen levels?

What are redbeds? What do they tell us about atmospheric oxygen levels? When did they begin forming?

Proterozoic Glaciations

M ultiple episodes of glaciation occurred during the Paleoproterozoic and Neoproterozoic eras. Glacial conditions are so much a part of the sedimentary record of the middle part of the Neoproterozoic that the interval between 850 and 630 million years ago has been named the Cryogenian Period (from the Greek *kryos* meaning "ice" and *genesis* meaning "birth").

The first glacial episode started near the beginning of the Paleoproterozoic Era, shortly after the cratons had amalgamated. Impressive evidence of this glaciation comes from the Gowganda Formation of southern Ontario, which has layers of distinctive greenish-gray **tillites** (lithified tills), often containing large boulders plucked by glacial ice from the recently formed Laurentian cratonic bedrock (**FIGURE 9.7**).

For the most part, the tillites represent **moraine** deposits, or rocks carried along in the lower part of the ice stream and then dumped at the glacial margin. Tillite layers are interspersed between successions of thin varves, or silt-clay couplets deposited in large, quiet water lakes near the glacial front. In places, dropstones have disturbed some of the thin sediment layers. The age of the Gowganda Formation is bracketed between 2.6 and 2.1 billion years ago because its sediments overlie crystalline (igneous and metamorphic) rocks of the Laurentian craton dated at 2.6 billion years, and they are crosscut by 2.1-billion-year-old intrusive igneous rocks. Tillites of the same age occur elsewhere in Laurentia, Baltica, Africa, and India, and they attest to the widespread nature of this Rhyacian Period (Paleoproterozoic Era) glacial episode.

Earth's earliest glaciation FIGURE 9.7

A Tillites of the Gowganda Formation (Rhyacian Period of the Paleoproterozoic Era) of northern Canada document the first-known glacial episode.

B In an ironic twist of events, large boulders of Gowganda Formation tillite were picked and scraped from the Proterozoic bedrock of Canada and carried southward by glacial advances during the last 2.6 million years, the most recent glacial interval. Gowganda boulders were then plunked down at the glacial margins and incorporated in Pleistocene till deposits now scattered over the U.S. Midwest.

SNOWBALL EARTH

During the Neoproterozoic, Earth experienced at least two episodes of global glaciation. The evidence includes tillites, dropstones, and glacially striated surfaces. One glacial episode was in the Cryogenian Period, between 745 and 725 million years ago, and the other was in the Ediacaran Period, between 590 and 550 million years ago. Neoproterozoic glacial sequences are difficult to correlate globally, but it seems likely that each episode represents at least two cycles of glacial advance and retreat.

The stratigraphic record of Neoproterozoic glaciation extends to most sizable areas of the Neoproterozoic world, even those close to the Equator. Paleomagnetic evidence shows that present-day Australia, for example, lay astride the Equator, yet it shows clear evidence of Cryogenian and Ediacaran glaciation. This peculiar condition is unmatched by anything we know from the Cenozoic Era. Not even during the Quaternary ice age did continental glaciation extend to the tropics. Even more peculiar, glacial deposits from the Cryogenian and Ediacaran periods are associated with banded iron formations and capped by carbonate rocks (see *What a Geologist Sees*).

> ■ **Snowball Earth hypothesis** The concept that during the Proterozoic Eon, the entire surface of Earth was repeatedly plunged into freezing conditions.

One explanation, offered by Joseph L. Kirschvink, for the unexpected assortment of Neoproterozoic glacial, banded iron, and carbonate cap sediments is the **Snowball Earth hypothesis**. The idea behind Snowball Earth is that the warming effects of greenhouse gases were rendered ineffective as **runaway freezing** quickly and repeatedly enveloped the planet's surface in ice.

Runaway freezing of Earth from the poles to the Equator requires that the warming effects of greenhouse gases had to be temporarily countered. The polar regions receive a lower dose of solar radiation than the tropics. Clustering of continental masses near the Equator during the Neoproterozoic may have left high-latitude regions vulnerable to extreme cooling. Once ice began spreading from the poles, even the addition of small amounts of greenhouse gases from volcanoes would have been insufficient to warm the planet enough to deter the spread of ice. After freezing of the surface, the high reflectivity (or **albedo**) of the ice would have increased the amount of solar radiation reflected from Earth. Because of this, cold conditions and reduced evaporation rates would have been maintained. With less water vapor in the atmosphere, the planet's ability to warm itself and melt the surface ice would have been diminished.

One line of evidence for repeated global freezing is rapid variation in the carbon-13: carbon-12 ($\delta^{13}C$) ratios preserved in carbonate deposits. Because photosynthesizing marine organisms tend to use carbon-12 in their life activities, limestones formed during times that photosynthetic cyanobacteria and algae "bloomed" were enriched in isotopically heavy carbon-13. Reversion to high carbon-12 isotopic levels in sediment has been interpreted as the result of deep freeze conditions that killed most photosynthetic life forms in the water and incorporation of their remains in the sediment.

Proponents of the Snowball Earth hypothesis maintain that banded iron formations are a consequence of an oxygen-depleted planet. Glaciation of global scale would be expected to seriously reduce photosynthesizers, and as a result, replenishment of atmospheric oxygen would be curtailed. In such a world, only nonoxidized or slightly oxidized iron-rich sedimentary deposits could form.

One puzzling aspect of Neoproterozoic stratigraphy is the carbonate rocks that cap the glacial sequences. Carbonates usually form in warm water, which implies the absence of ice. With Earth covered in ice, however, rocks would no longer experience significant chemical erosion involving carbonic acid (produced from carbon dioxide reacting with water). This would leave CO_2 to rapidly accumulate in the atmosphere. Eventually, enough warming should occur in the tropics to cause sea ice to melt. Because ice-free water has a lower reflectance than ice, more of the Sun's radiation should be absorbed, leading to further warming of the equatorial ocean and warming of the continental areas especially near the Equator. Rainwater should begin washing CO_2 from the atmosphere as a weak solution of carbonic acid, and the acid would then react with silicate rocks of the continental crust, liberating calcium, magnesium, and other ions. Those ions, after being washed into the ocean, could then recombine as layers of carbonate rocks deposited over glacial till.

Evidence for the Snowball Earth Hypothesis

Strata across the Cryogenian–Ediacaran boundary in Brachina Gorge, South Australia, demonstrate some of the evidence that Earth was entirely enveloped in ice during portions of the Neoproterozoic Era.

◀ **A** Reddish strata just below the base of the Ediacaran System include a thick Cryogenian tillite deposited by a continental glacier. This is a close-up photograph of the tillite. Paleomagnetic evidence shows that this region was close to the equator as the Cryogenian Period was coming to a close.

B Immediately above the Cryogenian tillite (lower-left corner of the photograph), tan-colored dolostone layers make up a "cap carbonate" sequence. The carbonate beds reflect a change toward warmer climatic conditions across the Cryogenian–Ediacaran boundary as Earth was emerging from a deep freeze (or "snowball") phase. ▶

Continued buildup of carbonate rocks would deplete enough CO_2 from the atmosphere that the world could freeze over again. This freeze–partial melt–carbonate deposition–refreeze cycle went on until continental masses near the equator had dispersed to higher latitudes. With landmasses no longer being strung out through the tropics, the most important prerequisite for operation of a Snowball Earth was eliminated.

Diversification of multicellular animals began in the Ediacaran Period, shortly after the last Neoproterozoic glacial episode. Whether melting of a Snowball Earth helped "pump" animal evolution during this time is uncertain. It is conceivable, though, that as melting of ice progressed, abundant CO_2 provided plenty of raw material to feed photosynthetic cyanobacteria and algae, which fueled an explosive population growth and rapid buildup of free oxygen (O_2) in the atmosphere. Higher O_2 levels, in turn, could have helped trigger the evolution of diverse and complex multicellular animals.

Was an earlier Snowball Earth–type scenario responsible for the glacial age 2.3 billion years ago, during the Paleoproterozoic Era? It is possible that the first appearance of atmospheric oxygen absorbed methane (a greenhouse gas) from the air. That, combined with a weaker radiation output from the Sun during the Paleoproterozoic, may have set glacial conditions in motion by forcing temperatures on Earth to plunge.

The concept of Snowball Earth is only one possible explanation for the evidence of equatorial glaciation during the Proterozoic. Another idea is that during this time, Earth's axis was tilted by as much as 60°. Even though continental areas were situated along the Equator, the net effect would be that Rodinia was in a "high-latitude" position, well away from the zone of greatest solar heating, so cooling could be expected. A second alternative to the Snowball Earth hypothesis involves true polar wander to around 60°. In either alternative scenario, glacial conditions may have developed in relatively lim-ited areas, similar to what we see in the world today. If either of the alternative explanations is correct, severe changes to the global climate do not need to be invoked.

CONCEPT CHECK STOP

What is the evidence for Proterozoic glaciation? When did glaciers develop during the Proterozoic Eon?

What is the Snowball Earth hypothesis?

What is the evidence favoring the Snowball Earth hypothesis? What are some of the alternatives to the Snowball Earth hypothesis?

Proterozoic Life Forms

NATIONAL GEOGRAPHIC

LEARNING OBJECTIVES

Outline the most important steps in the history of life on Earth.

Outline the Proterozoic history of prokaryotes and early eukaryotes.

Explain the origin of eukaryotes.

Summarize the evidence for the earliest animals.

Explain how evolutionary relationships of the Ediacaran fossils can be interpreted.

The Proterozoic Eon witnessed some of the most pivotal changes in the history of life on Earth. During the preceding Archean Eon, life was dominated by prokaryotes. As early as 3.5 billion years ago, organisms had adopted three major strategies for acquiring nutrients: chemosynthesis (in archeans), photosynthesis (in cyanobacteria), and heterotrophy (predation, scavenging, herbivory, and breakdown of detritus in eubacteria). Photosynthetic activity eventually led to evolution of an oxygenated atmosphere–ocean system. Heterotrophy gained in importance as the probable means by which the eukaryotic cell initially evolved (through the symbiotic association of predator and undigested prey). Later, heterotrophy became a major driving force in the evolution of skeletons and behavioral strategies in animals.

Simple eukaryotes may have first evolved during the Archean, but if so, they were a minor component of Earth's biota until well into the Proterozoic. Nevertheless, achievement of the eukaryotic condition set the stage for sexual reproduction and genetic recombination. This potent novelty opened the way for a rapid increase in biologic complexity and diversity. Multicellularity and development of organ systems (which are necessary for an effective division of labor within individual organisms) are early outcomes of the trend toward increasing complexity in organisms.

The first direct fossil evidence of eukaryotes appears in Proterozoic rocks. Multicellular eukaryotes (algae and animals) appeared for the first time during the Proterozoic. Some eukaryotic animals formed rigid skeletons, a condition that may have been selected for in part because of predation pressure. Later, during the early Paleozoic, increasing predation pressure, among other reasons, led to an explosive increase in skeletal types.

Most evolutionary innovations in biologic organisms that had ecosystem-scale ramifications had occurred by the end of the Proterozoic. In the succeeding Phanerozoic Eon, the evolutionary steps that fundamentally altered ecosystems were mostly ones involving changes in ecologic strategy such as invasion of the land, invasion of the air, and development of social structure.

PROTEROZOIC PROKARYOTES

At the start of the Proterozoic, prokaryotes were the dominant life forms on Earth. Molecular clock evidence suggests that origination times for most groups of eubacteria and archaebacteria were before the end of the Archean Eon, although we do not have much supporting data from fossils.

Two types of physical evidence for early (Archean to Proterozoic) prokaryotes are organic-walled bodies preserved in layers of sedimentary chert (silica) and biogenic-sedimentary structures constructed by bacterial consortia. The earliest prokaryotic specimens are preserved in chert and comprise small rounded cells and filaments formed of linked cells. They have been reported from Archean strata as old as 3.5 billion years. Similar fossils occur in chert layers of the Proterozoic Eonothem (FIGURE 9.8).

Biogenic-sedimentary structures constructed by cyanobacteria include stromatolites (mounds having layered structure) and thrombolites (mounds having clotted structure). Their proliferation in the Proterozoic Eon followed an increase in the size of continental masses and expansion of continental shelves, places where

Prokaryotic fossils FIGURE 9.8

This photomicrograph shows individual spheres and more complex strands of cells, inferred to be cyanobacterial fossils, from the 1.9-billion-year-old Gunflint Chert of Ontario, Canada.

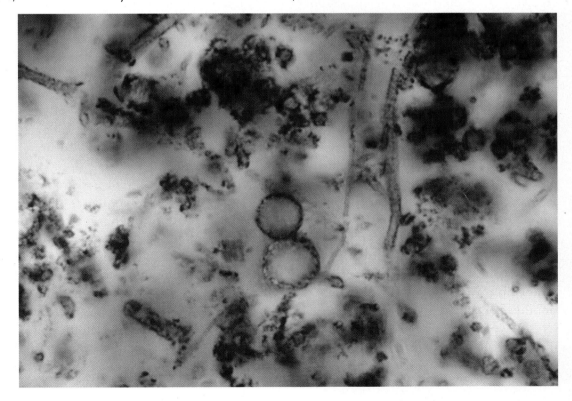

Coated grains FIGURE 9.9

Coated grains include spherical ooids and irregularly shaped oncolites. Both are multilayered structures normally formed in current-agitated carbonate environments. Precipitation of carbonate layers on ooids and oncolites is mediated by the life activities of cyanobacteria. This piece of limestone from the Neoproterozoic Era of Alaska shows both oncolites and giant ooids. Weathering has exposed the concentric layering of some grains. The largest grains are about 1 cm in diameter.

bacterial consortia apparently prospered. Mesoproterozoic and Neoproterozoic biogenic-sedimentary mounds could reach enormous proportions: some columnar stromatolites and thrombolites are 2 m or more in height.

Other biogenic structures found in Proterozoic strata that attest to the life activities of photosynthetic cyanobacteria are **coated grains**—ooids and oncolites (**FIGURE 9.9**). **Ooids** are spherical structures formed of concentric calcareous laminae, and **oncolites** are similar except that they are irregularly shaped. Coated grains form in modern carbonate environments through the accretion of thin calcareous layers mediated by gelatinous sheaths of cyanobacterial consortia. They are typical of shallow, high-energy areas where current action can roll the structures around, continuously exposing all sides of the growing objects to sunlight. Usually the core of an oncolite is a discernible object such as a calcareous mud chip (or in modern environments, a mollusk shell or other skeletal fragment).

Normally, ooids are small (smaller than 2 mm in diameter), but in the Neoproterozoic Era, ooids sometimes reached impressive proportions. Ooids up to 2 cm in diameter, known as **giant ooids**, were a common, and

essentially unique, feature of Neoproterozoic carbonate environments. Why ooids reached large sizes in the Neoproterozoic may be partly related to the existence of a high tidal range. Astronomical calculations suggest that the Neoproterozoic was a time when the Earth and Moon were closer to one another than they are today, and the distance between them has been increasing since the end of the Proterozoic Eon. If the Moon were closer to Earth, it would exert greater gravitational attraction on Earth, and its effect would be felt in a larger tidal range. In expansive, shallow, tropical-shelf seas, tidal energy would have provided considerable current agitation and, with it, a strong flow of nutrients to ooid-producing cyanobacteria, encouraging them to grow to "giant" size.

EARLY EUKARYOTES

Eukaryotes are known from rocks as old as the Proterozoic Eon, although molecular clock evidence suggests that they originated during the Archean Eon. Distinguishing the cells of simple, single-celled eukaryotic organisms from similar-appearing prokaryotic organisms is not always easy. Size and structure are the basic means of identifying cell type. Prokaryotic cells typically range up to 10 micrometers (μm) in diameter, whereas **eukaryotic cells** are usually larger than 10 μm. Inside some eukaryotic cells are distinct masses presumed to be remains of cell nuclei or other organelles. Others have thick, complex external walls that are unknown among prokaryotes.

How did eukaryotic cells evolve? The most commonly held idea, championed by Lynn Margulis, is that they arose through a form of **symbiosis** called **endosymbiosis**, in which one organism resides

eukaryotic cell
A cell that has a true nucleus.

symbiosis
A condition in which two or more dissimilar organisms live together.

within another (**FIGURE 9.10**). Nearly all eukaryotic cells contain **mitochondria**, which allow cells to extract energy from food. Interestingly, the DNA and RNA in mitochondria are different from the DNA and RNA in the cells in which they reside. This leads to the possibility that mitochondrial precursors were once independent prokaryotic organisms captured by other cells but resis-

Symbiotic origin of eukaryotes FIGURE 9.10

The eukaryotic cell type may have arisen through a symbiosis of prokaryotic cells in which neither of the cells in the association was consumed or destroyed.

The original, mitochondria-bearing eukaryotes were probably animal-like protozoans.

Plant-like protoctists probably evolved through another endosymbiotic association of cells. An ingested but undigested cyanobacterial cell could have transformed inside a protozoan into a chloroplast (a cell organelle in which photosynthetic activity occurs).

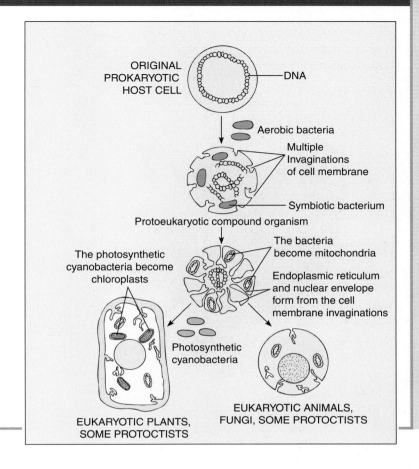

ORIGINAL PROKARYOTIC HOST CELL — DNA

Aerobic bacteria

Multiple Invaginations of cell membrane

Symbiotic bacterium

Protoeukaryotic compound organism

The photosynthetic cyanobacteria become chloroplasts

The bacteria become mitochondria

Endoplasmic reticulum and nuclear envelope form from the cell membrane invaginations

Photosynthetic cyanobacteria

EUKARYOTIC PLANTS, SOME PROTOCTISTS

EUKARYOTIC ANIMALS, FUNGI, SOME PROTOCTISTS

---

tant to digestion inside the predator cells. Minor alteration was required for the captured cells to adapt to life as endosymbionts and, ultimately, as cell organelles. The earliest eukaryotes were protoctists; protoctists became the root stock of fungi, plants, and animals.

Photosythetic eukaryotes probably arose from **protozoans** (animal-like protoctists) through a second endosymbiotic association with cyanobacterial cells (forming chloroplasts). Just like mitochondria, chloroplasts have their own distinctive DNA and RNA. They also have a structure that is little changed from that of cyanobacteria: **chlorophyll**, the pigment that absorbs sunlight, resides on layered membranes in both cyanobacteria and chloroplasts. Plant-like protoctists may have evolved multiple times when protozoans preyed on cyanobacteria.

The oldest-known multicellular eukaryotes are algae resembling modern seaweed. Coiled, ribbon-like *Grypania* fossils (FIGURE 9.11), some as long as 50

cm, occur in rocks estimated to be 1.4 to perhaps 2.1 billion years old. Accompanying the multicellular condition is the implication of sexual reproduction, although algae can also reproduce asexually.

An early multicellular eukaryote FIGURE 9.11

Grypania, which resembles seaweed, is the oldest-known multicellular eukaryote. This specimen was found in Michigan.

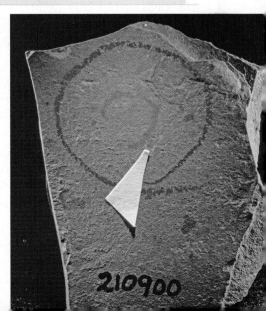

210900

Skiagia, an acritarch that lived on Earth from the Neoproterozoic Era to the Cambrian Period, is thought to be the resting stage of a planktonic green alga. The diameter of *Skiagia* is about 100 μm.

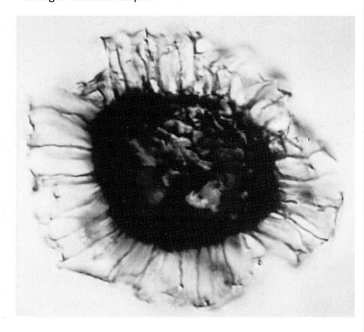

In strata younger than 2 billion years, spherical to spinose organic-walled microfossils called **acritarchs** (FIGURE 9.12) become increasingly abundant. They range in size from about 10 to 100 μm and seem to be of heterogeneous origin. Some are probably resting stages, or cysts, of **dinoflagellates** (a group of planktonic, eukaryotic green algae). Others may be prokaryotic cyanobacterial cells or spores of eukaryotic fungi. The diversity and abundance of planktonic acritarchs increased dramatically through the Proterozoic and early Paleozoic, and some species are useful as guide fossils. Near the end of the Ediacaran Period, acritarchs declined sharply in diversity. What caused them to become extinct in large numbers is unknown.

The final piece of evidence for eukaryotic life early in the Proterozoic Eon comes from biomarkers. Organic compounds known as **steranes**, which are present in eukaryotes, have been reported from rocks as old as 1.7 billion years.

THE EDIACARAN BIOTA AND THE RADIATION OF ANIMALS

Until the Ediacaran Period, most organisms were microscopic. That changed about 570 million years ago, with the appearance of a remarkable collection of Neoproterozoic organisms (FIGURE 9.13) that mark an important step toward the multicellular eukaryote-dominated world of the Phanerozoic.

The Ediacaran Period derives its name from a conspicuous assortment of fossils that have flattened zipper-like, concentric, frond-like, radial, and other miscellaneous shapes. Most are between 1 cm and 30 cm in size, but a few reach lengths of 50 cm or more. These fossils, collectively referred to as the **Ediacaran biota**, are so named for an early discovery site in the Ediacara Hills of South Australia. Today Ediacaran-type fossils are known from localities on all modern continents except Antarctica.

> **Ediacaran biota**
> Fossils dating from the Ediacaran Period, including the earliest putative animals.

Conventionally, the Ediacaran fossils are interpreted as multicellular animals, or **metazoans**. One zipper-like fossil, an ovate segmented creature called *Dickinsonia* (FIGURE 9.13A), has been interpreted as a flatworm. *Spriggina* (FIGURE 9.13B), a narrow, elongate, segmented fossil that has a horseshoe-like "head" may be an early annelid (polychaete) worm. Possible cnidarians are represented by some round concentric fossils that may be jellyfish and frond-like fossils that resemble modern sea pens (a group of colonial octocorals). *Tribrachidium* (FIGURE 9.13C), a disk-shaped fossil that has three curved radial arms superimposed on it, bears an intriguing resemblance to some Paleozoic echinoderms.

> **metazoan**
> A multicellular animal.

The exact nature of the Ediacaran fossils has been controversial. Not all paleontologists agree that they are the earliest animals. In one interpretation, Ediacaran fossils are regarded as members of an extinct kingdom characterized by a quilted architecture. The fossils have also been considered, variously, as protozoan colonies, fruiting bodies of fungi, and lichens. Some of the round fossils have even been considered trace fossils.

A diverse biota of organisms appeared in marine environments during the Ediacaran Period. In present-day South Australia, Ediacaran body fossils are preserved as molds in sandstone. Many of the fossils have been interpreted as ancestors of Phanerozoic animals. The unusual shapes of the fossils, however, have led some scientists to speculate that Ediacaran organisms have different evolutionary histories.

A *Dickinsonia* may be a primitive flatworm. It shows a subtle asymmetry where the divisions of the left and right sides meet at the longitudinal midline to form a zipper-like arrangement.

B *Spriggina* shows a resemblance to polychaete worms. In addition to having a segmented body, it appears to have a distinct head and tail region.

C *Tribrachidium* shows a radial morphology reminiscent of echinoderms.

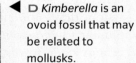

◀ **D** *Kimberella* is an ovoid fossil that may be related to mollusks.

E The possible life attitudes of *Dickinsonia*, *Spriggina*, *Tribrachidium*, *Kimberella*, and some frond-like organisms are shown in this reconstruction of a shallow Ediacaran Period seafloor. The large frond-like organisms may be early sea pens, a group of soft corals. ▼

One way of testing competing ideas about the affinities of Ediacaran life forms is to determine whether they share homologies, and therefore share genealogical relationships, with Phanerozoic organisms. A convincing case for Ediacaran sponges has been made with the discovery of articulated specimens retaining siliceous spicules just like those in Cambrian to Holocene glass sponges.

If most of the Ediacaran organisms represent a kingdom of creatures that became extinct at the end of the Neoproterozoic, we should not find Cambrian or younger representatives of the same biologic groups (clades). Perhaps the most emblematic fossils of the Ediacaran biota are frond-like fossils (Figure 9.13E on page 271). Once thought to have gone extinct before the beginning of the Cambrian Period, several examples of probable relatives are now recognized from Cambrian rocks. The best examples show polyps or polyp chambers, and this strengthens the case for a close evolutionary relationship with some frond-like sea pens, a group of "soft corals" that still inhabit the ocean. It is possible, though, that the frond-like shape evolved independently in multiple groups of organisms. Similar reasoning has been used to infer close relationships between Ediacaran organisms and Phanerozoic arthropods, annelids, flatworms, jellyfish, and mollusks.

The peculiar nonbiomineralized organisms of the Ediacaran biota were preserved under extraordinary sedimentary circumstances—mostly as impressions on the undersides of sandstone layers where they overlie more easily weathered and eroded shales or siltstones. Most of the fossils are impressions formed in relatively coherent sediment stabilized by sticky microbial colonies. Such **microbial mats**, which in the modern world are typified by slimy, greenish mud that develops mostly at the margins of ponds and ephemeral pools (Figure 9.14), apparently were a pervasive aspect of sediment–water interfaces in the Proterozoic Eon. Unlike today, where microbial mat-stabilized sediments are relatively restricted in their distribution, during the Neoproterozoic Era, microbial mats may have stabilized sedimentary surfaces across the continental shelves.

The beginning of a trace fossil record coincides with the beginning of a body fossil record of animals (Figure 9.15). Scratches on the mud surface, near-surface horizontal traces, and simple tubular burrows are about all that Ediacaran strata have to offer as evidence of animal behavior. However, these traces demonstrate that multicellular life forms had achieved a significant level of morphologic organization and nervous system development by at least 570 million years ago. Similar to *Dickinsonia*, *Spriggina*, and other typical Ediacaran body fossils, trace fossils on sediment surfaces were preserved through impression or replication in microbial mat–stabilized sediments.

> ■ **microbial mat**
> A layer of microscopic bacteria and fungi that grows at the sediment surface.

A microbial mat Figure 9.14

A greenish-colored mat composed of microbes stabilizes modern sediment at the margin of a pool of water in Yellowstone National Park, Wyoming. Similar microbial mats are thought to have been pervasive in water-laden sediments during the Neoproterozoic Era.

Ediacaran trace fossil FIGURE 9.15

This wiggle mark from Ediacaran strata of Australia is interpreted as a trace fossil formed on a microbial mat–stabilized sediment surface. The relatively complex nature of the trace indicates that it was constructed by a multicellular animal.

Early shell-like skeletons FIGURE 9.16

Calcareous tubes of *Cloudina* from the Ediacaran Period are some of the earliest biomineralized skeletons known from the fossil record. These fossils, which are only a few millimeters long, are from Namibia.

Most Ediacaran animals had relatively soft, pliable, but fairly durable external coverings. These soft "skins" may have been useful for locomotion and other functions but offered little protection from predators. Before the end of the Ediacaran Period, one group of organisms, the cloudiniids, evolved small conical skeletons of hard calcium carbonate (FIGURE 9.16). In part, these early biomineralized skeletons may have been a way for their owners to rid themselves of excess calcium buildup during metabolism, but they also reduced the threat of predation. Evidence shows that biomineralization was not a complete deterrent to predation, however, because some small cloudiniid tubes have tiny boreholes left by an unknown drilling organism.

One of the most remarkable indications of animal life during the Ediacaran comes from carbonate strata deposited in China and elsewhere just before the end of the period. Dissolution of the carbonate matrix in weak acid leaves residues containing microscopic eggs and embryos coated with thin crusts of acid-resistant phosphate. The embryos show spiral cleavage patterns typical of cnidarians, arthropods, and other **protostomes** or show radial cleavage patterns typical of echinoderms and chordates (the **deuterostomes**).

CONCEPT CHECK STOP

What are the most important steps in the history of life on Earth? Which steps had been reached by the end of the Proterozoic Eon?

What does the fossil record tell us about the abundance and diversification of prokaryotes and early eukaryotes during the Proterozoic Eon?

What is the most commonly accepted origin of eukaryotes?

What is the evidence for the earliest animals?

What types of organisms are present in the Ediacaran biota? What hypotheses have been used to explain their body forms? How can the hypothesized relationships of the Ediacaran fossils be tested?

1 Evolution of the Cratons and Proterozoic Supercontinents

1. The beginning of the Proterozoic Eon marks the beginning of the **platform phase** of Earth history. Tectonic processes operated in a modern way, and cratonic areas no longer expanded significantly. Subsequent modification occurred through accretion at the margins, suturing during collisional events, separation along rifts, and sedimentary deposition.

2. The first supercontinent, **Rodinia**, was assembled in the Mesoproterozoic Era. Rodinia was composed of Laurentia, the North American–Greenland craton; the Australian and East Antarctic cratons; Baltica, the Scandinavian craton; Siberia, the Siberian craton; the North China and South China cratons; and other pieces of continental crust. An extensive tectonic belt, the **Grenville orogenic belt**, developed in the interior of Rodinia between 1.3 and 1.0 billion years ago and today can be used as means of reconstructing the main crustal components of the supercontinent.

3. Rodinia began fragmenting 800 to 700 million years ago. The split between Laurentia and Australia opened up the Pacific Ocean. Following Rodinia's fragmentation, and continuing into the early Paleozoic, small crustal pieces were sutured together in a tectonic event called the **Pan-African orogeny** to form the craton of **Africa**.

4. During the Neoproterozoic Era, continued expansion of the Pacific Ocean may have brought the leading edges of Laurentia and eastern Gondwana together on opposite sides of the African craton, forming a second Proterozoic supercontinent, or cluster of close-lying crustal pieces, called **Pannotia**.

5. Pannotia separated at the end of the Neoproterozoic Eon. Laurentia, Baltica, and Siberia separated to become independent continents. This left the present-day Southern Hemisphere continents assembled as the continent **Gondwana**.

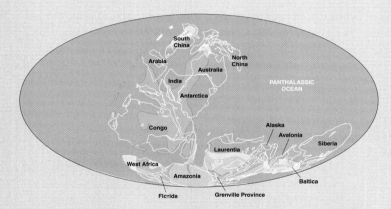

2 Oxygenation of the Atmosphere–Ocean System

1. Earth's early atmosphere probably had small quantities of free oxygen (O_2) released from the breakdown of water vapor in the upper atmosphere by the Sun's radiation. By about 3.5 billion years ago, photosynthetic cyanobacteria began releasing oxygen to the atmosphere–ocean system, and, in time, this was the primary source of oxygen input to the system.

2. **Banded iron formations**, which contain weakly oxidized iron minerals, show that Earth's atmosphere remained relatively depleted in oxygen until about 1.9 billion years ago. After this time, **redbeds**, which contain strongly oxidized, iron-bearing sediments and indicate a well-oxygenated atmosphere, appear in the stratigraphic record.

3 Proterozoic Glaciations

1. Proterozoic strata record glacial episodes in the Paleoproterozoic and Neoproterozoic eras.

2. During the Neoproterozoic Era, Earth experienced at least two distinct episodes of global glaciation, and the **Snowball Earth hypothesis** has been posed to explain evidence suggesting that the entire surface of Earth was frozen during those times.

4 Proterozoic Life Forms

1. Evolutionary steps that had ecosystem-scale consequences included the capacities for chemosynthesis, photosynthesis, and heterotrophy; endosymbiosis and development of the **eukaryotic cell**; sexual reproduction and genetic recombination; development of multicellularity and organ systems; development of resistant skeletons; invasion of the land; invasion of the air; and development of social structure. Most of these evolutionary steps had occurred by the end of the Proterozoic Eon. Chemosynthesis, photosynthesis, and heterotrophy first evolved in the Archean Eon, about the time that life first appeared on Earth. The eukaryotic condition probably evolved in the Archean, but the first evidence for it appeared in the Proterozoic. The first signs of sexual reproduction, the multicellular condition, and hard skeletons also appeared in the Proterozoic.

2. Much of the biologic record of the Proterozoic is composed of direct or indirect evidence of prokaryotes. Through the middle of the Paleoproterozoic Era, fossils seem to be exclusively of unicellular organisms. Biogenic-sedimentary structures constructed by prokaryotic cyanobacteria include stromatolites, thrombolites, ooids, and oncolites.

3. Eukaryotes are widely thought to have arisen through endosymbiosis of prokaryotic cells.

4. The first clear evidence of multicellular eukaryotes comes from seaweed-like algae in rocks ranging in age from about 1.4 to 2.1 billion years.

5. Some of the most abundant body fossils in strata younger than 2 billion years are organic-walled microfossils called acritarchs. These fossils probably have varied origins as resting stages of planktonic, eukaryotic green algae, prokaryotic cyanobacterial cells, or spores of eukaryotic fungi.

6. Biomarker evidence of eukaryotes has been reported from rocks as old as 1.7 billion years.

7. Until the Ediacaran Period, the terminal period of the Proterozoic Eon, most organisms were microscopic in size. The **Ediacaran biota**, which included multicellular eukaryotes capable of sexual reproduction, marks the appearance of more common macroscopic organisms. Trace fossils, presumably constructed by animals, also occur with Ediacaran body fossils.

8. The macroscopic, nonbiomineralized, Ediacaran organisms were mostly preserved as impressions in relatively coherent sediments stabilized by **microbial mats**. Microbial mat–stabilized sediments were typical of sediment–water interfaces in the Proterozoic.

9. During the Ediacaran Period, a group of organisms called cloudiniids evolved biomineralized skeletons of calcium carbonate.

10. Microscopic eggs and embryos of multicellular animals, or **metazoans**, have been recovered from acid residues of late Ediacaran carbonate rocks. They confirm that animals had evolved before the beginning of the Phanerozoic Eon.

KEY TERMS

1. What geologic evidence would you look for to test whether there were one or two times of supercontinent assembly during the Proterozoic?

2. Do you think banded iron formations could form anywhere in the world today? If so, where might you go to find modern analogs?

3. Could the world plunge into a Snowball Earth phase in the near future? What geologic and atmospheric conditions would make it possible?

4. Is it possible that the eukaryotic condition arose more than once in Earth history? Could more than one eukaryotic "experiment" have survived in living things to the present day? How would you test this possibility?

5. What conditions do you think were responsible for inhibiting the diversification of eukaryotes until late in the Proterozoic Eon?

6. How did the biologic innovations of the Proterozoic pave the way for eukaryotic life forms to later undergo adaptive radiation in the early Phanerozoic?

What is happening in this picture ?

A Fossil Embryo

This microscopic fossil was preserved with a coating of phosphate in limestone of the Doushantuo Formation (Ediacaran) of China. The fossil has been interpreted as a metazoan embryo.

■ What features of the fossil indicate that it is an embryo?

■ This delicate embryo is preserved three-dimensionally. How much time do you think it took for the specimen to fossilize?

SELF-TEST

1. What major phase of Earth history was used to mark the beginning of the Proterozoic Eon?

2. What was Rodinia? When did it exist?

3. How do the ages of Proterozoic mountain belts help in reconstructing early supercontinents?

4. What was Pannotia? What time interval does it signify?

5. What is a supercontinent cycle?

6. What hypothesis has been used to explain evidence of glaciation in tropical latitudes during the Proterozoic?

7. What was the origin of free oxygen in the atmosphere?

8. When did the atmosphere make the switch from being oxygen depleted to being well oxygenated?

9. What hypothesis has been used to explain the evolution of eukaryotes?

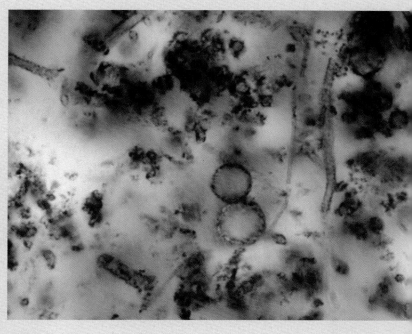

10. What is a metazoan?

11. What is the first evidence of animals in the fossil record?

12. What types of possible animals are represented by body fossils in the Proterozoic?

13. When did the first biomineralized skeletons appear in the fossil record?

14. What is a microbial mat, and when did microbial mats dominate sediment surfaces?

Paleozoic World

The next great "chapter" in our planet's chronicle is the Phanerozoic Eon. Although comprising only 12% of geologic history, this eon is full of intricate scene, character, and plot changes. This planet's features, and its living things, become increasingly familiar as we ascend the stack of stratigraphic "pages," like the layers of rock shown here, representing the last 542 million years of time. The story development however, is by no means simple or methodical. Predictable cycles of glacial buildup and melting, sea level rise and fall, and carbon burial and exhumation, are punctuated by extraordinary twists and turns—disasters strike unexpectedly, tectonic episodes reshape the margins of continents, and organisms become modified in myriad ways.

The word *Phanerozoic* is derived from Greek roots meaning "visible life," and it attests to the importance of life forms in the eon. Biologic events are the fundamental basis for marking the passage of time in the Phanerozoic. Three great episodes, or three great metaphoric scene changes—the Paleozoic, Mesozoic, and Cenozoic eras—characterize the Phanerozoic Eon. The means of identifying these three intervals is strikingly evident in the fossil assemblages left behind in the stratigraphic layers.

This chapter addresses the Paleozoic Era. Most Early Paleozoic life forms were marine. As ecologic interactions among species increased in complexity, new life forms evolved to fill open niches. As the Paleozoic wore on, waves of organisms invaded land environments and opened up a new world of ecologic and evolutionary possibilities.

NATIONAL GEOGRAPHIC

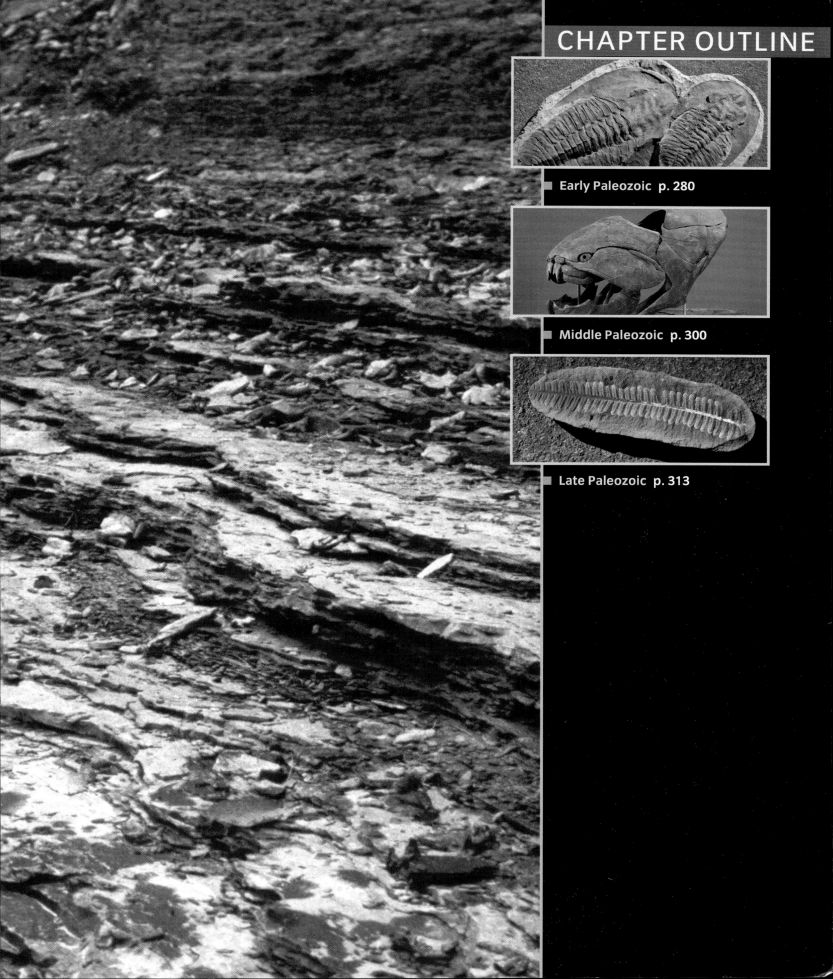

Early Paleozoic

LEARNING OBJECTIVES

Summarize early Paleozoic paleogeography and tectonic events.

Understand how biologic organisms and sediment surfaces changed during the early Paleozoic.

Explain how exceptional preservation of fossils has increased our understanding of ancient life.

Explain the sea level changes of the early Paleozoic and how they are related to biodiversity patterns.

he Paleozoic Era is the time of "ancient animal life." By the beginning of the era, the oceans were rich enough in oxygen to support advanced animal life. Commensurate with this, the genetic basis for diversification of body forms had been established sometime in the Proterozoic. Weathering and erosion of the cratons released calcium, magnesium, phosphorus, sulfur, silica, and other ions to the

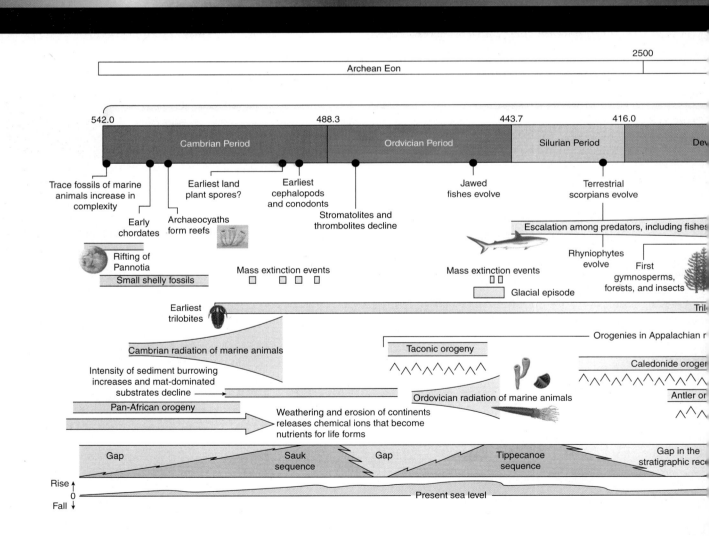

ocean, where they would act as nutrients for organisms, or where they would combine to form compounds for skeletons. Finally, sea level rise onto the continental shelves provided additional living space for marine-dwelling life forms. This combination of biologic, chemical, and physical conditions set the stage for the diversification of multicellular animals, and for preserving their remains as fossils.

From the Ediacaran to the Early Ordovician, eustatic sea level rose in stepwise fashion, primarily because of the rifting of Pannotia and addition of warm new crust at the mid-ocean ridges. Relatively high sea level resulted in the flooding of continental shelves by shallow seas

epeiric sea A shallow sea formed when marine water covers part of the continental crust. Also called epicontinental sea.

called **epeiric seas** (or **epicontinental seas**). Epeiric seas submerged most paleocontinents in the Cambrian and left behind rich marine sedimentary records. In tropical regions, such as Laurentia and Siberia, thick siliciclastics and limestones were deposited. In high-latitude areas such as Baltica, successions tend to be thin (condensed) organic-rich black shales. Most Paleozoic fossils—the bulk of the Paleozoic evolutionary record—come from the lithified marine sediments now blanketing the continental shelves.

Global sea level was relatively high through most of the Paleozoic Era. As a result, thick successions of

Visualizing

Timeline of Paleozoic events FIGURE 10.1

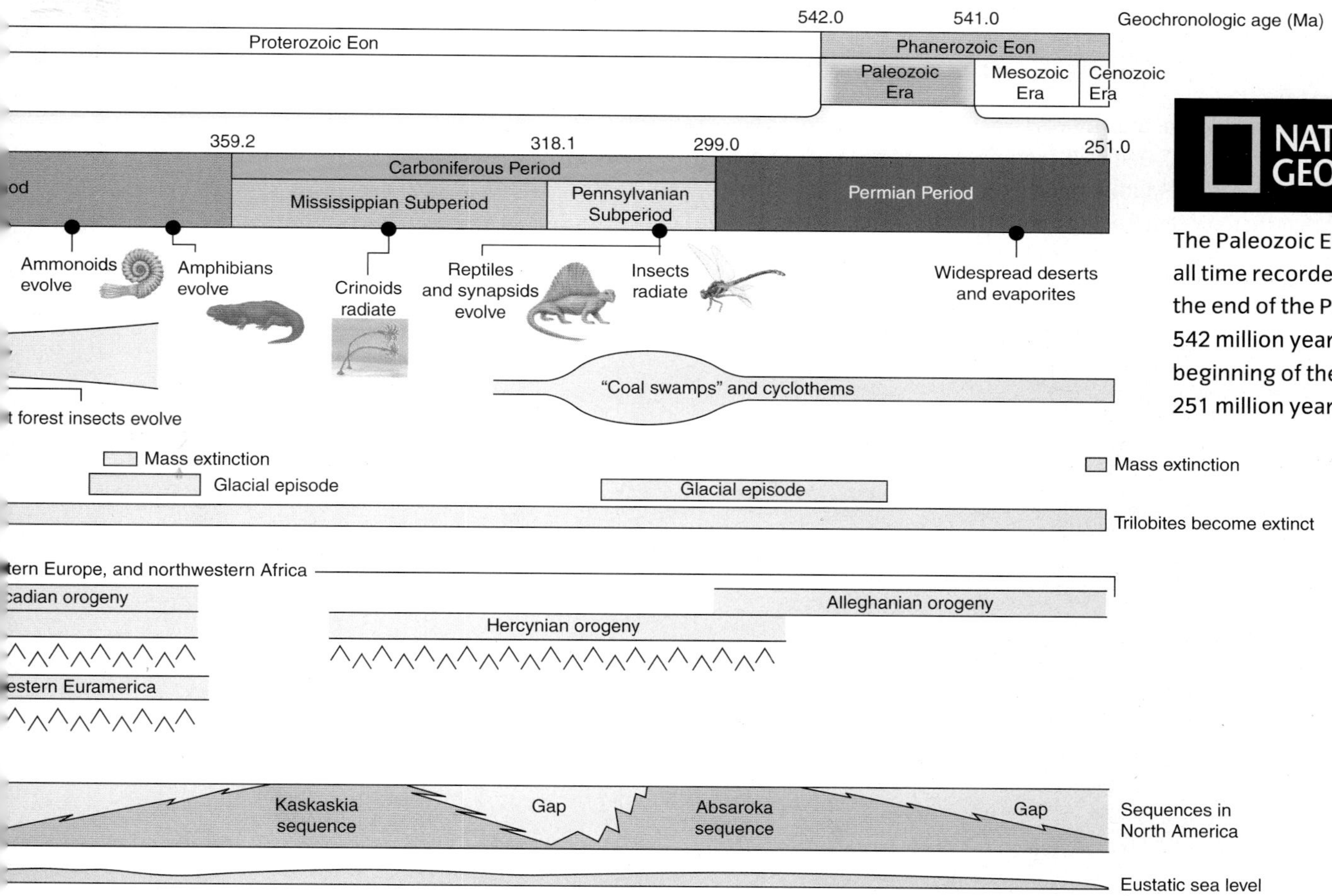

NATIONAL GEOGRAPHIC

The Paleozoic Era encompasses all time recorded on Earth from the end of the Proterozoic Eon 542 million years ago (Ma) to the beginning of the Mesozoic Era 251 million years ago.

marine sediments were deposited on continental shelves. The initial Paleozoic deposits have been referred to in North America as the **Sauk sequence**, and equivalent deposits occur worldwide. Subsequent Paleozoic deposits, all separated by regional unconformities reflecting drops in eustatic sea level, have been called the **Tippecanoe sequence**, **Kaskaskia sequence**, and **Absaroka sequence** (FIGURE 10.1 on pages 280–281).

At the start of the Paleozoic Era, Earth's tectonic plates were in a transitional stage between the breakup of Pannotia and reassembly as Pangea. Major cratonic pieces (FIGURE 10.2) were Laurentia, Baltica, Siberia, and Gondwana. Laurentia and Siberia were located in the tropics. Baltica and Avalonia were in mid-latitudes of the Southern Hemisphere. Gondwana stretched about halfway around the world. **Avalonia** was a smaller block that lay between southern Gondwana and Baltica. Kazakhstan was part of a chain of volcanic islands (an island arc) that stretched between Siberia and Gondwana.

Paleogeography of the Cambrian world FIGURE 10.2

At the start of the Paleozoic Era, the world's largest continental blocks were Laurentia (North America and Greenland), Baltica (Scandinavia), Siberia, and Gondwana (North and South China, Australia, Africa, South America, Antarctica, India, and Madagascar). Another small but significant tectonic block in the Cambrian was Avalonia (comprising England, Wales, New England, and Nova Scotia). Kazakhstan was part of a chain of islands.

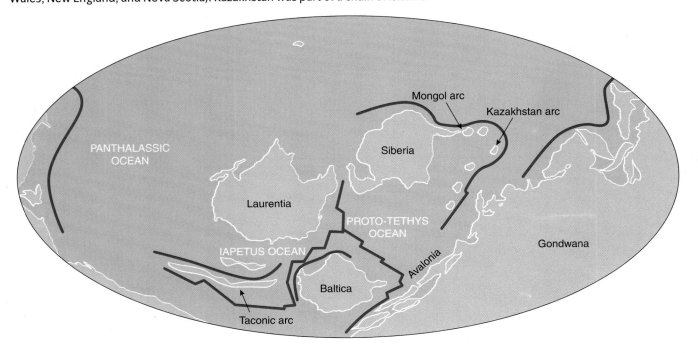

Cambrian radiation FIGURE 10.3

During a span of about 50 million years, across the Ediacaran–Cambrian transition, most animal groups represented in the fossil record made their appearance. Trace fossils and sediment burrowing increased greatly about the same time that trilobites first appeared. The record of nonbiomineralizing organisms coincides with the occurrence of deposits of exceptional preservation.

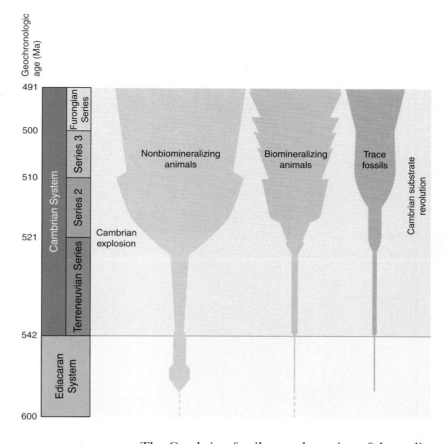

EVOLUTION'S BIG BANG

Cambrian explosion The appearance of numerous taxonomic groups in the Cambrian. It reflects a widespread evolutionary diversification, especially among multicellular animals.

The Cambrian Period marks a momentous episode in the history of life: it is the time that a good fossil record begins (**FIGURE 10.3**). Some animal groups, including sponges, cnidarians, flatworms, and annelid worms, can possibly be traced to the Ediacaran Period, but the first definitive fossil records of most forms, and the first records of complex behavior, are in the lower half of the Cambrian System.

The early evolutionary radiation of multicellular organisms is commonly called the **Cambrian explosion**. Not only did numerous organisms emerge during this time, but a variety of animals evolved resistant coverings or skeletons that could be preserved as fossils. Some groups reinforced their skeletons through biomineralization, which greatly increased their chances of preservation.

The Cambrian fossil record consists of three distinct assemblages that record major steps in the evolution of animals. Trace fossils indicating a varied assemblage of nonbiomineralized bottom-dwelling animals represents the first set of Cambrian fossils (**FIGURE 10.4**).

Earliest Cambrian fossil FIGURE 10.4

Trace fossils recording the complex behavior of multicellular animals signal the beginning of Cambrian time. The appearance of this form, called *Trichophycus pedum*, is used to identify the base of the Cambrian System.

small shelly fossils Phosphatic skeletal fossils, usually 1 or 2 mm, common in deposits of the lower half of the Cambrian System.

Tiny skeletal fossils preserved by phosphate, called **small shelly fossils** (FIGURE 10.5), comprise the second assemblage of Cambrian fossils. They also represent the earliest diverse assemblage of biomineralized body fossils. Small shelly fossils consist of a varied association of tiny skeletal pieces (or **sclerites**) that look like hats, honeycombs, pretzel twists, bugles, jacks, tubes, claws, teeth, and horns. A number of them may have originally been phosphate, but others had aragonite or calcite compositions and were secondarily replaced by phosphate in the fossilization process. In places, these phosphatic fossils are so numerous that they form a significant component of layers thick enough to be of economic value. The phosphate is mined to produce fertilizer and other useful products.

Small shelly fossils FIGURE 10.5

Tiny phosphatized skeletal parts from the lower part of the Cambrian are called small shelly fossils. These disarticulated pieces of organisms make up the first diverse biota of Cambrian body fossils.

A *Microdictyon* was originally known only from small honeycomb-like plates. Here *Microdictyon* plates are shown attached along the trunk of a marine relative of modern velvet worms (onychophorans).

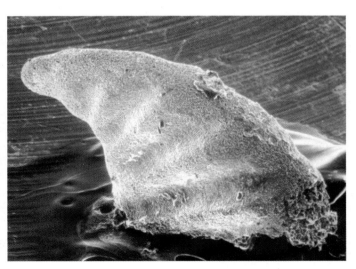

B *Helcionella* is a primitive mollusk, often small in size.

What animals formed the small shelly fossils has been determined in only a few instances. The origin of most small shelly fossils is uncertain because they were parts of skeletons that separated into pieces that water currents later concentrated in large aggregations. One rock can contain the remains of many species. Complicating matters further, various species secreted more than one type of skeletal piece when they were alive. Separating the species, and matching all the different pieces secreted by each of them, is not easy.

archaeocyath
A calcareous sponge of Cambrian age that secreted a double-walled skeleton.

The third and final assemblage of Cambrian fossils consists of skeletons of large animals. Soon after small shelly fossil–secreting animals passed their peak diversity, a peculiar group of reef-forming sponges called **archaeocyaths** appeared. Archaeocyaths (**FIGURE 10.6**) were unique because they secreted a continuous, pore-riddled, double-walled calcitic skeleton rather than making their skeletons of spicules as do most other sponges.

Archaeocyaths were a short-lived group. They diversified quickly and took up residence on most tropical marine shelves, but they became extinct midway through the Cambrian. As a group, they make good guide fossils to the first half of Cambrian time.

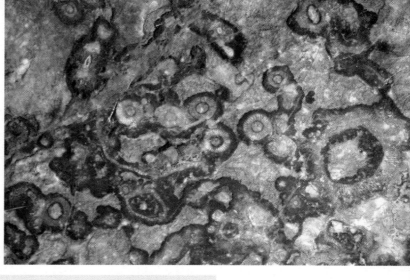

Archaeocyaths FIGURE 10.6

Archaeocyaths are a group of sponges that formed double-walled calcite skeletons. In this cross-section of a Cambrian reef, the sponges look like wagon wheels, with the inner and outer walls being connected by partitions that resemble spokes. The appearance of these archaeocyaths is enhanced by dark calcitic coatings secreted by microbes.

trilobite A Paleozoic marine arthropod characterized by a calcified exoskeleton divided lengthwise into three lobes.

Partway through the Cambrian, the first **trilobites** appeared (**FIGURE 10.7**). These arthropods (marine relatives of spiders, scorpions, and horseshoe crabs) left an excellent fossil record through the Paleozoic Era because of the calcite in their exoskeletons.

Early trilobites FIGURE 10.7

Trilobites, a distinctive group of Paleozoic arthropods, had biomineralized exoskeletons and left an excellent fossil record. The Cambrian Period, often referred to as the "Age of Trilobites," is not only the time when trilobites first evolved but also when they reached their peak diversity. *Cambropallas* and *Paradoxides* are large trilobites from the Cambrian of Morocco.

Cambropallas

Paradoxides

Early predators FIGURE 10.8

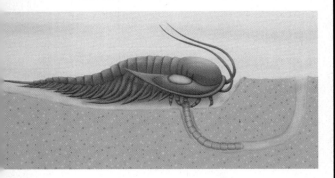

A Early trilobites preyed on worms and other creatures. Many captured their prey by burrowing into the sediment.

B This compound trace fossil shows numerous burrows left by Cambrian worms. At the end of one of the traces is a larger bilobed trace (left side of photo) made by a trilobite as it burrowed into the sediment to feed on the worm.

Trilobites quickly adopted a variety of lifestyles, but many were active predators or opportunistic scavengers. Like one group of modern relatives, the horseshoe crabs, they were strong swimmers, and many were also equally good burrowers. Some tiny ones formed enormous planktonic swarms, but trace fossils in sediment show that larger trilobites dug into the sediment searching for worms (**FIGURE 10.8**) or other prey. Some burrowing trilobites probably sifted through sediment layers in search of tiny creatures living in the spaces between sediment grains.

> **brachiopod** An invertebrate animal with two bilaterally symmetrical valves composed of calcite or chitinophosphate and a lophophore for food gathering.

Other animals that left good fossil records in Cambrian rocks are **brachiopods**, **mollusks**, **hyoliths** (a now-extinct group of animals characterized by conical shells), **echinoderms**, and **siliceous sponges** (**FIGURE 10.9**). All these forms could biomineralize their skeletons. Except for the hyoliths, each of these forms later became important characters in the story of life on Earth.

A number of animals had modest beginnings in the Cambrian but came to prominence later in time. Among them are the **conodonts**, a group of primitive **chordates** that secreted phosphatic toothlike structures; tabulate **corals**, colonial corals that were major reef-formers in the mid-Paleozoic; and **graptolites**, colonial animals distantly related to the chordates. In the latest Cambrian, **cephalopods**, a group of swimming carnivorous mollusks, most of which had shelly skeletons, first appeared in the fossil record.

> **conodont** An extinct early chordate that secreted phosphatic tooth-like structures along the pharynx.

> **chordate** An animal with a notochord and pharyngeal gill slits.

> **graptolite** An extinct hemichordate animal, or chordate relative, that had an organic skeleton.

The magnitude and impact of evolutionary diversification in the Cambrian was of a scale Earth would never see again, but it was only part of the ecologic change that occurred during the time. With the appearance of new animals, unexploited niches were soon

Apart from trilobites, the dominant animal fossils in Cambrian rocks are brachiopods (**A**), mollusks (**B**), hyoliths (**C**), siliceous sponges (**D**), and echinoderms (**E**). Conodonts (**F**) are locally common in deposits of the upper part of the Cambrian System; this specimen is 1.5 mm long. Later in the Paleozoic they became much more numerous and diverse. All these forms secreted hard, biomineralized skeletal parts that helped preserve them as fossils.

A Brachiopods

B Mollusks

C Hyolith

D Siliceous sponge

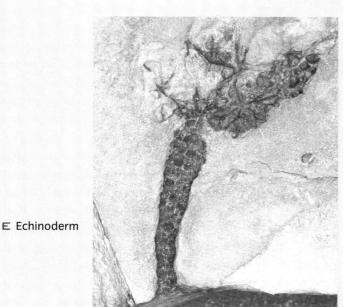

E Echinoderm

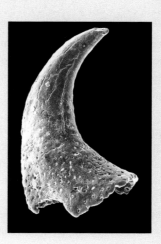

F Conodont

filled. The most thoroughly exploited area was the substrate, or sediment. While some animals continued to merely scratch at the sediment surface, others found it an area of opportunity. Many animals sought refuge in the sediment layers, some used it as a dwelling place, some sifted it for detritus, and ultimately, predators recognized the sediment as home to their quarry. Changes in marine sediment conditions, sometimes called the **Cambrian substrate revolution**, were driven in part by the diversification of animals, including burrowing trilobites and their prey (Figure 10.8 on page 286). By the end of the Cambrian, microbial mat-stabilized marine substrates typical of the Neoproterozoic had given way to burrowed and more fluidized sediments (see *What a Geologist Sees*).

Collectively, the astonishing radiation of organisms and the prodigious burrowing of sediment brought about a sweeping reorganization of marine ecosystems called the **Early Paleozoic marine revolution**. What caused these changes? As with most other events that had far-reaching effects on the biosphere, it is probably far too simplistic to invoke a single cause. A synergy of factors seems to have been responsible for opening the way to new body forms and new ways of making a living. Increased oxygenation made animal metabolism possible, nutrient input to the ocean provided the raw materials for constructing skeletons, and sea level rise provided living space on continental shelves. Much of the driving force for the biologic changes, however, probably came from ecologic interactions.

Burrowing the Sediment Layers Away

The Cambrian was a time of transition between mat-stabilization of sediment surfaces characteristic of the Neoproterozoic Era and more fluidized, burrowed marine sediments characteristic of most of the Phanerozoic Eon.

A Thinly laminated, virtually unburrowed sediment layers like these are common in the Ediacaran System. Sedimentary layers were mostly undisturbed by burrowing animals until the later half of the Cambrian.

C Marine sediments deposited after the Cambrian Period are often highly burrowed. This is a Devonian limestone that has been used as a building stone. The burrows in the limestone are clearly evident as a mottled, latticed texture after about 150 years of weathering. Since the early Paleozoic, layers that escaped burrowing had often been inhospitable to burrowing animals due to low oxygen availability, hypersalinity, or some other reason.

B By the end of the Cambrian Period, deeply burrowing animals transformed many thinly laminated shallow marine sediments into mottled beds, like these from the latest part of the period.

Anomalocaris was a large, voracious Cambrian predator that preyed on trilobites.

A The large grasping appendages of *Anomalocaris* are found at many localities in the world. This specimen is from the Burgess Shale of British Columbia.

B In most places, *Anomalocaris* remains are associated with fossils of bitten trilobites, and the bite marks match the shape of *Anomalocaris* mouth parts. This trilobite shows a healed bite mark on the right side.

C In this reconstruction, *Anomalocaris* is shown preying on a trilobite in much the same way as a shark preys on other animals in modern seas.

NATIONAL GEOGRAPHIC

During the Early Paleozoic marine revolution, evolving body forms and the arrival of new organisms on the scene meant changes in predators and their prey. As predators developed more efficient prey-capture or immobilization structures such as grasping appendages and streamlined bodies, prey species responded. Responses ranged from prey species concealing themselves within the sediment, evolving body forms that were more efficient for escape, reproducing at high rates, and developing predation-resistant spines, to developing thickened or mineral-reinforced skeletons. Prey species were not the only ones burrowing sediment; predators used the sediment as a hunting ground (Figure 10.8 on page 286). The ecologic escalation continued as predators and prey maneuvered for advantage in a great ecologic version of an "arms race." Animals quickly discovered that evolutionary advantage went to the species that could survive the predation pressure.

The Cambrian seas were apparently teeming with predators. One fearsome creature was *Anomalocaris*, a large streamlined arthropod with enormous grasping appendages, a mouth rimmed by sharp plates, and large stalked eyes (FIGURE 10.10). *Anomalocaris* could sometimes reach lengths exceeding 2 m, and one of its favorite meals was the trilobite. Bite marks this arthropod inflicted on trilobites record partly successful feeding attacks, but even more common in Cambrian strata are broken trilobite skeletal pieces, apparently the result of fully successful predation. Sometimes these pieces are found in **coprolites** (or fossil fecal matter) left by large predators.

Early Paleozoic 289

EXCEPTIONALLY PRESERVED FOSSILS AND CAMBRIAN BIODIVERSITY

It was not just the strengthening of skeletons that improved the quality of the fossil record during the Cambrian Period. A large and important share of the preserved record of Cambrian biodiversity comes from organisms lacking biomineralized skeletons.

Roughly coinciding with the first appearance of trilobites are some of the earliest Phanerozoic deposits containing exceptionally preserved fossils (or fossils retaining skin, muscle, internal organs, chitinous or pro-

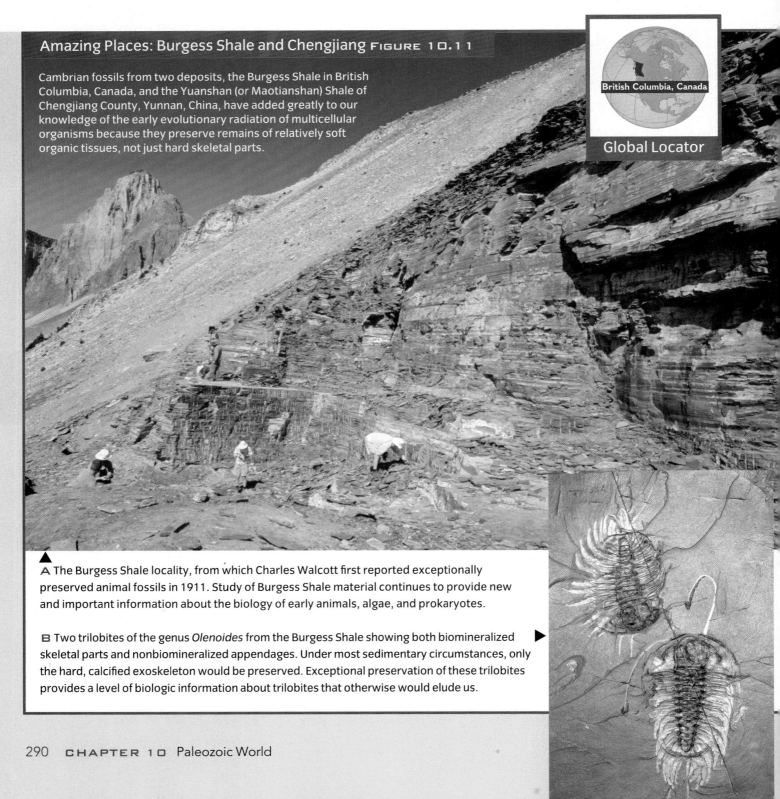

Amazing Places: Burgess Shale and Chengjiang FIGURE 10.11

Cambrian fossils from two deposits, the Burgess Shale in British Columbia, Canada, and the Yuanshan (or Maotianshan) Shale of Chengjiang County, Yunnan, China, have added greatly to our knowledge of the early evolutionary radiation of multicellular organisms because they preserve remains of relatively soft organic tissues, not just hard skeletal parts.

British Columbia, Canada

Global Locator

▲ A The Burgess Shale locality, from which Charles Walcott first reported exceptionally preserved animal fossils in 1911. Study of Burgess Shale material continues to provide new and important information about the biology of early animals, algae, and prokaryotes.

B Two trilobites of the genus *Olenoides* from the Burgess Shale showing both biomineralized skeletal parts and nonbiomineralized appendages. Under most sedimentary circumstances, only the hard, calcified exoskeleton would be preserved. Exceptional preservation of these trilobites provides a level of biologic information about trilobites that otherwise would elude us. ▶

teinaceous coverings, or cellulose, in addition to bio-mineralized fossils). Cambrian strata include an unusually large number of such deposits, the best-known of which are the Burgess Shale of British Columbia, Canada, and the Chengjiang deposit of Yunnan, China (FIGURE 10.11).

Why are deposits containing non-hard-part-bearing fossils important? Much of the answer has to do with the fact that most Cambrian organisms had non-biomineralized external coverings. The Cambrian world was dominated by an extraordinary number and diversity of arthropods, and except for trilobites, most had

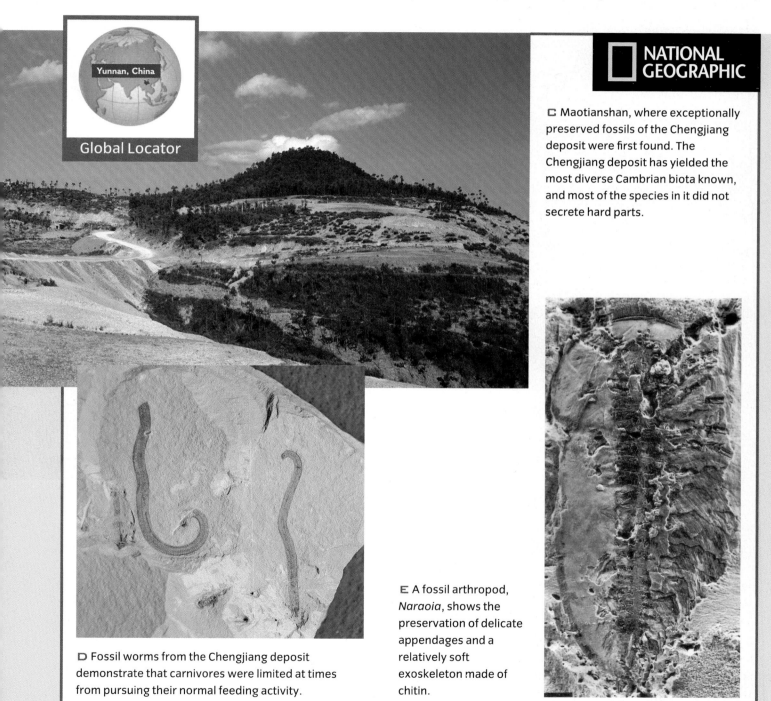

Global Locator

Yunnan, China

NATIONAL GEOGRAPHIC

C Maotianshan, where exceptionally preserved fossils of the Chengjiang deposit were first found. The Chengjiang deposit has yielded the most diverse Cambrian biota known, and most of the species in it did not secrete hard parts.

D Fossil worms from the Chengjiang deposit demonstrate that carnivores were limited at times from pursuing their normal feeding activity.

E A fossil arthropod, *Naraoia*, shows the preservation of delicate appendages and a relatively soft exoskeleton made of chitin.

A diorama illustrating a Cambrian ecosystem based on fossils from the Burgess Shale of British Columbia. Most organisms in the ecosystem were nonbiomineralizing, and under ordinary sedimentary circumstances would not have been preserved as fossils.

exoskeletons made of chitin. Although chitin affords protection, it is made of sugar compounds and can make an appetizing meal. Among the other nonbiomineralizing life forms were worms, primitive chordates, some sponges, algae, and macroscopic bacterial colonies. Together, exceptionally preserved fossils—the organisms without hard parts—comprise 90% or more of the macroscopic fossils in some Cambrian deposits. This implies that where only hard-part-bearing fossils are preserved, most of the original biodiversity was filtered out of the fossil record. This circumstance, in which the most abundant and diverse organisms are usually represented by a weak fossil record, is

the rule rather than the exception for most of the Phanerozoic and is known as the **preservation paradox**.

Having opportunities to study deposits containing nonbiomineralized fossils helps us compensate for the preservation paradox. These deposits provide unusually complete views of Phanerozoic ecosystems (FIGURE 10.12). In the Cambrian, these deposits provide remarkable glimpses into the evolution of all types of marine organisms at a critical stage in their history. The fossils help us fill gaps in our understanding of the relationships between major animal groups that otherwise would remain unfilled.

CAMBRIAN EXTINCTIONS

About 10 million years after trilobites first appeared in the oceans, something happened periodically to cause ecosystem collapse in the epeiric seas covering tropical areas of Laurentia and Gondwana. In the last 15 million years of Cambrian time, mass extinctions of trilobites and other animals occurred at least three times, at intervals of 4 to 6 million years. The extinctions were followed by evolutionary recoveries. These evolutionary "packages" delimited by extinctions are called **biomeres** (FIGURE 10.13). The biomere extinctions are associated with the cooling of marine water on tropical continental shelves. In one case (at the end of the Marjumiid biomere), extinction was associated with the burial of large volumes of organic matter, as recorded in the onset of a large positive excursion in the carbon-13:carbon-12 ratio. By the end of the Cambrian Period, trilobites were close to the brink of annihilation; many species were decimated, but the entire group did not become extinct.

Marine life of the Ordovician and afterward would have a much different character because of the Cambrian extinction pulses. Biomere events had the effect of "pruning" the evolutionary "tree" just after animal life went through its greatest period of experimentation and adaptive radiation. Many of the interesting and bizarre, nonbiomineralizing animals of the Burgess Shale and Chengjiang biotas (such as the large anomalocaridids) did not survive these biomere events. Luckily for us, some early chordates did survive the extinction events.

Cambrian extinction events FIGURE 10.13

Cambrian biomeres are evolutionary "packages" characterized by extinctions followed by recoveries. Here, the stratigraphic ranges of trilobites from Laurentia illustrate biomeres in the middle to late part of the Cambrian Period.

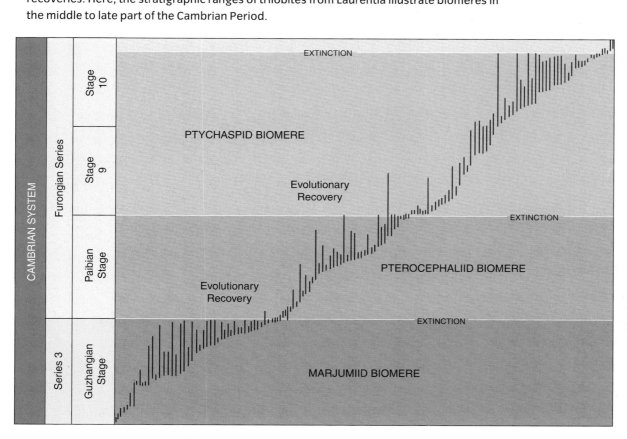

ORDOVICIAN SEA LEVEL AND TECTONICS

During the Ordovician, global sea levels reached some of the highest positions ever. The sea level rise of the Ediacaran–Cambrian continued into the earliest Ordovician, but before the end of the Early Ordovician, it dropped precipitously, which ended the Sauk sequence. Erosion and deep incision of shelf sediments ensued in many areas of the world. In Laurentia, **karst** (cave-related) topography developed in carbonate rocks of the Knox Group and equivalent strata. Later in the Paleozoic, some of the pores were filled with petroleum that migrated through the strata.

Following the great shallowing event of the Early Ordovician, eustatic sea level rose again and remained high until the final stage of the Ordovician. Shallow seas covered vast areas of Laurentia, Siberia, Baltica, and Gondwana. Ordovician carbonate and mixed carbonate-siliciclastic strata from tropical regions are among the most fossiliferous on Earth. In the tropics, relatively small wave-resistant reef communities, anchored by corals and coralline sponges, took hold in shallow shelf environments.

Iapetus The proto-Atlantic Ocean.

Shifting plates delivered big changes to the Ordovician world (**FIGURE 10.14**). Laurentia rotated counterclockwise across the Equator, and Siberia moved northward. Subduction between Laurentia, Baltica, and Gondwana collapsed a proto-Atlantic ocean called **Iapetus**. Closure of the Iapetus Ocean closed the distance between Laurentia and Avalonia and brought tectonism to eastern Laurentia and environs. A series of volcanic islands sprouted along the length of Iapetus as oceanic crust was subducted and remelted. Eruptions of these volcanoes

Ordovician paleogeography FIGURE 10.14

This map shows one interpretation of continental configurations in mid-Ordovician time. Closure of the Iapetus Ocean between Laurentia and Baltica resulted in Taconic orogenic events, the first of three major mountain-building episodes in what is now the region of the Atlantic Ocean.

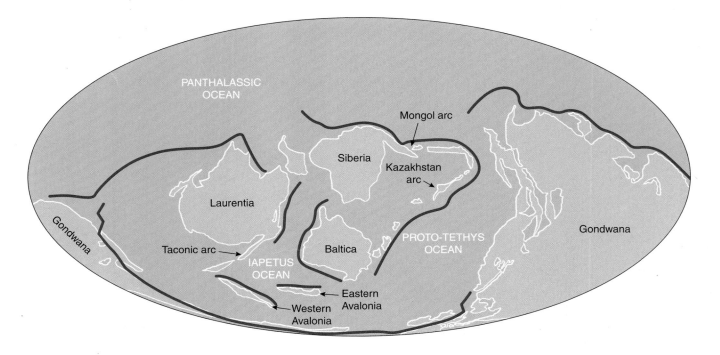

were the largest of the entire Phanerozoic. Two eruptive episodes spewed thick layers of ash over eastern Laurentia, Baltica, and the intervening ocean basin.

Orogenesis, or mountain building, on the Laurentian mainland resulted in rise of the Taconic Mountains. This event, the **Taconic orogeny**, was the first of three mountain-building episodes along the Appalachian margin of Laurentia during the Paleozoic.

> **Taconic orogeny**
> Orogenic activity during the Ordovician along the Appalachian margin of Laurentia.

Development of the Taconic Mountains is recorded through a combination of tectonic and sedimentary patterns. During the latest Cambrian and Early Ordovician, carbonate deposits extended broadly over the Appalachian margin of Laurentia. In the Middle Ordovician, crustal adjustment to collision along the continental margin caused the downdropping of a long **foreland basin** paralleling the continental margin. The first sediments in the deep, rapidly downdropped basin were organic-rich black shales. Turbidite deposits followed as earthquakes set unconsolidated sediment in motion downslope in response to gravity. This sedimentary pattern (black shales and turbidites) associated with early phases of orogeny is referred to as **flysch**.

Plate collision caused the thrusting of large wedges of flysch over shallow carbonate deposits. In places, slices of basaltic ocean floor, or **ophiolites**, were thrust onto the craton as well.

Closure of Iapetus caused the collision of island volcanoes and other crustal pieces, including Avalonia, with eastern Laurentia. These additions to the continental margin are referred to as **tectonostratigraphic terranes**, or **terranes** (FIGURE 10.15). Today they can be differentiated from cratons by their distinctive stratigraphic, paleontologic, and tectonic character, as well as by bounding faults.

As subduction continued along eastern Laurentia, the foreland basin pushed westward, and a **peripheral bulge** developed along the outer rim of the basin. This bulge produced low islands that were eroded, leaving distinctive disconformities in Ordovician strata.

Eventually, as the rising Taconic Mountains were eroded and sediment was being supplied to the basin faster than the basin could subside, flysch deposits were

Tectonostratigraphic terranes FIGURE 10.15

Much of the eastern seaboard of North America is composed of microcontinental fragments accreted to the continent in the Paleozoic Era during the Taconic, Acadian, and Alleghanian orogenies.

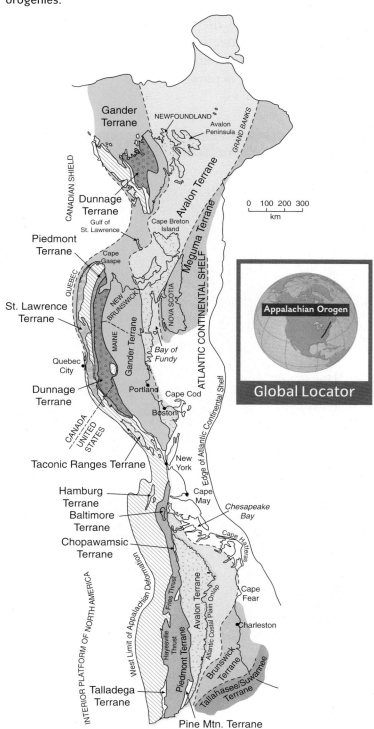

replaced by **molasse** (shallow marine and nonmarine detrital sediments). Molasse deposits, from a series of coalescing deltas at the mouths of rivers draining from the mountains, formed an enormous **clastic wedge** that thinned to the west. Today, the reddish-brown and green siltstones, sandstones, and shales of the Queenston Group exposed in the Niagara Gorge along the New York–Ontario border are a visible reminder of the enormous volume of sediment eroded from the Taconic Mountains and deposited in the Queenston Delta complex.

ORDOVICIAN MARINE LIFE

The Ordovician Period marks the appearance of an assemblage of marine life characteristic of the post-Cambrian Paleozoic (FIGURE 10.16). Extinctions in the late part of the Cambrian took a toll on animal groups, but the recovery was unexceptional. Trilobites, mollusks, and brachiopods diversified at a fairly slow rate. Few niche spaces were newly reopened to be exploited.

Cephalopods became the top predators in marine environments around the beginning of the Ordovi-

Ordovician life forms FIGURE 10.16

Ordovician seas were teeming with life forms characteristic of much of the Paleozoic Era. Fossils typical of Ordovician tropical platforms include trilobites, brachiopods, bryozoans, clams, snails, cephalopods, corals, and coralline sponges. This general assemblage dominated marine shelf areas through the end of the Paleozoic.

cian. The cephalopods, represented today by the chambered nautilus, are agile swimmers that capture prey or scavenge the sediment surface with tentacles. Some straight-shelled conical forms reached lengths of 2 m or more in the Ordovician.

Fishes were a threat to marine life of the Cambrian and Ordovician, although the extent of their impact is difficult to assess. Most fishes lacked true jaws until the mid-Paleozoic. Before that time, they may have been restricted to feeding on relatively small, soft prey (zooplankton, for example) and perhaps also plants and algae.

Graptolites, a now-extinct group of colonial animals related to the chordates, had their origins in the Cambrian but became much more diverse in the Ordovician and Silurian (FIGURE 10.17). Graptolites were composed of hundreds of tiny individual animals, called zooids, living as a colony enclosed in a resistant organic skeleton. Most of the colonies were planktonic, so individual graptolite species were widely distributed in Ordovician and Silurian oceans. That, combined with rapid

evolution in the group, makes them excellent biostratigraphic markers.

As sea level rose in the Middle Ordovician, marine organisms radiated quickly. Big evolutionary winners were the calcite-shelled articulate brachiopods, which evolved into hundreds of species in the Ordovician. An efficient tooth-and-socket arrangement at one end of the shell allowed the valves to open and close for filter-feeding on microplankton. Some brachiopods attached to rocks or shells by a fleshy stalk, and others developed thickened valves that allowed them to rest in place on the seafloor.

In the Ordovician, echinoderms (represented today by starfish, brittle stars, sea urchins, sea cucumbers, and crinoids) showed an important evolutionary expansion. Stalked crinoids (or sea lilies), which look much like flowers, as well as starfish and brittle stars, appear first in Ordovician rocks. During the Carboniferous, they were to become some of the dominant marine animals.

Graptolites FIGURE 10.17

Graptolites, distant relatives of the chordates, were common in Ordovician and Silurian seas. They are now important guide fossils for Ordovician and Silurian strata.

Another group whose evolutionary success story begins in the Ordovician is the **bryozoans** (or ectoprocts). Bryozoan colonies, usually made up of thousands of tiny zooids that fed on microplankton, built an array of hard (mostly calcium carbonate) skeletons that look like small bushes, twigs, fans, and many other shapes. They are among the most common marine fossils of the Ordovician through Devonian. Relatives of the Paleozoic bryozoans have survived to the present.

Mollusks other than cephalopods made some evolutionary gains in the Ordovician, but they were setting the groundwork for more spectacular developments later. One of the most significant outcomes of mollusk evolution in the early Paleozoic was a substantial decline in marine stromatolites and thrombolites. Grazing snails, chitons, and helcionellids, all voracious herbivores, scraped rocky surfaces clean of their cyanobacteria and algal coatings. This trend began in the Cambrian and intensified in the Ordovician. By the end of the Ordovician, cyanobacterial mounds were a rarity in normal marine environments. Since that time, stromatolites have largely been restricted to environments hostile to grazing herbivores, primarily freshwater and hypersaline environments.

EARLY FORAYS ONTO THE LAND

Live organisms have been washed ashore since early in Earth's history. Trace fossils from beach sandstones suggest that animals occasionally crawled on dry land as early as the Cambrian. They may have left the water by accident or even deliberately—perhaps the shore was a place to bury eggs (similar to the strategy used by modern horseshoe crabs and sea turtles). Jellyfish strandings on the shoreline occurred in the Cambrian and Ordovician just as they do today. At times, algae also must have washed ashore and become stranded.

The first suggestion that plants made the transition to land comes from the Cambrian of the Grand Canyon (**FIGURE 10.18**). Tidal flat deposits of the Bright Angel Shale in the eastern part of the canyon yield spores that have a distinctive pattern of grooves, called a trilete pattern, on the outer surface. Spores of primitive land plants such as mosses also have the same pattern of grooves, which leads to the conclusion that the Cambrian spores are from primitive land plants or their precursors. The earliest land plants may have evolved from **charophytes**, a group of aquatic green algae. Spores are the only known evidence of terrestrial plants older than the Silurian, when stems of land plants became fossilized.

Early land plant? FIGURE 10.18

A tiny spore from the Cambrian of Arizona is the earliest suggestion that plants invaded the land. Spores are the only evidence of possible land plants during the Cambrian and Ordovician periods.

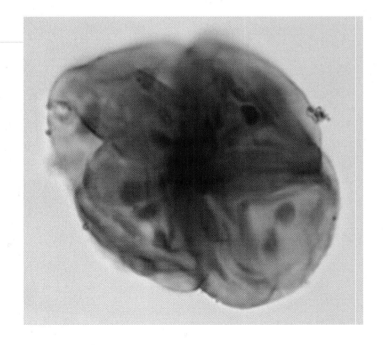

ORDOVICIAN GLACIATION AND EXTINCTION

As the Ordovician was coming to a close, global temperatures dropped and the ice cap in southern Gondwana expanded. In Gondwana, tillites, dropstones, and striated pavements record the glacial event. Elsewhere in the world, a sudden drop in sea level left disconformities in sedimentary strata.

Strong isotopic shifts provide further evidence of Ordovician climate change. An isotopic shift toward relatively heavy carbon-13 in sediments indicates that glaciation was triggered by a reduction of Earth's greenhouse warming effect. Burial of isotopically light (carbon-12-enriched) organic matter left isotopically heavier CO_2 in the atmosphere and ocean, where it would be recorded in sediments. Burial of large volumes of organic carbon reduced atmospheric CO_2 levels, and this sharp decline in greenhouse gases opened the way for rapid cooling. Change in the carbon isotopic composition of the ocean was mirrored in an isotopic shift toward heavy oxygen-18 in carbonate sediments as lighter oxygen-16 became locked up in glacial ice.

The effect of climate change on the biosphere was quick and catastrophic: it precipitated one of the most severe mass extinctions in Earth history. As glaciation reached a maximum, benthic brachiopod, bryozoan, and coral species disappeared in large numbers, as did nektobenthic trilobites, conodonts, and nautiloids. Higher in the water column, graptolites and acritarchs likewise showed declines in species numbers.

The end-Ordovician extinction shows a curious two-stage pattern. The first pulse of extinction coincided with the expansion of Gondwanan continental glaciers. A sea level drop quickly exposed wide areas of the continental shelves, wiping out niches occupied by lots of shallow marine creatures. Many tropical reef-forming coral and sponge species were eliminated when cold water covered the shelves. Biogeographic shifts also occurred as animals adapted to life in colder water expanded their ranges into shallow tropical areas. The second pulse of extinction happened as sea level rose during the initial melting of the glacial ice cap. Animals adapted to cold water were then eliminated from shallow low-latitude areas when warm water returned to the tropical continental shelves.

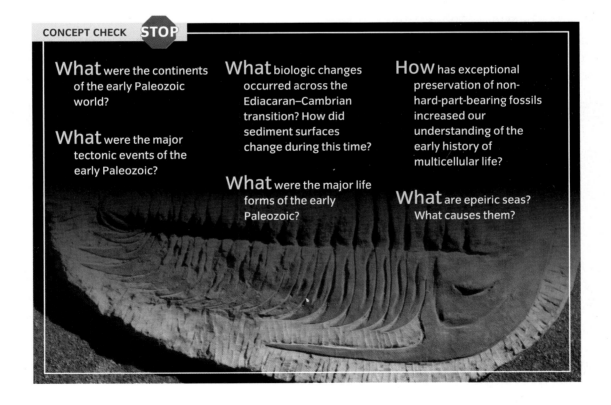

CONCEPT CHECK STOP

What were the continents of the early Paleozoic world?

What were the major tectonic events of the early Paleozoic?

What biologic changes occurred across the Ediacaran–Cambrian transition? How did sediment surfaces change during this time?

What were the major life forms of the early Paleozoic?

How has exceptional preservation of non-hard-part-bearing fossils increased our understanding of the early history of multicellular life?

What are epeiric seas? What causes them?

Middle Paleozoic

LEARNING OBJECTIVES

Discuss the formation of the Acadian Mountains and their influence on Devonian sedimentation patterns.

Explain the tectonosedimentary cycle.

Discuss the major biologic innovations of the Silurian and Devonian.

Explain the invasion of land by plants and animals.

ajor physical and biologic changes characterize middle Paleozoic time. Ancestral North America collided with parts of ancestral Europe to produce a spectacular mountain chain and related environments. While marine ecosystems were undergoing reorganization as new and more efficient predators evolved, animals and plants were also taking hold on dry land.

> **Caledonian orogeny** Silurian-Devonian orogenic activity that affected western Europe from the British Isles through Scandinavia.

SILURIAN-DEVONIAN TECTONICS AND SEA LEVEL

By the early part of the Silurian Period, Taconic mountain building had ceased in eastern Laurentia, and erosion left few remnants of the mountains. During this time, the eastern Laurentian coast became a passive margin as Baltica rifted eastward for a while and the Iapetus Ocean expanded.

In the mid-Silurian, renewed subduction and closure of Iapetus caused Baltica to collide with Laurentia (FIGURE 10.19A), principally along what is now eastern Greenland and Norway–Scotland–Ireland. This collisional event, known as the **Caledonian orogeny**, was really the first phase of a longer-term and more substantial orogenic event.

Apart from collision in the Caledonian belt, the Silurian was a time of relative quiescence on the cratonic platforms. Global sea level was high after melting of the ice that gripped the south polar cap of Gondwana during the latest Ordovician. On the North American shelf, sediments recording the Late Ordovician–Silurian sea level rise are called the Tippecanoe sequence (Figure 10.1 on pages 280–281). Eustatic sea level experienced a sharp and short-lived drop at the end of the Silurian but then

Silurian–Devonian paleogeography FIGURE 10.19

B In the Devonian, Acadian–Caledonian orogenic events brought Laurentia and Europe together as a single land mass, Euramerica, and the line of suturing was marked by a chain of spectacular mountains. ▼

A In the mid-Silurian, Baltica collided with Laurentia along the Greenland coast, initiating the Caledonian orogeny.

rose again through the Early and Middle Devonian. In North America, this rise is recorded by sediments of the Kaskaskia sequence (Figure 10.1 on pages 280–281). Carbonate deposits accumulated all the way into interior regions of Laurentia and other tropical areas during much of the Devonian, and huge reefs formed in many places. On carbonate islands not more than 1 or 2 m above sea level, land plants were taking hold.

As Caledonian orogenic activity proceeded southward through the Iapetus region, this tropical paradise was disrupted. Avalonia rammed into eastern Laurentia, and sutured to it, in the Middle Devonian. Baltica was next to come in contact with Laurentia. Mountains formed through thrusting, explosive volcanic activity increased, and the carbonate platform subsided into a foreland basin. The mountains born in the east were the Acadian Mountains, and their highest peaks rivaled the heights of the modern Himalayan summits. The larger landmass formed from these events is called **Euramerica** (FIGURE 10.19B), and in North America the Devonian phase of collision is called the **Acadian orogeny**.

The basin formed to the west of the rising Acadian Mountains would be known as the Appalachian Basin. It was a large, elongate depression extending from New York to Maryland (FIGURE 10.20A on page 302). The basin axis migrated westward, and eventually detrital sediments derived from the Acadian Mountains made their way as far west as Ohio.

The first indication of orogenic activity on the Appalachian margin of Laurentia was in the form of volcanic eruptions. Next, a blanket of quartz sand spread out over the eastern Appalachian Basin, snuffing out the carbonate platform for a short time. Carbonates returned, but then shallow-water sedimentation gave way to sudden basin deepening and the arrival of black shales (flysch deposits).

Uplift of the Acadian-Caledonian Mountains partially blocked access to the open ocean and restricted circulation within the epeiric seas covering Euramerica.

Euramerica A composite continent formed by collision of Laurentia with Avalonia and Baltica during the Devonian Period.

Acadian orogeny Orogenic activity during the Devonian along the Appalachian margin of Laurentia.

tectonosedimentary cycle The sedimentary record of continental collision: passive margin sedimentation followed by foreland basin deposits and then shallow water to nonmarine deltaic deposits.

Antler orogeny An orogeny that deformed Paleozoic rocks in the Great Basin during the Devonian and early Carboniferous.

At times, the water became stratified, with denser, more saline and oxygen-depleted water residing in the deeper parts. As a result, Devonian basins became sites of widespread black shale deposition.

Rivers carried detrital sediments eroded from the Acadian Mountains to their mouths along the shoreline of the Appalachian Basin and then deposited their loads in an ever-thickening clastic wedge called the Catskill Delta (FIGURE 10.20B on page 302). Before long, the rate of sedimentation exceeded the rate of basin subsidence, and the basin became filled with shallow marine-to-nonmarine silts, sands, and gravels, including redbeds (molasse deposits).

With the arrival of molasse deposits to the Appalachian Basin, the **tectonosedimentary cycle** was complete. The Acadian Mountains would become subdued through erosion in the Late Devonian and early part of the Carboniferous, but before the end of the Paleozoic, the Appalachian region would experience yet another major mountain-building episode.

On the eastern side of the Acadian-Caledonian Mountains, a parallel basin received sediments from the Silurian Period through the Carboniferous Period (Mississippian Subperiod). The Old Red Sandstone of the British Isles records deltaic sedimentation of sediments eroded from the Caledonian highlands.

Western Euramerica was covered in warm shallow seas in the Middle and Late Devonian. As elsewhere in the tropics, carbonate sediments were deposited, and reef communities thrived. A volcanic island chain, called the Klamath Arc, lay offshore. Subduction between the Klamath Arc and western Euramerica, which is called the **Antler orogeny**, caused the thrusting of deep-sea sediments more than 150 km onto the craton. This was the first major orogenic episode along the Cordilleran margin of the continent, and it continued into the Carboniferous. Today, results of Antler events can be seen in the Great Basin region (Nevada and adjacent areas) of the western United States.

At the start of the Devonian Period, the eastern edge of Laurentia was a relatively quiescent passive margin. Subduction of the ocean basin between Laurentia and Baltica, though, eventually led to collision of the two continents and rise of the Acadian-Caledonian Mountains during the Middle Devonian. Related thrust loading of the cratons caused subsidence of an elongate foreland basin on either side of the mountains, and the basins filled with sediments eroded from the mountains.

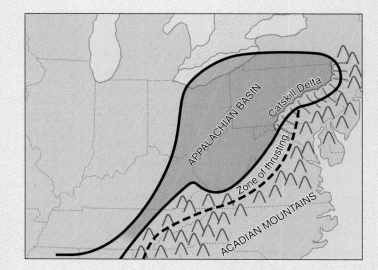

A Acadian orogenic events are reflected in the development of the Appalachian Basin to the west of the mountains and by infill of the basin with sediments of the Catskill Delta.

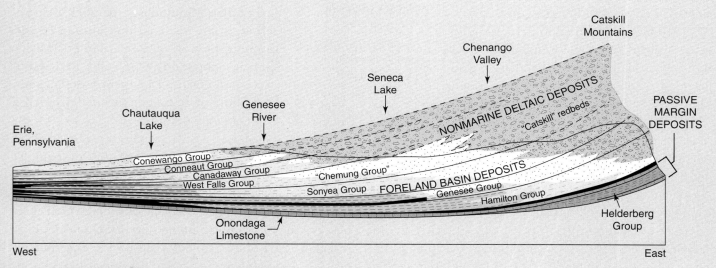

B The sedimentary record of the Catskill Delta shows a classic tectonosedimentary cycle: a passive margin gave way to development of a foreland basin, followed by deposition of shallow nonmarine, deltaic sediments.

C Sediments deposited during the Early Devonian passive margin phase include thick and widespread limestone deposits, like those shown here, and thin quartz sandstone deposits.

E The final phase of Appalachian Basin fill includes shallow marine and nonmarine sediments of the Catskill Delta. These are redbeds deposited at the top of the delta.

D Subsidence of the Appalachian Basin beginning in the Middle Devonian is reflected in the rapid changeover from passive margin deposits to deeper marine deposits, including these dark gray and black shales.

MID-PALEOZOIC MARINE LIFE

Glaciation at the end of the Ordovician decimated many shallow marine invertebrates, but recovery came rather soon as marine water flooded shelf areas, opening new habitats. For the most part, Silurian marine biotas mirrored those of the Ordovician. The end-Ordovician extinction vacated many niches, and in the Silurian, animals simply refilled them. Brachiopods, bryozoans, mollusks, and stalked echinoderms (crinoids and others) formed a large percentage of the shelly fossils. Trilobites were still important in marine ecosystems, but they were never as diverse after the Ordovician extinction as they were before it.

Corals, especially the massive, fast-growing tabulate corals, and coralline sponges (**stromatoporoids**), both refilled vacant niches and diversified into unfilled niche space. They formed numerous reefs on tropical carbonate platforms of Laurentia and Gondwana. Where there were breaks in the slope of the shelf topography, such as along structural arches and basins developing on the cratons, extensive reef tracts sprung up.

Coral-stromatoporoid reefs were a great ecologic and evolutionary success. They took hold sporadically in Ordovician seas, but they became prominent features of carbonate shelves in the Silurian and lasted well through the Devonian (FIGURE 10.21). The reefs attracted other life forms, including bryozoans, brachiopods, clams and snails, echinoderms, horn-shaped rugose corals, and algae, all of which contributed in various ways to the reef ecosystem. Behind the reefs, quieter water lagoons or inter-reef areas were places of carbonate mud deposition and homes to other communities of organisms, including brachiopods, mollusks, and trilobites.

High in the water column, graptolites, which had nearly disappeared at the end of the Ordovician, quickly diversified. Their numbers were never greater than in the Silurian Period.

Starting in the Silurian, marine life underwent its second great ecologic reorganization, the **Mid-Paleozoic marine revolution**. As a variety of new predators made an appearance, a new and intensified ecologic "arms race" between predators and prey ensued.

One of the most notable evolutionary events of the mid-Paleozoic was the rise of jawed fishes (FIGURE 10.22). Jaws, which first evolved in the Ordovician, made it possible for fishes to bite into prey and even to crush shells or skeletons. The earliest jawed fishes were **acanthodians**, or "spiny sharks," small, scale-covered fishes with numerous paired fins supported by spines. Jawed vertebrates are also known as **gnathostomes**. Acanthodians inhabited both marine and freshwater

acanthodian
A type of early, extinct jawed fish; characterized by fixed paired fins supported by spines.

gnathostome
A vertebrate animal with true jaws.

Middle Paleozoic reefs FIGURE 10.21

A In Silurian and Devonian time, reefs were rich in biodiversity and biomass. As shown in this reconstruction, these reefs were almost like "oases" in shallow tropical seas. Reef-dwelling organisms built carbonate structures that projected above the seafloor into the wave-agitated zone near the top of the marine water column. This is a reconstruction of a Silurian reef community.

B This is a Silurian reef exposed in a roadcut in upstate New York.

Jawed fishes FIGURE 10.22

Jawed fishes evolved in the Ordovician and diversified dramatically in the Devonian.

A Acanthodians, or "spiny sharks," like this one were small and had paired fins supported by spines. They represent the oldest vertebrate animals with true jaws.

B Fish teeth and plates are common in some Devonian rocks. Only the bony head shield of this giant fish, *Dunkleosteus*, normally preserves as fossils.

environments. Before becoming extinct in the late Paleozoic, they gave rise to more advanced jawed fishes.

In the Devonian, fishes had diversified so greatly that the period is often called the "Age of Fishes." Members of one group, the **placoderms**, had fearsome heads covered by bony plates. Some species reached lengths of 6 m or so. Placoderms survived only until the Carboniferous Period. More successful in the long term were the sharks, rays, lungfishes, coelacanths, and ray-finned fishes, all of which made their first appearance in the Devonian and have survived to the present.

A new group of cephalopods, the **ammonoids** (FIGURE 10.23), appeared in the Middle Devonian. Externally, they looked much like the modern nautilus.

> **placoderm** An extinct jawed fish that had jointed dermal armor (bony plates) covering the head and anterior trunk.

The inside of the shell, however, had walls, called septa, that changed from relatively simple folds early in their evolutionary history to complex, ornate structures later in their history. With their powerful jaws, the ammonoids rapidly became some of the most significant marine predators.

Two types of ammonoid cephalopods inhabited Paleozoic seas. Ammonoids with simple goniatitic sutures existed from the Middle Devonian to the Late Permian. Ammonoids with ceratitic sutures, which had small serrations on the sutures pointing toward the center of the shell, evolved from goniatites in the Middle Permian. Only the ceratites survived the end-Permian extinction event, and in the Triassic Period, they experienced a remarkable adaptive radiation.

Evolution of Paleozoic ammonoids FIGURE 10.23

Ammonoid cephalopods first appeared in the Middle Devonian. These swimming mollusks had powerful jaws and were important marine predators until their extinction at the end of the Mesozoic Era.

A Shells of the early ammonoids, like this one, had relatively simple folds in the septal walls. The line along which an internal folded wall meets the outer shell is called a suture. This, the most primitive suture type, is called a goniatitic suture, and it occurs in ammonoids ranging in age from the Middle Devonian to the latest Permian. ▶

B Ammonoids with goniatitic sutures gave rise to ones with more complex (slightly serrated) patterns, called ceratitic sutures, in the Middle Permian. Ceratitic ammonoids survived the end-Permian extinction event. The ceratites diversified in the Triassic Period after the goniatites became extinct. ▶

A Eurypterids, or "sea scorpions," were predatory marine arthropods whose remains are most abundant in Silurian dolostone deposits. They may have occasionally ventured onto land.

B A Devonian trilobite with large compound eyes composed of many separate facets capable of light collection in dimly lit places. This form may have been capable of nocturnal hunting.

Arthropods developed new predaceous forms in the mid-Paleozoic. **Eurypterids** (or "sea scorpions;" FIGURE 10.24A), which originated in the Ordovician Period, became much more numerous and diverse in the Silurian–Devonian. Eurypterids were agile swimmers, and each had two pairs of eyes. Their appendages included a set of strong pincers near the mouth. True scorpions, the earliest of which were marine, also evolved in the Silurian Period. Trilobites, which were both predators and prey, underwent changes in the mid-Paleozoic Era that improved their chances for success. One innovation was a pair of large compound eyes (FIGURE 10.24B) that could collect light during the night, which is when some may have hunted for food.

How did prey respond to the challenges posed by new or improved predators? Clams, snails, and brachiopods thickened their calcium carbonate shells (FIGURE 10.25A). Some clams burrowed into the sediment, and some brachiopods adapted to cryptic life habits. Spines (FIGURE 10.25B) were a predation deterrent that trilobites, brachiopods, clams, snails, and crinoids all evolved. Conceivably, some of these prey may have also developed poisons or foul tastes.

Predation resistance FIGURE 10.25

A Mid-Paleozoic strata are rich in remains of shelly fossils. This Devonian brachiopod has a thickened shell that was probably a deterrent to predation.

B The spines of this Devonian trilobite may have helped ward off predators.

LIFE ON LAND

Plants and animals appeared on land in several phases during the Paleozoic. A complete transition to land appears to have been successful by the end of the Silurian Period, and adaptive radiations were under way in the Devonian Period.

Life on land has certain basic requirements that are different from the requirements for life in water. Plants and animals overcame these challenges in different ways (TABLE 10.1). Once they did, life forms were free to exploit the wide-open world beyond the water's edge. The primary problems associated with the transition to land are (1) gas exchange with the outside environment, (2) internal fluid flow and maintenance of osmotic balance, (3) maintenance of support for the body, (4) internal temperature regulation and prevention of desiccation, and (5) protection of developing embryos.

Animals and plants had to overcome similar problems in making the transition from marine water to land. Groups making the transition at different times evolved in a variety of ways to meet the challenges.

To survive on land, terrestrial plants developed several new structures. Small pores called **stomata**, which can open and close, allowed plants to regulate the input of CO_2 and the release of O_2 after photosynthesis. Acquisition of roots in higher plants contributed to water and nutrient uptake and also provided anchoring.

The development of vascular tissues served multiple functions in land plants. First, these tissues formed a pipeline for delivering water, food, and other nutrients to all vital parts of the body. Second, they provided support. Marine plants could count on water to provide buoyancy and support their bodies, but on land, external support for the body was lost, and the need had to be filled by internal structures. Vascular tissues filled this need by enclosing water, a non-compressible liquid that provides turgor, within the body. Some higher plants further strengthened their bodies with **lignin**, which forms woody tissue.

Land plants minimized the loss of internal water partly by limiting gas exchange to the outside using the stomata. Plant transpiration releases not just O_2 but also water vapor, so closure of the stomata slows water loss. A waxy coating on the surfaces of leaves (the photosynthetic organs) and stems provides another way of limiting water loss. Most terrestrial plants have an epicuticle layer, and trees have bark. Both of these outer layers slow the loss of precious internal fluids to the outside.

Invasion of the land: problems and solutions TABLE 10.1

| Problems to Be Solved | Solutions | |
	Animals	Plants
Exchange of gases (O_2, CO_2)	Lungs Tracheae	Stomata
Maintaining osmotic balance	Blood (internal saline system)	Water-filled vascular network (with uptake of water by roots)
Maintaining support of the body in the absence of buoyancy provided by water	Internal skeleton (strong bone-ligament-tendon-muscle framework) External skeleton or covering (with muscle sets)	Vascular tissues (filled with water) Lignin
Protection against desiccation and temperature regulation	Skin Cuticle Hair Scales Feathers	Waxy epicuticle Bark
Protection of developing embryos	Eggs with resistant coverings Amniotic egg Internal fertilization Internal embryonic development Nesting	Seed covers Cones Seeds enclosed in ovaries

An organism is most vulnerable in its early stages, and it is no surprise that living things invest considerable metabolic energy in protecting their gametes and developing embryos. Primitive land plants tend to rely on moisture in the environment to bring gametes in contact. More advanced forms have specialized means of fertilization within reproductive structures, and some even use other organisms in the reproductive cycle. Land plants have adopted several strategies for shielding their spores or seeds until they are ready to be released. Spores or seeds may be hidden from the Sun's heat under leaves. Seeds may be covered in special reproductive chambers (seed covers, seed cones, or ovaries) to protect embryos from drying out and from other harmful effects.

■ **rhyniophyte**
A leafless, rootless, vascular plant, apparently the first true vascular land plant.

The earliest indication that plants made their way ashore comes from Cambrian spores. The spores may belong to bryophytes (mosses, liverworts, and their kin), which are non-vascular plants that live in moist places.

The first definitive vascular land plants were Late Silurian **rhyniophytes** (see *What a Geologist Sees*). These plants were essentially small, forked, leafless stems. Vascular tissue allowed them to transport food and water, but they lacked roots, so they were confined to marshy areas and had limited growth potential. They reproduced using spore organs on the tips of their stems. Rhyniophytes survived only to the Carboniferous Period, but before disappearing, they gave rise to other vascular plants.

Early land plant evolution

What a Geologist Sees

◄ A Before the evolution of vascular land plants in the late part of the Silurian, cyanobacteria, algae, fungi, lichens, and possibly spore-bearing mosses may have been the only vegetative matter in moist terrestrial environments. Rhyniophytes like this one had true vascular tissue for carrying water and nutrients through the body, an essential step toward covering land areas with greenery.

B Evolution of vascular land plants was rapid, and by the mid-Paleozoic, trees 8 m or more in height had already begun forming the canopy for early forests. These stumps of a progymnosperm are from a Middle Devonian wetland deposit, part of the Catskill Delta, near Gilboa, New York.

The first really efficient land plants, ones with roots, leaves, and voluminous vascular systems within their stems, were the spore-bearing **lycopods**, which made their appearance in the Devonian Period. Tiny club mosses are the only lycopods that have survived to modern times, but in the Paleozoic, they were much more diverse in marshy areas. Some Carboniferous lycopods developed woody tissue and grew into trees.

Once plants developed the means for survival and reproduction in the terrestrial realm, they quickly experimented with new forms. In addition to the lycopods, three other spore-bearing plants groups—the **sphenophytes** (scouring rushes or horsetails), the ferns (or **pteridophytes**), and the **progymnosperms**—all emerged in the Devonian. Stumps of progymnosperms from New York and Belgium show that these plants had developed into trees, the basis for forests, by the Middle Devonian (see *What a Geologist Sees*). Progymnosperms gave rise to the first seed-bearing plants, or **gymnosperms**.

gymnosperm
A plant, such as a conifer, cycad, ginkgo, or seed fern, bearing uncovered seeds.

Seed ferns (or **pteridospermophytes**), the earliest gymnosperms, first appeared in the Devonian. With the innovation of a seed, plants were free from their dependence on moist habitats for reproduction, and they could invade drier upland areas.

The spread of forests onto the landscape had at least three important consequences. One was stabilization of soil by the plants' root systems. This affected the courses of river systems. Meandering river systems with broad floodplains would become more common. The second effect of forestation was an increase in the rate of biochemical weathering on land. Plant roots, and associated fungi, release organic acids that weather bedrock to soil. Terrestrial soil formation increased in the Devonian with the advent of forests. The third major effect of early forestation was that it opened new habitats for animals as they ventured onto the land.

Animals had to overcome the same basic bioengineering hurdles as plants in order to survive on land. In water, O_2 and CO_2 exchange was easy: it could occur over gills, or, in some examples, it could occur directly across the body wall. To adapt to life on land, animals had to use either internal lungs or **tracheae** (sets of openings along the body leading to internal gas-exchange organs). Lungs evolved independently in terrestrial arthropods, snails, and vertebrates, whereas tracheae are restricted to arthropods such as insects and scorpions.

To nourish the vital tissues and provide osmotic balance, animals use a blood system. Blood is an internal saline fluid, essentially a replacement for the ocean water that supports marine animals.

Structural support in terrestrial animals is mostly afforded by a skeletal system that includes muscles and ligaments. Land-dwelling arthropods and most mollusks have external skeletons, and vertebrates have internal skeletons. Both types of skeletons were **preadaptations** for life on land (features already present in the marine ancestors). A coelom and blood system in terrestrial animals provides additional support for the body.

To help regulate internal temperature and prevent desiccation, land animals have an outer skin or cuticle. Secretion of an oily or waxy coating, or even sweat, helps to stabilize internal temperatures or limit water loss. Reptilian scales, mammalian hair or fur, and avian feathers also regulate temperature and minimize internal drying. Probably the greatest source of water loss among vertebrate animals occurs through the mouth as water vapor passes to the outside along with CO_2. To slow the loss of water vapor during breathing, vertebrates have developed **choana**, or tubes connecting the nasal passages to the back of the throat, allowing them to breathe through their noses and exhale far less water vapor than they would if breathing directly through their mouths.

Finally, land animals use various measures to protect their developing embryos from drying out and from other harmful effects. Many arthropods and some vertebrates have nests to keep their eggs safe. Amphibians still use water to keep their eggs safe and healthy. Fully terrestrial animals have almost universally adopted internal fertilization to maximize reproductive success. Typically, embryonic development is at least partly internal as well. Development of the **amniotic egg**, with its protective, semipermeable membrane, was a key factor in the evolutionary success of land vertebrates. Amniotic eggs

amniotic egg
A vertebrate egg that has a large yolk and is covered by a shell lined with extra-embryonic membranes, including an amnion, to conserve water and allow gaseous exchange.

are either held internally within the female until they come to term or are covered with a resistant leathery or mineralized shell and laid carefully outside the body.

Terrestrial scorpions of the latest Silurian and Early Devonian are among the earliest animals to make the transition to land. They developed tracheae to replace the gills of their marine ancestors.

Another wave of arthropod invasion of the land is recorded in exceptionally preserved Devonian fossils from the Rhynie Chert of Scotland. In this deposit, freshwater pools, enriched in silica from a hydrothermal vent, quickly encased the bodies of early land plants, fungi, and animals, including the oldest-known six-legged terrestrial arthropods (**hexapods**).

The evolution of winged **insects** (FIGURE 10.26) from non-wing-bearing terrestrial hexapods around the Middle Devonian marks the beginning of the animals' greatest evolutionary success story. By the end of the Carboniferous Period, insects had undergone an al-

most unimaginably large adaptive radiation, propelled by a combination of wings for flight, a resistant but adaptable outer skeleton, and a high reproductive rate leading to rapid speciation. Before the end of the Paleozoic Era, insects had conquered almost every conceivable terrestrial habitat and become the most diverse animals on Earth. They have remained that way ever since, especially after going through additional major radiations in the early Mesozoic associated with early radiations of dinosaurs and mammals, and again in the Cretaceous–Paleogene associated with the diversification of flowering plants. At some relatively early point in their evolutionary history, insects such as termites, ants, and bees developed complex social behavior. This innovation also contributed to their astounding success. According to conservative estimates, at least 80% of all animal species alive today are insects.

> **insect** An arthropod with three pairs of legs and wings, at least primitively.

Animals with backbones, the vertebrates, made the transition to land by the Late Devonian (FIGURE 10.27). The first of the four-legged vertebrate animals (also called **tetrapods**) was an **amphibian**. Modern amphibians include small frogs, toads, and salamanders. Amphibians have not fully made the leap to land. Their eggs are laid in water, and they also live in water as juveniles. Most species undergo metamorphosis, or change, from the juvenile stage to the adult stage and emerge from the water as air-breathing adults.

Early amphibians show features that clearly indicate that their ancestors were lobe-finned fishes. The modern lobe-finned fishes, or **crossopterygians**, include coelacanths and lungfishes. They have lungs, by which they can breathe air, or swim bladders. They also have four stout fins formed of an array of bones, including bones that are homologous with the humerus, radius, ulna, and other limb bones of amphibians. Supporting structures of the shoulder and pelvic girdles in lobe-finned fishes are homologous with bones in the shoulder and pelvic girdles of amphibians. Modern African lungfishes have been observed to crawl from one ephemeral freshwater pool to another, making use of their lungs and powerful fins. Devonian lobe-finned fishes could also make occasional jaunts across

> **crossopterygian** A lobe-finned fish, such as a coelacanth or lungfish.

Early insect FIGURE 10.26

Insects, the most successful of all animal groups, got their start on land during the Devonian Period. By the end of the Carboniferous Period, they had become the most diverse land animals. This is a Carboniferous cockroach.

The first vertebrate animals to leave the water, ultimately becoming the forerunners of true land animals, were lobe-finned fishes, or crossopterygians. The lobe-finned fishes were "preadapted" to spending short time intervals on land because they had stout fins and could breathe air through their lungs. The fins of these fishes evolved into legs as the animals made the transition to amphibians.

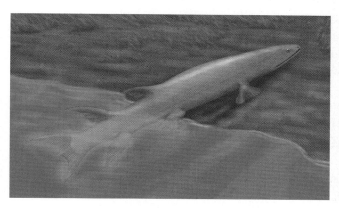

B *Ichthyostega* was one of several known Devonian amphibians that evolved from lobe-finned fishes. This is an artist's reconstruction of *Ichthyostega*.

A This is *Eusthenopteron*, a Devonian lobe-finned fish that had lungs with which it could intermittently breathe air and stout leg-like paddles that allowed it to crawl on land.

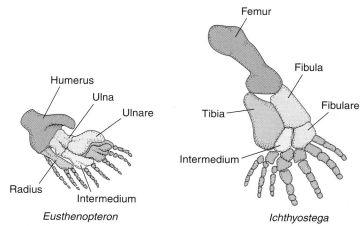

Eusthenopteron

Ichthyostega

C Comparing a limb of *Eusthenopteron*, a lobe-finned fish, with a limb of *Ichthyostega*, an early amphibian, shows a smooth transition from fish to land-dweller. The homologous arrangement of bones in the forelimb of the amphibian demonstrates a sister-group relationship with lobe-finned fishes, a clear example of an evolutionary transition in animals.

Process Diagram

dry land. Lobe-finned fishes and amphibians show similar complex tooth structure, which is another indication of the evolutionary pathway to amphibians.

Upper Devonian strata of eastern Greenland have yielded an assemblage of early amphibians, including one, *Ichthyostega*, that has a morphology intermediate between lobe-finned fishes, such as *Eusthenopteron* (known from the Devonian of Quebec), and more advanced amphibians (Figure 10.27 on page 311). *Ichthyostega* had four legs like an amphibian, and a fishlike skull. It also had a long fishlike tail that was useful for swimming but was probably an encumbrance on land.

DEVONIAN GLACIATION AND BIOTIC CRISIS

Late in the Devonian, close to the Frasnian Stage–Famennian Stage boundary, the polar ice cap in Gondwana expanded during a time when Earth was entering a cooling phase. At the same time, marine organisms declined in species diversity. More than 50% of genera known from fossils disappeared in a time span of perhaps 3 million years. Extinction affected many shelly invertebrates but nearly wiped out entirely the placoderm fishes and acritarchs. Only a few species of placoderms survived into the Carboniferous, and acritarchs never diversified again after the Devonian Period. Brachiopods, trilobites, conodonts, ammonoids, and jawless fishes all suffered major losses in the Late Devonian.

Multiple factors, including global cooling, may have caused the Late Devonian biotic crisis. Glacial deposits have been found in Brazil, which was near the South Pole in the Devonian. Glaciation triggers a fall of worldwide sea level, and this would have reduced the extent of shallow epeiric seas on continental shelves. The earliest phase of extinction coincides with two anoxic events, signaled by two widespread facies shifts toward dark, organic-rich shales, that stressed marine ecosystems. The final wave of extinction coincides with low eustatic sea level, glacial deposits in Gondwana, and isotopic shifts of carbon recorded in carbonate sediments and fossils.

An indication that the Late Devonian crisis was related to global cooling is in the pattern of extinction. Most species affected by the crisis inhabited warm tropical waters. Nearly unaffected were animals living in cool,

mid- to high-latitude waters covering areas of high-latitude Gondwana that are now the areas of South America, South Africa, and Antarctica.

The stimulus for the Devonian glacial interval is speculative, but one interesting hypothesis holds that global cooling was triggered by a reduced greenhouse effect stemming from growth of the first forests. Initial expansion of forests could have diminished atmospheric CO_2 levels through the burial of organic carbon (plant leaves and wood) and through biochemical weathering by carbonic acid on land. Declining concentrations of greenhouse gases would have opened the way to global cooling and the onset of an icehouse world.

Another reason for the biotic crisis was the way tropical marine species refilled niches vacated by extinct ones. In other words, the biotic crisis may have been due more to a decrease in speciation than to an increase in extinction. New species commonly arise when the geographic ranges of ancestral species are divided. The physical distance between newly separated populations leads to genetic divergence. This concept is known as **vicariance**.

In the Late Devonian tropical oceans, vicariance speciation rates, especially among brachiopods and clams, dropped significantly, so replacement of species that became extinct occurred slowly. Low vicariance rates were ultimately tied to expansion of the geographic ranges of species that survived extinction. The ranges of these species expanded as a consequence of small sea level rises that promoted larval dispersal between formerly separated basins.

CONCEPT CHECK   STOP

How did the Acadian Mountains form? What influence did erosion of the mountains have on Devonian sedimentation patterns in Euramerica?

What is a tectonosedimentary cycle?

What was the Mid-Paleozoic marine revolution?

When did plants invade the land? When did animals invade the land? What adaptations increased the odds of success on land?

Late Paleozoic

LEARNING OBJECTIVES

Discuss the assembly of Pangea and its effects on global climatic patterns.

Understand the origin of calcite seas and aragonite seas and explain how the magnesium:calcium ratio of seawater affects carbonate-secreting organisms.

Explain the causes and effects of late Paleozoic glaciation.

Understand the major late Paleozoic animals and plants.

Explain the end-Permian extinction.

The late Paleozoic was a time of important tectonic change, and tectonic events influenced global climatic and sea level changes. Continental masses began assembling into the late Paleozoic supercontinent Pangea, a process that was completed in the Triassic.

CARBONIFEROUS–PERMIAN TECTONICS

During the Carboniferous, Gondwana moved northward and collided first with small terranes that today make up southern Europe, and then with Euramerica (**FIGURE 10.28A**). In Europe and northwestern Africa, this

Carboniferous–Permian paleogeography FIGURE 10.28

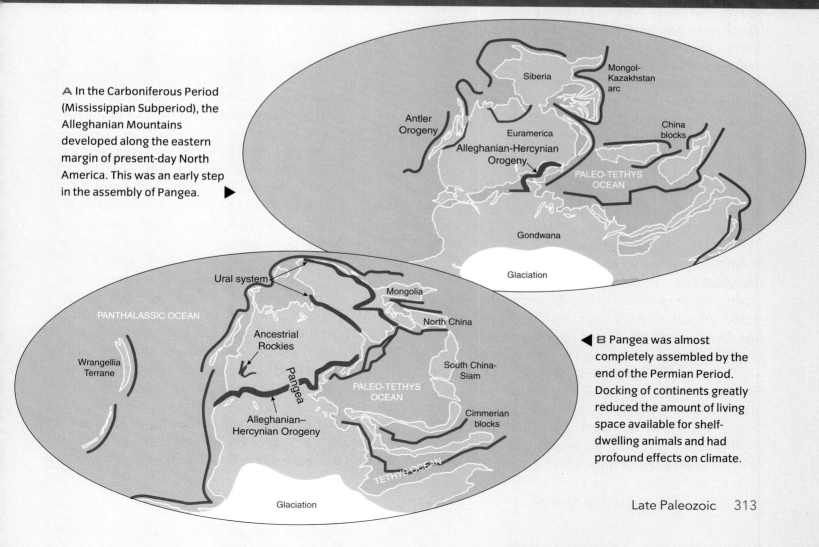

A In the Carboniferous Period (Mississippian Subperiod), the Alleghanian Mountains developed along the eastern margin of present-day North America. This was an early step in the assembly of Pangea. ▶

B Pangea was almost completely assembled by the end of the Permian Period. Docking of continents greatly reduced the amount of living space available for shelf-dwelling animals and had profound effects on climate.

mountain-building event is called the **Hercynian** (or **Variscan**) **orogeny**. In North America, the orogenic episode is called the **Alleghanian** (or **Appalachian**) **orogeny**, the third and final phase of Paleozoic mountain building in the Appalachian region. Each of the mountain-building events (the Taconic orogeny in the Ordovician Period, the Acadian orogeny in the Devonian Period, and the Alleghanian orogeny in the Carboniferous and Permian periods) was followed by opening of an ocean basin. The cyclic recurrence of plate-tectonic spreading and basin closure through subduction has been referred to as the **Wilson cycle**.

Alleghanian mountain building was focused along the southeastern margin of present-day North America. Collision with Gondwana formed the fold-and-thrust belt (or Valley and Ridge Province) of the central and southern Appalachians by squeezing crustal slices together, sliding them along thrusts faults, and crumpling them into mountains. To the east, in the Blue Ridge Province, Proterozoic rocks originally deformed during the Grenville orogeny were metamorphosed and deformed again because they were so close to the zone of suturing (the Piedmont Province). Collision sutured large chunks of Gondwanan crust to Euramerica. Most of Florida and parts of the Carolinas are among the Gondwanan terranes now attached to North America (Figure 10.15, see page 295).

The rising Alleghanian Mountains shed detrital sediments westward into the Appalachian Basin, forming a clastic wedge. Unlike the two previous tectonostratigraphic cycles of the Appalachian region (the Taconic and Acadian orogenies), the Alleghanian deposits only show deltaic molasse sediments because the craton had not subsided much since the Devonian. Westward-flowing streams deposited coarse sediment in deltaic areas as they wound across broad marshy floodplains where coal seams were forming.

Erosion of the Alleghanian Mountains down to their roots has left this former grand mountain range in a state of **inverse topography**, where the cores of former

■ **Hercynian orogeny** Late Paleozoic orogeny that affected Europe and northwestern Africa during the Carboniferous and Permian periods. Also known as the Variscan orogeny.

■ **Alleghanian orogeny** Orogenic activity during the Carboniferous and Permian periods along the Appalachian margin of Laurentia. Also known as the Appalachian orogeny.

■ **Wilson cycle** The successive recurrence of plate-tectonic spreading and convergence.

■ **Ouachita orogeny** Mostly Carboniferous (Pennsylvanian) mountain building along the southern margin of Euramerica, related to collision with Gondwana.

mountains are now deeply eroded valleys, and the hills are ridges of resistant sandstones and other rocks that once flanked the peaks. The mountain roots, which are of lower density than the mantle below, have risen isostatically (accounting for their relatively high position today) as weathering and erosion have slowly and continuously shaved material from the surface.

Important events were also taking place in western North America during the Carboniferous and Permian periods. Orogenic activity, a westward extension of the mountain building that produced the Appalachian Mountains when Gondwana collided with Euramerica, was responsible for folded rocks of the **Ouachita orogeny**. Deformation of Ordovician through Carboniferous strata took place in a deep foreland basin where black shales were accumulating as folding proceeded. Several small crustal pieces were involved in the tectonic collision.

Tectonic movement, possibly related in some way to the Ouachita orogeny, occurred to the north and west of the Ouachita Mountains during the Carboniferous Period (Pennsylvanian Subperiod). Two of the uplifted areas, the Front Range and Uncompahgre uplifts, show Precambrian basement rocks that were forced upward through Paleozoic limestones and other marine sediments. The uplift produced a series of islands, called the ancestral Rocky Mountains, that projected from an epeiric sea covering a large portion of western North America.

In west Texas and adjacent areas, shallow epeiric seas filled a subsiding basin to the north of the Ouachita Mountains. When the Ouachita Mountains were being uplifted in the Carboniferous Period, a small fault block developed that eventually led to separation of the basin into two smaller ones, the Delaware Basin to the west and the Midland Basin to the east. These basins have yielded large volumes of petroleum trapped in the subsurface. Large reefs grew around the margins of the Delaware Basin, and the Midland Basin filled with sediment. The large Capitan

Reef in the Guadalupe Mountains of Texas exemplifies the reef communities that once rimmed the Delaware Basin.

In the late Paleozoic, the western margin of North America extended lengthwise through Nevada. West of this position, a volcanic island arc was developing, and sediments were shed into adjacent marine areas. Deformation in Nevada caused by closure of the basin west of the continental margin and east of the island arc is referred to as the **Sonoman orogeny**. From the Late Permian to the Early Triassic, deep-sea sediments were thrust onto the craton. Also, crustal pieces were welded to the continent crust. The Sonoman orogeny was responsible for great westward expansion of western North America.

By the end of the Permian Period, Pangea was almost completely assembled, and it formed a continuous landmass that stretched from the north polar region to the South Pole (**FIGURE 10.28B**). Suturing of Siberia to eastern Europe along the Ural Mountains left only parts of eastern Asia to be incorporated in the supercontinent during the early Mesozoic.

LATE PALEOZOIC CLIMATIC CHANGES AND SEA LEVEL HISTORY

The Carboniferous Period is divided into two subperiods, the Mississippian and the Pennsylvanian. Strata of the Mississippian and Pennsylvanian subperiods are separated by a major regressive episode, often marked by a disconformity, that represents the dividing line between the Kaskaskia and Absaroka sequences of North America.

Climatic conditions of the Carboniferous were a study in contrasts. A greenhouse world in the early half of the period (the Mississippian Subperiod) gave way to a series of rapid glacial cycles, reflected in sea level fluctuations, in the later half (the Pennsylvanian Subperiod).

Rapid pulses of sea level change in the Carboniferous Period (Figure 10.1 on pages 280–281) were closely tied to glacial–interglacial cycles. In the Mississippian Subperiod, when the world was warm, sea level rose, flooding continental shelves with marine water. In the mid-Carboniferous, Earth switched from a greenhouse world to an icehouse world. Tectonic motion shifted present-day Antarctica over the South Pole (**FIGURE 10.28B** on page 313), and the south polar ice cap quickly expanded. Southern Gondwana became blanketed in a sheet of glacial ice, referred to as the **Gondwanide glaciation**, during this time, and sea level fell sharply. Glaciation continued through multiple cycles of ice advance and retreat well into the Permian Period.

The late Paleozoic climatic cycles were ultimately related to intervals of warming and cooling brought on by changes in Earth's orbital patterns with respect to the Sun. These cycles are referred to as **Milankovitch cycles** after the Serbian astrophysicist Milutin Milankovitch (1879–1958), who described their periodicity. Shorter intervals (approximately 19,000 and 23,000 years each) are related to **precession**, or slight wobble of Earth's axis over time, which causes Earth to tip away from the Sun's radiant energy from time to time. Longer intervals (approximately 41,000-year cycles) resulted from **obliquity**, or the tilt of Earth's axis which affects where and how much sunlight falls on the globe. Finally, the longest cycles, 100,000 to 120,000 years, are related to **eccentricity**, or deviation over time in Earth's elliptical orbit around the Sun. Less of the Sun's energy arrives on Earth when Earth is at a distant point in its orbital path than when it is closer to the Sun. When the effects of two or more of these cycles coincide, the cooling effect can be sudden and dramatic.

Our best estimate of the onset of Gondwanide glaciation comes from sedimentation patterns in low paleolatitudes that were linked to the waxing and waning of Southern Hemisphere glacial sheets. In marginal marine settings across much of Euramerica, Carboniferous–Permian shifts in sea level are recorded in sedimentary cycles called cyclothems. Classically, **cyclothems** show a change up-section from continental deposits including redbeds, fluvial sediments, and coal beds; through marginal-marine and shallow marine sandstones,

Carboniferous cyclothems FIGURE 10.29

A Cyclic sedimentary deposits, referred to as cyclothems, are common in cratonic deposits of continents that were in low paleolatitudes during the Carboniferous. These are cyclothem deposits from the Carboniferous of eastern Kansas. They record geologically rapid sea level changes along a marine coastline.

Approximate thickness (m)

Generalized lithologic members

Sea level curve

← Fall

Outside Shale

Upper Limestone

Core Shale — Black Shale

Middle Limestone

Outside Shale — Coal bed

Regression

Maximum transgression

Transgression

Rise ⟶

B Generalized model of a Carboniferous cyclothem. Nonmarine deposits in the lower part of the succession give way to shallow marine shales and limestones. Black shales record the deepest water conditions and represent the inflection point in the transgressive–regressive cycle.

siltstones, and limestones; to deeper-water gray and black shales (FIGURE 10.29)315. The shales mark an inflection point as sea level was shifting from a transgressive (deepening) phase to a regressive (shallowing) phase, so overlying sedimentary layers look like a mirror-image of those below. The deeper-water shales are replaced by shallow marine deposits, which are replaced in turn by marginal-marine and then continental deposits.

Sea level changes in coastal areas well away from the influence of major river systems and local tectonic movements are often related to climatic changes. In the Carboniferous, the buildup of glaciers in southern Gondwana forced sea level position to fall worldwide. Melting of the glaciers allowed sea level to rise.

Global sea level was high at the beginning of the Permian Period but dropped close to present levels midway through the Permian. With sea level at a relatively low level by Paleozoic standards, vast continental areas were exposed. The enormous Pangean supercontinent altered ocean circulation patterns and had a major impact on climate. Climatic gradients along the length of Pangea were strong, ranging from continental ice sheets and sea ice in polar regions to expansive deserts with shifting sands and large dunes in low latitudes. Evaporates accumulated in the dry trade wind belt near the Equator. Much of the world's reserves of salt come from Permian strata. Lush swamps where coal was forming still persisted in China, which was mostly a series of large tropical islands.

CALCITE AND ARAGONITE SEAS

The Carboniferous was a time of transition from a **calcite sea** to an **aragonite sea**. Calcite and aragonite are the two forms of calcium carbonate precipitated from seawater with the help of marine organisms. Fluctuations in the magnesium: calcium ratio in seawater control whether carbonate cements or carbonate-secreting organisms mostly form skeletons of low-magnesium calcite (during times of calcite seas) or of aragonite and high-magnesium calcite (during times of aragonite seas).

Earth experienced a long interval of calcite seas lasting from the Cambrian to the Early Carboniferous (FIGURE 10.30). Many reef-formers during this time (such as tabulate corals and stromatoporoid sponges) secreted calcite skeletons. The switchover to an aragonite sea occurred in the Carboniferous Period, when algae and other aragonite-formers dominated reef communities. This aragonite sea lasted until midway through the Jurassic Period, when there was a return to calcite seas. The most recent switch to an aragonite sea occurred in the Cenozoic Era, close to the Paleogene–Neogene boundary.

The magnesium:calcium ratio of seawater seems closely related to the volume of mid-ocean ridges. When ridge volumes were largest, magnesium:calcium ratios fell because chemical reactions involving warm seawater and newly formed oceanic crust along the ridges sped up, resulting in preferential precipitation of calcite. This occurred following the dispersion of Pannotia in the Ediacaran–Cambrian and following the breakup and dispersion of Pangea beginning in the Jurassic Period.

Calcite and aragonite seas FIGURE 10.30

The magnesium:calcium ratio of seawater has changed through time, and this has influenced which form of calcium carbonate was precipitated in the oceans at any given time point. Organisms secreting skeletons of low magnesium-bearing calcite were favored during times of calcite seas, and organisms secreting skeletons of aragonite and high magnesium-bearing calcite were favored during times of aragonite seas.

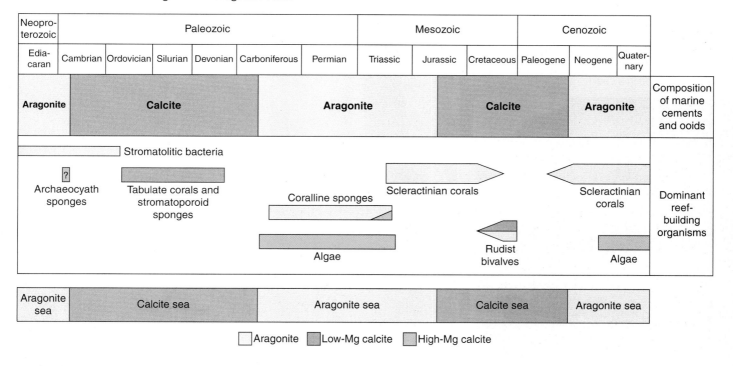

LATE PALEOZOIC MARINE LIFE

Marine life of the Carboniferous was similar to that of the Devonian. Except for placoderm fishes, reef-formers, and acritarchs, most organisms rebounded after the end-Devonian crisis. The seas were rich in brachiopods, bryozoans, and mollusks. Trilobites were no longer diverse, but some species were locally abundant.

Tabulate corals and stromatoporoids had all but ceased to form reefs in the Carboniferous. Instead, green algae, crinoids, and bryozoans assumed important positions as carbonate reef-forming or reef-dwelling organisms in the early Carboniferous.

One striking feature of early Carboniferous (Mississippian) deposits is the abundance of crinoid-rich limestones or **encrinites**. Crinoids and other stalked echinoderms formed impressive "meadows" that stretched for hundreds of kilometers across the warm, shallow, epeiric seas of Euramerica and Gondwana, contributing enormous quantities of carbonate grains to the sediment when they died.

Permian marine faunas were much like those of the late Carboniferous: brachiopods, bryozoans, rugose corals, and mollusks were especially common. In some areas, an aberrant group of productid brachiopods that developed large spines projecting from their shells became reef-formers. A group of foraminifera known as **fusulinids**, which secreted small tests that look like rice grains, evolved rapidly through the late Carboniferous and Permian. Their widespread and abundant remains,

Coal is formed in swampy areas where peat accumulates in great quantities. Organic matter from dead vegetation is added to the swamps at such a high rate that it cannot be fully oxidized and decomposed in water. Anoxic water leads to preservation of the organic material. In time, with the addition of overlying sediment, the water is squeezed from the peat, and heat and pressure associated with burial forces volatiles (gases) from it, leaving behind carbon. Later, metamorphism often increases the purity of the carbon in coal.

B Pennsylvanian coal swamp floras were diverse, and included lycopods, conifers, sphenophytes, ferns, seed ferns, progymosperms, and others.

A In the late part of the Carboniferous Period (Pennsylvanian Subperiod), lush coastal wetlands produced thick accumulations of coal in many low-latitude areas. This leaf was preserved in a coal layer.

together with short species ranges, make them excellent biostratigraphic tools.

LATE PALEOZOIC TERRESTRIAL LIFE

The most conspicuous terrestrial life forms of the Carboniferous and Permian were plants. During warmer intervals, especially in the late part of the Carboniferous, forests flourished in extensive coastal wetlands of the tropics. These areas of massive vegetative production are sometimes referred to as **coal swamps** (FIGURE 10.31). Much of the world's coal reserves come from thick peat deposits, which were lithified to coal seams and deposited in these settings. It was the prominent coal beds of Britain that gave rise to the name "Carboniferous." Coal was a much sought-after natural resource in the early 1820s, as it fueled Britain's Industrial Revolution, a time when mechanization was powered mostly by the burning of coal.

In low-latitude wetland areas, lycopods rose high above the ground cover to form the forest canopy. In places, progymnosperms and early conifers (called **cordaites**) also grew into trees. Cordaites, which were a seed-bearing (gymnosperm) group, were freed from living only in moist areas because their seeds were protected by cones. This adaptation allowed them to form woodlands much like modern pine forests. Shrubby ferns,

> **coal swamp** An ancient wetland where vegetative matter produced in massive quantities became lithified to coal.

Visualizing

Coal swamps and their fossils FIGURE 10.31

Oxygen-poor conditions in coal swamp environments are conducive to the exceptional preservation of fossils. Carboniferous coal swamps were commonly sites of the formation of siderite (iron carbonate) concretions.

D This fern or seed fern leaf was beautifully preserved in a siderite concretion.

C Today, swamps are widespread in low-relief-area coastal plains and deltas, and they make excellent modern analogs for ancient coal swamps. This is a swamp on the Atlantic coastal plain of South Carolina. Many of the trees are cypress. The dark water is poor in oxygen because of the high rate of input of dead vegetative matter.

E Much of our record of nonbiomineralizing Carboniferous arthropods comes from siderite concretions formed in coal swamp settings. This horseshoe crab was preserved in a concretion.

seed ferns, cycads, and lycopods filled in much of the undergrowth in forests. Along riverbanks, scouring rushes commonly grew into shrubs and small trees.

Climatic conditions of the Permian in most of Pangea shifted from moist to dry as the supercontinent assembled and most areas lay distant from the moderating effects of ocean waters. Plants adapted to dry conditions—such as true conifers, cycads, ginkgos, and other gymnosperms—tended to replace other forms that had once thrived in wetland environments.

In high southern latitudes of Pangea, the greenery was much less diverse than that living at lower latitudes. During the late Paleozoic, *Glossopteris*, which was adapted to relatively arid conditions, was a dominant component of the flora.

In coal swamps where meandering streams carried freshwater into brackish or marine areas, concretions of siderite (iron carbonate) often formed around decaying organisms. These concretions, which are common in North American and European coalfields, are notable for their exceptionally preserved fossils (FIGURE 10.31D, FIGURE 10.31E on page 319) of terrestrial, freshwater, and marine animals and plants.

Significant among the fossils preserved in the iron carbonate concretions are arthropods, including insects. Exceptional preservation of fossils in concretions shows that insects had, by the Carboniferous, already diversified to become the dominant land animals. Primitive groups (ones unable to fold their wings back over their bodies, represented by dragonflies and mayflies), as well as more advanced groups (represented by beetles, flies, and cockroaches, all of which have foldable wings), are known from Carboniferous rocks.

In a few cases, insects, along with some myriapods (the group that includes millipedes and centipedes) had reached scary dimensions: flying cockroaches measured up to 8 cm in length, dragonflies had wingspans of 20 cm or more, and myriapods 2 m long snaked along the forest carpet. Most land-dwelling arthropods of the Carboniferous were, however, of sizes comparable to their modern relatives.

Freshwater habitats became better populated with animals during the Carboniferous. Clams and snails were more abundant in streams and lakes. Small, bivalved crustacean arthropods, such as ostracodes, soon dominated many quiet freshwater pools. By the Carboniferous, freshwater lakes and streams were inhabited by numerous species of fishes, especially ray-finned fishes and sharks.

Until the Permian Period, amphibians were the dominant terrestrial vertebrates. They occupied a variety of niches and varied widely in size and shape. Some species resembled snakes, and others looked much like crocodiles.

Early reptile FIGURE 10.32

The first completely terrestrial land vertebrates, reptiles, first appeared in the late part of the Carboniferous. They were both numerous and diverse in the Permian Period. This is the skull of an early reptile.

Mammalian ancestors FIGURE 10.33

Primitive synapsids, forerunners of Triassic mammals, made their first appearance on land in the Carboniferous Period and then increased in number during the Permian Period.

A This skull of an early synapsid has only one bone in each side of the jaw, just like modern mammals.

B Pelycosaurs, or sail-backed "mammal-like reptiles," were an unusual group of Carboniferous–Permian synapsids. The sails on their backs may have helped the animals regulate their body heat. This is a skeleton of the Permian pelycosaur *Dimetrodon*.

Reptiles (FIGURE 10.32) appeared in the late Carboniferous, and in the Permian they began diversifying into the groups we know as **squamates** (lizards and snakes), **archosaurs** (crocodiles, dinosaurs, and flying reptiles), and others. Big evolutionary advantages they had over their amphibian ancestors were internal fertilization and an amniotic egg.

Another important group of terrestrial animals that first appeared in the late Carboniferous and diversified in the Permian was the **synapsids**. Early synapsids (FIGURE 10.33) had reptilian skeletons, but their skulls showed features characteristic of mammals, leading to the moniker "mammal-like reptiles." Synapsids, including the more derived mammals (which appeared in the Mesozoic), have only one bone in each side of the lower jaw. In contrast, reptiles have two or three bones in the jaw. Synapsids also have a ball-and-socket articulation between the back of the skull and the neck. One group of primitive synapsids is known as the **pelycosaurs**, or sail-bearing "mammal-like reptiles." These animals carried large "sails" on their backs that were supported by rodlike extensions from the vertebrae. The sails may have acted as thermoregulatory devices to help control body heat.

synapsid
A vertebrate animal having a skull that has one opening located behind and below the eye socket; the clade includes mammals and "mammal-like reptiles."

pelycosaur
A synapsid having a "sail" supported by rodlike extensions of the vertebrae extending from the back.

The most devastating mass extinction in geologic history occurred at the end of the Permian Period.

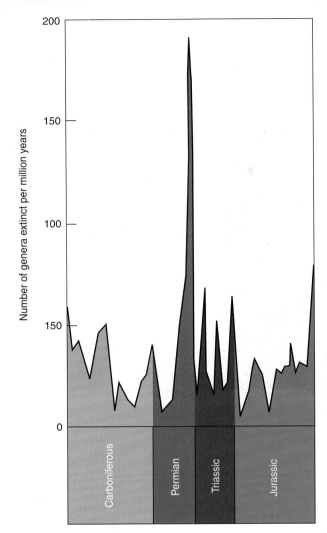

This graph shows the rate of extinction of genera from the Carboniferous Period through the Jurassic Period. There was roughly a three-fold increase in the disappearance of genera at the end of the Permian as compared to the background rate of extinction.

PERMIAN EXTINCTION

More than 80% of all marine species are estimated to have become extinct in the waning stages of the Permian Period. By all estimations, the events of the late Permian amount to the most catastrophic events that life has ever faced (FIGURE 10.34). Tropical marine life forms such as corals, bryozoans, echinoderms, and brachiopods were especially hard-hit. Extinction in the seas actually came in two waves. About 9 million years before the end of the period, reef-formers declined. They collapsed entirely close to the end of the period. The last of the trilobites disappeared in this event, as did rugose corals, productid brachiopods, fusulinids and lacy (fenestrate) bryozoans. Only a few species of ammonoid cephalopods survived to the Triassic. On the land, extinction also took place, perhaps most notably in some amphibian, reptilian, synapsid, and insect groups, as well as among the progymnosperms and gymnosperms (cordaites). Extinction was so severe among terrestrial vegetation that coal seams, which result from the accumulation of peat, are absent from strata of the Early Triassic.

What caused the Permian biotic crisis? Clues abound, but a definitive answer is still elusive. In all likelihood, a combination of changes dealt the lethal blow to so many species. As the Permian Period came to a close, the continents were still moving together and fusing into the supercontinent Pangea. This reduced the total area of shallow epeiric seas and the seaways between continental masses. Adding to the crisis, sea level dropped roughly 100 m during the last 2 million years of the Permian Period. Shallow shelf seas, where marine life was densely concentrated, was increasingly limited to the narrow rims of continents. With shelf areas disappearing, many of the life forms were simply squeezed out of their habitats.

Fusion of continental pieces also had profound effects in terrestrial environments. The large interior areas of Pangea formed widespread deserts because of their great distances from the ocean. These deserts experienced extreme temperature fluctuations. Mountainous areas created by continental collision would have become sites of very cold temperatures, and perhaps glacial buildup. Seasonality increased worldwide. Aridification of vast areas led to widespread evaporite deposition. All these changes must have had serious long-term consequences for animals and plants living on land.

Marine and terrestrial ecosystems of the Late Permian were clearly in a state of crisis caused by continental amalgamation, sea level change, and changing climate. Further events, though, may have exacerbated the problems. A large continental fissure eruption in Siberia 251 million years ago spread flood basalts, called the Siberian traps, across a wide area. There was a similar eruption in China about the same time. Emissions from the eruptions may have blocked sunlight from reaching Earth or added CO_2 and water vapor to the atmosphere, enhancing the greenhouse effect. Carbon isotopes in sediments became enriched in carbon-12 close to the end of the Permian Period, and organic-rich sediments attest to anoxic conditions in the deep sea. At times, upwelling in the seas may have brought oxygen-poor water to the surface, poisoning creatures living in the water column. Another suggested cause of extinction is the release of poisonous methane produced by methanogenic bacteria and archeans in oceanic sediments. Methane hydrates can be released when the temperature rises quickly or the pressure on methane-containing sediment drops quickly. Finally, the rate of extinction at the end of the Permian Period may have been accelerated by the impact of a meteorite or comet. Upper Permian strata are unusually enriched in soccer-ball-shaped carbon molecules, called fullerenes, that contain argon and helium-3 atoms. Most terrestrial helium is of the isotope helium-4, and helium-3 is largely of extraterrestrial origin.

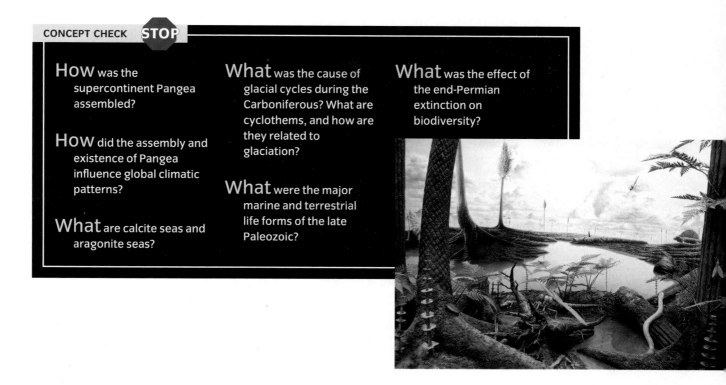

CONCEPT CHECK STOP

How was the supercontinent Pangea assembled?

How did the assembly and existence of Pangea influence global climatic patterns?

What are calcite seas and aragonite seas?

What was the cause of glacial cycles during the Carboniferous? What are cyclothems, and how are they related to glaciation?

What were the major marine and terrestrial life forms of the late Paleozoic?

What was the effect of the end-Permian extinction on biodiversity?

SUMMARY

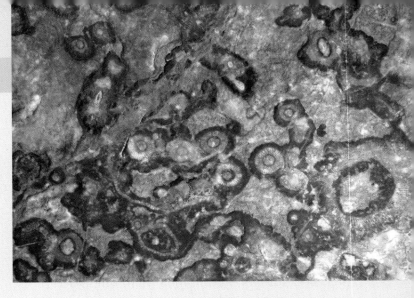

1 Early Paleozoic

1. The major paleocontinents at the start of the Paleozoic were **Laurentia, Baltica, Siberia**, and **Gondwana**. Smaller tectonic pieces included **Avalonia** and the **Kazakhstan Arc**.

2. Sea level was high through most of the Paleozoic, resulting in flooding of the continental shelves with shallow seas called **epeiric seas**, or **epicontinental seas**.

3. The **Cambrian explosion** was the time when most marine animal groups originated or first radiated. Many forms developed resistant skeletons. **Biomineralization** in some groups led to the advent of a good fossil record.

4. In the early Paleozoic, burrowing animals converted mat-stabilized sediment surfaces to ones that were more fluid rich. This ecological change is called the **Cambrian substrate revolution**.

5. The Cambrian fossil record is characterized by three distinct fossil assemblages: (1) diverse **trace fossils**; followed by (2) diverse **small shelly fossils**; and (3) **archaeocyaths** and other **sponges, trilobites, brachiopods, mollusks, hyoliths**, and other remains.

6. Some Cambrian deposits are notable for their **exceptionally preserved fossils**—fossilized remains of nonbiomineralized or lightly biomineralized cuticle, gut contents, and internal soft parts. These deposits provide unusually complete views of ancient ecosystems.

7. **Trilete spores** indicate that plants may have made the transition to land in the Cambrian Period.

8. Several **mass extinctions** in Cambrian time resulted in quite different post-Cambrian communities of marine organisms.

9. The **Taconic orogeny** (Ordovician), due to closure of the Iapetus Ocean between Laurentia and Baltica, was the first of three mountain-building episodes in the Appalachian region of North America.

10. In the Ordovician, predatory **cephalopods** and **fishes** became prevalent. Other important faunal elements included **graptolites, bryozoans, starfishes, brittle stars, brachiopods**, and **mollusks**.

11. Glaciation occurred at the end of the Ordovician Period. As glaciation reached a maximum, marine organisms suffered extinctions, although major groups were not eliminated.

2 Middle Paleozoic

1. Silurian marine invertebrates mostly filled niches vacated in the end-Ordovician biotic crisis. Two exceptions were **tabulate corals** and **stromatoporoid sponges**, which became reef-formers.

2. The first **jawed fishes** appeared in Ordovician oceans.

3. The earliest definitive vascular land plants are Silurian in age. Arthropods also invaded the land by the Silurian Period.

4. Continental collision between Laurentia and Baltica initiated the **Caledonian orogeny** in the Silurian–Devonian and the **Acadian orogeny** in the Devonian. Suturing of the continents resulted in a larger continent called **Euramerica**. Sediments shed to the west of the Acadian Mountains filled the **Appalachian Basin**, and sediments shed to the east comprise the Old Red Sandstone and its equivalents.

5. The Devonian Period was a time of important biotic changes in marine and terrestrial ecosystems. Jawed fishes diversified, **ammonoid cephalopods** appeared, forests were established on land, land-dwelling **amphibians** appeared, and **insects** evolved from wingless terrestrial hexapods.

6. Many marine species were eliminated in a biotic crisis near the end of the Devonian Period.

3 Late Paleozoic

1. The Carboniferous Period was a time of climatic transition from a greenhouse world in the early half of the period to an icehouse world in the later half.

2. Assembly of **Pangea** began in the Carboniferous Period. Mountain building in Europe and northwestern Africa is ascribed to the **Hercynian** (or **Variscan**) **orogeny**. In North America, orogenic activity called the **Alleghanian** (or **Appalachian**) **orogeny** comprises the third and final phase of mountain building in the Appalachian region.

3. The Carboniferous Period was a time of transition from a **calcite sea**, which had been in existence since the Cambrian, to an **aragonite sea**. Aragonite sea conditions lasted until the Jurassic Period, when calcite sea conditions returned. The most recent switch to an aragonite sea occurred in the mid-Cenozoic.

4. Climate cycles of the late Carboniferous were brought on by changes in Earth's orbital patterns referred to as **Milankovitch cycles**. Sudden, intense cooling can result when the effects of two or more orbital cycles coincide.

5. Cycles of glacial expansion and contraction in polar Gondwana are reflected in sea level fluctuations recorded in **cyclothems**.

6. In the early Carboniferous (Mississippian), epeiric seas left widespread carbonate deposits, including deposits rich in **crinoids**, in tropical shelves.

7. In the late Carboniferous (Pennsylvanian), enormous wetlands stretched along tropical coasts. The massive production of vegetative biomass turned coastal wetland areas into **coal swamps**.

8. Carboniferous coal swamps were home to diverse insects and other arthropods, amphibians, and the earliest **reptiles**.

9. In the Permian Period, the **supercontinent** Pangea was a continuous landmass that altered ocean circulation patterns and experienced strong latitudinal climatic gradients.

10. Late Paleozoic marine faunas were rich in typical Paleozoic forms, especially brachiopods, bryozoans, rugose corals, and mollusks. **Fusulinid foraminifera** evolved rapidly through the late Carboniferous and Permian.

11. Land-dwelling reptiles, including lizards and **archosaurs**, diversified in the Permian Period. **Synapsids** appeared for the first time in the Permian.

12. More than 80% of marine species became extinct by the end of the Permian Period. Among the groups that disappeared were trilobites, rugose corals, productid brachiopods, fusulinids, and lacy bryozoans.

KEY TERMS

- **epeiric sea (epicontinental sea)** p. 281
- **Sauk sequence** p. 282
- **Tippecanoe sequence** p. 282
- **Kaskaskia sequence** p. 282
- **Absaroka sequence** p. 282
- **Cambrian explosion** p. 283
- **small shelly fossils** p. 284
- **archaeocyath** p. 285
- **trilobite** p. 285
- **brachiopod** p. 286
- **conodont** p. 286
- **chordate** p. 286
- **graptolite** p. 286

- **Iapetus** p. 294
- **Taconic orogeny** p. 295
- **Caledonian orogeny** p. 300
- **Euramerica** p. 301
- **Acadian orogeny** p. 301
- **tectonosedimentary cycle** p. 301
- **Antler orogeny** p. 301
- **acanthodian** p. 304
- **gnathostome** p. 304
- **placoderm** p. 305
- **rhyniophyte** p. 308
- **gymnosperm** p. 309
- **amniotic egg** p. 309

- **insect** p. 310
- **crossopterygian** p. 310
- **Hercynian orogeny** p. 314
- **Alleghanian orogeny** p. 314
- **Wilson cycle** p. 314
- **Ouachita orogeny** p. 314
- **Sonoman orogeny** p. 315
- **Milankovitch cycles** p. 315
- **cyclothem** p. 315
- **coal swamp** p. 319
- **synapsid** p. 321
- **pelycosaur** p. 321

CRITICAL AND CREATIVE THINKING QUESTIONS

1. Why do Paleozoic marine deposits crop out over so many areas of the continents today? What does this tell us about the current position of sea level relative to its positions in the Paleozoic? Where is the additional supply of water today? What could cause sea level to again reach the positions it reached in the Paleozoic?

2. What groups of animals and plants successfully made the transition to land? What made some groups, such as insects, more successful as land-dwellers than other groups, such as snails?

What is happening in this picture ?

Carboniferous Crinoids

Stalked crinoids rose to prominence in the early part of the Carboniferous (Mississippian). This slab is from Indiana.

How many different pieces are there in one specimen?

Are the pieces all the same size and shape?

If specimens were always found disarticulated, how would you be able to determine which pieces belonged to the same individual and reconstruct what the animal looked like in life?

1. What were the major paleocontinents of the Paleozoic?

2. What was the Cambrian explosion?

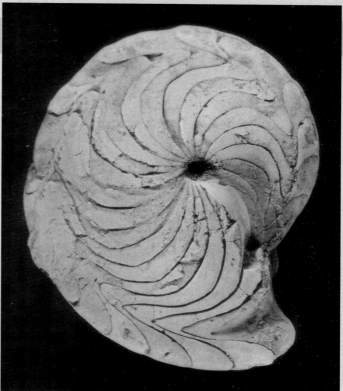

3. How did animals influence changes in marine sediments of the early Paleozoic?

4. What is exceptional preservation of fossils, and how has it helped us better interpret life of the past?

5. How has sea level history contributed to our understanding of Paleozoic marine life?

6. What three orogenic events were responsible for geologic developments in the Appalachian region of North America? How were those events mirrored in Europe and Africa?

7. What is the tectonosedimentary cycle?

8. What is a clastic wedge?

9. How did the rise of jawed fishes and ammonoid cephalopods change the dynamics of marine ecosystems?

10. How did plants and animals cope with the new challenges posed by life on land?

11. When was Pangea assembled?

12. How did the assembly and existence of Pangea influence climate history and the diversity of marine life?

13. What is a calcite sea? What is an aragonite sea?

14. What is a cyclothem?

15. What aspect of Earth history spurred scientists to name a period the Carboniferous?

16. What was the sequence in the evolution of fully terrestrial vertebrate animals?

17. What events led to the mass extinction at the end of the Permian Period?

Mesozoic World

The Mesozoic Era, or time of "middle animal life," consists of three periods, the Triassic, Jurassic, and Cretaceous. It begins with most of the world's continents united in one giant landmass, Pangea, with sea level at a low point, and with biodiversity at its lowest levels since the Cambrian. In the 185.5-million-year-span of the Mesozoic, the large island areas of present-day Asia finally sutured to Pangea, and significant biotic changes reshaped marine and terrestrial ecosystems. In the oceans, new predators and predatory strategies appeared on the scene, and prey evolved to meet the challenges. Land environments became dominated by archosaurian reptiles, including dinosaurs such as *Allosaurus* (shown here). Remains of dinosaurs, both large and small, as well as trackways they left behind, are a common feature of nonmarine rocks around the world. In the late Mesozoic, angiosperms (flowering plants) overtook gymnosperms as the dominant land plants. Mammals made their first appearance in the early Mesozoic, and insects diversified.

In the mid-Mesozoic, Pangea began rifting apart, extensive flood basalts spilled out along rift systems, and sea level rose to record levels shortly thereafter. The era closed with a dramatic mass extinction that took a heavy toll on marine and terrestrial life forms.

Triassic Period

G lobal sea level at the beginning of the Triassic Period was relatively low and close to its modern position. It rose to a level perhaps 100 m above its present position and then fell again at the end of the period. In North America, marine deposits of the Triassic Period constitute the upper part of the Absaroka sequence (**FIG-URE 11.1**).

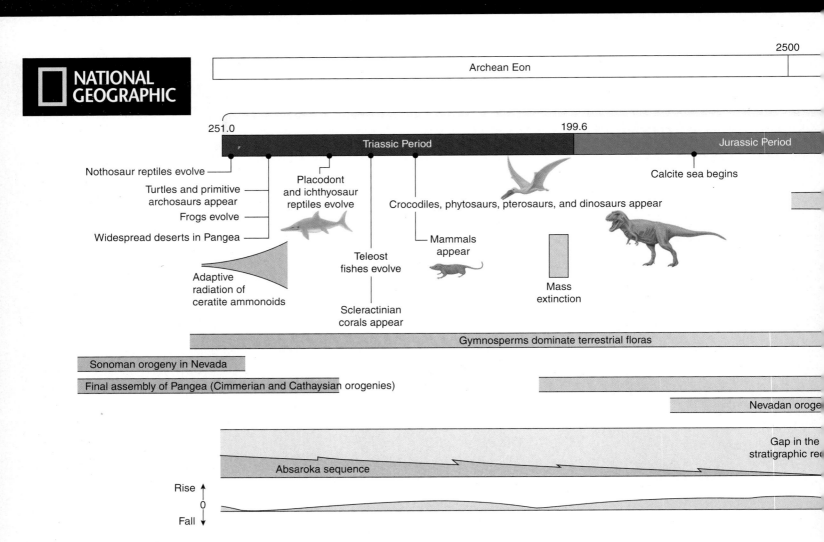

NATIONAL GEOGRAPHIC

2500

Archean Eon

251.0 — 199.6

Triassic Period — Jurassic Period

Nothosaur reptiles evolve

Turtles and primitive archosaurs appear

Frogs evolve

Widespread deserts in Pangea

Placodont and ichthyosaur reptiles evolve

Adaptive radiation of ceratite ammonoids

Teleost fishes evolve

Scleractinian corals appear

Mammals appear

Crocodiles, phytosaurs, pterosaurs, and dinosaurs appear

Calcite sea begins

Mass extinction

Gymnosperms dominate terrestrial floras

Sonoman orogeny in Nevada

Final assembly of Pangea (Cimmerian and Cathaysian orogenies)

Nevadan oroge

Gap in the stratigraphic rec

Absaroka sequence

Rise ↑
0
Fall ↓

Orogenies in Eurasia (the Cimmerian orogeny in southern Eurasia and the Cathaysian orogeny along the southeastern margin of Eurasia) completed the assembly of Pangea. East of Pangea was the Tethys Ocean, and west of it was the Pacific Ocean (FIGURE 11.2 on page 332). Large portions of Pangea, areas far from ocean waters, formed arid deserts. The high peaks of the Appalachian Mountains, which arose in the late Paleozoic, were subdued through erosion in the Triassic Period. By the latest Triassic, Pangea began to split, forming elongate rift basins in eastern North America that enlarged through the Jurassic Period, eventually forming the Atlantic Ocean.

TRIASSIC MARINE LIFE

Extinction at the end of the Permian Period wiped out great numbers of marine organisms, leaving a new world to be exploited by those surviving the crisis. Major groups that failed to survive were the trilobites, fusulinid foraminifera, rugose corals, productid brachiopods, and lacy bryozoans. Other groups suffered substantial declines but were not completely annihilated.

Recovery from the Permian extinction event was a protracted process for most marine animals other than ammonoid cephalopods and conodonts. Ammonoids, which were reduced to two genera at the end of the

Visualizing

Timeline of Mesozoic events FIGURE 11.1

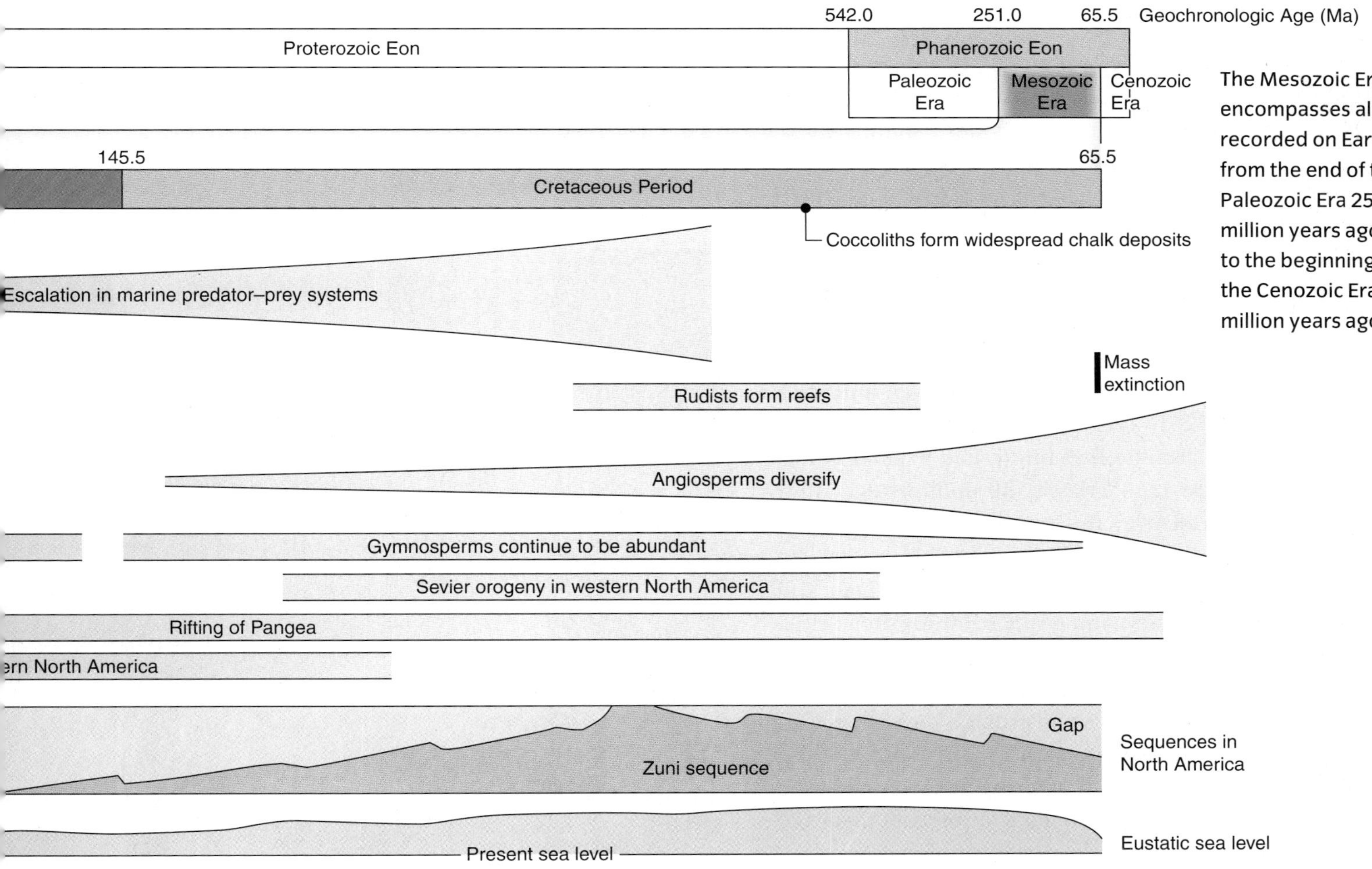

The Mesozoic Era encompasses all time recorded on Earth from the end of the Paleozoic Era 251 million years ago (Ma) to the beginning of the Cenozoic Era 65.5 million years ago.

Triassic paleogeography FIGURE 1 1 .2

Pangea was fully assembled during the Triassic Period. To the east was the Tethys Ocean, and to the west was the Pacific Ocean.

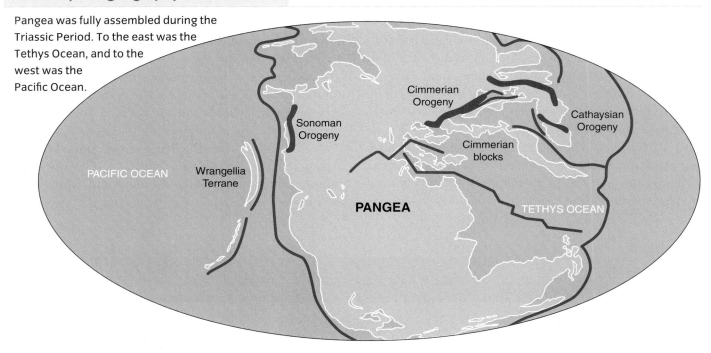

Permian, increased to more than 100 genera in the first 6 million years of the Triassic Period. Those ammonoids had ceratitic sutures, patterns that were slightly more complex than the simple goniatitic sutures of their Paleozoic Era forebears. Conodonts, which secreted phosphatic toothlike microfossils, evolved quickly in the Triassic Period. As the Triassic wore on, clams, snails, and brachiopods slowly recovered from the end-Permian crisis. Clams and brachiopods were common in the Early Triassic, but their diversity was limited. Grazing snails were decimated in the Permian Period, and for a brief time in the Early Triassic, without the threat of significant grazing, stromatolites multiplied in shallow marine waters. By the Late Triassic, 50 million years after the Permian extinction, the oceans were once again filled with abundant life forms.

At the start of the Triassic Period, algae and sponges filled the reef-forming niche, but by the Middle Triassic, scleractinian corals (descendants of Paleozoic tabulate corals) dominated in that setting. By the latest Triassic, scleractinian corals constructed mounds towering tens of meters above the seafloor (**FIGURE 1 1 .3**). Rapid growth of the reefs was probably facilitated by a symbiotic association of green algae in the soft tissues of the corals, just as it is in modern reef-forming corals.

Reef-forming corals FIGURE 1 1 .3

Reef-forming scleractinian corals first appeared in the Triassic Period. After a decline in the Jurassic, they reappeared as major reef-formers in the Cenozoic Era. This close-up view of a coral skeleton is from a late Cenozoic reef.

A Modern-type teleost, or bony, fishes evolved in the Triassic Period. This is a specimen of Jurassic age from Germany.

B Teleost fishes of the Mesozoic Era shared the water with a variety of animals, including reptiles of various kinds. The teleosts were largely predators, but they also served as prey for carnivorous reptiles such as these plesiosaurs (left) and ichthyosaurs (right).

Bony fishes called **teleosts** (FIGURE 11.4) were introduced in the Triassic Period, and they soon overtook ray-finned bony fishes in abundance and diversity. Early teleost fishes had skeletons composed partly of cartilage, rather than of bone like their modern descendants. Similar to their Paleozoic Era relatives, their tails were asymmetrical. Many of the early forms had simple jaw structures and had swim bladders to control buoyancy. As time passed, the diamond-shaped scales of early teleost fishes gave way to more hydrodynamic rounded scales. Today, most living fishes are teleosts, and their feeding habits are almost as varied as their body forms.

Sea urchins expanded into a variety of benthic niches in the Triassic and Jurassic periods. Originally surface dwellers, these deposit-feeders evolved the capability to burrow into the sediment, some of them reaching depths of many centimeters below the sediment surface.

Reptiles first invaded the sea during the Triassic Period. The lizardlike **nothosaurs**, which have limbs modified into paddles, appeared in the Early Triassic. A related group, the **placodonts**, which had wide, armored bodies and blunt teeth used for crushing mollusk shells, appeared later in the Triassic. Neither reptile group lived beyond the Triassic, but **plesiosaurs**—large, sleek, carnivores descended from nothosaurs in the Middle Triassic—were among the most successful oceangoing reptiles of the Mesozoic.

In a fascinating example of convergence (homoplasy), a group of marine reptiles called the **ichthyosaurs** (literally "fish lizards") evolved a general shape

plesiosaur
A Mesozoic marine reptile with a euryapsid skull type, a broad body, and large, paddlelike limbs.

ichthyosaur A Mesozoic marine reptile with a dolphin-like body.

A Ichthyosaurs are marine reptiles that appeared in the Triassic Period and became even more prevalent in the Jurassic Period.

B From specimens preserved in black shale with outlines of the carbonized skin, we know ichythyosaurs had crescentic, vertical tails reinforced in the lower part by a downwardly bent backbone.

similar to that of a fish or a dolphin (FIGURE 11.5). Ichthyosaurs, which evolved in the Triassic Period, had long snouts for catching food, paddlelike limbs, large eyes, and sleek, hydrodynamic bodies.

Near the end of the Triassic Period, mass extinction ripped through marine ecosystems. The timing of extinction coincides with floral evidence of a climatic shift to more arid conditions in terrestrial environments of Gondwana, and it coincides with a large drop in sea level. Roughly 20% of marine families disappeared in a two-step extinction during the last 15 million years of the Triassic. Conodonts, nothosaurs, and placodonts were eradicated. Most other marine animal groups suffered losses but managed to survive into the Jurassic Period.

TERRESTRIAL LIFE OF THE TRIASSIC

Land plants were little affected by extinction at the end of the Paleozoic Era. As the world entered the Mesozoic Era, gymnosperms such as ferns, conifers (FIGURE 11.6), cycads, and ginkgos dominated the landscape. Seed ferns survived into the Triassic Period but became extinct before the end of the period.

Petrified logs FIGURE 11.6

Conifer logs, petrified by colorful silica, from Triassic strata of the Colorado Plateau, Arizona. Gymnosperms like these dominated Mesozoic terrestrial forests until the Cretaceous Period.

Phylogeny of the reptiles and their close relatives FIGURE 11.7

Reptiles and their sister group, the synapsids, show four skull types that help to characterize phylogenetic lines. Turtles have anapsid skulls, which apparently arose homoplasically. Placodonts, ichthyosaurs, and plesiosaurs show the euryapsid skull form. Dinosaurs and other archosaurs have the diapsid skull type. Primitive synapsids (or "mammal-like reptiles") and mammals have synapsid skull types.

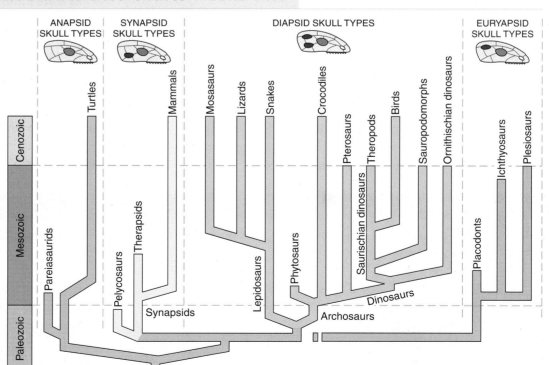

A Early dinosaurs, including *Rioarribasaurus*, were relatively small, sleek predators. This skeleton is from Triassic rocks of New Mexico.

B An artist's reconstruction of *Rioarribasaurus*.

Three land vertebrates appeared during the Early Triassic—an amphibian group (frogs) and two reptilian groups (turtles and primitive archosaurs). Primitive archosaurs (or **thecodonts**) stood or walked upright on two legs. Prior to their evolution, reptiles such as lizards showed only a squat, sprawling posture requiring four legs. Evolution of an upright posture opened the way for archosaurs to develop greater speed and agility.

The group known as **archosaurs** (FIGURE 11.7) includes **crocodiles, phytosaurs, pterosaurs** (flying reptiles), and **dinosaurs**. All these reptiles appeared in the Late Triassic.

Two archosaur groups developed similar body forms, another interesting example of convergence (or homoplasy). Crocodiles and phytosaurs both had long, lizardlike bodies, but they walked on four legs carried under the body, rather than out to the sides in lizardlike fashion. Both reptiles lived as predators in swampy environments. Phytosaurs, though, evolved from a different lineage of primitive archosaurs than did the crocodiles, and they were distinguished from their look-alikes primarily by the position of the nostrils. Whereas crocodiles have nostrils at the front of the snout, phytosaurs had them on the top of the head, just in front of the eyes. Unlike crocodiles, phytosaurs did not survive past the Triassic Period. Some Mesozoic Era crocodiles reached lengths several times the size of their modern relatives.

The earliest dinosaurs were small bipedal animals (FIGURE 11.8). During the Jurassic Period, dinosaurs diversified and some reached large proportions.

archosaur A reptile having a diapsid skull type with teeth set in sockets. This group includes the crocodiles, pterosaurs, dinosaurs, and birds.

A *Cynognathus*, a primitive synapsid of the Triassic Period, had a reptile-like body and a mammal-like skull.

B Early mammals were small and looked like rodents. Multituberculate mammals, like the one shown here, appeared in the Triassic Period.

Two types of synapsids are known from Triassic strata (FIGURE 11.9). Primitive synapsids (or "mammal-like reptiles") persisted from the Late Paleozoic. Fossils of one Triassic therapsid, *Lystrosaurus*, were helpful in matching continents fragmented from Pangea. **Mammals**, represented first by a group called the multituber-culates, appeared in the Late Triassic. The early mammals remained small and inconspicuous through the entire Mesozoic Era. Most were reminiscent of modern rodents in size and shape. Late in the Mesozoic, monotremes and marsupials evolved, but placental mammals did not appear until the Paleogene Period.

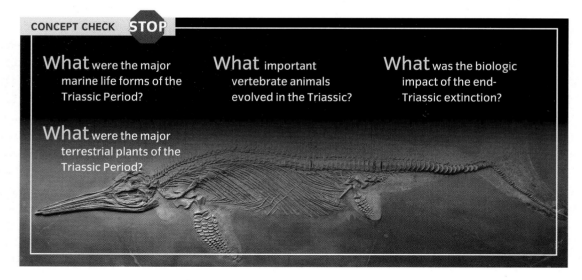

CONCEPT CHECK STOP

What were the major marine life forms of the Triassic Period?

What important vertebrate animals evolved in the Triassic?

What was the biologic impact of the end-Triassic extinction?

What were the major terrestrial plants of the Triassic Period?

Jurassic Period

he Jurassic Period was a time of great change tectonically and biotically. The period opened with global sea level at one of its lowest points in geologic history. Rising sea level beginning in the Early Jurassic Period resulted in deposits referred to in North America as the **Zuni sequence**. Pangea was beginning to rift apart, and both marine and terrestrial ecosystems were being reshaped. Separation of Pangea led not only to the development of the modern continents but also to the evolution of separate biotas associated with them. As rifting proceeded, sea level rose (only to fall quickly near the end of the period). The rising sea level ushered in a new time of calcite seas. The switchover to calcite seas happened in the Middle Jurassic.

Most major life forms of the Jurassic Period were already in existence by Triassic time, but extinction near the end of the Triassic Period reshuffled the evolutionary cards and gave the upper hand to groups that previously did not have ecologic prominence. For the remainder of the Mesozoic Era, reptiles would be the most formidable animals on land, in the water, and in the air.

EARLY–MIDDLE MESOZOIC TECTONIC EVENTS

Modern continents have their origin in tectonic events beginning in the latest Triassic–Middle Jurassic (**FIGURE 11.10**). Assembly of Pangea was completed in the Triassic Period. It did not last as a unified supercontinent much longer though. By the latest Triassic, splits had begun to develop along the contact zone between the Americas and Europe to Africa. It can be

Jurassic paleogeography FIGURE 11.10

Pangea underwent splitting in the Jurassic Period. Initial rifting formed a series of small basins that filled with sediment. Eruptions of basaltic lavas also accompanied the early phases of rifting.

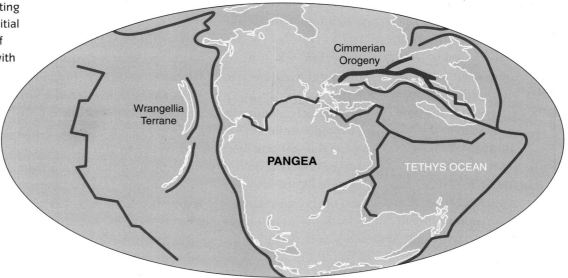

inferred that Pangea shifted to a position overlying a line of hot spots that caused crustal expansion and cracking, similar to that occurring today in the East African rift zone. By the Early Jurassic, a line of elongate, fault-bounded rift basins lined either side of the incipient Atlantic Ocean.

The Atlantic rift basins, extending from eastern Canada southward through New York City and environs, to the Gulf Coast, opened in the latest Triassic Period through the Jurassic Period (FIGURE 11.11). More than 5 km of nonmarine sediments, assigned to the Newark Supergroup, accumulated in some basins. Arkoses (poorly sorted sandstones and conglomerates, rich in minimally weathered feldspar minerals) were rapidly deposited in alluvial fans. Shallow lakes partly filled the basins. Evaporite minerals, indicative of a warm, arid climate, precipitated around the margins of lakes. Also along the lakeshores, dinosaur populations left footprints in what became reddish paleosols.

Accompanying the expansion of rift basins was the upwelling of basaltic magmas. Vertical dikes fed hot magma to the surface, where it spilled out in widespread sills. As the magma cooled and contracted, it formed distinct columnar joint patterns. The Palisades, exposed along the Hudson River near New York City, is a nice example of one of these sills. The vertical columns, resembling a palisade, or large wooden fence, represent the weathered faces of columnar joints of basalt (FIGURE 11.11C).

Jurassic splitting of Pangea was not confined to the Atlantic region. Similar rifts, often with basaltic outpourings, developed in Antarctica, extending all the way to Tasmania and into southern Africa. The "backbone" of the great Transantarctic Mountains is mostly formed of Jurassic basaltic rocks filling an extensive rift system. Along the margins of the present-day continents are a number of failed rifts whose presence is made obvious by river systems such as the Mississippi and the Amazon.

In the Middle to Late Jurassic, one rift arm split North America from South America and opened up the Gulf of Mexico. At times, seawater made its way into this newly formed rift basin. Evaporation of the seawater led to deposition of widespread evaporite deposits. Salt has a lower density than other sediments and flows under pressure. Jurassic salt deposits surrounding the Gulf of Mexico now have numerous salt domes that formed from overpressurized and mobilized salt. Many of the salt domes have become petroleum traps.

Tectonics of the Jurassic to Early Cretaceous considerably altered the western margin of North America. Accretion of a complex series of tectonostratigraphic terranes, consisting of microplates and volcanic islands, greatly increased the size of the North American continent by additions extending from Alaska, through western Canada, to California.

One microcontinent, called Sonomia, comprising parts of Nevada plus California and Oregon, sported a volcanic arc (called the Golconda arc) on its eastern margin, as it neared North America in the late Paleozoic. The distance was progressively closed through subduction, which provided remelted igneous magma to the island arc. The deformation associated with this collisional event is referred to as the **Sonoman orogeny**. Early in the Triassic Period, Sonomia sutured to North America. The accretionary wedge on the leading edge of the microcontinent was squeezed along the contact zone, forming what we now call the Golconda terrane.

Evidence of the former existence of a subduction zone along western North America comes from rocks called the Franciscan sequence. These rocks, exposed in the vicinity of San Francisco Bay, consist of deep-water sediments (fine siliciclastics, cherts, and some limestones), as well as volcanics, all of which formed an accretionary wedge that was metamorphosed and deformed at high pressures but low temperatures through collision with North America.

Rising plumes of magma, formed from oceanic crust remelted along subduction zones in the Pacific Ocean, thickened the continental crust of North America and formed new igneous rocks. During the time of the **Nevadan orogeny** (Jurassic to Early Cretaceous), folding and metamorphism occurred in the western part of the North American Cordillera. Also, emplacement of the great batholiths of the Sierra Nevada Range began during this time. The batholiths continued to form through the Late Cretaceous. Today, granitic rocks of the Sierra Nevada are spectacularly exposed in Yosemite National Park, California.

> **Nevadan orogeny** An orogenic episode that occurred in western North America during the Jurassic and Early Cretaceous.

Pangea began rifting apart during the latest Triassic Period through the Jurassic Period. Spectacular evidence of that rifting is preserved in a series of basins formed along the Atlantic coastline of North America.

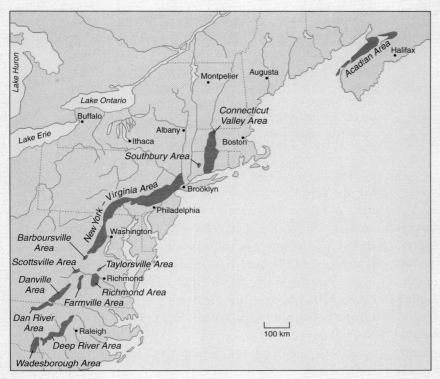

A Locations of rift basins produced by block faulting in the Atlantic region during the early stages of the rifting of Pangea. The basins filled with lake and stream sediments, as well as reddish soils.

C As the North Atlantic Ocean opened up, fissures fed by dikes formed in the rift basins. When magma spilled out close to the surface, large sills were formed. One such sill is the Palisades along the Hudson River of New York and New Jersey.

Stage 1

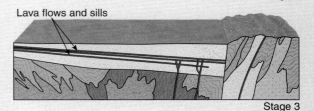

Triassic deposits

Stage 2

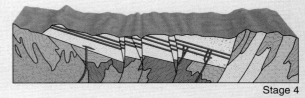

Lava flows and sills

Stage 3

Stage 4

B Stages in the development of the Connecticut Valley rift basin

Stage 1: The topography and complex structures inherited from the Alleghanian orogeny were eroded.

Stage 2 represents a low-profile plain that had developed by the Late Triassic.

Stage 3: A large, trough-like fault basin opened up and became the site of sedimentary deposition. Basaltic dikes, sills, and flows also partly filled the basin.

Stage 4: During the Early Jurassic, the Connecticut Valley basin was complexly broken through normal faulting. This is the origin of the modern structure of the region.

NATIONAL GEOGRAPHIC

A Belemnoids, which are Mesozoic relatives of modern squids, are best known for having cigar-shaped internal skeletons.

B Oysters, a group of clams, became diverse in Mesozoic oceans. Many, like this specimen of *Exogyra*, had thickened, coiled shells. The large shell of this form helped stabilize the oyster in soft marine mud, and a small lid-like valve opened and closed for feeding and other life functions.

JURASSIC MARINE LIFE

In the Jurassic Period, marine predator–prey systems were in the midst of a third, and final, time of escalation, the **Mesozoic Marine Revolution**. This time, reorganization of marine ecosystems was facilitated by the extinction of many species and the reopening of once-occupied niche spaces near the end of the Triassic Period.

Swimming predatory mollusks included the ammonoids and relatives of the modern squids called the **belemnoids** (FIGURE 11.12A). These cephalopods enjoyed a varied diet, including crustacean arthropods, fishes, and mollusks. Ammonoids evolved rapidly in the Jurassic and Cretaceous periods, with most species having ranges no more than 1 million years or so. This makes their remains valuable as biostratigraphic tools today. Belemnoids first showed up in marine waters during the late Paleozoic, but they were not really prominent until the Mesozoic Era. The belemnoids are best known for having cigar-shaped, calcareous internal shells. As a measure of their prominence in Jurassic deposits, one of the analytical standards used for developing oxygen isotopic ratios from carbonate deposits is called PDB, which stands for Peedee belemnite, a collection of belemnoid fossils from the Peedee Formation of New Jersey.

Coiled oysters (FIGURE 11.12B) appeared in the Jurassic. Their shells were commonly thickened, making them difficult prey for shell-crushing predators.

Fishes of the Mesozoic Era were mostly carnivorous. Sharks and rays, holdovers from the Paleozoic Era, increased in number in the Jurassic and Cretaceous periods. Some teleosts had sharp, daggerlike teeth adapted for catching other fishes (FIGURE 11.13) or

Fishing in the Mesozoic FIGURE 11.13

Teleost fishes, most of them predatory, radiated in the Mesozoic Era. In a few instances, a fish's last meal has been fossilized in the gut of a larger fish. This specimen comes from Cretaceous strata of Kansas.

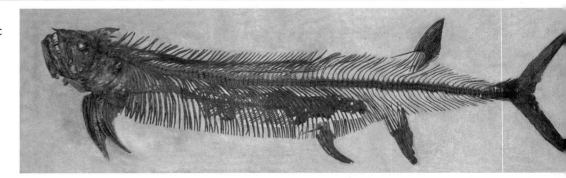

Swimming and flying reptiles FIGURE 11.14

B Ichthyosaurs were swimming reptiles that gave live birth in the water.

C Pterosaurs were flying archosaurs that lived only during the Mesozoic Era. This specimen of *Pterodactylus* is from a Jurassic deposit, the Solnhofen Limestone of Germany.

A Plesiosaurs were large carnivorous reptiles that inhabited Mesozoic seas.

cephalopods, and others had blunt teeth adapted for crushing clam and snail shells.

In addition to fishes, a number of other vertebrate predators patrolled the seas. Plesiosaurs (FIGURE 11.14A) had developed into a major threat, and they apparently had a large vertical range in the water column. Some skeletons show strange bony ossifications, suggesting that they could dive deeply—at times so deeply that they developed the bends from failing to return slowly to the ocean surface and repressurize properly.

Ichthyosaurs (FIGURE 11.14B) were among the top predators of Mesozoic seas. Their large eyes apparently helped them identify food under low light conditions, including at night, and their fishlike bodies allowed them to cut through water quickly. From some remarkably preserved remains of female ichthyosaurs containing skeletons of fetal individuals, we know that mother ichthyosaurs did not lay eggs like many terrestrial reptiles do; instead, they gave birth to live young.

Some pterosaurs (FIGURE 11.14C), or flying reptiles, probably relied on fishes, crustaceans, and cephalopods swimming close to the water's surface for a large part of their nutrition. Pterosaurs had small bodies, forelimbs expanded into giant wings, and thin, light, hollow bones to facilitate flight. Their snouts were long and narrow. Some species were armed with sharp teeth, and others were toothless. These dragonlike monsters of the air had wingspans ranging from about 30 cm to more than 10 m.

In the Jurassic Period, echinoderms (FIGURE 11.15) expanded into more significant predatory roles. Most notable among them were the starfishes, which use the suction power of their tube feet to attach

Mesozoic echinoderms FIGURE 11.15

A Starfishes became important benthic predators in the middle part of the Mesozoic Era.

B Sea urchins, which were once mostly round, like this specimen, became more heart shaped in the Jurassic and Cretaceous periods as they adapted to burrowing in sediment.

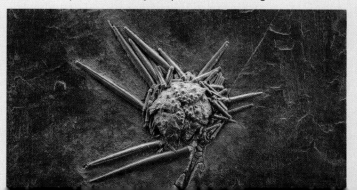

Jurassic lobster FIGURE 11.16

Crustaceans diversified greatly beginning in the mid-Mesozoic and became a significant group of predators.

to clams and open their shells. Sea urchins also increased in prominence, adapting to a variety of life habits, including sediment burrowing.

Other benthic or nektobenthic carnivores included some snails and crustaceans. Snails that could use their hard radulas to drill into prey, such as clams and other snails, became a much greater menace in the Mesozoic Era than they were in the Paleozoic Era. An adaptive radiation of aqueous crustaceans (FIGURE 11.16) began in the Jurassic Period and continued through the early Cenozoic. Lobsters and crabs acquired thickened, powerful claws with which they could quickly dispatch snails and other prey.

In response to the heightened threat from predators in and above the Mesozoic seas, prey species (many of them predators themselves) commonly armed themselves with thicker shells and with spines. Many prey species evolved cloaking strategies such as camouflage patterns or attachment of empty shells to their own shells. Squids developed glands that could release ink into the water when predators came near. Cephalopods and crustaceans could rapidly propel themselves backward in the water when threatened. Some sea urchins and clams evolved the ability to burrow deeply in sediment. Even skates and rays used the sediment for con-

cealment, evolving a flattened body form that allowed them to escape detection by enemies and prey alike.

By the Jurassic Period, brachiopods and stalked echinoderms fell into great decline, and they never again achieved more than a minor role in marine ecosystems. To a large extent, their decline was probably related to the expansion of marine predators during the Mesozoic Marine Revolution.

Not all marine life forms of the Jurassic Period were engaged in a vicious cycle of ecologic escalation. Algae comprising the group we call **calcareous nannoplankton** made their appearance during this period. Calcareous nannoplankton, including coccolithophorids, have been major contributors to the sediment of the deep sea, especially in the tropics, where they were most abundant through much of the Mesozoic and Cenozoic eras.

Another innocuous group of floating algae, called the **dinoflagellates**, which produce durable organic cysts, emerged in the Jurassic Period as important contributors to the phytoplankton biomass. Dinoflagellates have an evolutionary history that extends back to the Proterozoic Era (where some are referred to under the collective term "acritarchs"), but they achieved a more impressive diversity in the Jurassic Period. They have continued to remain diverse through to the present time.

Ancient and modern cycads FIGURE 11.17

A Gymnosperms were the dominant land plants of the early and middle parts of the Mesozoic Era. Cycads were so abundant during Jurassic time that the period is often referred to as the "Age of Cycads."

B Modern cycads like these live in tropical and subtropical areas of the world.

TERRESTRIAL LIFE OF THE JURASSIC

Gymnosperms were the most common land plants of the Jurassic Period. Cycads (FIGURE 11.17), conifers, ferns, and ginkgos dotted the landscape and provided much of the food for herbivorous dinosaurs and mammals.

Most terrestrial animal groups of the Triassic Period continued into the Jurassic Period. Synapsids declined in importance near the end of the Triassic, and phytosaurs do not seem to have survived beyond the end of the period. Dinosaurs became the most conspicuous land animals of the Jurassic Period, and they remained so through the Cretaceous Period.

One odd group of amphibians that appeared in the Jurassic is known as the **caecilians**. These snakelike animals have as many as 200 vertebrae, and their limbs are either reduced or lost altogether. Modern caecilians live in the tropics, seeking food in ponds or in the leaf litter of forests.

DINOSAURS: RULERS OF THE LAND

Few animals that have ever lived evoke more fascination than the dinosaurs. Although they clearly rank as some of the most successful of all land vertebrates, their success did not come instantaneously. **Dinosaurs** descended from primitive archosaurs in the Late Triassic. In the first 20 million years or so of their reign on Earth, they were overshadowed by another vertebrate group, the synapsids. However, the extinction of many synapsids in the Late Triassic gave the evolutionary advantage to dinosaurs. In the Middle and Late Jurassic, dinosaurs reached their greatest diversity and their greatest individual dimensions.

Dinosaurs (FIGURE 11.18 on page 344) are characterized by an upright posture, carrying their legs below the body, and by having a skull with two openings behind the eye. There are two major groups, each distinguished principally on the basis of hip bone structure. **Saurischian dinosaurs**, or "lizard-hipped" dinosaurs, have the pubis bone pointing forward and the ischium bone pointing backward. This is the same configuration as in the earliest archosaurs and in modern lizards. **Ornithischian dinosaurs**, or "bird-hipped" dinosaurs, have both the pubis and ischium pointing backward, as in modern birds.

dinosaur An archosaur characterized by either ornithischian or saurischian hip bones and upright posture.

saurischian dinosaur An archosaur characterized, at least primitively, by a lizard-like hip. The clade includes herbivorous sauropodomorphs, carnivorous theropods, and birds.

ornithischian dinosaur A Mesozoic archosaur characterized by a bird-like hip; most species were herbivores.

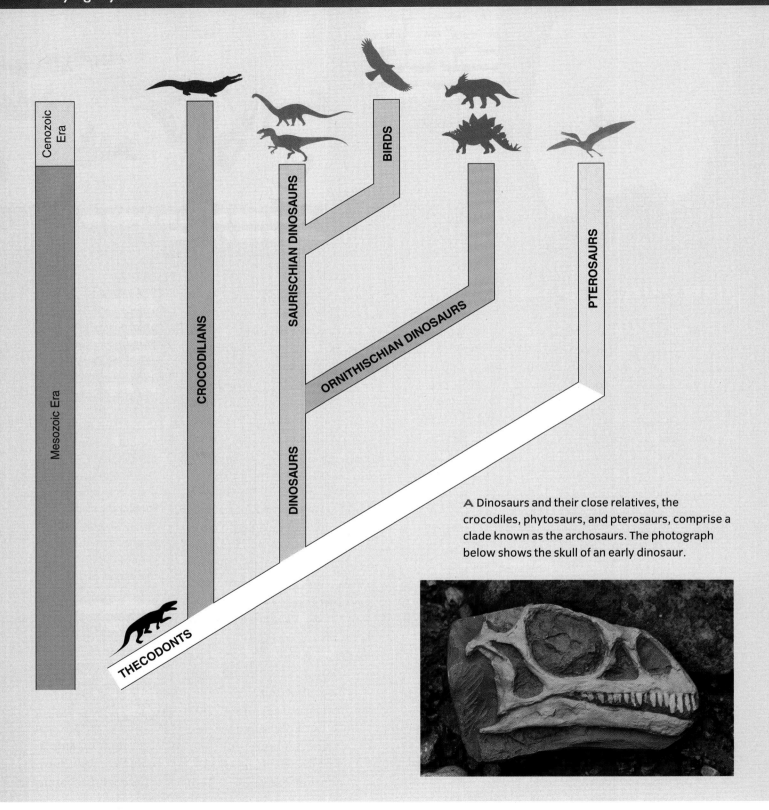

A Dinosaurs and their close relatives, the crocodiles, phytosaurs, and pterosaurs, comprise a clade known as the archosaurs. The photograph below shows the skull of an early dinosaur.

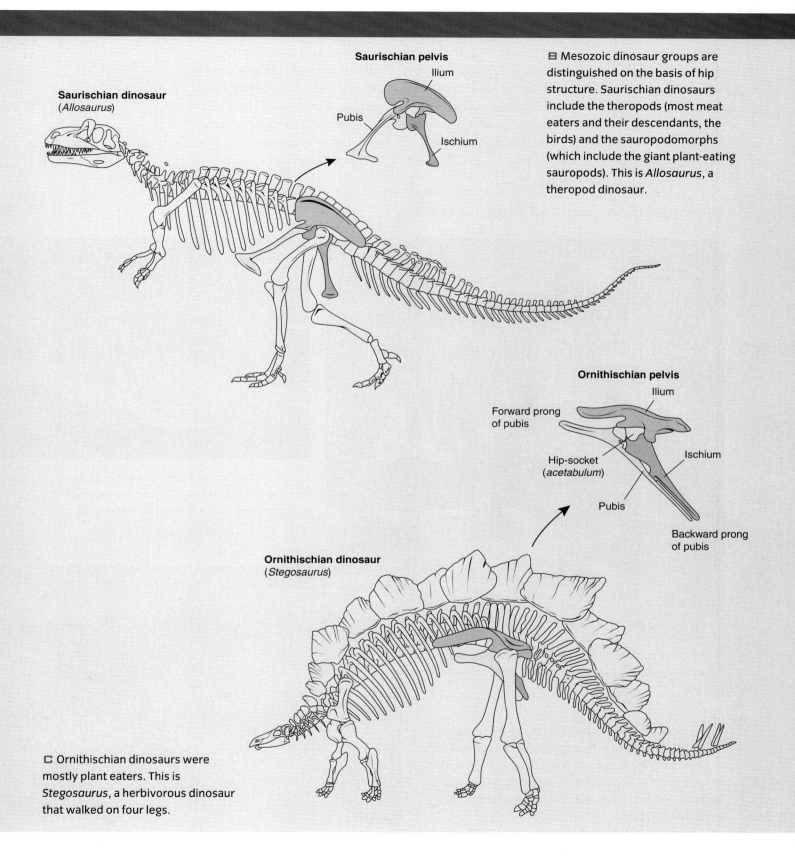

Saurischian pelvis

Ilium

Pubis

Ischium

Saurischian dinosaur
(*Allosaurus*)

B Mesozoic dinosaur groups are distinguished on the basis of hip structure. Saurischian dinosaurs include the theropods (most meat eaters and their descendants, the birds) and the sauropodomorphs (which include the giant plant-eating sauropods). This is *Allosaurus*, a theropod dinosaur.

Ornithischian pelvis

Ilium

Forward prong of pubis

Ischium

Hip-socket (*acetabulum*)

Pubis

Backward prong of pubis

Ornithischian dinosaur
(*Stegosaurus*)

C Ornithischian dinosaurs were mostly plant eaters. This is *Stegosaurus*, a herbivorous dinosaur that walked on four legs.

Saurischian dinosaurs are of two basic types: theropods and sauropodomorphs. **Theropods** (see *What a Geologist Sees*) were two legged (bipedal) and carnivorous. The earliest ones, Triassic in age, were relatively small, reaching barely 3 m in length. Each had three toes per foot and three fingers per hand, and pointed teeth with serrated edges. Three toes were retained through the entire theropod lineage, which includes *Allosaurus* (Jurassic Period) and *Tyrannosaurus* (Cretaceous Period). Some forms, including the giant (14-m-long) *Tyrannosaurus*, had only two fingers on each hand. *Ornithomimus*, *Struthiomimus*, *Oviraptor*, and similar Cretaceous theropods lost the teeth from their beaks, making them appear much like featherless ostriches and emus.

theropod
A clade of saurischian dinosaur having bipedal gait and, at least primitively, teeth adapted for carnivory. The clade includes birds.

Theropod Dinosaurs and Their Trackways

A Theropod dinosaurs, including the Cretaceous giant *Tyrannosaurus*, were active meat eaters that walked upright.

C This three-toed footprint, left on the shore of a Mesozoic lake, is attributable to a theropod tracemaker because of its close similarity in size and shape to the extremities of the hind limbs of dinosaurs such as *Ornithomimus*. The dinosaur that made this track was an active animal that could walk and run.

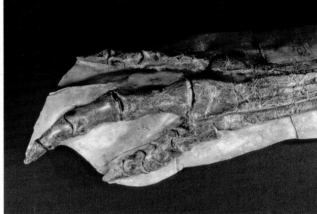

B *Ornithomimus* was a Cretaceous theropod that, like all other theropods, had three toes on each hind limb. This is a leg of *Ornithomimus*.

Feathered dinosaurs FIGURE 11.19

A A small theropod dinosaur fossil from a Cretaceous lake bed deposit in China shows exceptionally preserved feathers in places around the margin of the fossil.

B This fossil of the dinosaur *Sinosauropteryx* from China shows tiny hairlike feathers on the head and neck region. Feathers on this dinosaur were probably used for insulation.

Distinctive three-toed footprints left by theropod dinosaurs, combined with body fossil evidence, reveal interesting information about the physiology of these dinosaurs. Many species were not giants, but they did walk upright, carrying their tails high in the air. Theropods often had long strides, meaning they could move quickly. This is consistent with the interpretation that they were **endothermic**, or "warm blooded," rather than **ectothermic**, or "cold blooded," like present-day lizards and crocodiles. Numerous canals in the long bones of theropods were places where blood vessels resided, a feature usually associated with warm-blooded animals. Another indication of an active lifestyle is that some small Cretaceous dinosaurs covered their bodies with feathers (**FIGURE 11.19**) or similar structures adapted for insulation and display, although not for flight. *Sinosauropteryx* had hairlike "feather" structures resembling the down of modern birds, and *Caudipteryx* had true quilled feathers with vanes.

endothermic
Characterized by being "warm blooded," capable of generating internal heat.

ectothermic
Characterized by being "cold blooded," incapable of generating internal heat.

Sauropodomorphs (FIGURE 11.20) consist of the **prosauropods**, which evolved in the Late Triassic, and their descendants, the **sauropods**, which appeared in the Jurassic Period. Sauropodomorphs were herbivores, and basically quadrupedal (walking on all four legs). Prosauropods such as *Plateosaurus* were relatively small and had short necks. The more derived sauropods were the largest animals to ever live on land. They developed elongate necks for reaching leaves high in the treetops and had long tails for counterbalancing their enormous bodies. *Apatosaurus, Camarasaurus, Brachiosaurus,* and *Mamenchisaurus* were typical sauropods, each ranging between about 20 and 28 m in length and weighing perhaps 20 to 55 tons as adults. *Supersaurus* might have reached a length of 42 m. In comparison, African elephants, the largest land animals alive today, are 5 to 7.5 m long and weigh 5 to 7.5 tons.

Why did sauropods achieve such gargantuan sizes? One reason may be as a deterrent to predators. Many predators have a preference for smaller or weaker

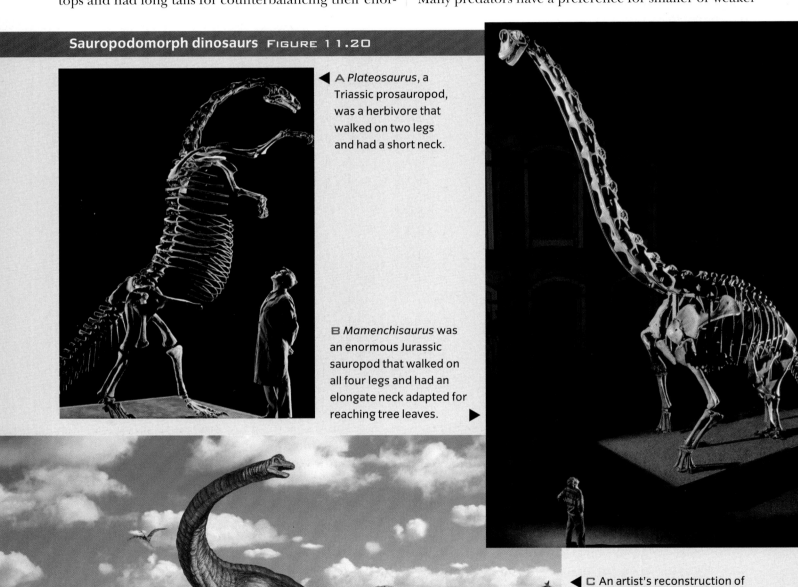

Sauropodomorph dinosaurs FIGURE 11.20

◀ A *Plateosaurus*, a Triassic prosauropod, was a herbivore that walked on two legs and had a short neck.

B *Mamenchisaurus* was an enormous Jurassic sauropod that walked on all four legs and had an elongate neck adapted for reaching tree leaves. ▶

◀ C An artist's reconstruction of *Mamenchisaurus*, an extremely long-necked sauropod dinosaur that lived during the Jurassic Period.

Clues to dinosaur habits FIGURE 11.21

B Gastroliths, or gizzard stones, are rounded stones often found in body cavities of dinosaurs. Dinosaurs apparently swallowed the stones and used them to help grind up their food.

A Trackways of large sauropod dinosaurs show that they were active, dry-land dwellers, probably much like modern elephants. They must have carried their long tails in the air because there is no indication that the tails were dragged behind as they walked.

prey, so growth to enormous size may have been a survival tactic for the sauropods. Another reason may have been related to their physiology. Large reptilian animals normally cannot warm their bodies quickly. However, extremely large size may limit an animal's heat loss because the surface area (where heat loss occurs) is relatively small compared to the total volume of the body. Animals that retain internal body heat because of their immense size are sometimes referred to as **gigantotherms**. Trackways made by sauropods (FIGURE 11.21A) reinforce the interpretation that they were active, living in herds like modern elephants and helping to raise each other's young to adulthood.

How were the giant herbivorous sauropods able to consume enough food to maintain their size and internal temperature? They used their relatively small mouths primarily for gathering and swallowing food, not chewing it much. Instead, they swallowed stones that

lodged in the gizzard, where they would help grind up plant material. Many modern birds operate the same way. Gizzard stones, or **gastroliths** (FIGURE 11.21B), have been found associated with the skeletons of a variety of dinosaurs, not just the sauropods.

Ornithischian dinosaurs, which range from the latest Triassic to the end of the Cretaceous Period, were mostly herbivores (plant eaters). The earliest known ornithischian, *Lesothosaurus*, was bipedal, was little more than 1 m in length, and had two separate types of teeth in the skull. The side teeth were sharp and broad, adapted for slicing leaves. The front teeth were sharp and pointed, most likely adapted for occasionally devouring small animal prey. In more derived ornithischians, teeth in the front of the skull were lost in favor of a beak that could be used for nipping at vegetation. Although the bipedal condition was common in ornithischians, lots of derived forms were quadrupedal.

One clade of four-legged ornithischians (FIG-URE 11.22) includes the ankylosaurs and the stegosaurs. Many of these dinosaurs had large plates and stout spines on their backs. In ankylosaurs (*Ankylosaurus* and its relatives), which ranged up to about 11 m in length, the plates and spines covered the backs and sides and mostly served a defensive function. The underside was exposed, though, so when attacked by predaceous theropods, ankylosaurs may have crouched low to the ground to survive. *Ankylosaurus* had a bony club at the end of its tail that could deliver a bone-breaking blow to a predator if timed just right.

In most stegosaurs (*Stegosaurus* and its relatives), which ranged up to 9 m in length, plates covered with grooves and canals were arranged along the back. One popular idea concerning the function of the grooved plates is that they held blood vessels and served as giant "radiators," regulating the internal body temperature of the animals. In addition to plates, stegosaurs had spikes that were useful for defensive purposes. In *Stegosaurus*, they were limited to the end of the tail, but in other genera, spikes substituted for plates in varying degrees. Other possible functions of the plates in stegosaurs may have been sexual display and camouflage.

Ornithischian dinosaurs FIGURE 11.22

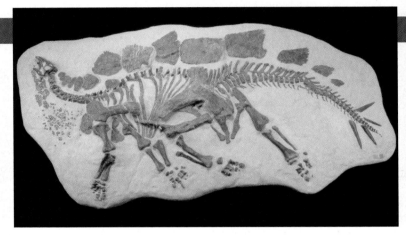

A Cretaceous anklyosaurs had bony plates on their backs. Some had clubs at the ends of their tails.

B Stegosaurs, including the Jurassic genus *Stegosaurus*, were ornithischian dinosaurs that had plates arranged along their backs. The plates may have functioned similar to radiators, allowing the animals to regulate their internal body temperatures.

C *Triceratops*, a Cretaceous ceratopsian, had a large bony shield covering its head. Spines projecting from the shield were used to ward off predators.

D *Parasaurolophus*, a Cretaceous ornithopod, had a wide toothless "bill" similar to that of a duck. Its teeth were along the sides of the mouth. This dinosaur also had a distinctive hollow crest on its skull.

Dinosaur nesting behavior FIGURE 11.23

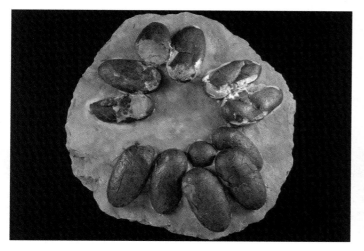

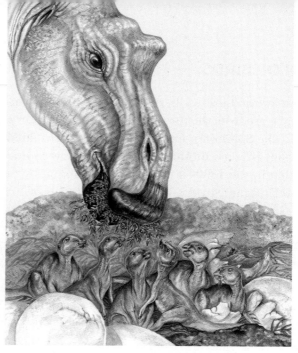

A Dinosaur egg clutches are known from many parts of the world. Here is a set of eggs laid by the Cretaceous dinosaur *Ornitholestes* from Cretaceous strata of the Gobi Desert of western China.

B Some dinosaur species apparently cared for their eggs and their young. *Maiasaura* adults have been found in association with eggs in their nests, as well as with babies.

Another clade of quadrupedal ornithischians was the ceratopsians, or horned dinosaurs. Genera include Cretaceous *Protoceratops*, *Triceratops*, and others, all of which had giant bony shields covering the head and neck. Except for *Protoceratops*, large spines also extended from the head shield. These dinosaurs ranged from about 2.5 to 9 m in length. The head shield surely had a defensive function, but like the plates of stegosaurs, also may have helped regulate the animals' body temperature.

The final major group of ornithischian dinosaurs is the bipedal ornithopods, including *Iguanodon* and the duckbills (hadrosaurs) *Hadrosaurus*, *Parasaurolophus*, *Corythosaurus*, and *Maiasaura*, among others. These dinosaurs, ranging from 2 to 15 m in adult length, showed little evidence of defensive traits. *Iguanodon* had a spike on each forelimb that might have deterred predators, but other ornithopods probably worked to avoid falling victims to theropods through speedy escape or herding behavior. Skeletons of ornithopods often show bite marks inflicted by large meat eaters.

Hadrosaurs developed wide, flat, toothless "bills" in their skulls, resembling those of ducks. Along the sides of the jaws were rasplike rows of teeth adapted for chewing coarse plants. Many hadrosaurs also had hollow bony crests on the skull. The crests extended the nasal passages, which may have improved the animals' ability to detect the smell of predators, potential mates, or babies. They also could have been used to produce distinctive sounds, enabling animals of the same species to communicate. Still another possibility is that they functioned in sexual display.

Nests and eggs laid by dinosaurs (FIGURE 11.23) are known from many areas of the world, and they show that certain species cared for both their unborn young and their young hatchlings. One of the most revealing discoveries is from Cretaceous strata of Montana, where nesting grounds of the hadrosaur *Maisaura* (literally "good mother lizard") were rapidly buried under shifting desert sands. Eggs were laid nearly vertically in bowl-shaped depressions in sand more than 1 m across. Skeletons of adult *Maiasaura*, 7 to 9 m in length, have been unearthed, along with the skeletons of babies about 1 m in length. The babies were perhaps a month old—too young and physically immature to live on their own when they were buried in sand. Their hip bones had not yet fused to their backbones, so they were unable to walk efficiently. Despite this, their teeth were worn from chewing coarse plant matter. This means that their parents brought their food to them while they were young.

ORIGIN OF BIRDS

The oldest known bird fossils, *Archaeopteryx* (literally "ancient wing"), are from the Jurassic of Germany (FIGURE 11.24). The Solnhofen Limestone, which has yielded exceptionally preserved bird and other fossils, is so fine grained that in the 1800s it was used for lithography (a printing technique). In the 1860s, the first fossils of *Ar-* *chaeopteryx* were discovered—first a single feather and later a skeleton with feathers attached.

Archaeopteryx (FIGURE 11.24B) had a skeleton with traits identifying its theropod dinosaur ancestry—sharp teeth, a long tail, and claws on the forelimbs. Birdlike hind limbs of *Archaeopteryx*, with a fourth, backward-facing toe adding to the three forward-facing toes of each foot, are a feature known in some theropods

Dinosaur–bird transition FIGURE 11.24

One impressive example of an evolutionary transition is the rise of birds from a theropod dinosaur ancestor. The transition took place during the Jurassic Period. The earliest bird, *Archaeopteryx*, shows a mix of primitive archosaurian features and more advanced avian features.

A *Compsognathus* was a small theropod dinosaur that closely resembled early birds. It coexisted with *Archaeopteryx*. This skeleton is from the Solnhofen Limestone of Germany.

B *Archaeopteryx*, the first bird, had skeletal features that mostly resembled those of theropod dinosaurs. Its feathers, elongate fingers, and wishbone (furcula), however, indicate that it is ancestral to modern birds.

(**FIGURE 11.24A**) and in birds. *Archaeopteryx* also had feathers. Unlike the "feathered dinosaurs," the feathers of *Archaeopteryx* were asymmetrical about the central shaft, which shows that they aided in flight. *Archaeopteryx* lacked a breastbone with a medial ridge, or keel, which implies that it had weak flying muscles. It was probably an adequate gliding animal but an ungainly flier.

Archaeopteryx turned out to be a classic missing link, in this case between warm-blooded theropod dinosaurs and modern birds (**FIGURE 11.24C**). More derived birds, which had appeared by the Cretaceous Period, evolved a keeled breastbone (making powered flight a reality), lost the clawed forelimbs, developed shorter tails, and eventually lost their teeth. They have kept their hind limb arrangement to the present day. Phylogenetically, birds are merely a derived group of theropod dinosaurs, a group that survived extinction at the end of the Mesozoic Era.

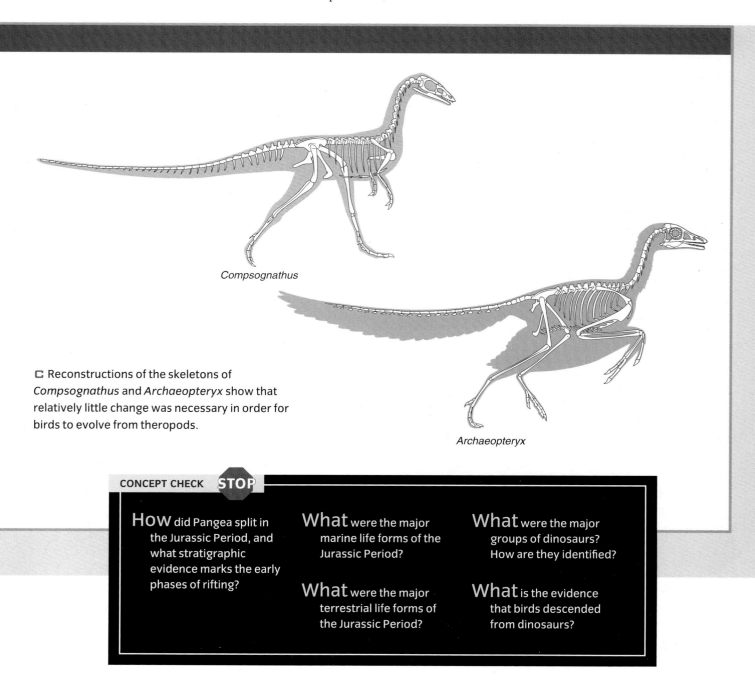

Compsognathus

Archaeopteryx

C Reconstructions of the skeletons of *Compsognathus* and *Archaeopteryx* show that relatively little change was necessary in order for birds to evolve from theropods.

CONCEPT CHECK STOP

HOW did Pangea split in the Jurassic Period, and what stratigraphic evidence marks the early phases of rifting?

What were the major marine life forms of the Jurassic Period?

What were the major terrestrial life forms of the Jurassic Period?

What were the major groups of dinosaurs? How are they identified?

What is the evidence that birds descended from dinosaurs?

Cretaceous Period

During the Cretaceous Period, Pangea continued fracturing into its modern drifting continents. High sea level associated with the spreading mid-ocean ridges caused flooding of the continental shelves. Greenhouse climatic conditions, driven partly by undersea volcanoes spewing their contents, warmed the oceans well beyond the tropics, and planktonic microorganisms bloomed in the seas until marine ecosystems collapsed at the end of the period.

Earth experienced an unusually long interval of normal polarity, lasting about 40 million years, during the mid-Cretaceous. In the 20 million years prior to this, quick polarity reversals, often at less than 1 million year intervals, were the rule, just as they had been through much of the Mesozoic Era. In the final 20 million years, prolonged magnetic polarity intervals gave way to quick reversals again, and this trend has continued to the present day.

CRETACEOUS TECTONICS AND OCEAN HISTORY

Supercontinent breakup and rifting accelerated in the Cretaceous Period. By the Early Cretaceous, splitting westward of the **Tethys Ocean** divided Pangea (**FIGURE 11.25**) into a Southern Hemisphere continental block (**Gondwana**) and a Northern Hemisphere block (**Laurasia**). By the Late Cretaceous, Gondwana had separated into South America, Africa, India, Antarctica, and Australia. Splitting of Laurasia resulted in the modern Atlantic margins of North America and Europe. Florida (derived from Gondwana) was left attached to North America. Similarly, pieces of Laurentia were left attached to continental Europe (in western Norway) and the British Isles (Northern Ireland and Scotland).

Cretaceous paleogeography FIGURE 11.25

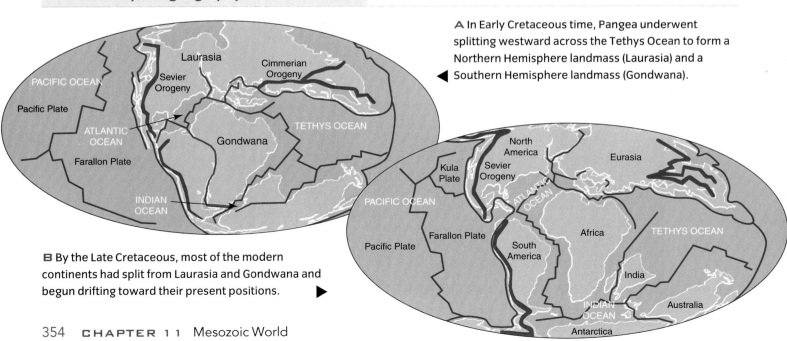

A In Early Cretaceous time, Pangea underwent splitting westward across the Tethys Ocean to form a Northern Hemisphere landmass (Laurasia) and a Southern Hemisphere landmass (Gondwana).

B By the Late Cretaceous, most of the modern continents had split from Laurasia and Gondwana and begun drifting toward their present positions. ▶

Western North American tectonic and sea level history FIGURE 11.26

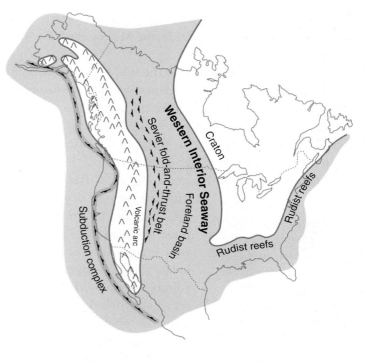

A The Western Interior Seaway was part of an enormous epeiric sea that stretched lengthwise along western North America in the Cretaceous Period.

B A large subduction zone in western North America during the Cretaceous Period and Cenozoic Era played an important role in driving collisional tectonics of the Nevadan, Sevier, and Laramide orogenies. To the east of the subduction, a volcanic arc complex (now batholiths exposed in California, Idaho, British Columbia, and other areas) was developed. Thrusting and foreland basin development extended as far east as Utah, Colorado, and Montana.

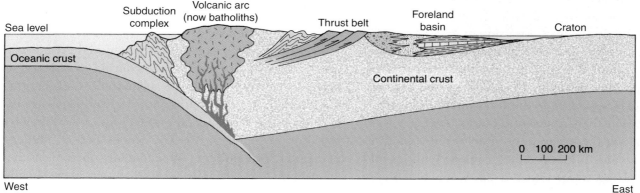

In western North America, three orogenic episodes caused deformation during the Cretaceous Period. The Nevadan orogeny extended from the Jurassic Period into the Early Cretaceous. The second orogenic episode, the **Sevier orogeny**, which culminated in the Late Cretaceous, involved folding and eastward thrusting in what is now the eastern Great Basin. Finally, the **Laramide orogeny** resulted in most of the structural features of the eastern Rocky Mountains. The modern landscape of the region, however, was largely formed through Cenozoic Era uplift and erosion.

Sevier orogeny
Folding and thrusting in the eastern Great Basin that culminated in the Late Cretaceous.

Laramide orogeny
Deformation in the Rocky Mountain region that occurred from the Late Cretaceous to the Paleocene.

The Laramide orogeny lasted from the Late Cretaceous through the Paleocene Epoch.

Rifting and drifting of fragments that were once part of Pangea caused ridge volumes to swell to dimensions previously unknown in the Phanerozoic Eon. In response, sea level around the world rose to high levels and flooded the continental shelves. One great epeiric sea, called the Western Interior Seaway, stretched lengthwise through western North America, leaving peaks of the rising Rocky Mountains protruding as islands (**FIGURE 11.26**). The Tethys Ocean was an unobstructed passageway that allowed trade winds to push warm tropical waters westward across Asia, the Middle East, Europe, and northern Africa.

As sea level peaked in the mid-Cretaceous, dark, organic-rich muds were deposited on continental shelves as a result of ocean circulation patterns that were different from those of the present. Today, cold, oxygen-rich water tends to sink in polar areas, after which it spreads out along the ocean bottom toward the Equator, mixing oxygen into the water as it goes. This leaves only a thin layer in the ocean, called the **oxygen-minimum zone**, poor in oxygen. Today, this zone is typically positioned just below the shelf edge, but warm climates of the mid-Cretaceous may have forced its expansion onto the shelves, resulting in the deposition of anoxic muds.

> **oxygen-minimum zone** An interval in a body of water in which the amount of dissolved oxygen is less than that above or below it.

In Cretaceous time, polar regions were warm enough that ocean water failed to cool enough to descend into the deep ocean. Emission of gases, especially carbon dioxide (CO_2) from rift zones, probably stimulated greenhouse warming. Enormous volumes of organic matter were being buried, which should have offset the input of some CO_2 to the atmosphere. However, it is hypothesized that hypersaline waters flowing toward the poles along the ocean bottom delivered warmth to those regions and influenced their climate. Without cold waters supplying oxygen to the lower reaches of the ocean, the seas became stagnant, and the oxygen-minimum zone expanded greatly. Anoxic conditions through most of the water column ensued, and organic-rich muds that lithified to black shales were deposited both in the deep sea and on continental shelves.

Organic materials in the Cretaceous muds, once they were heated through deep burial, often transformed into petroleum. Many of the world's largest oil and natural gas reserves were developed from Cretaceous rocks. Especially significant ones are in the Middle East (formerly the Tethys Ocean), the North Sea, and a band from Texas through Alberta to Alaska (formerly the Western Interior Seaway).

At the end of the Cretaceous Period, beginning about 67 million years ago, sea level dropped precipitously. Oxygen isotopic ratios recorded in marine foraminifera and a change in land plant communities in high latitudes point to cooling of the poles. A drop in sea level worldwide reduced the shallow seas where hypersaline waters could form, and the Tethys Ocean was severely constricted between Eurasia and Africa.

CRETACEOUS MARINE LIFE

Sea level reached its highest point during the Cretaceous, and warm conditions meant that the oceans were teeming with life. Microorganisms forming the bases of food webs were major contributors to marine sediments. Photosynthetic **diatoms**—phytoplankton that make siliceous skeletons—radiated beginning in the mid-Cretaceous. Diatoms account for a large proportion of the silica in deep-sea siliceous oozes, and their lithified equivalent, chert layers, in Cretaceous through Cenozoic strata.

> **diatom** Single-celled phytoplanktonic alga having an ornate microscopic skeleton (test) composed of two valves impregnated with opaline silica.

Other microorganisms that enjoyed great evolutionary success in the marine realm during the Cretaceous Period were dinoflagellates, planktonic and benthic foraminifera, and coccolithophorids. Dinoflagellates experienced a series of adaptive radiations in the Phanerozoic Era, including in the Cretaceous Period. Modern planktonic **foraminifera** (amoeboid microorganisms that live within tiny skeletons) arrived on the scene in the Jurassic Period. Calcareous forms experienced an explosive diversification in warm Cretaceous seas. They, along with **coccolithophorids**, a group of calcareous nannoplankton that formed spherical tests composed of tiny disklike calcite plates, blanketed the seafloors of "calcite seas" with countless skeletons, forming thick layers of soft limestone called **chalk** (FIGURE 11.27). Today, major deposits of chalk are found in the British Isles (where they are famously known as the White Cliffs of Dover), in neighboring areas of western Europe, in western Kansas, in Texas, in Alabama, and elsewhere. These strata inspired the period's name, Cretaceous, from Latin, *creta*, meaning chalk.

> **chalk** Soft marine limestone composed mostly of calcitic coccolithophorid plates.

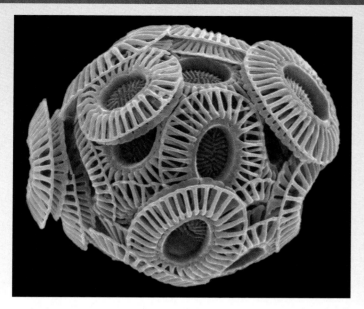

A Cretaceous marine microfossils included coccolithophorids, diatoms, and foraminifera. Coccolithophorids, like this example, were in life spherical objects formed of numerous disklike calcitic plates. Upon death, the plates disarticulated and fell to the seafloor in great numbers.

B Cretaceous chalk deposits, like these on Rugen Island in Jasmund National Park, Germany, are formed mostly of coccolithophorid tests.

Mollusks continued to evolve quickly in the Cretaceous. Coiled oysters were important constituents of benthic faunas, some of them forming large, weighty coiled shells by which they could anchor themselves in muddy substrates. Other bivalves (clams) adapted to burial deep in the sediment.

An unusual group of clams, called **rudists** (FIGURE 11.28), which had one long, curved valve and one tiny valve that formed a "lid," became a fixture of tropical seaways during the Late Cretaceous. They displaced corals as reef-formers by forming large clusters of wave-resistant structures, and their growth was probably enhanced by a symbiotic association with photosynthetic algae. Many rudist reefs preserved in subsurface strata are highly porous and permeable, and they serve today as petroleum reservoirs. In some instances, though, carbonate mud (now limestone) that is pink, reddish, or yellowish in color has filled the spaces between the clamshells. The col-

rudist A type of Mesozoic clam that had valves of unequal sizes and often formed reefs in the Cretaceous.

orful limestone has been used as ornamental building stone through Europe and North America.

Many families of modern snails appeared in the Cretaceous Period. In contrast to earlier snails, which were primarily herbivorous, the new wave of forms was

Rudist bivalves FIGURE 11.28

Rudist clams were important reef-formers, especially in the Tethys Ocean, during the Cretaceous Period.

dominated by carnivorous ones that fed on clams, snails, and worms.

Swimming carnivorous mollusks, the belemnoids, and especially the ammonoids, account for some of the most conspicuous Cretaceous fossils. Ammonoids evolved quickly during this period. Their shells had ornate fluted, or "corrugated," septa forming sutures where they joined the outer wall of the shell. These distinctive sutures are referred to as **ammonitic sutures**, and the animals possessing them are referred to as **ammonites** (FIGURE 11.29). The septa added structural strength to the ammonites' shells without substantially increasing their weight. It allowed them to dive to greater depths in the ocean. Today, ammonites are an important group of Cretaceous guide fossils. In some deposits, the original nacreous (mother-of-pearl) layer of their shells has been preserved unaltered; when it is polished, it is highly prized for jewelry or other ornamental purposes.

Teleost fishes, which are the dominant group of fishes today, became prevalent in Cretaceous seas. They

> **ammonite**
> An ammonoid cephalopod having ammonitic sutures.

included relatives of many modern forms, including carp, salmon, and eels. The largest known Cretaceous teleosts, 5 m in length, were a menacing presence in North American epeiric seas.

The Cretaceous Period witnessed the onslaught of a new and formidable arthropod predator in benthic marine ecosystems: the modern crabs. With their powerful claws, crabs can break open the shells or skeletons of their prey, operating much like a can opener, to get at the nutritious soft tissues inside. Crabs maintained their position as significant predators through the Cenozoic, and modern beaches are littered with the shells of snails and other animals showing breakage patterns attributable to these arthropods.

During the Cretaceous Period, the top of the marine food web was occupied mostly by reptiles. Ichthyosaurs and marine crocodiles had diminished in importance near the end of the Jurassic Period, but plesiosaurs were still common in the Cretaceous Period. Marine turtles of the Cretaceous grew to lengths of 4 m in

Ammonite cephalopods FIGURE 11.29

Ammonoids with ammonitic suture patterns evolved from ammonoids with ceratitic suture patterns in the Jurassic Period. Ammonites diversified quickly until becoming extinct at the end of the Cretaceous Period.

A This fossil ammonite from Cretaceous strata of Madagascar still has the original outer shell, which was made of aragonite, preserved in unaltered state. The subtly colorful reflectance of the specimen is caused by the nacreous, or mother-of-pearl, layer of the shell.

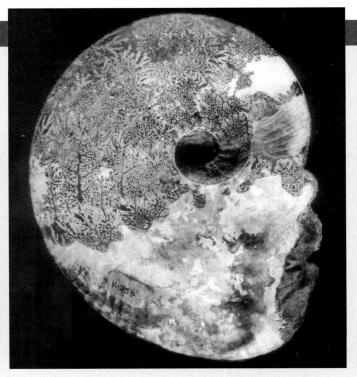

B This ammonite, from Cretaceous rocks of Seymour Island, Antarctica, has most of the outer shell stripped away, revealing the complex ammonitic suture pattern beneath.

Mosasaur and its prey FIGURE 11.30

A Mosasaurs were large oceangoing lizards of the Cretaceous Period.

B Mosasaurs preyed on ammonite cephalopods, which themselves could reach impressive sizes. Occasionally specimens like this one are found, showing puncture marks inflicted by the sharp teeth of mosasaurs.

some cases. Another marine reptile, the **mosasaur** (FIGURE 11.30), which was essentially a large swimming lizard, was a terrifying predator. Some mosasaurs reached lengths of 15 m, and they had large mouths full of sharp teeth, round in cross-section, with which they could pierce their prey. Shells of large ammonites have been found showing sets of round punctures, their arrangement clearly demonstrating that the ammonites had fallen victim to mosasaurs (FIGURE 11.30B).

TERRESTRIAL LIFE OF THE CRETACEOUS

Dinosaurs still ruled the land during the Cretaceous Period. The giant sauropods were all but gone, but ceratopsians and ornithopods expanded in numbers.

Other vertebrate animals held their own on land, or even diversified somewhat, but did not reach the level of prominence that dinosaurs enjoyed. By the Late Cretaceous, the two modern groups of mammals, placentals and marsupials, had evolved, but they remained small and inconspicuous. Flying reptiles and birds were the largest animals in the air.

Near the beginning of the Cretaceous, a gymnosperm species gave rise to the flowering plants, or **angiosperms** (FIGURE 11.31). Angiosperms, such as lilies, roses, magnolias, maples, and oaks, have a tremendous ecologic advantage because they enclose their seeds in a special reproductive chamber (an ovary). Within the ovary, flowering plants provide food to their seeds through a process called double fertilization. The first fertilization event produces a seed, and the second one produces a store of food for the seed.

angiosperm
A plant with true flowers in which seeds are enclosed in an ovary.

Early flowering plants FIGURE 11.31

A The earliest angiosperm fossils are from the earliest Cretaceous Period. This is a large angiosperm specimen from a Cretaceous lake bed deposit of Liaoning, China.

B In this detailed view of another angiosperm fossil from China, it is clear that the flower at the end of the stem has been modified from leaf bracts.

The starchy part of a corn kernel, a bean, or a pea represents the food supply for the seed. Gymnosperms do not provide food for their seeds, so they tend to have reproductive cycles of a year or more. Angiosperms, in contrast, release seeds with enough food that they can quickly grow to produce their own seeds, often within a few weeks.

The evolutionary success of angiosperms was bolstered by their development of the flower. Showy flowers—features modified from leaves—are attractive to animals, especially insects, but also birds and mammals. Insects are attracted to the nutritious nectar in flowers, and as they travel from one to the next, pollen (which is needed for plant fertilization) sticks to the insects' bodies, and they unwittingly carry it from flower to flower.

Insect species are often specific about the plants they will pollinate in the nectar-gathering process, and changes in the flowers can lead to rapid speciation through the rise of a new flower–insect association. Such a close evolutionary relationship between species is referred to as **coevolution**. The beneficial association of flowering plant and insect species led to an astonishing

> **coevolution** An evolutionary pattern in which species in two unrelated lineages profoundly influence each other's evolution so that the two evolve as an integrated complex.

adaptive radiation in both groups during the Cretaceous Period. By the end of the period, angiosperms became the dominant land plants, mostly by filling in open niche spaces rather than by displacing gymnosperms from theirs. Radiation of both angiosperms and insects, especially bees, wasps, butterflies, and moths, continued into the Cenozoic Era.

Still another contributor to the success of angiosperms was evolution of seed-containing fruits and berries. These reproductive structures are eaten by animals, especially mammals and birds, and the animals then disperse the seeds. During the Cretaceous Period, herbivorous dinosaurs also fed on fruits and berries and then helped to scatter the seeds they contained.

END-CRETACEOUS EXTINCTION

The Mesozoic Era came to a close during a time of biologic crisis that had been several million years in the making (FIGURE 11.32). The ichthyosaurs had disappeared well before the end of the Cretaceous Period,

Cretaceous extinction FIGURE 11.32

This graph shows the number of extinctions of genera through the Mesozoic Era and part of the Cenozoic Era. Although organisms had been in decline for several million years prior to the end of the Cretaceous Period, many were annihilated at the very end of the period.

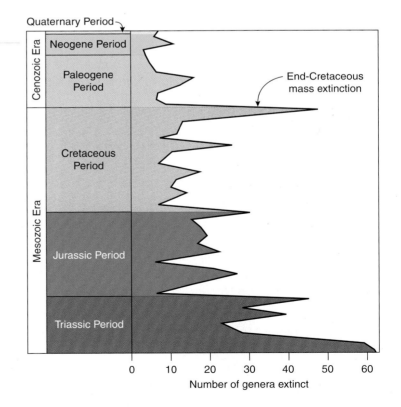

but various other groups were also declining in numbers toward the end of the period. Finally, the impact of a bolide from outer space at the end of the Cretaceous may have crippled ecosystems that were already in fragile condition.

The Late Cretaceous was a time of major volcanic activity, especially along spreading centers, where gases and dust were emitted. Increased dust levels may have placed stress on marine food webs from the bottom up by blocking sunlight from reaching photosynthetic plankton. With less sunlight reaching Earth's surface, global temperatures should have declined. Hydrogen sulfide and nitrous oxide gases vented from volcanoes, when combined with water, yielded sulfuric acid and nitric acid that may have poisoned some life forms.

Sea level dropped markedly at the end of the Cretaceous Period. Wide areas of the continental shelves were exposed, and shallow marine organisms were in peril. Global cooling, including the cooling of tropical seas, meant swift extinction of warm-water-adapted creatures such as the reef-forming rudists. Ammonites and belemnoids suffered from the shrinking of shallow marine habitats as the main source of their diets (crustaceans and other organisms) also declined. With food sources diminishing, large predators such as mosasaurs and plesiosaurs were the next to feel the impact of stressed marine ecosystems.

Terrestrial ecosystems were no less affected in the waning phases of the Cretaceous Period. If photosynthetic plant communities were stressed by reduced incident sunlight, plant eaters would have experienced the effects immediately, and meat eaters would have suffered next. Larger animals, ones requiring greater food supplies to maintain their metabolism, would have been less able to deal with rapidly changing conditions than smaller, more adaptable ones. Land animals that survived the end-Cretaceous drama were mostly small ones—insects, mammals, snakes, lizards, crocodiles, turtles, and birds, among others.

Major climatic, sea level, and biologic perturbations in the Earth system left the world in no way prepared for the effects of what was to come next. Streaking toward Earth 65.5 million years ago was a large extraterrestrial body, probably an asteroid that had fallen from

> **shocked quartz**
> Quartz grains showing distinctive parallel sets of welded microscopic planes, called shock lamellae.

its orbit, that finally crashed into the northern tip of the Yucatan Peninsula of Mexico (FIGURE 11.33 on page 362). The crater, now buried deeply in sediment layers but detectable through geophysical profiling, has a diameter of 170 to 300 km. Flaring of the crater to one side indicates that the trajectory of the body was toward the northwest when it struck Earth. The size of the rock that left such a massive crater, called the Chicxulub structure, is estimated to have been at least 10 km in diameter.

Impact of the Chicxulub bolide was felt globally. Shock waves reverberated from the impact site, and dust and debris were sent high into the atmosphere. The impact caused distinctive parallel sets of microscopic planes, called shock lamellae, to form in quartz grains at the impact site. These **shocked quartz** grains (FIGURE 11.33B on page 362), upon being blown out of the crater, were carried by winds around the world. Another immediate effect was the formation of microspherules when rocks at the impact site were instantaneously melted and then splashed into the atmosphere. The superheated globules cooled so quickly they were unable to crystallize and instead formed tiny glassy spheres. The extreme heat and pressure of the impact caused carbon in the rock to recrystallize into microscopic diamonds. Like shocked quartz grains, microspherules and microscopic diamonds also became widely distributed. Finally, anomalously high iridium levels were incorporated in sediments at the Cretaceous–Paleogene boundary (FIGURE 11.33D on page 363). High levels of this element are most commonly attributed to an extraterrestrial source. Together, the iridium-rich "boundary clay" at the base of the Paleogene, the shocked quartz grains, the microspherules, and the microscopic diamonds, make a distinctive set of correlation tools. They also speak clearly to a geologic catastrophe of proportions unknown at any other time in the Phanerozoic Eon.

Opinions differ about the effect of the Chicxulub impact on Earth's life forms. One possibility is that it sparked widespread wildfires. Dust blown into the atmosphere would have blocked sunlight from reaching Earth's surface for months, limiting photosynthesis. Earth, darkened from dust carried aloft, may have plunged into cold, winter-like conditions, and sulfuric

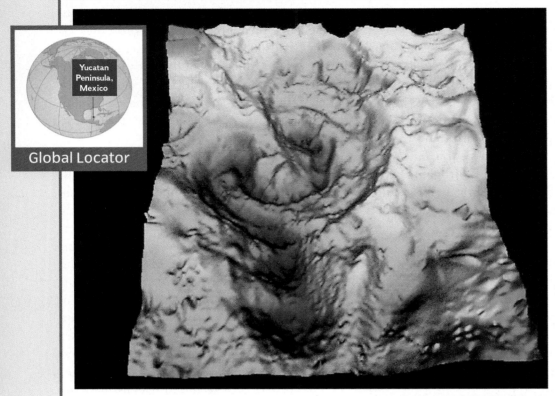

Global Locator

Yucatan Peninsula, Mexico

A The Chicxulub crater on the Yucatan Peninsula, Mexico, was caused by the impact of a bolide, probably at least 10 km in diameter, that crashed into Earth at the end of the Cretaceous Period. This seismic reflection profile shows the arrangement of strata in the subsurface of the Yucatan Peninsula, Mexico. With the overlying stratigraphic layers stripped away by computer visualization software, the structure of an impact crater is clearly visible.

NATIONAL GEOGRAPHIC

B Microscopic evidence of a bolide impact at the end of the Cretaceous Period: a shocked quartz grain. The distinct parallel lines in the grain are shock lamellae, or welded fractures, which indicate that the quartz grain had been subjected to extreme pressure.

acid–tainted rain may have poisoned life forms on the ground. After the dust settled from the air, greenhouse gases that remained would have stimulated a rapid warming of Earth.

The aftereffects of the Chicxulub impact, added to an already stressed ecosystem, may have been enough to drive many groups to extinction. Perishing completely from the seas were mosasaurs, plesiosaurs, ammonites, belemnoids, and rudist clams. Severely decimated about this time were sea urchins, bryozoans, planktonic foraminifera, and various phytoplankton. On land, ornithischian dinosaurs and pterosaurs were extinguished forever. Saurischian dinosaurs, with the exception of the derived theropods we recognize as birds, also vanished.

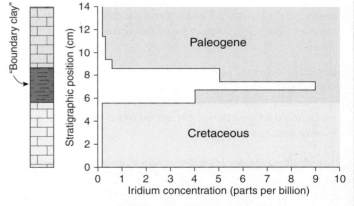

C The Cretaceous–Paleogene boundary is placed at the base of a clay layer rich in the element iridium (middle of the photo). The clay "boundary layer" records the effects of a large bolide impact at the close of the Cretaceous Period. This is a photograph of the boundary at Gubbio, Italy. The boundary clay separates white Cretaceous limestone (below) from pink Paleogene limestone (above).

D Iridium values recorded through strata across the Cretaceous–Paleogene boundary in Italy show unusually high levels in the boundary clay layer.

CONCEPT CHECK **STOP**

What was the origin of chalk and black shale in the Cretaceous? How did these facies relate to changes in the Earth system?

What were the major marine and terrestrial life forms of the Cretaceous Period?

How did flowering plants evolve? What gave them an evolutionary advantage over gymnosperms?

What are the possible causes of species extinction at the end of the Cretaceous Period?

SUMMARY

1 Triassic Period

1. In the Triassic Period, orogenies in Eurasia completed the assembly of Pangea. To the east of the supercontinent was the **Tethys Ocean**, and to the west was the **Pacific Ocean**.

2. Splitting of Pangea began in the Late Triassic to Jurassic. Rift basins enlarged through the Jurassic Period to form the **Atlantic Ocean**.

3. In the Triassic Period, ammonoid cephalopods diversified, **scleractinian corals** became reef-formers, and **bony fishes (teleosts)** first appeared.

4. Reptiles first invaded the sea during the Triassic Period. **Nothosaurs, placodonts, plesiosaurs**, and **ichthyosaurs** all evolved in the Triassic.

5. **Gymnosperms** were the predominant land plants of the Triassic Period.

6. Among terrestrial vertebrates, **frogs, turtles, archosaurs**, and **mammals** made their first appearance in the Triassic Period. The archosaurs included **crocodiles, phytosaurs, pterosaurs**, and **dinosaurs**.

7. **Conodonts**, nothosaurs, and placodonts became extinct at the end of the Triassic.

2 Jurassic Period

1. Beginning in the Jurassic Period, numerous tectonostratigraphic terranes were accreted along the western margin of North America.

2. Predators in Jurassic seas included ammonoid and **belemnoid cephalopods**, fishes, starfishes, crustaceans, plesiosaurs, and ichthyosaurs. Pterosaurs soared above the water. Prey species commonly thickened their shells, added spines, or evolved evasion strategies.

3. **Calcareous nannoplankton**, algae that contribute large volumes of deep-sea sediment, appeared in the Jurassic Period.

4. Gymnosperms, especially **cycads**, were the most common land plants of the Jurassic.

5. In the Middle and Late Jurassic, **dinosaurs** reached their greatest diversity. There are two major groups of dinosaurs: **saurischian**, or lizard-hipped, dinosaurs; and **ornithischian**, or bird-hipped, dinosaurs.

6. Carnivorous theropod dinosaurs gave rise to the **birds**.

3 Cretaceous Period

1. During the mid-Cretaceous, swollen mid-ocean ridges caused global sea level to reach its highest point. Black shale deposits record deposition of organic-rich muds on continental shelves.

2. Greenhouse climatic conditions warmed the oceans, and **coccolithophorids**, foraminifera, and **diatoms** bloomed in the seas. Thick layers of chalk were deposited in the Cretaceous "calcite sea."

3. Cretaceous **mollusks** evolved new ecologic strategies. Some **clams** (or **bivalves**) adapted to burrowing in sediment, and **rudist bivalves** became reef-formers. Many families of modern **snails** appeared, including carnivorous ones. Belemnoids and ammonoids having **ammonitic sutures** are some of the most conspicuous Cretaceous fossils.

4. Important predators in Cretaceous marine food webs were teleost fishes, crabs, and reptiles (plesiosaurs and mosasaurs).

5. Dinosaurs continued to be the prominent land-dwelling vertebrates of the Cretaceous Period.

6. The most important change in terrestrial biotas during the Cretaceous Period was the rise of **angiosperms**, or flowering plants.

7. A devastating extinction occurred at the end of the Cretaceous Period. The impact of a large asteroid in Mexico probably intensified the damage to ecosystems that had already been stressed.

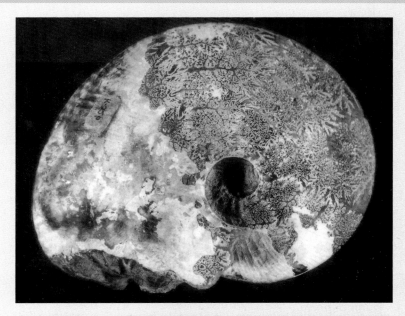

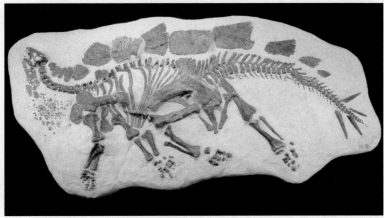

KEY TERMS

CRITICAL AND CREATIVE THINKING QUESTIONS

1. The earliest dinosaurs and mammals evolved at approximately the same time. Why do you think dinosaurs were better able to achieve prominence in the Mesozoic than were mammals?

2. Teleost fishes, sharks, ichthyosaurs, and other aquatic animals all developed broadly similar body forms. What characteristics of this general body form make it adaptive for life in the water? Have any flying animals evolved similar body forms, and if so, why?

3. What characteristics of the body and trace fossil records of dinosaurs might you expect to provide clues as to whether dinosaurs were slow, lumbering animals or more active ones?

4. Was the end of the Cretaceous Period the only time in Earth history when a large bolide struck Earth? What is the likelihood of it happening again? How might an asteroid impact affect the world today?

What is happening in these pictures ?

The ginkgo, a gymnosperm with ancient evolutionary roots.

Fossil ginkgo leaves (left). Ginkgos were common enough in the Mesozoic Era to serve as an important food source for animals.

Modern ginkgo leaves (right). Today, ginkgos are represented by only one species, *Ginkgo biloba*. Humans use the seeds as food.

What distinguishes the ginkgo as a gymnosperm?

How much have ginkgos changed since the Mesozoic Era?

1. What is the geologic evidence for the initial splitting of Pangea?

2. When did reptiles, which evolved on land, first invade the ocean?

3. What are the major groups of archosaurs?

4. Describe an example of convergent evolution (homoplasy) in Mesozoic reptiles.

5. When did mammals first evolve?

6. What was the main cause of global sea level rise in the Jurassic and Cretaceous?

7. What is the origin of the Sierra Nevada Range?

8. What were the dominant land plants of the Triassic and Jurassic?

9. What were the major marine predators of the Mesozoic?

10. What are the two major groups of dinosaurs, and how are they distinguished from one another?

11. What group of dinosaurs is thought to have given rise to the birds?

12. What microorganisms formed the Cretaceous chalk deposits?

13. What are angiosperms?

14. What evolutionary advantages did angiosperms have over gymnosperms?

15. What is the evidence for an asteroid impact on Earth at the end of the Cretaceous Period?

16. What organisms became extinct at the end of the Cretaceous Period?

Cenozoic World

The chapter of Earth history that brings us to the present is the Cenozoic Era. This time interval, 65.5 million years in duration, amounts to 12% of the Phanerozoic Eon and only 1.4% of time since Earth's beginning. The word *Cenozoic* means "new life," and it alludes to the increasingly modern appearance of the era's life forms. Among those modern life forms are fossils of hominids, our own closest relatives. Hominids, represented by this skull of an australopithecine, make their first appearance on the pages of Earth's history book at the very end of the very last chapter, less than 7 million years ago.

Following widespread extinction at the end of the Cretaceous Period, the biotic world changed greatly. On land, mammals, birds, flowering plants, and insects became prevalent. In the oceans, mollusks, crustaceans, corals, planktonic foraminifera, calcareous nannofossils, teleost fishes, sharks, and mammals diversified.

Three periods—the Paleogene (from the Greek for "old birth"), Neogene (from the Greek for "new birth"), and Quaternary (referring to the fourth division of geologic time, a carryover from the early 1800s)—comprise the Cenozoic Era. As we approach the present, the number and quality of stratigraphic tools at our disposal increases. For example, the beginning of the Paleogene Period is marked by the iridium-rich "boundary clay" that was deposited following the Chicxulub impact. Calcareous nannofossils help identify the beginning of the Neogene Period. The point also correlates with the beginning of a normal polarity interval. Finally, the beginning of the Quaternary Period coincides with the onset of major glaciation in the Northern Hemisphere. Sequence stratigraphy and calibration of sedimentary patterns to Milankovitch cycles allow for high-resolution stratigraphic correlation.

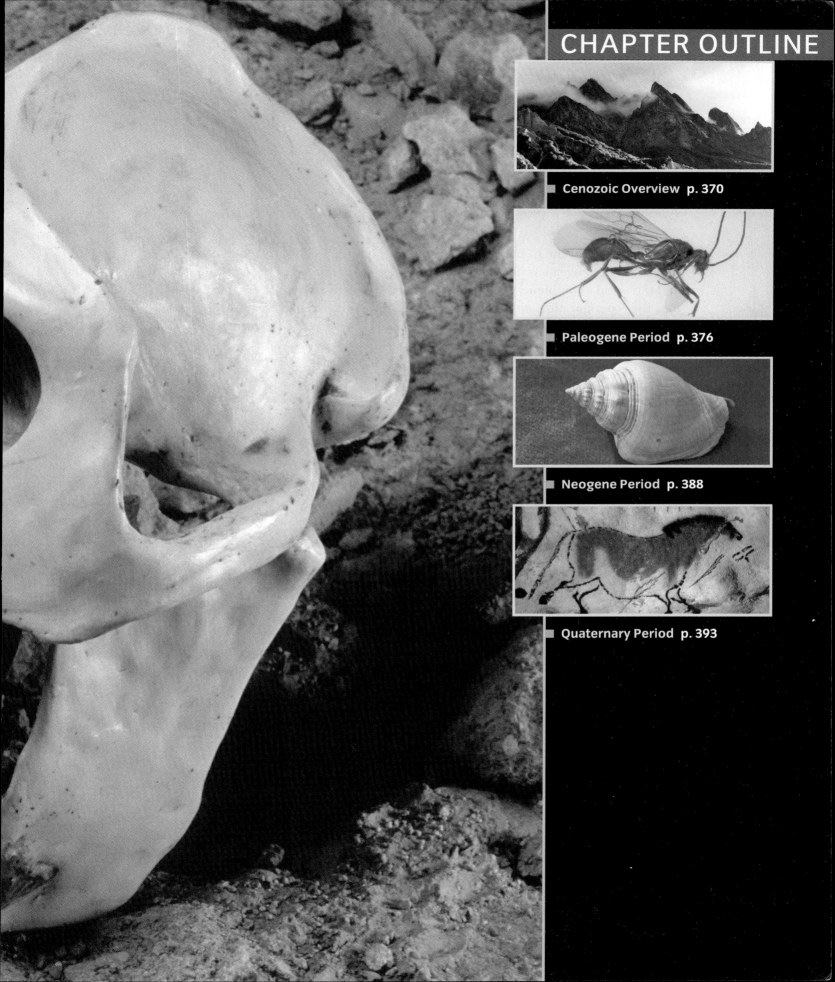

Cenozoic Overview

By the early part of the Cenozoic Era, fragmentation of Pangea was nearly complete, and the world's continents were approaching the configuration they have today (**FIGURE 12.1**). South America and Australia slid northward, isolating Antarctica. An extension of the North Atlantic rift pierced northward and separated the island of Greenland from North America. Later, active rifting switched to the east of Greenland, and the North Atlantic Ocean opened further as the island increased its

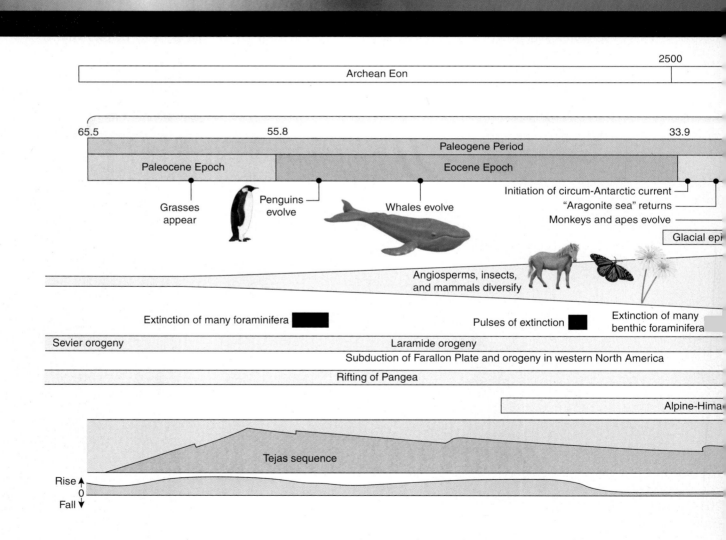

2500

Archean Eon

65.5   55.8   33.9

Paleogene Period

Paleocene Epoch   Eocene Epoch

Initiation of circum-Antarctic current
"Aragonite sea" returns
Monkeys and apes evolve

Grasses appear   Penguins evolve   Whales evolve

Glacial epi

Angiosperms, insects, and mammals diversify

Extinction of many foraminifera   Pulses of extinction   Extinction of many benthic foraminifera

Sevier orogeny   Laramide orogeny

Subduction of Farallon Plate and orogeny in western North America

Rifting of Pangea

Alpine-Hima

Tejas sequence

Rise
0
Fall

distance from Scandinavia. As rifting ensued, basaltic lavas were extruded in Ireland (as the Giant's Causeway), Scotland, Greenland, Baffin Island, and the Svalbard islands. This spreading created an open passageway from the North Atlantic Ocean to the Arctic Ocean and helped establish the modern conveyor-belt pattern of marine circulation. The Arctic Ocean could now supply oxygen-rich water to the North Atlantic because cold, dense Arctic waters would sink below warmer Atlantic surface waters and travel toward the Equator along the ocean floor.

Seafloor spreading associated with the rifting of Pangea has resulted in renewal of a significant portion of the ocean floor. Roughly half of the present ocean floor was formed during the Cenozoic Era. Spreading has been greatest in the Atlantic and Indian oceans.

Another region that experienced rifting during the Cenozoic Era was the Middle East and eastern Africa. The Red Sea began opening up in the late Paleogene Period as the Arabian Peninsula rifted away from Africa. Shortly afterward, during the Neogene Period, the Great Rift Valley of East Africa began forming through extensional faulting.

Supercontinent breakup played a role in biogeography. Separation of Gondwana, for instance, occurred at a critical phase in the history of mammals. Australia drifted from the rest of Gondwana just before the diversification of placental mammals, leaving the island continent to be inhabited by marsupials such as kangaroos, wallabies, and wombats. Elsewhere, the marsupials were largely outcompeted by placental mammals.

Visualizing

Timeline of Cenozoic events FIGURE 12.1

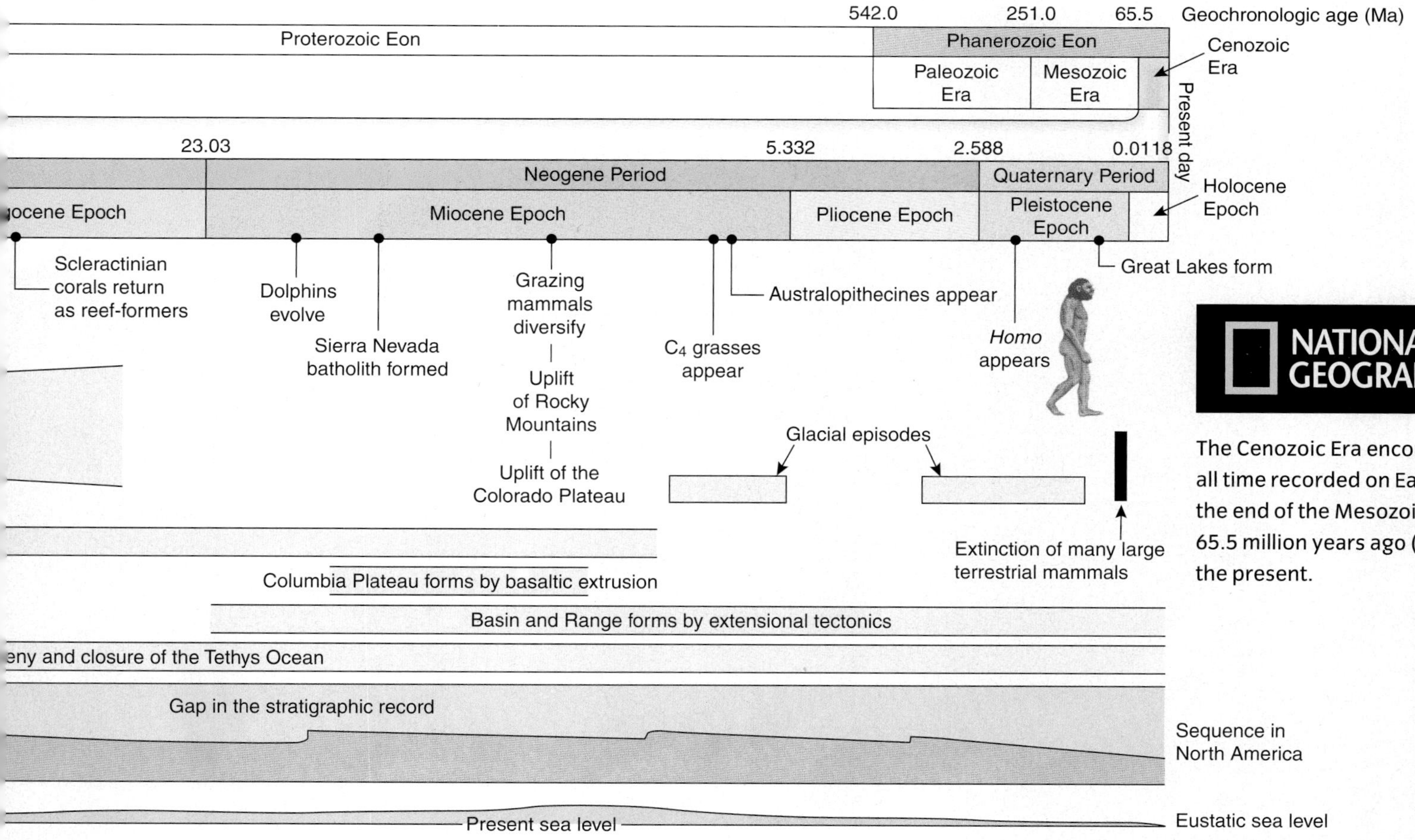

NATIONAL GEOGRAPHIC

The Cenozoic Era encompasses all time recorded on Earth from the end of the Mesozoic Era 65.5 million years ago (Ma) to the present.

One major exception to the drifting apart of continental blocks in the Cenozoic Era occurred on the Indian–Australian Plate. India and Australia are perched on opposite sides of the plate, with oceanic crust of the Indian Ocean separating them. As Gondwana dispersed and the Indian–Australian Plate moved northward away from Antarctica, it rotated, driven by subduction along the southern margin of Asia. In the Neogene Period, the Indian subcontinent rammed into Asia, producing an impressive array of geologic features that classically illustrate the effects of continent–continent collision: thrust slices and crustal thickening, metamorphic zones where suturing has occurred, an ophiolite sequence, and the highest mountain peaks on Earth today, the Himalaya Mountains (FIGURE 12.2).

The Himalayas were not the only great mountains to form through the movement of Gondwanan fragments. As the African Plate forged northward, it col-

Amazing Places: Himalaya Mountains FIGURE 12.2

In the Cenozoic Era, collisional tectonics produced a long string of mountains stretching across Europe, North Africa, and Asia, including the Pyrenees, Apennines, Alps, Atlas Mountains, and Himalayas.

A The Himalayan Mountains arose in the Cenozoic, through rotation of the Indian–Australian Plate followed by subduction of the oceanic crust that once separated India and southern Asia.

NATIONAL GEOGRAPHIC

lided with the southern margin of Europe in the Mediterranean region, and the Alps, Apennines, and other mountain ranges of southern and central Europe rose in a series of events called the Alpine orogeny. The long string of mountain chains arising from Spain and North Africa through Kashmir, India, and the Tibetan Plateau (Nepal and China), which is called the **Alpine–Himalayan orogenic belt**, nearly completely collapsed the once-enormous Tethys Ocean. The modern Medit-

erranean Sea, which now connects to the Atlantic Ocean through a narrow strait on its western margin, is one of the last vestiges of the Tethys.

The Pacific coast of the Americas was flanked by subduction zones through much of the Cenozoic. Collisional tectonics thrust large slices of crust onto the continental margins, loading the crust and inducing foreland basin development. Oceanic crust, remelted through subduction, later penetrated the crust and emplaced

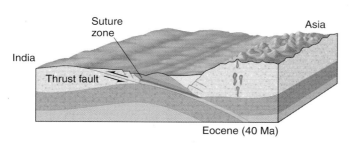

Eocene (40 Ma)

B Subduction shortened the distance between India and southern Asia. This cutaway view shows the evolving Himalayas during the Eocene Epoch (40 million years ago).

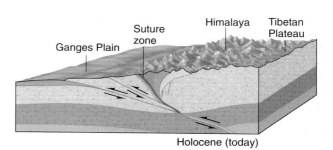

Holocene (today)

D After India came in contact with southern Asia, thrusting and magmatic activity thickened the crust. This is a cutaway view of the Himalayas today. The Himalaya Mountains are still tectonically active. The mountain ranges continue to undergo uplift.

C This map shows plate movement leading to the development of the Himalaya Mountains. During the Cenozoic Era, the Indian-Australian Plate shifted from its Cretaceous position (80 million years ago). The plate rotated and moved northward, bringing India into contact with southern Asia.

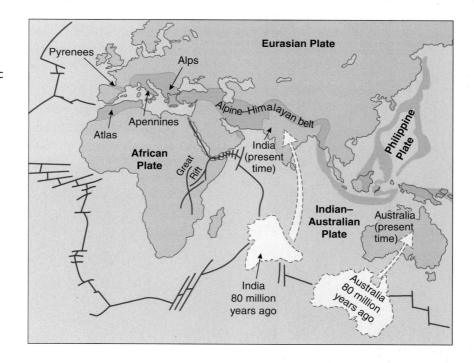

large intrusive bodies that fed a discontinuous chain of volcanoes, some of which are active to this day (FIGURE 12.3). South of the Canadian border, subduction along the Pacific coast of North America is now almost over, and compressional tectonics has given way in some areas to strike-slip faulting. Western California and Baja are currently sliding roughly northward along the San Andreas and related faults.

Sea level at the beginning of the Paleocene Epoch was low, but it rose through the epoch, only to fall again in the Oligocene as the circumpolar current became established around Antarctica. From that point forward, Cenozoic sea level history (FIGURE 12.4) has seen rapid and large fluctuations, driven mostly by the waxing and waning of continental glaciers in polar regions. The **Tejas Sequence** is used to describe shelf deposits of the Cenozoic Era in North America. Major glacial advances, and consequent drawdowns of sea level, occurred in the Oligocene, Miocene, and Pleistocene epochs.

Sea level changes in the Cenozoic Era had profound effects on the distribution of life forms. The Arctic Ocean remained mostly separated from the Pacific Ocean because of the narrow connection between Siberia and Alaska across the Bering Strait. In times of glacial maxima, when sea level was low, the Bering region emerged as a land bridge, allowing animals and plants to cross between Asia and North America.

Substantial glacial buildup in the Arctic and Antarctic regions occurred during the Oligocene, Miocene, and Pleistocene epochs. Each time, Milankovitch cyclity factors (obliquity, precession, and eccentricity of Earth) were the root causes of climatic shifts. Antarctica's position over the South Pole favored the buildup of continental glaciers. Without any land mass covering the North Pole in the Cenozoic Era, continental glaciers could only form on nearby continents (North America, Europe, and Asia) and islands (especially Greenland). Sea ice, which is only a couple meters thick, would have spread across the Arctic Ocean.

Cenozoic orogeny in western North America FIGURE 12.3

A During the Cenozoic Era, subduction and remelting of basaltic rock beneath western North America provided the magmas for volcanoes of the Cascade Range. As this map shows, by the Miocene Epoch, extensional tectonics resulted in development of the Basin and Range Province, and basaltic lavas had poured from a hotspot to form the Columbia Plateau.

B Cascade Range volcanoes today. Some of these peaks, such as Mount St. Helens, are still active.

Rapid fluctuations in global sea level through the Cenozoic Era, as interpreted from sequence stratigraphy, are proxy evidence of major tectonic movement and global climate change.

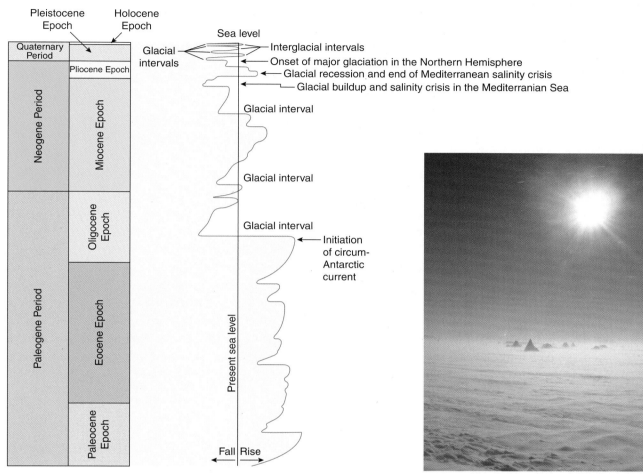

A This chart shows global sea level history through the Cenozoic Era. Sea level fall in the Oligocene was caused by initiation of the circum-Antarctic current. Subsequent climatic isolation of Antarctica led to a repeated buildup and melting of glaciers as Earth switched between icehouse and greenhouse conditions that stimulated sea level falls and rises.

B The world is now undergoing a transition to greenhouse conditions, and recession of large continental glaciers has led to rapid sea level rise. This is a field camp on the Antarctic glacial ice sheet. During the Pleistocene, extensive areas of the Northern and Southern Hemispheres probably looked much like this image.

CONCEPT CHECK **STOP**

How did rifting in the North Atlantic Ocean result in isolation of Greenland as an island?

How did the Himalaya Mountains develop?

How was the Mediterranean Sea formed? What ocean was its predecessor?

What major factors underlie the rapid climate and sea level changes of the late Cenozoic?

Paleogene Period

The Cretaceous–Paleogene boundary is one of the most clearly recognizable in the stratigraphic record because of the strong contrast in fossils on either side of it and because of the widely recognized "iridium spike" preserved in sediments at the base of the Paleogene System. Extinctions at the end of the Cretaceous Period opened the way for adaptive radiations among the survivors, and with those radiations came life forms of increasingly modern character.

The Paleogene Period is divided into three epochs: the Paleocene, Eocene, and Oligocene. They add up to about 42.5 million years of geologic time and span the interval from 65.5 million years ago to about 23 million years ago.

ISOLATION OF ANTARCTICA AND CLIMATE CHANGE

Global temperatures rose early in the Eocene Epoch, and the world entered a greenhouse state. In the Eocene, even the poles were warm, and lush forests grew within the Arctic Circle.

Greenhouse conditions were not to last long. As a consequence of the dispersal of Gondwanan continental fragments, Antarctica became isolated within the south polar circle (FIGURE 12.5). Antarctica has been an integral part of Gondwana since the late part of the Paleozoic Era, and its latitudinal position has changed little since that time. By the early part of the Oligocene Epoch, South America and Australia had

Paleogene paleogeography FIGURE 12.5

By the mid-Paleogene Period, the world's continents had achieved a nearly modern configuration. This map shows the arrangement of continents in the Eocene Epoch. At that time, India approached Asia as the oceanic crust separating them was subducted. Soon afterward, Antarctica became isolated, leading to development of the circum-Antarctic current.

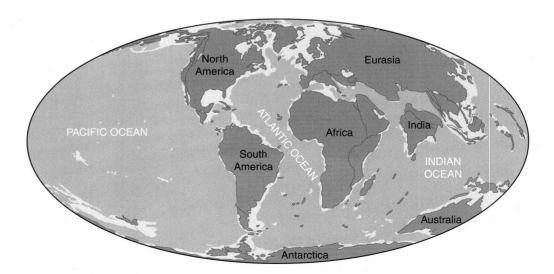

Rifting of Gondwana in the Paleogene Period led to separation of the modern Southern Hemisphere continents, progressive isolation of Antarctica, and development of the circum-Antarctic current.

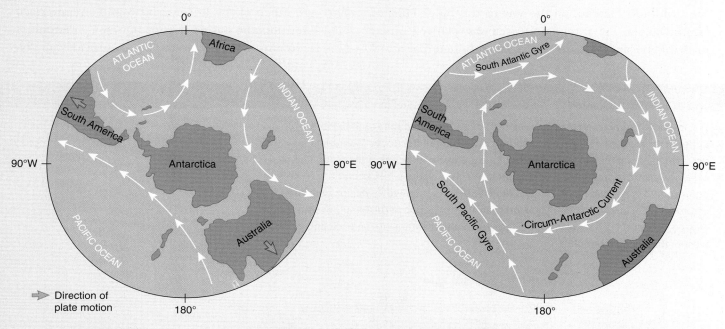

A In the Eocene Epoch, Australia and South America were just beginning to rift from Antarctica.

B In the early part of the Oligocene Epoch, Antarctica became fully separated from Australia and South America. Waters from the southern oceans became trapped in a circum-polar current, which resulted in climatic isolation of Antarctica and colder temperatures on the continent.

drifted far enough from Antarctica to isolate the southernmost continent with a continuous passageway and the new **circum-Antarctic current**. Warm surface waters of the South Atlantic, Indian, and South Pacific oceanic gyres, which had previously warmed the Antarctic shores, were ensnared in the circum-polar current and became progressively colder as they cycled around the continent (FIGURE 12.6). Antarctica became climatically isolated and much colder. Glaciers, which had been confined to relatively small areas before, were now free to expand over the land surface. By the mid-Oligocene, enough water from the world ocean had become locked up in Antarctic glacial ice to force a sharp decline in worldwide sea level. Since the Oligocene Epoch, glaciers have remained a persistent feature of Antarctica, although their volume has fluctuated with climatic cycles.

circum-Antarctic current A circum-polar ocean current system that flows around Antarctica and isolates it climatically from South America, Africa, and Australia.

TECTONICS OF WESTERN NORTH AMERICA

Orogenic activity in western North America continued almost unabated from the Cretaceous Period (Sevier orogeny) through the Paleogene Period. In the Paleogene, mountain building stretched along the Pacific margin from Canada to northern Mexico. Typical of the tectonic activity is that ascribed to the Laramide orogeny, which resulted in much of the uplift of the Rocky Mountains (**FIGURE 12.7**). Laramide events were related to descent of a large oceanic plate, the Farallon Plate, along a subduction zone at the Pacific continental margin. Large crustal blocks were uplifted east of the Colorado Plateau, all the way to the Black Hills of South Dakota. The intervening area was relatively quiet, probably because oceanic crust was subducted at a low angle. Large fold-and-thrust belts developed north and south of the basement uplifts.

By the end of the Eocene Epoch, basins formed in western North America through tectonic activity were filling with sediment fed by erosion of upland areas. Through much of the Eocene, lakes and streams occupied **intermontane basins** (basins between mountains). Fine sediments deposited in the basins provide valuable

Laramide orogeny FIGURE 12.7

A Laramide orogenic events were related to subduction of the Farallon Plate beneath the western margin of North America. Uplift of the Rocky Mountains occurred far from the margin because of a low angle of descent of the plate. This is a diagrammatic representation of the tectonics associated with Laramide deformation in the early part of the Eocene Epoch, about 55 to 45 million years ago. Later in the Eocene (45 to 35 million years ago), subduction of the Farallon Plate caused deflection of a mantle plume, leading to uplift, extension, and volcanism west of the Rocky Mountains.

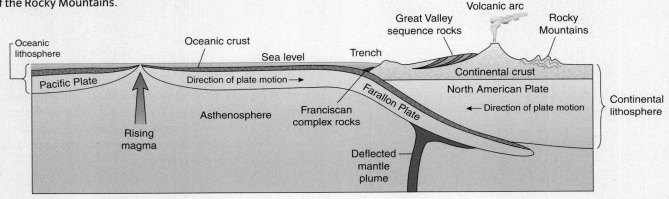

B The Rocky Mountains owe their grandeur to Laramide orogenic events, renewed uplift in the Neogene Period, and the effects of weathering and erosion, especially during the Pleistocene Epoch.

A Fine sediments of the Green River Formation filled intermontane lake basins of Wyoming, Utah, and Colorado. Spectacular fish fossils are common in the thinly laminated sediments.

B Palms and other plants thrived in the warm, moist climates of North America during the Eocene Epoch. This palm leaf was buried in lake sediments, along with fishes, where it became fossilized.

information about paleoclimate in the region and are now an energy source. A rich biota (**FIGURE 12.8**) of plants, insects, fishes, and other organisms attests to a warm, moist climate. Thinly laminated lake deposits are rich in algal remains transformed through burial into a low-grade petroleum product called **kerogen**. Because the kerogen can be processed to form oil, these strata are often called **oil shales**.

By the Oligocene Epoch, western North America was largely a flat plain broken sporadically by hills. The Badlands region of South Dakota, where flash floods have carved through relatively soft, poorly

oil shale Kerogen-bearing, finely laminated shale or siltstone that can be distilled to produce hydrocarbons.

consolidated Eocene and Oligocene sediments, reveal the nature of sediments that were once deposited widely through the region. The terrestrial sediments, rich in mammal remains and other fossils, are also rich in paleosols. Changes in soil types through time show that climate in the region changed from moist in the Eocene Epoch to arid in the Oligocene Epoch.

Renewed uplift in the Neogene Period produced the Rocky Mountains as we know them today. The subdued, almost flat, Paleogene plains are still evident today in the tilted flanks of the Rockies.

Following the Laramide orogeny, western North America experienced extension over a broad area from present-day Utah to California and Idaho to Mexico. This area is known as the **Basin and Range Province** (FIGURE 12.9). The name derives from flat-bottomed basins formed through downdropping along normal faults bounded by relatively high-standing mountain ranges.

Geographic provinces of North America FIGURE 12.9

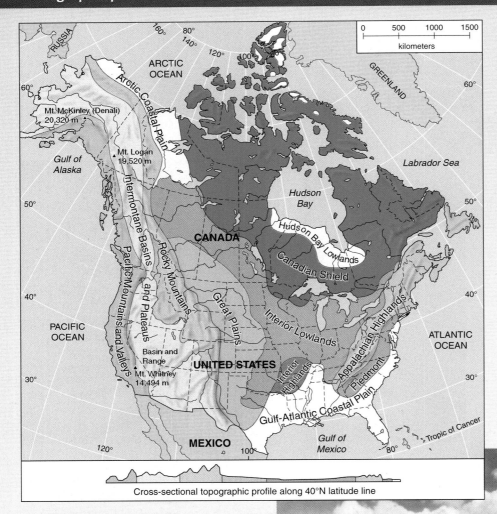

A The modern geographic provinces of North America took final form during the Cenozoic Era. In the eastern part of the continent, the physiography is dominated by the Appalachian highlands and coastal plains along the Atlantic seaboard and the Gulf of Mexico. Lowlands and the Great Plains dominate the central part of the continent. In the west, major features are the Rocky Mountains, intermontane basins and plateaus, and Pacific mountains and valleys. The Canadian Shield is exposed in the northeastern part of the continent, and the Arctic Coastal Plain rims part of the northern coast.

Cross-sectional topographic profile along 40°N latitude line

B The Basin and Range Province, which covers a wide area of western North America, experienced extensional tectonics during the Cenozoic Era. It consists of numerous flat basins delimited by mountain ranges. This is a view of the Basin and Range Province in western Nevada. The flat basin is in the foreground, and a range of fault-bounded mountains is in the distance.

PALEOGENE MARINE LIFE

Paleogene oceans were filled with life forms that are quite modern in appearance. Clams, snails, sea urchins, cheilostome bryozoans, teleost fishes, calcareous nanno-plankton, and both planktonic and benthic foraminifera radiated into many marine niches in the Paleogene. Also during this time, decapod crustaceans (shrimps, crabs, and lobsters) underwent their most impressive evolutionary expansion. As consumers of plankton and animal prey, as scavengers, and as food for many larger aqueous creatures, crustaceans (FIGURE 12.10) filled a vital position in Cenozoic food webs. Scleractinian corals took a long time to repopulate reef communities. Their return as reef-formers in the Oligocene Epoch coincided with a switchover of the world ocean from a calcite sea to an aragonite sea.

Some marine life forms changed little across the Cretaceous–Paleogene boundary, but they became more numerous and diverse. One example is the sharks, which replaced Mesozoic marine reptiles as top predators in Cenozoic seas.

Mammals multiplied on land during the Paleogene Period, but some left the land to become successful predators in the sea. Whales (FIGURE 12.11A), or **cetaceans**, made the transition from land to become

cetacean
A member of the mammal order Cetacea, which includes whales and porpoises.

Paleogene crustacean FIGURE 12.10

The diversity of marine crab, shrimp, and lobster species increased dramatically in the Paleogene Period. This is a fossil crab from the Miocene of New Zealand.

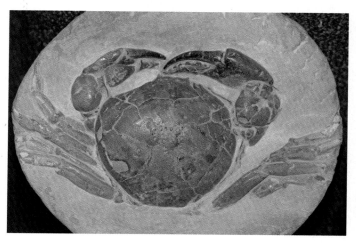

aqueous carnivores during the Eocene Epoch. Their limbs were reduced in length, with the front ones becoming powerful, stout propulsion devices and their rear limbs becoming simple rudders. Other mammals that made the transition to the ocean in the Cenozoic Era were the ancestors of modern manatees, seals, and dolphins.

Ocean-going mammals and birds FIGURE 12.11

A Whales are mammalian carnivores that adapted to life in the sea in the Paleogene Period. Some whales achieved dimensions greater than those of the largest Mesozoic dinosaurs.

B Penguins, the group of birds most well adapted to life at sea, evolved in the Eocene Epoch.

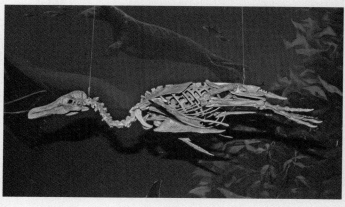

Birds also had a few representatives in aqueous settings. It is probably better to think of them as amphibious, however, because their usual life habit involves nesting on shore and seeking food in the water. Some birds developed wide feet with webbing between the toes that could function as paddles in the water. Others developed extremely long leg bones for wading in shallow water along shorelines.

Penguins (**FIGURE 12.11B** on page 381) are the birds adapted best to an amphibious lifestyle. They developed a more streamlined, torpedo-like body shape; decreased the size of their feathers (which would have reduced frictional response when swimming in water); transformed their forelimbs into large, curved, and powerful flippers; reduced the length of their rear limbs; and modified their light, air-filled, long bones into more substantial support structures. Penguin ancestors were adapted for flight in the air, but by the Eocene Epoch, penguins evolved the capacity for "flight" in water.

Foraminifera suffered pulses of extinction between the middle Eocene and the early Oligocene, in part related to changing climatic conditions. With the expansion of warm, oxygen-poor bottom waters in the early Eocene, some 70% of marine foraminifera became extinct. Later, benthic foraminifera suffered again with the final episode of Paleogene cooling.

PALEOGENE TERRESTRIAL LIFE

Three important groups of land-dwellers diversified in the Paleogene: mammals, angiosperms (flowering plants), and insects. Birds also experienced evolutionary expansion in the Cenozoic Era, and much of that diversification has its roots in the Paleogene Period.

Mammals emerged in the Paleocene Epoch as the inheritors of a world mostly wiped clean of large terrestrial animals (**FIGURE 12.13**). These energetic, warm-blooded animals were well poised to outcompete the reptiles, most of the birds, and the amphibians in the race to reoccupy niches vacated by the dinosaurs. Within 10 to 15 million years, the mammals had a toehold in most terrestrial ecosystems.

Although the earliest mammals of the Mesozoic looked rather rat-like, true rodents (a group of **placental mammals**) did not appear in the fossil record until the Paleocene. Molecular clock evidence suggests a Late Cretaceous origination time for rodents. Multituberculates, monotremes, and marsupials all persisted from the Mesozoic Era, but with the expansion of rodents, multituberculates declined and finally became extinct in the Oligocene Epoch. With few exceptions in the Cenozoic Era, monotremes (such as the echidna and duckbilled platypus) and marsupials (such as the kangaroo and wallaby) were confined to Australia, which was biogeographically separated from other continents after the Cretaceous Period.

Before the middle of the Eocene Epoch, most orders of mammals known today had appeared. In the Americas, Eurasia, and Africa, placental mammals such as rodents, rabbits, ungulates, elephants, **carnivores** (cats, dogs, bears, and their relatives), primates, and others roamed the land. By the end of the Eocene Epoch, bats fluttered through the air, and whales took to the sea.

At the end of the Eocene Epoch, there were nearly 100 families of mammals in existence, approximately the number that exist now. Many of the mammal families that appeared were **ungulates**, which are the mammals that have hooves, at least primitively (**FIGURE 12.14** on page 384). There are four principal groups of ungulates: odd-toed ungulates (**perissodactyls**), even-toed ungulates (**artiodactyls**), tethytherids (elephants and their close relatives), and cetaceans (whales, porpoises, and their close relatives).

placental mammal A mammal that nourishes its developing fetus with a placenta and gives lives birth.

carnivore As used for an order of mammals (order Carnivora), placentals belonging to the clade that includes cats, dogs, bears, hyenas, whales, seals, and dolphins.

ungulate A hoofed placental mammal.

perissodactyl An odd-toed ungulate mammal such as a horse, rhinoceros, hippopotamus, or tapir.

artiodactyls An even-toed ungulate mammal such as a deer, sheep, goat, pig, cow, camel, llama, or giraffe.

Mammal Evolution FIGURE 12.12

Mammals originated in the Mesozoic Era but diversified dramatically in the Cenozoic Era. Placental mammals have been especially successful.

Odd-toed ungulates include the horses, rhinoceroses, hippopotamuses, and tapirs. Horses (Figure 12.13) first appeared as small, multi-toed animals in the Eocene Epoch. By the end of the Eocene Epoch, horses vanished from Eurasia, but they survived in North America, where they underwent most of their subsequent evolution. Other odd-toed ungulates were well established by the Eocene Epoch, and some rhinoceroses reached immense proportions by the Oligocene Epoch.

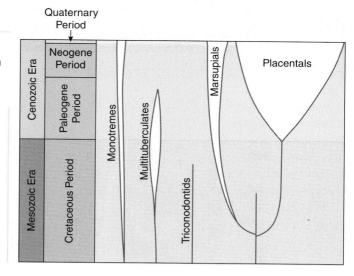

Evolution of the horses FIGURE 12.13

Early Cenozoic horses were small and multi-toed. Through time, horses increased in size and lost some of their toes. The molar teeth of horses also increased in length and surface complexity as horses evolved from browsers that fed on leaves of deciduous plants to grazers that feed on grasses.

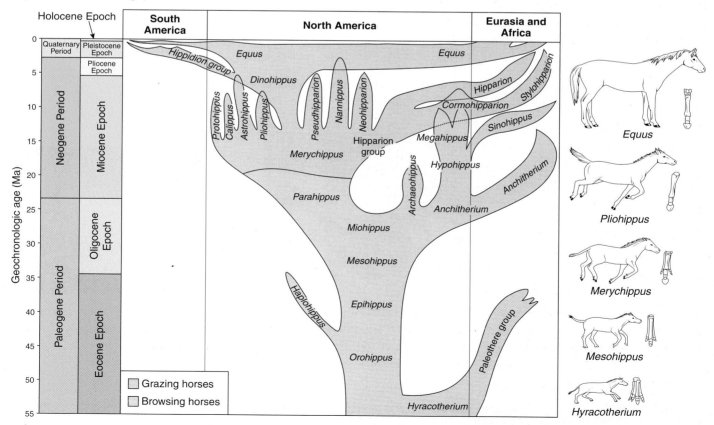

The ungulates are placental mammals that have hooves or that had hooves early in their evolutionary history. The ungulates first appeared in the Paleocene Epoch and quickly radiated into a range of terrestrial and marine habitats.

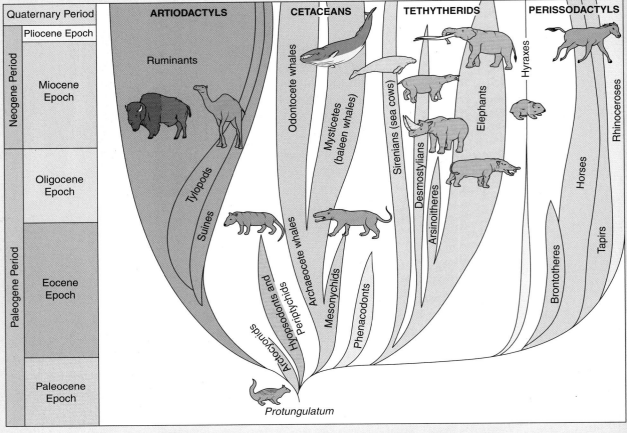

A The ungulates comprise four main groups (the artiodactyls, cetaceans, tethytherids, perissodactyls) and a number of smaller groups, of which one, the hyraxes, still lives today.

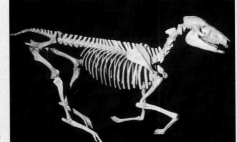

B Artiodactyls, such as camels, giraffes, sheep, and pigs, are ungulates that have two or four toes. This is the skeleton of a Pleistocene camel. Camels have two toes in their hooves.

C Tethytherids include elephants, and sirenians (sea cows). This is a Pleistocene mastodon, one of the elephants.

D Perissodactyls, such as horses, tapirs, and rhinoceroses, have an odd number of toes. This is a Miocene horse, *Miohippus*. This form had three toes.

Even-toed ungulates include deer, antelope, sheep, goats, pigs, bison, cattle, camels, llamas, and giraffes. These cloven-hoofed forms did not appear until the late Eocene. The even-toed ungulates, like the odd-toed ungulates, commonly showed phylogenetic size increase through the Cenozoic Era. One important group of even-toed ungulates is the **ruminants**, which are the cud-chewers, such as bison, cattle, and giraffes.

Elephants, which are members of the tethytherid ungulates, underwent a general size increase through time. They also showed progressively larger tusks and larger trunks, which were modified from their noses.

Early carnivores, the forerunners of modern wolves, dogs, cats, bears, pandas, minks, and others, appeared in the Eocene Epoch and radiated in the Oligocene Epoch. Some of the cats developed enlarged canine teeth, or sabers, that were used for dispatching prey (**FIGURE 12.15**).

Primates (**FIGURE 12.16**) were certainly in existence by the early part of the Paleogene Period, al-

Early carnivore FIGURE 12.15

The carnivore mammals became well adapted for predaceous food gathering early in their evolutionary history. This is a skull of *Dinictis*, an Oligocene saber-toothed cat.

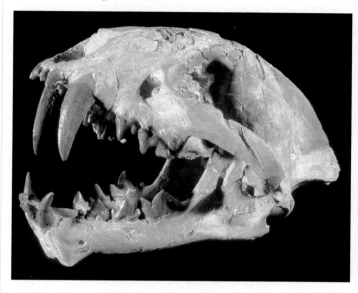

though the earliest lemurs may have evolved in the Cretaceous Period. The earliest forms were small and mostly arboreal (tree-dwelling). By the Oligocene Epoch, monkeys and larger apelike primates were living in forests of Africa and Eurasia.

Early primate FIGURE 12.16

Monkeys and apelike species appeared by the Oligocene Epoch. This is a skull of *Aegyptopithecus*, an early ape from the Oligocene of Egypt.

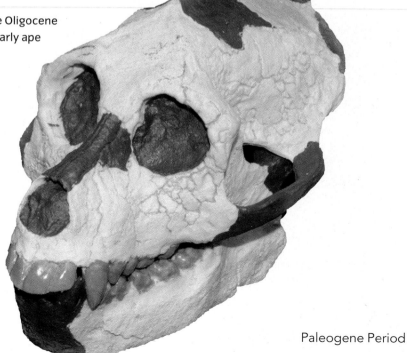

Angiosperm–Insect Coevolution

Flowering plants diversified in the Paleogene Period, in part spurred by coevolution with insects.

A A fossil flower from the Eocene of Washington. Without insects to assist in their reproduction, angiosperms may not have speciated at such a fast rate as they did during the Paleogene Period.

B An assemblage of Paleogene angiosperm leaves. Efficient reproductive strategies helped angiosperms become the dominant land plants of the Cenozoic Era.

C A fossil ant from the Eocene of the Dominican Republic, was one of many insect pollinators.

D Butterflies and moths have been important pollinators since the Paleogene Period. This is a fossil moth.

Climate change in the late Eocene apparently had an impact on terrestrial mammals, although their extinction may have been more directly tied to changes in land floras. Until the end of the Eocene Epoch, large areas of North America and Eurasia were covered with moist tropical and subtropical forests. With increasing aridity, forests were replaced by grass-covered savannas. Mammals having teeth adapted to chewing on soft leaves were increasingly replaced by ones having teeth capable of chewing coarse grasses.

The biggest innovation among Paleogene land plants was evolution of the grasses, an angiosperm group, in the Paleocene Epoch. The earliest grasses had discontinuous growth habits, like most plants. However, by the late Oligocene or early Miocene, they had evolved a pattern of continuous growth. This important novelty enabled grasses to quickly recover from grazing by herbivores and enabled herbivores to colonize upland areas.

Reproduction in many angiosperms has depended in part on insect pollinators. As flowering plants diversified, so did insects (see *What a Geologist Sees*). Hymenopterids (bees, wasps, and ants) and lepidopterans (butterflies and moths) probably expanded at rates greater than those of most other animals.

The fossil record of birds in the Paleogene Period includes large flightless forms (FIGURE 12.17). Some species reached the dimensions of modern ostriches and emus, and they developed powerful clawed feet and large beaks with which they could subdue and devour prey. Small mammals, lizards, snakes, amphibians, and arthropods were probably the favorite food sources of these giant flightless birds. Either flying birds were less numerous than they are today or, perhaps because of their life habits, the fossil record is biased against their preservation.

A large flightless bird FIGURE 12.17

Large flightless birds, evolved in places such as North America and South America during the Paleogene Period. This is a skeleton from Argentina.

CONCEPT CHECK STOP

What were the major episodes of tectonic deformation in the Cenozoic of western North America?

What major changes occurred in marine ecosystems during the Paleogene Period?

What major changes occurred in terrestrial ecosystems during the Paleogene Period?

What types of mammals evolved in the Paleogene Period? How did mammals evolve following the rise of grasses?

Neogene Period

The Neogene Period embraces two epochs, the Miocene and Pliocene, which together lasted more than 20 million years. The beginning of the Miocene Epoch and the Neogene Period is dated at close to 23 million years ago, and the end of the Pliocene Epoch is dated at 2.588 million years ago. During this period, Earth essentially reached its modern plate configuration (FIGURE 12.18), developed much of its modern biota, and witnessed the magnificent Himalaya Mountains and the Rocky Mountains reaching skyward.

Continued subduction along the Pacific margin of North America supplied magma emplaced, mostly during the Miocene Epoch, as the Sierra Nevada. Beginning in the Miocene, basaltic magma sporadically erupted from fissures in western North America, producing the Columbia River Plateau (Washington), the Snake River Plateau (Washington and Idaho), and the Black Rock Desert (Utah and Arizona).

Active tectonism has continued into the Quaternary Period around the margin of the Pacific Ocean. Subduction of oceanic lithosphere into the Peru–Chile Trench supplies the volcanoes of the Andes Mountains with newly melted magma. Similarly, subduction along the Japan Trench has ensured continuing tectonic activity along the Japanese island arc system.

Uplift of the Colorado Plateau in the western United States stimulated downcutting through the stratigraphic layers by the Colorado River. Beginning its formation in the Miocene Epoch, the Grand Canyon has reached its present shape in less than 20 million years.

Climatic change, often linked to biotic change, was one of the dominant themes of Neogene history. Milankovitch cyclicity is distinctly reflected in oceanic sediments and oxygen isotopic ratios. Early in the Neogene, the Mediterranean Sea dried out, and it later refilled. Aridification of many areas helped grasses, herbs, and weeds fan out across continental interiors. Silica-rich

Miocene paleogeography FIGURE 12.18

In the Neogene Period, continents shifted to what are essentially their present positions. Major areas of tectonism, ones that have remained active into the Quaternary Period, include the Pacific rim and the Himalayas. By the Miocene, the Tethys Ocean had become nearly fully closed, leaving the Mediterranean Sea and other small basins. Hotspots are responsible for formation of the Hawaiian Islands, the East African rift system, and other features.

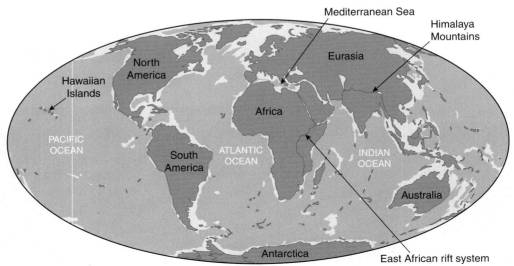

grasses favored grazing mammals with long teeth (such as some horses, antelope, and deer). In the savannas of Africa, humans evolved from an early form of apes.

Several large asteroid impacts occurred during the Neogene Period, and they have left their mark in widespread fields of impact-related glass droplets called **tektites**. One such impact involved a binary asteroid that crashed into Earth approximately 14.5 million years ago (in the mid-Miocene Epoch). The impact left two craters in southern Germany and a field of glassy tektites spread over parts of Germany, Austria, and the Czech Republic.

SALINITY CRISIS IN THE MEDITERRANEAN SEA

The Mediterranean Sea is the largest remaining vestige of the once-enormous Tethys Ocean. By the mid-Miocene, collision of the African Plate and India with Eurasia had almost fully constricted the Tethys (Figure 12.18), leaving behind the Mediterranean and a few other small basins (the Black Sea, the Caspian Sea, and the Aral Sea).

In the latest Miocene (Messinian Age), about 6 million years ago, global sea level had fallen by an estimated 50 m, enough to restrict water flow into the Mediterranean from the Atlantic Ocean at the Strait of Gibraltar. At that time, the Mediterranean region was warm, evaporation rates were high, and the sea shrank considerably in size. Evaporite minerals, especially salt (halite) and anhydrite, were deposited through much of the basin.

Rivers flowing into the Mediterranean all have deeply incised valleys filled with Pliocene sediment. The valleys of the Nile River (Egypt), the Rhône River (France), and the Po River (Italy) were cut dramatically in the latest Miocene, when their **base level**, or the lowest level to which they can erode, was defined by the lowest level ever reached by the Mediterranean. At the start of the Pliocene Epoch, 5.3 million years ago, sea level rose just enough to allow Atlantic waters over the sill at Gibraltar. As Mediterranean waters rose, the rivers feeding the basin rapidly filled their valleys with sediment up to the new base level.

NEOGENE MARINE LIFE

One of the most striking aspects of Neogene to Quaternary marine fossils is their increasingly modern species composition (**FIGURE 12.19**). Recognition of this fact is even encoded in the names given to some of the epochs of the Neogene Period by Sir Charles Lyell in the 1830s and 1850s. Lyell defined the Miocene and Pliocene epochs of the Neogene Period, as well as the Pleistocene Epoch of the Quaternary Period, using percentages of marine mollusk species still alive today. The Miocene (from the Greek *meion*, meaning "few") was defined as having more than 8% extant mollusks. Of the mollusks in the Pliocene (from Greek, *pleion*, meaning "more," and originally spelled Pleiocene), 50%–90% are still living today. Finally, of the mollusk species in the Pleistocene (a modification of the term "post-Pleiocene"), more than 90% are still living today.

Neogene marine fossils FIGURE 12.19

Marine fossils of the Neogene Period are essentially indistinguishable from a modern assemblage of marine animals. They include clams (A), snails (B), sand dollars (C), crabs (D), and sharks' teeth (E).

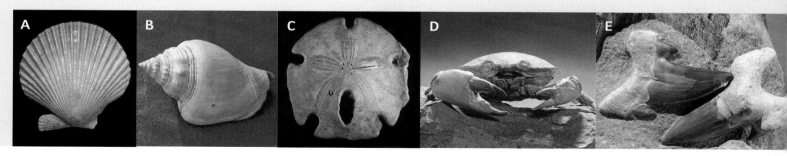

Planktonic microorganisms formed integral links in marine food webs of the Neogene Period, and they have continued to be important into the Quaternary Period. Planktonic foraminifera, whose diversity diminished in the Paleogene Period, made a strong comeback in the early Neogene Period. Their remains are useful for determining ages of strata, indicating oceanic circulation patterns, and estimating ocean water temperature. Because they are sensitive to water temperature, their biogeographic distribution parallels latitudinal position on the globe. Some species change their direction of coiling according to water temperature, and this adds to the oxygen isotopic evidence of greenhouse-to-icehouse climatic shifts, as reflected in ocean temperature changes (**FIGURE 12.20**).

Neogene and Quaternary ocean temperature FIGURE 12.20

Two ways of assessing paleotemperature changes in the ocean are using isotopic analyses and studying foraminiferan tests.

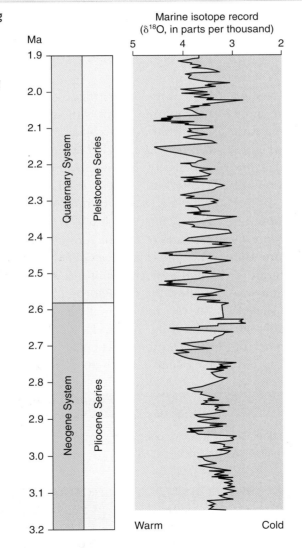

A This is a record of oxygen isotopes ($\delta^{18}O$) from a sediment core in the subtropical Atlantic Ocean.

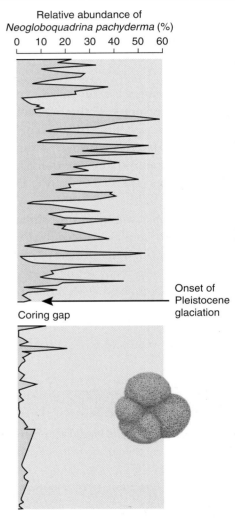

B This graph depicts the relative abundance of the foraminiferan *Neogloboquadrina pachyderma*, a species that favors cold marine water. Its abundance in sediments increases during intervals of cooler climate.

Other microfossils that are significant components of Neogene assemblages are calcareous nannoplankton, diatoms, radiolaria, and dinoflagellates. In the deep sea, the tests of these microorganisms often form a large fraction of the sediment (FIGURE 12.21) draping the ocean floor.

In shallow tropical areas, scleractinian corals and carbonate-secreting (or coralline) sponges emerged as important reef-formers. The corals, with their algal symbionts, were able to compete and flourish even in places where food resources were slim. Most Quaternary reef communities are also dominated by scleractinian corals and coralline sponges.

By the Miocene Epoch, vertebrate life in the sea was distinctly modern in character. Whales increased their species numbers, and a new specialized form, the dolphins, appeared. Like their terrestrial ancestors, most whales were carnivores that fed on live prey, including fishes, cephalopods (such as squid and octopuses), and crustaceans. A few species strained zooplankton from the water. Marine fishes of the Neogene Period were mostly teleosts, but sharks, rays, and even rare holdovers from much earlier times remained in the seas. Marine vertebrates changed little, in general terms, during the Quaternary Period.

NEOGENE TERRESTRIAL LIFE

Two major factors influenced changes in the terrestrial biota of the Neogene Period: changing climatic conditions and changes in food supplies. Animals that were better adapted to the changing conditions thrived, but those that could not adapt were not so lucky.

Two groups of plants, the herbaceous plants, or herbs, and grasses, proliferated in response to changing climate during across the Paleogene–Neogene transition. Herbaceous plants are non-woody annuals that die after releasing their seeds and include daisies, marigolds, sunflowers, and so-called weeds. As cooler, drier temperatures arrived in the late Oligocene Epoch and continued into the Miocene Epoch, forests shrank in size, and herbaceous plants and grasses began to spread.

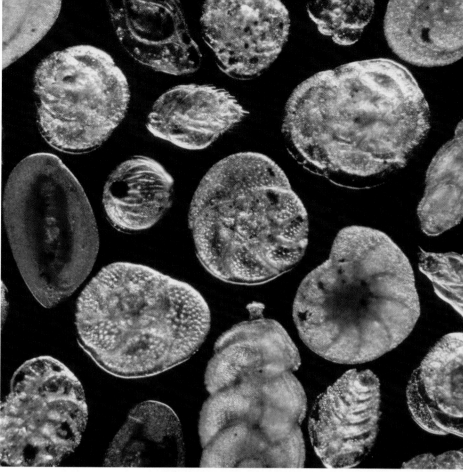

Microfossils in ocean sediment FIGURE 12.21

Foraminifera and other microfossils recovered from sediment cores in the world's oceans provide detailed information about the ages of strata. Foraminiferans are among the best guides to the relative ages of Cenozoic marine strata.

Their preference for open habitats and their resistance to relatively dry conditions, even drought, made them effective competitors in the new climatic regime.

In the waning phases of the Miocene Epoch, between 7 and 6 million years ago, grasses referred to as C_4 grasses partly replaced the previously dominant form of grasses, called C_3 grasses. These grasses have different physiologies: C_4 grasses extract more carbon-13 from the atmosphere than do C_3 grasses. Because of this, carbon-12/carbon-13 isotopic ratios left in terrestrial sediments make a sudden shift toward carbon-13 enrichment at the time that C_4 grasses dispersed across most continents.

Another difference between C_4 and C_3 grasses is that C_4 grasses incorporate more silica in their stems and leaves, and this had an effect on grazing mammals. By

Grasses, which form a large part of the diet of many terrestrial mammals, including humans, have their origins in the Paleocene Epoch. They underwent other major changes later in the Cenozoic Era.

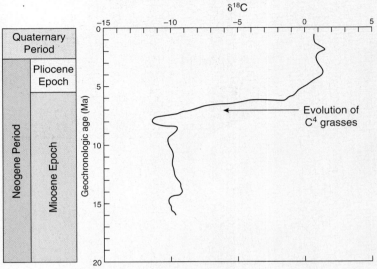

A In the Miocene Epoch, modern-type C_4 grasses evolved. This graph shows the trend of carbon-13/carbon-12 ratios ($\delta^{13}C$) from Neogene paleosols. The large positive shift reflects the spread of C_4 grasses in the late Miocene, between 7 and 6 million years ago.

B A field of grass that is being grown for hay. The evolution of C_4 grasses like these had an effect on the evolution of grazing mammals. In horses, longer teeth were selected for following the appearance of these silica-rich grasses.

the Miocene Epoch, many modern types of grasses (FIGURE 12.22) were in existence and a major component of the diets of herbivores. Mammals, such as horses, that grazed on grasses suffered a rapid decline near the end of the Miocene Epoch. Those that were selected for tended to have longer teeth than those that went extinct. Silica is harder than the phosphatic teeth of vertebrate animals, so the more silica there is in grasses, the more that grazing animals will grind down their teeth. The rise of silica-rich C_4 grasses may have contributed to the extinction of grazing mammals having short teeth.

In the Miocene Epoch, species of even-toed ungulates (artiodactyls) increased, even as odd-toed ungulate species (perissodactyls) declined. Deer, antelopes, cattle, sheep, pigs, camels, giraffes, and their kin—all having long teeth—were beneficiaries of the expansion of silica-rich grasses into open plains.

The rise of new herbivorous ungulates was probably connected to changes among carnivores. Dogs (including wolves) and cats appeared in the Paleogene Period, and bears and hyenas appeared in the Miocene Epoch.

Frogs, rodents, songbirds, and some snakes (mostly the poisonous vipers) all radiated during the Neogene Period. Increasing species diversity in these groups may have been related to changes in terrestrial food webs. Songbirds (FIGURE 12.23) and rodents largely de-

Cenozoic bird FIGURE 12.23

Songbirds radiated in the Neogene Period. Their evolution was fueled in part by the availability of grains and other seeds produced by angiosperms. This exceptionally preserved specimen, retaining feathers, was found in Germany.

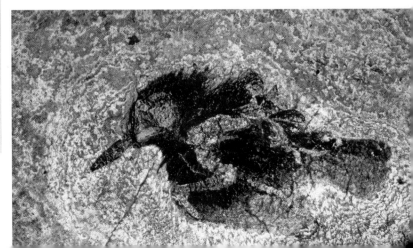

pended on angiosperms, including grasses, for food. Some bird species may have eaten grains and other seeds, but others certainly consumed insects, which underwent a coevolutionary radiation with the angiosperms. Insects also made up a large part of the diet of frogs. For snakes, major food sources were frogs and rodents.

Much of the evolution of primates took place during the Neogene Period. Monkeys originated in the Oligocene Epoch, apparently in Africa or Eurasia. These are known as the "Old World monkeys." By the end of the Oligocene Epoch, monkeys had spread to South America and founded a group called the "New World monkeys." These monkeys are best known for their prehensile tails,

which they use in climbing trees. Apes, native only to the Old World, were numerous in the mid-Paleogene but then declined in the late part of the period.

CONCEPT CHECK **STOP**

How did global climate change through the Neogene Period? What was its effect on the distribution and evolution of animals and plants?

Why did the Mediterranean Sea nearly dry out?

Quaternary Period

LEARNING OBJECTIVES

Explain the evolution of humans and their close relatives.

Explain the effects of Quaternary glaciation on the landscape and terrestrial life forms.

he most recent 2.588 million years of Earth history is dubbed the Quaternary Period. It consists of two epochs, the Pleistocene, or most recent ice age; and the Holocene (or "recent"), the present time. The Quaternary Period is characterized by a series of glacial advances and retreats in the Northern Hemisphere (**FIGURE 12.24**). Glaciers sculpted the landscape of North America and

Pleistocene glaciation and its effects FIGURE 12.24

A In the Quaternary Period, the Northern Hemisphere experienced a series of glacial advances and retreats. This map shows the maximum extent of glacial advance during the Pleistocene Epoch, when continental glaciers covered as much as 30% of Earth's surface. In the Northern Hemisphere, glaciers extended south to about the 40th parallel.

CHINA
RUSSIA
Siberia
Eurasian Ice Sheet
ARCTIC OCEAN
Innuitian Ice Sheet
Beringia
ICELAND
EUROPE
Greenland Ice Sheet
Alaska
Sea Ice
Cordilleran Ice Sheet
Laurentide Ice Sheet
UNITED STATES

B Pleistocene glaciers reshaped the landscape of North America and Eurasia. In the Yosemite Valley of California, granitic batholiths of the Sierra Nevada were carved into spectacular peaks, U-shaped valleys, and hanging waterfalls. The batholiths were once kilometers below the surface and later uplifted and uncovered through the combination of Neogene tectonics and erosion.

Eurasia and mantled the land with rocky sediment (till and outwash). The Great Lakes and the Finger Lakes were carved out during this time, and exposure of the once-deeply buried igneous rocks of the Sierra Nevada left them vulnerable to glaciers that sculpted them, especially in the Yosemite Valley of California.

Just as in the Neogene Period, climate change has been a dominant theme of Quaternary history. Climate change has influenced landscape evolution, the geographic distributions of organisms, and the evolution of organisms. Arid conditions that developed concurrently with climatic cooling resulted in expansion of the Sahara Desert across the northern part of Africa. Similar

expansion of the Gobi Desert occurred across Mongolia and western China. During glacial episodes, some large mammals dispersed across the Bering Strait from Eurasia to North America. In the savannas of Africa, the hominid lineage evolved from primitive apes during the Neogene and Quaternary periods. The genus *Homo* appeared in the Quaternary Period.

Glaciers of continental scale expanded and contracted multiple times across North America and Eurasia. Floral and faunal migrations according to temperature conditions were common. Falling sea level related to glacial buildup opened dispersal routes across land bridges between continents, and rising sea level related

Amazing Places: California Tar Pits FIGURE 12.25

In the heart of Los Angeles, California, lies the site of one of the richest Pleistocene fossil discoveries—the Rancho La Brea Tar Pits. In the Pleistocene, when the Los Angeles area was a savanna with sporadic watering holes, mammals, birds, insects, and other creatures were trapped in sticky asphalt and became fossilized in it. The animals may have been first lured to the tar pits by water floating on the surface of the asphalt.

Los Angeles, California

Global Locator

NATIONAL GEOGRAPHIC

to glacial melting closed those dispersal pathways. Close to the end of the last glacial maximum, large mammals adapted to cold-weather conditions in the Northern Hemisphere died out. Some may have been hunted into extinction by humans.

Asteroid impacts, recorded in part by tektite fields, continued into the Quaternary Period. One significant impact occurred in the Pleistocene Epoch, 79,000 years ago, when an asteroid slammed into southeast Asia. Tiny tektites were scattered across a wide region including China, the Pacific Ocean, and the Indian Ocean. The microtektite layer left by the impact is distinctive and can be used for detailed stratigraphic correlation in the Indo-Pacific region.

QUATERNARY LIFE

Marine strata of the Pleistocene Epoch show marine biotas that are essentially fully modern in species and generic composition. More than 90% of Pleistocene mollusk species, for example, are still living today.

Terrestrial vertebrates of the Pleistocene Epoch have close relatives still living today, but some of the species seem strange by modern standards (FIGURE 12.25) or had geographic distributions that are different from those of their present-day relatives. The Pleistocene dire wolf was about twice the size of a modern wolf. *Smilodon*, a saber-toothed cat, had greatly elon-

◀ A Even today, tar, distilled from petroleum bubbling up from underlying sediment layers, continues to form pools at the surface of the La Brea Tar Pits. Sculptures of Pleistocene animals at the tar pits depict events that occurred during the ancient past.

Among the Pleistocene animals whose bones and teeth have been excavated from the Rancho La Brea Tar Pits of California are giant ground sloths (B), saber-tooth cats (C), and dire wolves (D). Dire wolves are especially common fossils.

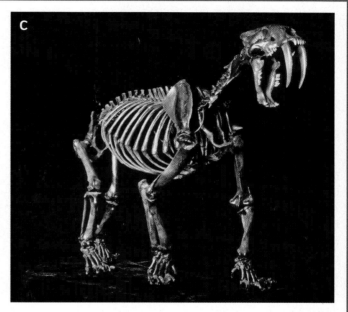

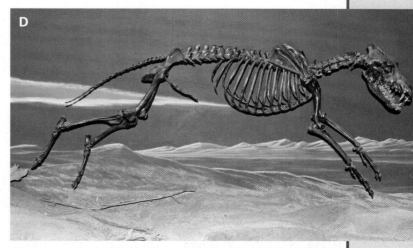

gated, dagger-like canine teeth. Large elephants (woolly mammoths and mastodons) roamed in the Northern Hemisphere. Giant ground sloths and glyptodonts (which are much like giant armadillos) were impressive herbivorous mammals that lived in the Americas.

Until late in the Cenozoic Era, North and South America maintained dissimilar terrestrial faunas. The two continents were separated since the Cretaceous Period by the Caribbean Ocean. As subduction took place at the eastern and western margins of the Caribbean Ocean, though, a sizable microcontinent (now Nicaragua and Panama) slid into a position between the two Americas. A narrow land bridge—the Isthmus of Panama—thus emerged, allowing the migration of Pliocene and Pleistocene mammals. Giant ground sloths, armadillos, glyptodonts, capybaras, monkeys, and opossums all made

their way to North America. South America received a larger assortment of mammals, including elephants, camels, deer, cats, foxes, wolves, bears, peccaries, and rabbits, in what has been called the "Great American Faunal Interchange" (FIGURE 12.26).

HUMAN EVOLUTION

Hominids and their close relatives have a short evolutionary history, geologically speaking. The hominids, or species belonging to the family Hominidae, date back to the late Miocene Epoch (6 to 7 million years ago). The hominids include the extinct australopithecines (the genera *Sahelanthropus, Orrorin, Kenyapithecus, Ardipithecus, Australopithecus,* and *Paranthropus*) and the genus *Homo* (FIGURE 12.27).

> **hominid**
> The primate group that includes australopithecines and species in the genus *Homo*.

Great American faunal interchange
FIGURE 12.26

Movement of a thin slice of continental crust into the space between North and South America during the Neogene Period opened the way for a great faunal exchange between the two continents during the Neogene and Quaternary Periods. Giant ground sloths, glyptodonts, and others migrated to North America, and cats, wolves, camels, and others migrated to South America.

Hominids first appeared in Africa between 6 and 7 million years ago. They evolved separately from the modern apes, sharing with them a primitive ape as a common ancestor.

A Approximate geologic ranges of hominid species. Hominids consist of the australopithecine genera plus species in the genus *Homo*.

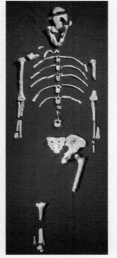

B "Lucy," a partial skeleton of a Pliocene australopithecine (*Australopithecus afarensis*) from Ethiopia.

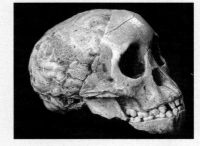

C This is the skull of a juvenile of *Australopithecus africanus*, a specimen known as the "Taung baby."

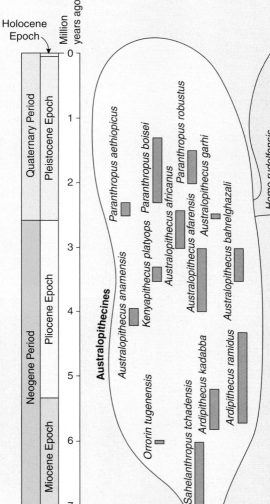

D A skull of *Paranthropus boisei*, an australopithecine with a robust skull.

E *Homo erectus* was a geographically widespread species belonging to our genus.

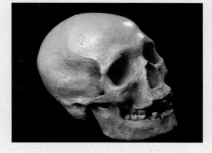

G Skull of a Cro-Magnon, an early example of our species, *Homo sapiens*.

F Skull of a Neanderthal, *Homo neanderthalensis*. In many ways, Neanderthals resembled our own species, but there are distinct genetic differences between the two forms.

The modern apes, a sister group to the hominids, include the bonobos, chimpanzees, gorillas, and orangutans. Fossil evidence shows that modern humans did not evolve from any of these apes. Instead, humans and modern apes evolved separately from a common ancestor (a primitive type of ape) that existed around the early Miocene, roughly 20 million years ago.

Within 5 million years of their emergence in Africa, early apes had spread through parts of Africa and Eurasia. The number of middle and late Miocene ape species exceeded the number alive today, and some reached great sizes. One genus is appropriately named *Gigantopithecus*.

The first australopithecines appeared in the Miocene Epoch. The early australopithecines essentially represent an intermediate step between ancient apes and more recent hominids. *Australopithecus* had an apelike skull, with a small brain, heavy ridges over the eyes, and a large jaw. The skull was also shorter than that of modern humans. Postcranial skeletal remains, such as the skeleton nicknamed "Lucy," show that austalopithecines had wide hips like modern humans. Fossil trackways found in association with australopithecine remains demonstrate that they walked upright (**FIGURE 12.28**).

Species of the genus *Homo* appeared in the Pleistocene Epoch, about 2.4 million years ago. In the Pleistocene, the genus had multiple species, some of which were coeval. The skull of the earliest *Homo* was proportionately larger than that of any ape, and it had smaller teeth. These early hominids were well on their way to becoming modern humans.

In places, *Homo* and *Australopithecus* overlapped in geographic range. With their larger brain capacities, though, early species of *Homo* had a competitive advantage.

Early hominid species, including early *Homo* species and perhaps some australopithecine species, learned how to make simple tools by chipping flakes from stones and producing sharp edges. Early *Homo* species probably had diets consisting largely of plants, but tools helped them acquire meat to add to their diets. The oldest tools belong to what we call the Oldowan culture (**FIGURE 12.29A**), after Olduvai Gorge in Kenya, where many important early hominid discoveries were made.

Homo erectus (**FIGURE 12.27E** on page 397), the ancestor of our own species, appeared about 1.8 million years ago. This species migrated over a wide area, including Africa, Europe, and Asia, and existed until about 200,000 years ago. With a larger, more powerful brain than its predecessors, *Homo erectus* produced more refined stone tools than earlier hominids.

Homo erectus had a different skull shape than our species. The skull was more elongate and had a low, sloping forehead. In addition, compared to our species, the skull had more pronounced brow ridges, a larger jaw, and a forward-projecting mouth.

The brain size of *Homo erectus*, at 775 to 1300 cc, was smaller than that of modern *Homo sapiens* (which has a brain size of 1200 to 1500 cc). The pelvis (hip bone)

Hominid trackway FIGURE 12.28

Australopithecine footprints left in a Pliocene volcanic ash layer on the Laetoli plain, Tanzania, show that these early hominids could walk upright.

was also narrower. The size of a baby's head at birth is limited by the size of the opening in the mother's pelvis because the baby's head must pass through the opening during birth. Because *Homo sapiens* has a wider pelvis, a baby's brain can grow to larger size even before birth, and this gives infants an advantage in reaching maximum brain size in adulthood.

Neanderthals (**FIGURE 12.27F** on page 397) show even more similarities to modern humans than does *Homo erectus*. DNA evidence from Neanderthal bones, however, shows strong genetic dissimilarities, so they are now regarded as a separate species, *Homo neandethalensis*. Both advanced species of *Homo* seem to have arisen from *Homo erectus*.

Evolving hominid culture FIGURE 12.29

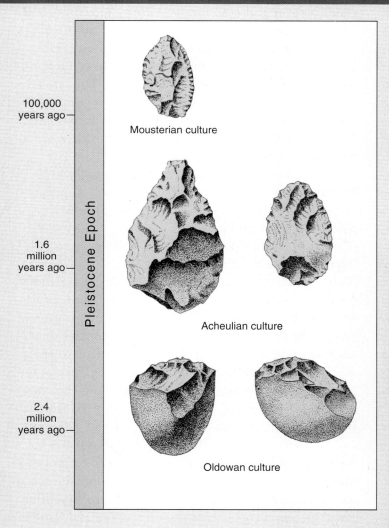

A Stone tools made by hominids became increasingly sophisticated through time. Tools of the Oldowan culture were crude, formed from chipping pebbles to form sharp edges. Oldowan tools were made by early species of *Homo* and perhaps also by australopithecines. Stone implements belonging to the Acheulian culture were probably flaked by *Homo erectus*. Tools of the Mousterian culture were the work of Neanderthals.

B A cave painting from Europe, evidence of the sophisticated level of Cro-Magnon culture.

Neanderthal fossils first appear in Pleistocene sediments dated at about 200,000 years. Molecular clock evidence, though, suggests that their origins could have been as much as 500,000 years ago. They coexisted with *Homo sapiens* for a long time until becoming extinct about 30,000 years ago. Neanderthal skulls have heavy brow ridges and large, chinless jaws, but otherwise these cousins of our species appear much like modern humans. Even their brain capacity equaled or exceeded that of modern *Homo sapiens.*

Neanderthals were distributed through much of the Old World, and they developed a rather sophisticated culture that included well-fashioned stone tools. Many Neanderthals sought refuge in caves, especially in Europe, and they were avid hunters who brought down prey mostly using sharpened sticks as spears. Neanderthals buried their dead and apparently cared for the sick and injured while they were living.

Homo sapiens (FIGURE 12.27G on page 397) evolved in Africa during the Pleistocene Epoch, approximately 200,000 years ago. In this case, molecular clock evidence and fossil evidence are nearly in accord. The earliest modern human remains from Europe are about 33,000 years old. It was close to this time that Neanderthals became extinct. Whether *Homo sapiens* helped drive *Homo neanderthalensis* into extinction is speculative. The culture of *Homo sapiens* that developed first in Europe is called the Cro-Magnon culture. It was much more sophisticated than the culture of any previous species of *Homo* and included finely fashioned stone tools, elaborate artwork left on cave walls (FIGURE 12.29B on page 399), clay models, decorated bones, and jewelry formed from animal teeth and shells.

NEOGENE–QUATERNARY ICE AGES

The Miocene–Pliocene transition was a time of dramatic climate shift (Figure 12.4 on page 375). In the latest Miocene, when polar areas were cold, glacial buildup intensified and sea level dropped. Erosion occurred along many shelf areas, and river valleys were deeply incised, resulting in disconformities that can be correlated through much of the world. Warming on a global scale at the beginning of the Pliocene Epoch brought glacial retreat, a sea level rise of 50 m or more, and renewed sedimentation on the shelves. Warm conditions ensued for about 2.7 million years.

Evidence of the onset of the most recent Ice Age in the Northern Hemisphere marks the beginning of the Pleistocene Epoch. Glacial advance occurred about 2.588 million years ago. Since that time, global climate has alternated between cold episodes of glacial expansion and warmer **interglacial** episodes when glaciers receded. Warm and cold intervals are reflected on both large and small scales in oxygen isotopic ratios ($\delta^{18}O$) preserved in layers of ice cored from glaciers in Greenland, Antarctica, and alpine areas; in the tests of planktonic foraminifera; and in coral skeletons. They are also reflected in other ways, such as in the relative abundances of warm- and cold-water-adapted foraminifera fossils, in changes to terrestrial flora, and in evidence of ice rafting of sediment to the deep ocean by icebergs (called Heinrich events).

> **interglacial**
> A relatively warm time interval between successive glacial episodes.

At times of maximum glacial cover, as much as 30% of Earth's surface may have been covered by ice. In the Northern Hemisphere, continental glaciers expanded southward across North America, Europe, and Asia to around the 40th parallel. Extensive areas were enveloped by glacial sheets up to 3 km thick. A zone of permafrost stretched a few hundred kilometers beyond the glacial margin. The weight of the dense, compact ice depressed the underlying continental crust, and each time glaciers retreated, the continents experienced isostatic rebound (FIGURE 12.30).

In the Southern Hemisphere, glacial pulses have been reflected in the waxing and waning of continental glaciers across Antarctica. The Andes Mountains were ice covered, and an ice cap extended from Argentina. Small glaciers were also present in New Zealand, Tasmania, and the mountains of Africa.

Beyond the glacial margins, large lakes developed. In the Basin and Range Province of the western United States, dozens of basins filled with water mostly derived from increased precipitation and reduced

Crustal rebound FIGURE 12.30

Thick, dense continental glaciers are extremely heavy, and they cause depression of the crust. When the glaciers recede, the weight is released, and the crust is allowed to rebound isostatically.

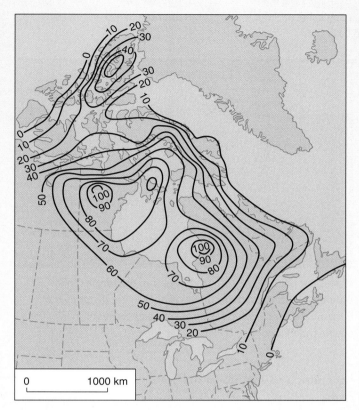

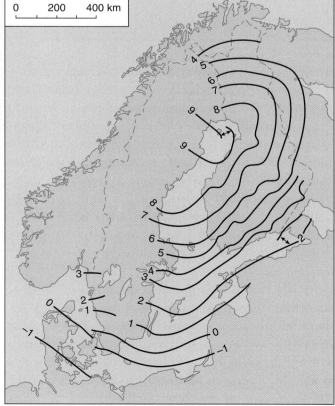

A A map of North America showing the amount of isostatic rebound, or crustal uplift, in meters since 6000 BP, once continental glaciers had fully receded. Two centers of uplift reflect the positions of thick ice bodies that lasted until late in the process of glacial recession.

B A map of northern Europe today, showing isostatic rebound of the crust following recession of the Pleistocene glacial sheets. The contour lines show rates of uplift in millimeters per year compared to present sea level.

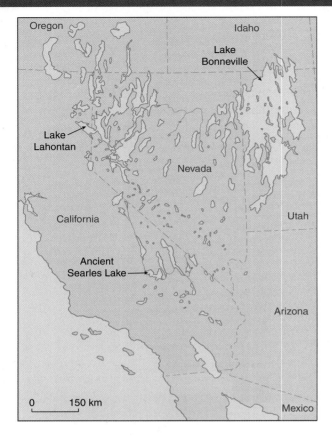

A During the Pleistocene Epoch, increased precipitation and decreased evaporation rates led to the formation of pluvial lakes beyond the glacial front. Some of the former pluvial lakes survive today as much smaller entities. Examples are the former Lake Bonneville, which survives as the Great Salt Lake, and the former Lake Lahontan, which survives as Pyramid Lake.

B This flat plain in western Utah is part of the former Lake Bonneville. About midway up the ranges in the distance is a horizontal terrace, which reflects a time of standstill in lake level during the Pleistocene Epoch.

evaporation to become **pluvial lakes** (FIGURE 12.31). The region occupied today by the Great Salt Lake (Utah) and the Bonneville Salt Flat (Utah–Nevada) was submerged as Lake Bonneville during the Pleistocene Epoch. Blockage of the lake's outlet in Idaho contributed to a buildup of water. Glacial meltwaters were also largely responsible for water buildup across a large area of north-central North America. This feature, called Lake Agassiz, survives today as the considerably smaller Lake Winnipeg. In Eurasia, expansive lakes also developed from glacial runoff.

Associated with global cooling was diminished evaporation from the oceans and lakes. Deserts became drier and more extensive. Prior to about 2.588 million years ago, the Sahara Desert was relatively small, and tropical forests were growing along its southern margin. Increased aridity in the Pleistocene Epoch drove expansion of the desert across northern Africa. About the same time, an extensive **loess** plain spread out over much of China, as dry winds carried silt particles from western Asian desert areas toward the Pacific Ocean.

Continental glaciers expanded and contracted through the entire Pleistocene Epoch, and by all indications, the Holocene Epoch is an interglacial phase. At least four major glacial advances can be identified from the sedimentary record in the Northern Hemisphere, although the ages of glacial deposits in separate regions may not have been exactly synchronous. In the United States, the glacial stages are called Nebraskan, Kansan, Illinoisan, and Wisconsin (or Wisconsinan), and the interglacial stages are called Aftonian, Yarmouthian, and Sangamon (or Sangamonian). In the Alpine region of Europe, comparable glacial cycles have been given the names Günz, Mindel, Riss, and Würm. The Alpine interglacial cycles are called Günz–Mindel, Mindel–Riss, and Riss–Würm (TABLE 12.1).

> **pluvial lake**
> A lake formed during an interval of exceptionally heavy rainfall.

Glacial and interglacial stages of the Pleistocene Epoch TABLE 12.1

| North America | | Europe | | Years before present |
Glacial stages	Interglacial stages	Glacial stages	Interglacial stages	
Wisconsin		Würm		75,000
	Sangamon		Riss-Würm	125,000
Illinoisan		Riss		265,000
	Yarmouthian		Mindel-Riss	300,000
Kansan		Mindel		435,000
	Aftonian		Günz-Mindel	500,000
Nebraskan		Günz		1,800,000

As Pleistocene glaciers advanced, they eroded the land surface, polishing resistant rock surfaces and leaving behind striations where rock fragments were dragged along the ice–bedrock interface, scouring V-shaped stream valleys into broader U-shaped valleys, and depositing rocks of variable sizes (erratics) in moraines, eskers, drumlins and outwash plains, often at great distances from their origins (FIGURE 12.32). In northern Canada, glaciers left bare igneous and metamorphic rocks of the Archean core of North America, which we call the Canadian Shield. The Baltic Shield of Scandinavia was similarly exposed by glacial action. South of the Canadian Shield, in the American Midwest and Plains, glacial till contains pebbles, cobbles, and boulders carried from as far away as Hudson Bay. Glacial till also mantles much of northern Europe and Asia.

Streams and valleys that existed before glacial advance were permanently changed as continental glaciers passed over them. Along the Canada–United States border, a series of valleys was scoured out by glaciers to form

Glacial imprint on the landscape FIGURE 12.32

A Glacial striations on the surface of a Devonian limestone in northeastern New York point southward, the direction of glacial movement during the Pleistocene Epoch.

B The Finger Lakes of New York are former stream valleys that were scoured out by Pleistocene glaciers. The valleys, which were originally V-shaped, have changed to become U-shaped through glacial scouring.

C Large drumlins, or fluted hills of glacial till, litter the Finger Lakes area of central New York. This is a drumlin near Waterloo, New York.

D This relief map shows how extensive the drumlin field is across central New York. The drumlins are all aligned, which reflects shaping of the mounds during glacial advance southward.

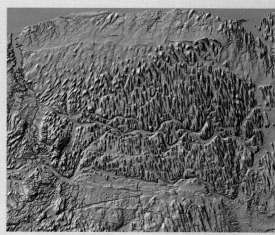

the string of lakes now known as the Great Lakes (FIG-URE 12.33). Glacial meltwaters then filled the deep scours with freshwater. In central New York, small stream valleys were glacially scoured to form the Finger Lakes. Before the glaciers reached the American Midwest, part of the Ohio River flowed to the northeast and emptied into the Gulf of St. Lawrence. Part of the Missouri River flowed northward and emptied into Hudson Bay. The northern segments of both rivers, where they met the glacial front,

were redirected and forced to flow along the glacial margin until finding outlets to the south. Today, the Ohio and Missouri rivers continue to drain to the south.

In the northwestern United States, the glacial front dammed water flowing along the Clark Fork River, and this formed a long, narrow lake, called Lake Missoula, that extended diagonally across western Montana. About 18,000 years ago, as glacial recession was proceeding, the ice dam was repeatedly breached, each time

Evolution of the Great Lakes FIGURE 12.33

The Great Lakes, along the Canada–United States border, began as a series of valleys that were later carved and enlarged by Pleistocene glaciers. This sequence of maps illustrates the hypothesized glacial drainage patterns in the Great Lakes region, with positions of the future Great Lakes indicated by dotted lines. The maps show the beginning of basin infilling by water as glaciers were starting to recede over the region 14,000 years ago (A); redirection of drainage along the glacial front about 13,000 years ago (B); and the appearance of the lakes after the glaciers had completely receded about 10,000 years ago (C).

A

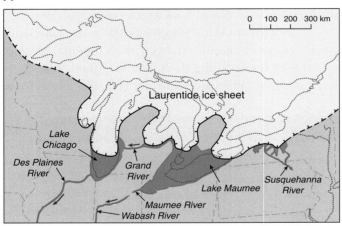

B

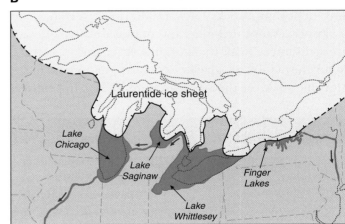

C

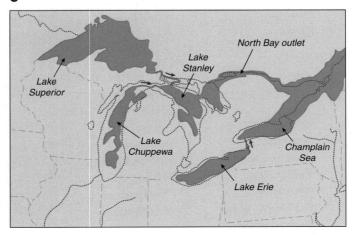

Niagara Falls FIGURE 12.34

Retreat of Pleistocene glaciers exposed an escarpment of resistant dolostones belonging to the Lockport Group between Lake Erie and Lake Ontario. The Niagara River, which flows between the two lakes, reached the escarpment and formed a waterfall at the edge of the cliff. Water rushing over the falls causes the strata below the resistant caprock to erode quickly, which undermines the falls and has led to its recession upstream, toward Lake Erie.

A A view of Niagara Falls from the Canadian side of the Niagara River.

B From the top of the Niagara Escarpment, the drop-off to the plunge pool at the base of the falls is sudden and dramatic. The distance to the base of Niagara Falls is about 55 m at this point.

spilling vast amounts of water into eastern Washington. The currents must have been torrential because they eroded huge channels and left behind enormous volumes of gravel, cobbles, and boulders. The resulting landscape, called the Channeled Scablands, seems like something from another world.

One of the great exposures of Paleozoic strata in North America and a major tourist destination, Niagara Falls (FIGURE 12.34) has its origin with Pleistocene glaciation. The falls was formed along the Niagara River, which flows northward from Lake Erie to Lake Ontario. Retreating ice of the most recent glacial episode (the Wisconsin glaciation) uncovered an **escarpment**, called the Niagara Escarpment, that was capped by resistant Silurian dolostones of the Lockport Group. Relatively weak Silurian and Ordovician shales underlying the Lockport Group are continuously eroded by the great rush of water over the 55 m cliff, which undermines the **caprock** and leads to upstream retreat of the falls.

In addition to the glaciers themselves, wind currents beyond the limits of the Pleistocene ice sheets played a role in landscape evolution. After Lake Michigan was scoured out and filled with glacial meltwater, sands originally deposited as outwash were carried eastward by winds and then redeposited in large dunes along the southeastern shore of the lake. These survive today in such places as Indiana Dunes National Lakeshore and Michigan's Sleeping Bear Dune. Easterly winds of the Pleistocene also picked up and transported large amounts of silt from outwash plains and floodplains. The wind-blown silt was redeposited in thick layers of loess, which now blanket a large area of the central United States.

Where glacial sheets met the ocean, icebergs calved off and floated out to sea. Icebergs carried rocks frozen in their lower reaches, and in time they were released as dropstones as the icebergs melted. Pulses of iceberg formation (called Heinrich events) culminated in small-scale cooling cycles, perhaps 5,000 to 10,000 years

Bering land bridge FIGURE 12.35

During times of low sea level in the Quaternary Period, a land bridge linked Asia and North America across the Bering Strait and provided a conduit for the migration of land animals between the two continents. This map shows the Bering land bridge (Beringia) at 18,000 years before present (BP).

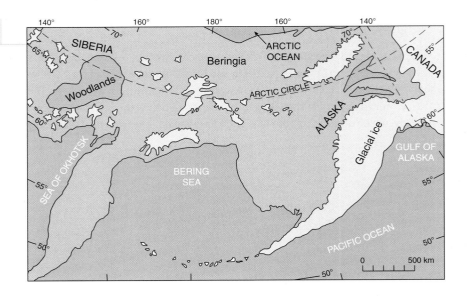

in duration, and were followed by abrupt shifts to warmer temperatures.

Each major glacial expansion resulted in eustatic sea level change and changes in biotic patterns. During glacial maxima, sea level fell on the order of 100 to 120 m, enough to expose many continental shelves all the way to the shelf edge. Species migration routes hinging on connections between regions were affected by sea level fluctuations. During glacial episodes, the Bering Strait turned into a land bridge, called Beringia, between Asia and North America (FIGURE 12.35). This corridor opened the way for many mammals, including humans, to enter the Americas.

The end of the most recent glacial episode was associated with the extinction of many large mammals, especially in the Northern Hemisphere (FIGURE 12.36). Large elephants, both mammoths and mastodons, that once roamed close to the glacial margin in North America and Eurasia, vanished. Woolly rhinoceroses, cave bears, Irish "elk," and others disappeared from Eurasia, and giant ground sloths, glyptodonts, saber-tooth cats, dire wolves, and others perished from the Americas.

Pleistocene extinction FIGURE 12.36

A Large mammals, including mammoths and mastodons, became extinct at the end of the Pleistocene Epoch. B Extinction of Pleistocene elephants may have been related in part to overhunting by early humans.

A

B

Humans probably played a role in the extinctions of some Ice Age mammals. Mammoth and mastodon skeletons from peat bogs of the American Midwest and Canada commonly show evidence of hunting and butchering. Stone implements are occasionally found in association with the skeletal remains, but even more commonly the bones show surficial scratches where they were sliced by flint tools. Tailbones are rarely preserved with these elephant skeletons, probably because the tails were removed when the hides were stripped. Many bogs were places where ice blocks lodged in ground moraines or other soft sediments. As the blocks slowly melted, they formed ponds, but the ice melted slowly enough that the water stayed cool for years. Early humans apparently used these ice-filled ponds as natural "refrigerators" for storing their elephant meat.

EARTH'S PAST, PRESENT, AND FUTURE

Earth's long, intricate story is not yet complete. The narrative you have just read is merely a progress report. The Earth system continues to evolve and to diligently record the outcome of physical, chemical, and biologic processes in its stratigraphic "pages."

As you turn to the "page" of Earth's history "book" representing the present time, it is worth recalling the relevance of Earth history to our lives and to the lives of generations yet to come. Charles Lyell's principle of uniformitarianism teaches us that Earth processes are ongoing, and they will continue to operate in the future as they have in the past. The geologic past therefore provides lessons for our planet's future.

Humans have, and no doubt will continue to have, a significant influence over Earth systems (FIGURE 12.37). Exactly how our planet evolves, and what gets written on the next stratigraphic "pages," will be partly in your hands. Your life activities, the choices you make, will help shape our planet's future. Any perturbation to Earth systems will have consequences. If we are to ensure that Earth evolves in ways that will continue to support our species long into the future, we must treat it responsibly, with respect for its rich, complex, and fascinating history.

The world today and tomorrow FIGURE 12.37

Humans have significantly influenced Earth systems, and changing technologies are accelerating the pace of change.

A For about 2200 years, the most conspicuous sign of the impact of humans on the land was the Great Wall of China, which was consolidated during the Qin Dynasty. Although an impressive edifice, Earth systems were affected little by it.

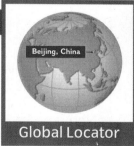

Beijing, China

Global Locator

B Today, cities such as Beijing are growing. They have enormous demands for energy and face concerns about air and water quality, as well as waste disposal.

CONCEPT CHECK STOP

What is the evolutionary history of humans? What are our nearest relatives?

How did Quaternary glaciation affect landscape evolution?

What caused extinction of large mammals in the late Pleistocene?

SUMMARY

1 Cenozoic Overview

1. In the Cenozoic Era, the modern continents drifted toward their present positions, and the modern conveyor-belt ocean circulation pattern was established in the Atlantic Ocean. Eurasia witnessed uplift of the Himalaya Mountains, the Alps, and the Apennines, and closure of the Tethys Ocean. Subduction and orogeny occurred along the Pacific coast of the Americas.

2. Sea level fluctuated rapidly in the Cenozoic Era, driven mostly by glacial–interglacial cycles. Sea level changes had profound effects on the geographic distribution of animals and plants. Emergence of the Bering land bridge allowed migration between Asia and North America.

2 Paleogene Period

1. Antarctica became geographically and climatically isolated in the Oligocene Epoch.

2. Subduction of the Farallon Plate beneath western North America led to orogenic activity that resulted in uplift of the Rocky Mountains. Later, extensional tectonics formed the Basin and Range Province.

3. Following the end-Cretaceous extinction, marine organisms refilled open niche spaces. Coinciding with a switchover of the world ocean from a calcite sea to an aragonite sea, scleractinian corals regained their position as reef-formers.

4. Mammals, especially **placental mammals**, proliferated on land during the Paleogene Period. Some, such as whales, left the land to become successful predators in the sea.

5. Birds radiated in the Paleogene Period. Large flightless birds appeared on land.

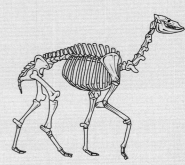

6. Angiosperms and insects radiated, in part because of **coevolution**, in the Paleogene Period. Grasses first appeared during this time.

3 Neogene Period

1. During the Neogene Period, the Sierra Nevada batholiths were emplaced in western North America, the Colorado Plateau was uplifted, and the Grand Canyon was cut by the Colorado River.

2. Aridification in the Neogene Period caused expansion of deserts across northern Africa and western Asia. Forests declined, and grasses, herbaceous plants, and weeds spread across continental interiors.

3. In the late Miocene Epoch, the Mediterranean Sea largely dried out, and evaporate minerals were deposited through much of the basin.

4. Marine organisms acquired a progressively modern aspect during the Neogene Period. Planktonic foraminifera, calcareous nannoplankton, diatoms, radiolaria, and dinoflagellates became important contributors to marine sediment. Reefs were built by scleractinian corals and sponges. Teleost fishes and mollusks were important components of marine faunas.

5. On land, herbaceous plants and grasses flourished, particularly following the arrival of generally cooler, drier temperatures in the latest Paleogene to early Neogene. In the Miocene Epoch, silica-rich C_4 grasses evolved.

4 Quaternary Period

1. The beginning of the Quaternary Period and the Pleistocene Epoch is characterized by evidence of the onset of major glaciation in the Northern Hemisphere.

2. Mammalian faunal interchange between North and South America occurred when the Isthmus of Panama opened a land bridge between the two continents.

3. Evolution of **primates** took place primarily during the Neogene and Quaternary periods. Monkeys originated in the Oligocene Epoch, and apes became numerous in the mid-Paleogene.

4. **Hominids** (australopithecines and *Homo*) first appeared in the Miocene Epoch. The genus *Homo* evolved in the Pleistocene Epoch. *Homo sapiens* evolved in Africa approximately 200,000 years ago. The earliest modern human remains are from Europe, dated at about 33,000 years old.

5. Since the beginning of Pleistocene glaciation, global climate has alternated between colder **glacial episodes** and warmer **interglacial episodes**. At least four major glacial advances are recorded in sediments of the Northern Hemisphere.

6. Glacially related conditions have had important effects on landscape evolution, including the development of **pluvial lakes**, scouring of river valleys and lakes basins, rerouting of river courses, deposition of till and outwash deposits, and deposition of loess deposits.

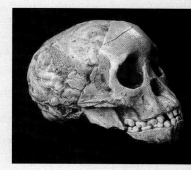

7. Close to the end of the last glacial episode, large mammals adapted to cold-weather conditions in the Northern Hemisphere died out. Some may have been hunted into extinction by humans.

KEY TERMS

- **circum-Antarctic current** p. 377
- **oil shale** p. 379
- **cetacean** p. 381
- **placental mammal** p. 382
- **carnivore** p. 382

- **ungulate** p. 382
- **perissodactyl** p. 382
- **artiodactyl** p. 382
- **ruminant** p. 385
- **primate** p. 385

- **hominid** p. 396
- **interglacial** p. 400
- **pluvial lake** p. 402

CRITICAL AND CREATIVE THINKING QUESTIONS

1. What factors probably contributed to the great evolutionary success of mammals during the Cenozoic?

2. Evolution of C_4 grasses played a vital role in the evolution of terrestrial biotas. What animals depend on these grasses for survival? How have hominids been influenced by C_4 grasses?

3. The graph shows global temperature change over the past 1,000 years. What could be the cause of the rapid rate of temperature change since 1900? How would temperature changes of this sort affect global sea level? How would you expect coastal regions of the world to be affected by temperature change and consequent changes in sea level?

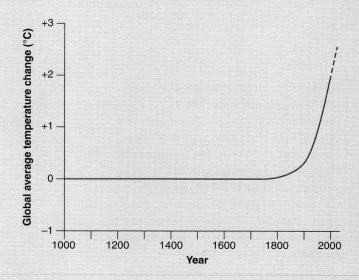

CRITICAL AND CREATIVE THINKING QUESTIONS

4. These two photos of reefs are from the same island in The Bahamas (San Salvador). Today, reefs encircle the island in shallow water. In the island's interior, Pleistocene and early Holocene reefs are exposed above sea level. What does this tell you about sea level history in the Bahamian islands? What could have changed the elevation of sea level from the late Pleistocene to the present? How would facies patterns change if sea level were to rise quickly from its present position?

What is happening in these pictures ?

Quaternary glaciation

The Holocene Epoch is considered to be an interglacial interval. What evidence supports this hypothesis?

Given the present configuration of continents and oceans, is it likely that the Antarctic glaciers will fully melt in the near future? What geologic, climatologic, or other factors would be necessary for their complete disappearance?

In Antarctica, we find excellent analogs for the enormous glacial sheets that existed on Earth during the Pleistocene Epoch. Here are two views of Antarctic glaciers: the Beardmore Glacier, which flows past the Transantarctic Mountains (A), and the Taylor Glacier, a large valley glacier close to the margin of the continent (B). An end moraine around the margin of the Taylor Glacier indicates that the glacier once extended farther in the direction of the Ross Sea.

1. What happened to cause destruction of the Tethys Ocean?

2. How did the Himalaya Mountains form?

3. How did Antarctica become separated from other Southern Hemisphere continents?

4. How were the Rocky Mountains formed?

5. How did the Basin and Range Province form?

6. What biologic groups experienced great evolutionary radiation in the Cenozoic?

7. How did the Sierra Nevada Range form?

8. What oceanic condition made it possible for scleractinian corals to return as reef-formers in the Cenozoic?

9. What conditions led to formation of the Grand Canyon?

10. When and why did the Mediterranean Sea nearly dry out?

11. How did evolution in grasses affect the evolution of terrestrial mammals?

12. When did the first hominids evolve?

13. When did *Homo sapiens* evolve?

14. What areas of the world were ice-covered during the Pleistocene ice age?

15. Extinction at the end of the Pleistocene Epoch mostly affected what types of animals?

Appendix A Geologic Time Scale

Table 1 (Cenozoic – Mesozoic)

Eonothem Eon	Erathem Era	System Period	Series Epoch	Stage Age	Geochronologic Age (Ma)
Phanerozoic	Cenozoic	Quaternary	Holocene		0.0118
			Pleistocene	Upper	0.126
				Middle	0.781
				Lower	1.806
				Gelasian	2.588
		Neogene	Pliocene	Piacenzian	3.600
				Zanclean	5.332
			Miocene	Messinian	7.246
				Tortonian	11.608
				Serravallian	13.65
				Langhian	15.97
				Burdigalian	20.43
				Aquitanian	23.03
		Paleogene	Oligocene	Chattian	28.4 ±0.1
				Rupelian	33.9 ±0.1
			Eocene	Priabonian	37.2 ±0.1
				Bartonian	40.4 ±0.2
				Lutetian	48.6 ±0.2
				Ypresian	55.8 ±0.2
			Paleocene	Thanetian	58.7 ±0.2
				Selandian	61.7 ±0.2
				Danian	65.5 ±0.3
	Mesozoic	Cretaceous	Upper	Maastrichtian	70.6 ±0.6
				Campanian	83.5 ±0.7
				Santonian	85.8 ±0.7
				Coniacian	89.3 ±1.0
				Turonian	93.5 ±0.8
				Cenomanian	99.6 ±0.9
			Lower	Albian	112.0 ±1.0
				Aptian	125.0 ±1.0
				Barremian	130.0 ±1.5
				Hauterivian	136.4 ±2.0
				Valanginian	140.2 ±3.0
				Berriasian	145.5 ±4.0
		Jurassic	Upper	Tithonian	150.8 ±4.0
				Kimmeridgian	155.7 ±4.0
				Oxfordian	161.2 ±4.0
			Middle	Callovian	164.7 ±4.0
				Bathonian	167.7 ±3.5
				Bajocian	171.6 ±3.0
				Aalenian	175.6 ±2.0
			Lower	Toarcian	183.0 ±1.5
				Pliensbachian	189.6 ±1.5
				Sinemurian	196.5 ±1.0
				Hettangian	199.6 ±0.6
		Triassic	Upper	Rhaetian	203.6 ±1.5
				Norian	216.5 ±2.0
				Carnian	228.0 ±2.0
			Middle	Ladinian	237.0 ±2.0
				Anisian	245.0 ±1.5
			Lower	Olenekian	249.7 ±0.7
				Induan	251.0 ±0.4

Table 2 (Permian – Cambrian)

Eonothem Eon	Erathem Era	System Period	Series Epoch	Stage Age	Geochronologic Age (Ma)
Phanerozoic	Paleozoic	Permian	Lopingian	Changhsingian	251.0 ±0.4
				Wuchiapingian	253.8 ±0.7
			Guadalupian	Capitanian	260.4 ±0.7
				Wordian	265.8 ±0.7
				Roadian	268.0 ±0.7
			Cisuralian	Kungurian	270.6 ±0.7
				Artinskian	275.6 ±0.7
				Sakmarian	284.4 ±0.7
				Asselian	294.6 ±0.8
		Carboniferous	Penn-sylvanian Upper	Gzhelian	299.0 ±0.8
				Kasimovian	303.9 ±0.9
			Penn-sylvanian Middle	Moscovian	306.5 ±1.0
			Penn-sylvanian Lower	Bashkirian	311.7 ±1.1
			Missis-sippian Upper	Serpukhovian	318.1 ±1.3
			Missis-sippian Middle	Visean	326.4 ±1.6
			Missis-sippian Lower	Tournaisian	345.3 ±2.1
		Devonian	Upper	Famennian	359.2 ±2.5
				Frasnian	374.5 ±2.6
			Middle	Givetian	385.3 ±2.6
				Eifelian	391.8 ±2.7
			Lower	Emsian	397.5 ±2.7
				Pragian	407.0 ±2.8
				Lochkovian	411.2 ±2.8
		Silurian	Pridoli		416.0 ±2.8
			Ludlow	Ludfordian	418.7 ±2.7
				Gorstian	421.3 ±2.6
			Wenlock	Homerian	422.9 ±2.5
				Sheinwoodian	426.2 ±2.4
			Llandovery	Telychian	428.2 ±2.3
				Aeronian	436.0 ±1.9
				Rhuddanian	439.0 ±1.8
		Ordovician	Upper	Hirnantian	443.7 ±1.5
				Stage 6	445.6 ±1.5
				Stage 5	455.8 ±1.6
			Middle	Darriwilian	460.9 ±1.6
				Stage 3	468.1 ±1.6
			Lower	Stage 2	471.8 ±1.6
				Tremadocian	478.6 ±1.7
		Cambrian	Furongian	Stage 10	488.3 ±1.7
				Stage 9	~ 492.0
				Paibian	~ 496.0
			Series 3	Guzhangian	501.0 ±2.0
				Drumian	~ 503.0
				Stage 5	~ 506.5
			Series 2	Stage 4	~ 510.0
				Stage 3	~ 517.0
			Terreneuvian	Stage 2	~ 521.0
				Fortunian	~ 534.6
					542.0 ±1.0

Table 3 (Precambrian)

Eonothem Eon	Erathem Era	System Period	Geochronologic Age (Ma)
Precambrian Proterozoic	Neo-proterozoic	Ediacaran	542
		Cryogenian	~630
		Tonian	850
	Meso-proterozoic	Stenian	1000
		Ectasian	1200
		Calymmian	1400
	Paleo-proterozoic	Statherian	1600
		Orosirian	1800
		Rhyacian	2050
		Siderian	2300
Archean	Neo-archean		2500
	Meso-archean		2800
	Paleo-archean		3200
	Eoarchean	Lower limit is not defined	3600

Appendix B
Units and Their Conversions

Commonly Used Units of Measure

Length

Metric Measure

1 kilometer (km)	= 1000 meters (m)
1 meter (m)	= 100 centimeters (cm)
1 centimeter (cm)	= 10 millimeters (mm)
1 millimeter (mm)	= 1000 micrometers (μm)
	(formerly called microns)
1 micrometer (μm)	= 0.001 millimeter (mm)
1 angstrom (Å)	= 10^{-8} centimeters (cm)

Nonmetric Measure

1 mile (mi)	= 5280 feet (ft) = 1760 yards (yd)
1 yard (yd)	= 3 feet (ft)
1 fathom (fath)	= 6 feet (ft)

Conversions

1 kilometer (km)	= 0.6214 mile (mi)
1 meter (m)	= 1.094 yards (yd)
	= 3.281 feet (ft)
1 centimeter (cm)	= 0.3937 inch (in)
1 millimeter (mm)	= 0.0394 inch (in)
1 mile (mi)	= 1.609 kilometers (km)
1 yard (yd)	= 0.9144 meter (m)
1 foot (ft)	= 0.3048 meter (m)
1 inch (in)	= 2.54 centimeters (cm)
1 inch (in)	= 25.4 millimeters (mm)
1 fathon (fath)	= 1.8288 meters (m)

Area

Metric Measure

1 square kilometer (km²)	= 1,000,000 square meters (m²)
	= 100 hectares (ha)
1 square meter (m²)	= 10,000 square centimeters (cm²)
1 hectare (ha)	= 10,000 square meters (m²)

Nonmetric Measure

1 square mile (mi²)	= 640 acres (ac)
1 acre (ac)	= 4840 square yards (yd²)
1 square foot (ft²)	= 144 square inches (in²)

Conversions

1 square kilometer (km²)	= 0.386 square mile (mi²)
1 hectare (ha)	= 2.471 acres (ac)
1 square meter (m²)	= 1.196 square yards (yd²)
	= 10.764 square feet (ft²)
1 square centimeter (cm²)	= 0.155 square inch (in²)
1 square mile (mi²)	= 2.59 square kilometers (km²)
1 acre (ac)	= 0.4047 hectare (ha)
1 square yard (yd²)	= 0.836 square meter (m²)
1 square foot (ft²)	= 0.0929 square meter (m²)
1 square inch (in²)	= 6.4516 square centimeter (cm²)

Volume

Metric Measure

1 cubic meter (m³)	= 1,000,000 cubic centimeters (cm³)
1 liter (l)	= 1000 milliliters (ml)
	= 0.001 cubic meter (m³)
1 centiliter (cl)	= 10 milliliters (ml)
1 milliliter (ml)	= 1 cubic centimeter (cm³)

Nonmetric Measure

1 cubic yard (yd³)	= 27 cubic feet (ft³)
1 cubic foot (ft³)	= 1728 cubic inches (in³)
1 barrel (oil) (bbl)	= 42 gallons (U.S.) (gal)

Conversions

1 cubic kilometer (km^3)	= 0.24 cubic miles (mi^3)
1 cubic meter (m^3)	= 264.2 gallons (U.S.) (gal)
	= 35.314 cubic feet (ft^3)
1 liter (l)	= 1.057 quarts (U.S.) (qt)
	= 33.815 ounces (U.S. fluid) (fl. oz.)
1 cubic centimeter (cm^3)	= 0.0610 cubic inch (in^3)
1 cubic mile (mi^3)	= 4.168 cubic kilometers (km^3)
1 acre-foot (ac-ft)	= 1233.46 cubic meters (m^3)
1 cubic yard (yd^3)	= 0.7646 cubic meter (m^3)
1 cubic foot (ft^3)	= 0.0283 cubic meter (m^3)
1 cubic inch (in^3)	= 16.39 cubic centimeters (cm^3)
1 gallon (gal)	= 3.784 liters (l)

Mass

Metric Measure

1000 kilograms (kg)	= 1 metric ton (also called a tonne) (m.t)
1 kilogram (kg)	= 1000 grams (g)

Nonmetric Measure

1 short ton (sh.t)	= 2000 pounds (lb)
1 long ton (l.t)	= 2240 pounds (lb)
1 pound (avoirdupois) (lb)	= 16 ounces (avoirdupois) (oz) = 7000 grains (gr)
1 ounce (avoirdupois) (oz)	= 437.5 grains (gr)
1 pound (Troy) (Tr. lb)	= 12 ounces (Troy) (Tr. oz)
1 ounce (Troy) (Tr. oz)	= 20 pennyweight (dwt)

Conversions

1 metric ton (m.t)	= 2205 pounds (avoirdupois) (lb)
1 kilogram (kg)	= 2.205 pounds (avoirdupois) (lb)
1 gram (g)	= 0.03527 ounce (avoirdupois) (oz) = 0.03215 ounce (Troy) (Tr. oz) = 15,432 grains (gr)
1 pound (lb)	= 0.4536 kilogram (kg)
1 ounce (avoirdupois) (oz)	= 28.35 grams (g)
1 ounce (avoirdupois) (oz)	= 1.097 ounces (Troy) (Tr. oz)

Pressure

Metric Measure

1 pascal (Pa)	= 1 newton/square meter (N/m^2)
1 kilogram-force square centimeter (kg/cm^2 or kgf/cm^2)	= 1 technical atmosphere (at)
	= 98,067 Pa
	= 0.98067 bar
1 bar	= 10^5 pascals (Pa)
	= 1.02 kilogram-force/square centimeter (kgf/cm^2 or at)

Nonmetric Measure

1 atmosphere (atm)	= 1.01325 bar = 14.696 lb/in^2 (psi)
1 pound per square inch (lb/in^2 or psi)	= 68.046 × 10^{-3} atmospheres (atm)

Conversions

1 kilogram-force/ square centimeter (kgf/cm^2 or at)	= 0.96784 atmosphere (atm)
	= 14.2233 pounds/square inch (lb/in^2 or psi) = 0.98067 bar = 98,067 Pa
1 bar	= 0.98692 atmosphere (atm)
	= 10^5 pascals (Pa)
	= 1.02 kilogram-force/square centimeter (kgf/cm^2)
1 Pa = 10^{-5} bar	= 10.197×10^{-6} kgf/cm^2 (or at)
	= 9.8692×10^{-6} atm
	= 145.04×10^{-6} lb/in^2 (or psi)
1 atm	= 101,325 Pa = 1.01325 bar

Temperature

Metric Measure

0 degrees Celsius (°C)	= freezing point of water at sea level
100 degrees Celsius (°C)	= boiling point of water at sea level
0 degrees Kelvin (K)	= −273.15°C = absolute zero
1°C	= 1K (temperature increments)
273.15K	= 0.0° C

Nonmetric Measure

Fahrenheit (°F)	= (K · 9/5) − 459.67
Fahrenheit (°F)	= (°C · 9/5) + 32

Conversions

degrees Kelvin (K)	= °C + 273.1
degrees Celsius (°C)	= K − 273.15
degrees Fahrenheit (°F)	= (°C · 9/5) + 32
degrees Celsius (°C)	= (°F − 32) · 5/9

Appendix C
Tables of the Chemical Elements and Naturally Occurring Isotopes

Alphabetical list of the elements

Element	Symbol	Atomic Number	Crustal Abundance, Weight Percent	Element	Symbol	Atomic Number	Crustal Abundance, Weight Percent
Aluminum	Al	13	8.00	Lithium	Li	3	0.0020
Antimony	Sb	51	0.00002	Lutetium	Lu	71	0.000080
Argon	Ar	18	Not known	Magnesium	Mg	12	2.77
Arsenic	As	33	0.00020	Manganese	Mn	25	0.100
Barium	Ba	56	0.0380	Mercury	Hg	80	0.000002
Beryllium	Be	4	0.00020	Molybdenum	Mo	42	0.00012
Bismuth	Bi	83	0.0000004	Neodymium	Nd	60	0.0044
Boron	B	5	0.0007	Neon	Ne	10	Not known
Bromine	Br	35	0.00040	Nickel	Ni	28	0.0072
Cadmium	Cd	48	0.000018	Niobium	Nb	41	0.0020
Calcium	Ca	20	5.06	Nitrogen	N	7	0.0020
Carbon[a]	C	6	0.02	Osmium	Os	76	0.00000002
Cerium	Ce	58	0.0083	Oxygen[b]	O	8	45.2
Cesium	Cs	55	0.00016	Palladium	Pd	46	0.0000003
Chlorine	Cl	17	0.0190	Phosphorus	P	15	0.1010
Chromium	Cr	24	0.0096	Platinum	Pt	78	0.0000005
Cobalt	Co	27	0.0028	Polonium	Po	84	Footnote[d]
Copper	Cu	29	0.0058	Potassium	K	19	1.68
Dysprosium	Dy	66	0.00085	Praseodymium	Pr	59	0.0013
Erbium	Er	68	0.00036	Protactinium	Pa	91	Footnote[d]
Europium	Eu	63	0.00022	Radium	Ra	88	Footnote[d]
Fluorine	F	9	0.0460	Radon	Rn	86	Footnote[d]
Gadolinium	Gd	64	0.00063	Rhenium	Re	75	0.00000004
Gallium	Ga	31	0.0017	Rhodium[c]	Rh	45	0.00000001
Germanium	Ge	32	0.00013	Rubidium	Rb	37	0.0070
Gold	Au	79	0.0000002	Ruthenium[c]	Ru	44	0.00000001
Hafnium	Hf	72	0.0004	Samarium	Sm	62	0.00077
Helium	He	2	Not known	Scandium	Sc	21	0.0022
Holmium	Ho	67	0.00016	Selenium	Se	34	0.000005
Hydrogen[b]	H	1	0.14	Silicon	Si	14	27.20
Indium	In	49	0.00002	Silver	Ag	47	0.000008
Iodine	I	53	0.00005	Sodium	Na	11	2.32
Iridium	Ir	77	0.00000002	Srontium	Sr	38	0.0450
Iron	Fe	26	5.80	Sulfur	S	16	0.030
Krypton	Kr	36	Not known	Tantalum	Ta	73	0.00024
Lanthanum	La	57	0.0050	Tellurium[c]	Te	52	0.000001
Lead	Pb	82	0.0010	Terbium	Tb	65	0.00010

Element	Symbol	Atomic Number	Crustal Abundance, Weight Percent	Element	Symbol	Atomic Number	Crustal Abundance, Weight Percent
Thallium	Tl	81	0.000047	Vanadium	V	23	0.0170
Thorium	Th	90	0.00058	Xenon	Xe	54	Not known
Thulium	Tm	69	0.000052	Ytterbium	Yb	70	0.00034
Tin	Sn	50	0.00015	Yttrium	Y	39	0.635
Titanium	Ti	22	0.86	Zinc	Zn	30	0.008-2
Tungsten	W	74	0.00010	Zirconium	Zr	40	0.0140
Uranium	U	92	0.00016				

Source: After K. K. Turekian, 1969.

[a]Estimate from S. R. Taylor (1964).

[b]Analyses of crustal rocks do not usually include separate determinations for hydrogen and oxygen. Both combine in essentially constant proportions with other elements, so abundances can be calculated.

[c]Estimates are uncertain and have a very low reliability.

[d]Elements formed by decay of uranium and thorium. The daughter products are radioactive with such short half-lives that crustal accumulations are too low to be measured accurately.

GLOSSARY

These and additional terms are defined online at www.wiley.com/college/Babcock.

Absaroka sequence   The name applied to a large scale, unconformity-bounded sequence of sediments deposited on the craton of Laurentia during the mid-Carboniferous through Middle Jurassic periods.

Acadian orogeny   Orogenic activity during the Devonian Period along the Appalachian margin of Laurentia; a result of the convergent-margin tectonism and collision of exotic terranes (such as Avalon) along the eastern margin of Laurentia. Equivalent in part to the Caledonide orogeny of western Europe.

acanthodian   A member of a group of extinct fishes that includes the oldest known jawed vertebrates (gnathostomes); characterized by fixed paired fins supported by spines. Stratigraphic range: Silurian to Permian.

active margin   A continental margin that is also a plate margin undergoing significant seismicity and deformation.

adaptation   Modification, as the result of natural selection, of an organism or its parts so that it is better fitted for survival in the conditions of its environment.

adaptive breakthrough   Modification of the traits of an organism, through natural selection, so that descendant species can survive under significantly different environmental conditions than the ancestral species.

adaptive radiation   Rapid evolutionary diversification of organisms to fill unoccupied ecologic niches.

age   1. When used as a formal geochronologic (geologic time) unit, and signified by combination with a formally ratified age name from the standard geologic time scale, the unit below epoch in rank. Its chronostratigraphic equivalent is a stage. 2. When used informally, age refers to the position or time of duration of anything in the geologic time scale.

Alleghanian orogeny   Orogenic activity during the Carboniferous-Permian periods along the Appalachian margin of Laurentia; a result of the collision between Laurentia and the African sector of Gondwana. Equivalent to the Hercynian (Variscan) orogeny of Europe. Synonyms: Alleghenian orogeny, Appalachian orogeny.

allele   One of the possible mutational states of a gene, and distinguished from other alleles by phenotypic effects.

alluvial fan   A relatively low, outspread, sloping mass of loose sediment, shaped like an open fan, deposited by a stream where it issues from a narrow mountain valley onto a broad plain; usually alluvial fans occur in semi-arid regions and open onto playas.

amino acid   Any of the subunit building blocks of organic molecules that are covalently linked to form proteins.

ammonite   An ammonoid cephalopod having ammonitic sutures.

ammonoid   A type of extinct cephalopod (mollusk) characterized by an external shell that is usually coiled in one plane and

symmetrical; internally, partitions called septa are present. Stratigraphic range: Devonian to Cretaceous.

amniotic egg   A vertebrate egg that (at least primitively) has a large yolk and is covered by a shell lined with extra-embryonic membranes, including the amnion, to conserve water and allow gaseous exchange. The membranes are shed at hatching or birth.

analogous structures   Morphological structures in organisms having the same function but different evolutionary origins.

angiosperm   A plant with true flowers in which seeds, resulting from double fertilization, are enclosed in a ovary (the fruiting structure).

angular unconformity   An unconformity in which sedimentary strata are present below and above the erosion surface but the strata below the unconformity are positioned at an angle compared to those above the erosion surface.

Animalia   In biologic classification, the animal kingdom, which consists of diploid eukaryotes that develop from an embryo (blastula) formed by the fusion (fertilization) of a haploid egg and a sperm.

Antler orogeny   An orogeny that deformed Paleozoic rocks in the Great Basin (Nevada and adjacent areas) during the Devonian and early Carboniferous.

Appalachian orogeny   See Alleghanian orogeny.

aragonite sea   The oceanic condition in which the magnesium:calcium ratio of seawater favors the precipitation of aragonite and high-magnesium calcite shells and marine cements.

Archaea (pronounced "ar-kee-a")   In biologic classification, the domain of prokaryotes that includes methanogenic, halophilic, and thermoacidophilic archaebacteria.

Archaebacteria (pronounced "ar-kee-back-tee-ree-ya")   In biologic classification, the only kingdom in the domain Archaea. The common name is archaebacteria (singular, archaebacterium).

archaeocyath (pronounced "ark-ee-oh-sy-ath")   A member of an extinct group of calcareous sponges that secreted double-walled conical or vasiform skeletons, and commonly reef-forming. Stratigraphic range: Cambrian.

archosaur (pronounced "ark-oh-sore")   A reptile having a diapsid skull type with teeth set in sockets; includes the crocodiles, pterosaurs, dinosaurs, and birds. Primitive archosaurs are sometimes referred to as thecodonts.

arthropod   An invertebrate animal characterized by a chitinous exoskeleton, jointed appendages, and a segmented body. Some arthropods, such as trilobites, added biominerals to their exoskeletons. Arthropods are the most diverse animals. Examples of arthropods are anomalocaridids, trilobites, insects, crustaceans (including ostracodes, shrimps, crabs, lobsters, and barnacles), myriapods (millipedes and centipedes), spiders, scorpions, horseshoe crabs, mites, ticks, velvet worms, and tardigrades.

artiodactyl　An even-toed ungulate mammal such as a deer, antelope, sheep, goat, pig, bison, cow, camel, llama, or giraffe.

asthenosphere　The layer within the upper mantle and below the lithosphere, approximately 100 to 350 km below the surface, where rocks are relatively ductile and easily deformed.

aulacogen　A sediment-filled continental rift formed where extensional tectonics have ceased; often formed as the failed arm of a three-armed spreading center.

australopithecine　A hominid belonging to the genus *Australopithecus* or *Paranthropus*.

autapomorphy　A derived character unique to a taxon, usually a single species.

autotroph　An organism that uses raw inorganic materials to make its own food. Synonym: primary producer.

autotrophy　"Self-feeding" by means of either harvesting light energy from the Sun or from oxidation of inorganic compounds to make organic molecules.

Avalonia　A Paleozoic microcontinent that was caught up in Acadian orogenesis during the Devonian, sutured to Laurentia, then fragmented with the later breakup of Pangea. Today, pieces of Avalonia comprise such areas as southeastern Newfoundland, eastern Massachusetts, and parts of the British Isles such as North Wales.

Bacteria　In biologic classification, the domain that includes all prokaryotic microorganisms other than those in the Archaea. They are the most widespread and abundant organisms on Earth, and include decomposers, symbionts, and pathogens. Bacteria are important in the mediation of mineral formation and other reactions.

Baltica　The Paleozoic continent consisting mostly of the craton underlying Scandinavia.

banded iron formation　A sedimentary rock composed of thin chert (cryptocrystalline quartz) bands interlayered with iron oxide minerals (commonly hematite and magnetite). Almost all banded iron formations are Archean or Proterozoic in age. Acronym: BIF.

barrier island　A long, narrow, sandy coastal island above high tide level and parallel to the shoreline.

basalt　Mafic (dark colored) igneous rock having fine crystal size and composed largely of calcic plagioclase feldspar and pyroxene; the finely crystalline equivalent of gabbro.

basement　As used in geology, Archean and Proterozoic crystalline rock (igneous and metamorphic rock) forming the craton.

basin　A relatively depressed area of the crust that receives sedimentary deposition.

batholith　An irregularly shaped intrusive (plutonic) igneous rock body having an aerial extent of at least 100 km^2 and no known base.

bedding　Layering, normally in sedimentary rocks.

bedrock　The rock at the Earth's surface or immediately underlying soil or other unconsolidated sediments.

benthic　The term used for an organism that lives on or in the substrate in an aqueous environment.

BIF　The acronym for banded iron formation.

binominal nomenclature　The technique of identifying species of biologic organisms using a two-part name, a genus name followed by a species name. The names are written in Latin and separated from ordinary text by italics or underlining. Synonym: binomial nomenclature.

biodegrader　An organism that breaks down organic tissues through predation, scavenging, boring, decay, or other means.

biofacies　A sedimentary unit defined on the basis of an assemblage of fossils contained within it.

biofilm　A consortium of microorganisms including bacteria and fungi involved in the biodegradation of organic matter.

biologic evolution　Change through time in biologic organisms. The term is usually construed to mean macroevolution, or evolution at the species level.

biological species concept　The concept that members of a species can interbreed and produce fertile offspring, and that members of a single species are distinguished from other species by reproductive isolation.

biomarker　An organic compound having a structure that can be related to a particular type of organism, and that serves as a "fingerprint" to that type of organism.

biomineral　A mineral of organic origin (secreted or mediated by biologic activity).

biomineralization　Secretion of minerals by biologic organisms; usually involves secretion as bones, teeth, shells, external coverings, or secretion within internal organs.

biostratigraphic unit　A stratigraphic unit defined by the ranges of fossils. A biozone (or zone) is the fundamental biostratigraphic unit.

bioturbation　Reworking of sediment by organisms.

biozone　A stratigraphic interval defined by its fossil content, and usually given the name of a characteristic fossil present in that interval; it is the fundamental unit of biostratigraphy. Synonyms: zone, biostratigraphic zone.

bird　A feathered theropod; usually with forearms modified into wings, a breastbone, and a pelvic structure modified to a bifurcate form.

body fossil　Direct or altered remains of an ancient organism.

bolide　An extraterrestrial body that has struck Earth, another planet, or a moon; may be a meteorite, an asteroid, a comet, or some other body.

brachiopod　An invertebrate animal characterized by two bilaterally symmetrical valves composed of calcite or chitinophosphate, and a lophophore for food gathering.

brackish Slightly salty water; salinity is intermediate between that of fresh water (0‰ salt) and normal marine water (35‰ salt).

breccia (pronounced "brech-ee-ah") Rock composed largely of angular particles, the largest of which are larger than 2 mm in size; may have an origin as a primary sedimentary deposit, through dissolution and collapse of rock material, through breakage of rock along a fault, or through explosion.

bryophyte A small, nonvascular land plant that may have stems and leaves but not true roots. Bryophytes include mosses, liverworts, and hornworts.

bryozoan A "moss animal," a colonial animal comprising numerous tiny zooids, each having a tentacled lophophore for gathering food.

calcite sea The oceanic condition in which the magnesium:calcium ratio of seawater favors the precipitation of low-magnesium calcite shells and marine cements.

Caledonian orogeny Silurian-Devonian orogenic activity that affected western Europe from the British Isles through Scandinavia, creating the Caledonides orogenic belt, and roughly corresponding to the Acadian orogeny.

Cambrian explosion The appearance of numerous taxonomic groups in the Cambrian sedimentary record; it reflects a widespread evolutionary diversification especially among multicellular animals.

carbonate A mineral that contains a carbonate ion, CO_3^-. Important carbonate minerals are calcite and aragonite (calcium carbonate, $CaCO_3$), dolomite (magnesium-calcium carbonate, $(Mg,Ca)(CO_3)_2$), and siderite (iron carbonate, $FeCO_3$).

carbonization A fossilization process in which the original organic compounds have been reduced to a thin film of carbon.

carnivore 1. An animal that feeds on animal tissues. 2. A member of the placental mammalian group that includes cats, dogs, bears, hyenas, whales, seals, and dolphins.

cephalopod A type of mollusk, usually carnivorous, having the foot developed into a head with advanced sensory functions including eyesight. Examples include octopi, squids, chambred nautilus, and their extinct relatives the endoceratoids, nautiloids, belemnoids, and ammonoids.

cetacean A member of the mammal order Cetacea, which includes whales and porpoises.

chalk Soft, fine-textured limestone of marine origin, composed mostly of calcitic coccolithophorid plates; normally skeletons of calcareous nannoplankton are also present.

chemoautotroph An autotrophic organism that uses chemical energy released through oxidation of inorganic compounds to synthesize organic molecules.

chemostratigraphy The branch of stratigraphy dealing with chemical signals such as cycles of chemical isotopes.

chemosynthesis The process of producing nutrients or synthesizing organic molecules by breaking down compounds such as hydrogen sulfide and methane rather than by photosynthesis.

chordate An animal possessing a notochord and pharyngeal gill slits.

chromosome A genetic structure by which hereditary information is physically transmitted from one generation to the next. In eukaryotic cells, chromosomes consist of long threads of DNA associated with protein.

chronostratigraphic unit A time- rock, or time-stratigraphic unit; the tangible representation of a geologic time (geochronologic) unit. A system is the fundamental chronostratigraphic unit.

circum-Antarctic current A circumpolar ocean current system that flows around Antarctica and isolates it climatically from South America, Africa, and Australia.

clade A phylogenetically related lineage consisting of an ancestral species and all its descendants. Synonyms: monophyletic group, natural group.

cladistics The method in biologic classification in which ancestor-descendant relationships among taxa are analyzed according to the distribution of synapomorphies (shared derived characters). Synonym: phylogenetic analysis.

cladogram A branching diagram illustrating a phylogenetic hypothesis.

clastic Sediment containing accumulated particles of broken rocks and the skeletal remains of organisms. Also referred to as detrital sediment.

clastic wedge A wedge-shaped deposit of sediments shed from an active thrust belt and filling a foreland basin.

clay Sediment having a grain size smaller than 1/256 mm.

clay mineral Any potassium-aluminum phyllosilicate mineral that forms clay deposits, including the minerals kaolinite, montmorillonite, clay mica, and chlorite.

coal swamp An ancient wetland where vegetative matter produced in massive quantities became lithified to coal.

coevolution An evolutionary pattern in which species in two unrelated lineages profoundly influence each other's evolution so that the two evolve as an integrated complex.

concretion A rounded mineral body (commonly calcite, siderite, pyrite, or quartz) enclosed in sedimentary rock and apparently formed as the result of biogenically mediated precipitation, usually microbial decomposition of a decaying organism.

conglomerate Sedimentary rock composed largely of rounded particles, the largest of which are larger than 2 mm in size. The lithified equivalent of gravel.

conifer A gymnosperm having needlelike or scalelike leaves and naked seeds carried in cones.

conodont A member of a group of early chordates that secreted phosphatic tooth-like structures along the pharynx.

consumer An organism that derives nutrients by feeding on or decaying the organic tissues of other organisms.

contact metamorphism Localized changes in texture and mineralogy of a rock associated with the intrusion or extrusion of an igneous magma; heat and hydrothermal fluids may be involved in the metamorphic activity.

continental crust The solid, outer part of the Earth underlying the continents and continental shelves; composed largely of felsic igneous rocks (especially granitic and rhyolitic rocks).

continental drift The hypothesis advocated by Alfred Wegener that the Earth's continents moved to their present positions after fragmentation of a larger landmass in the geologic past. This concept was later amalgamated with aspects of seafloor spreading to develop the theory of plate tectonics.

continental shelf The part of the continental margin between the shoreline and the continental slope; characterized by a very low slope of about 0.1°.

continental slope The part of the continental margin between the continental shelf and the continental rise if there is one; characterized by a relatively steep slope of 1.5° to 6°.

convergent evolution See homoplasy.

convergent plate boundary A boundary between two plate that are moving toward each other.

coprolite Fossilized excrement.

cordaite An early form of conifer, characterized by long strap-like leaves.

correlation In stratigraphy, the matching of strata from one location to another. Matching can potentially be based on any observable or measurable characteristics.

craton The core of a continent. Part of the Earth's continental crustal areas that has attained relative stability and received little deformation for at least 1 billion years. A craton includes shield areas, where Archean and Proterozoic rocks are exposed, and platform areas, where the craton is overlain by a thin layer of Phanerozoic strata.

cross-bedding Inclined strata, normally more than 1 cm thick, formed through the rippling of sediment, and developed at different angles through changes in current strength or direction.

crossopterygian A lobe-finned fish. Modern lobe-finned fishes include coelacanths and lungfishes.

crust The outermost layer of the Earth, defined by density, composition, and a seismic velocity difference from the underlying mantle; the layer overlying the Mohorovicic discontinuity.

crystal A solid composed of atoms and molecules having a regular internal structure and an external form defined by flat faces.

cyclothem A series of beds deposited during a single sedimentary cycle, usually a transgressive-regressive sequence. It is an informal lithostratigraphic unit.

daughter (nuclide) Isotope formed from the radioactive decay of another (parent) isotope.

deep time Time, as expressed on a geologic scale.

degassing See outgassing.

delta Sedimentary prism or wedge that develops near the mouth of a stream or river as sediment is rapidly deposited in relatively quiet oceanic or lake water.

depositional sequence See sequence.

descriptive classification Classification of rock type according to their texture, fabric, and composition.

detrital Particles of broken rocks and the skeletal remains of organisms accumulated at a depositional site following transportation.

diagenesis All the chemical, physical, and biologic changes that sediments undergo (excluding weathering and metamorphism) between the time of deposition and the time of lithification.

diatom A type of aquatic protoctist characterized by a microscopic, ornate skeleton (called a test) composed of two valves made of organic matter impregnated with opaline silica (SiO_2). Diatoms are single-celled phytoplankton, and may form simple filaments or colonies.

dinoflagellate An aquatic alga, usually single-celled, having two dissimilar flagella at some point in the life cycle. Some dinoflagellates secrete relatively resistant organic-walled tests.

dinosaur An archosaur characterized by either ornithischian or saurischian hip bones and upright posture.

disconformity An unconformity in which strata below and above the erosion surface are parallel.

divergent plate boundary A tectonic boundary between two plates that are moving apart.

DNA Deoxyribonucleic acid. A macromolecule consisting of two helical polynucleotide chains held together by hydrogen bonds. DNA is the basic storage vehicle for hereditary information.

dysoxic The condition in which oxygen is deficient but not completely absent.

Earth dynamics The study of Earth materials, features, and processes operating on and within the Earth. Synonym: physical geology.

Earth history The study of the origin and development of Earth, including its life forms, through time. Synonym: historical geology.

Earth system The sum of the physical, chemical, and biologic processes operating on and within the Earth.

eccentricity A deviation over time in the Earth's elliptical orbit around the Sun.

echinoderm An invertebrate deuterostome animal characterized by radial symmetry, a water-vascular system, and an endoskeleton formed of calcitic ossicles, plates, or sclerites. Echinoderms include starfish, sea urchins, brittle stars, crinoids, and sea cucumbers.

ectothermic The "cold-blooded" condition, in which animals are incapable of generating internal heat.

Ediacaran biota Fossils dating from the Ediacaran Period of the Neoproterozoic Era; the biota includes the earliest putative animals.

embryology   The study of unborn or unhatched offspring.

endothermic   The "warm-blooded" condition, in which animals are capable of generating internal heat.

eolian   A term applied to wind-blown deposits such as sand dunes or loess.

eon   The largest geochronologic (geologic time) unit; longer in time duration than an era. Its chronostratigraphic equivalent is an eonothem.

epeiric sea   A shallow sea formed when marine water covers part of the continental crust. Synonym: epicontinental sea.

epibenthic   A term applied to an organism that lives on the surface of an aqueous substrate.

epicontinental sea   See epeiric sea.

epifaunal   A term applied to an animal that lives on the surface of an aqueous substrate.

epoch (pronounced "eh-peck")   The geochronologic (geologic time) unit lower in rank than a period and higher than an age. Its chronostratigraphic equivalent is a series.

era   The geochronologic (geologic time) unit lower in rank than an eon and higher than a period. Its chronostratigraphic equivalent is an erathem.

erosion   The general process in which Earth materials are loosened, dissolved, worn away, and transported. Erosion usually includes weathering, dissolution, corrosion, and transportation, but usually excludes mass wasting.

Eubacteria   In biologic classification, the only kingdom in the domain Bacteria. The common name is eubacteria. Synonym: Monera.

Eukarya   The domain name applied to eukaryotic organisms, or organisms having nucleated cells.

eukaryote   An organism that has a nucleated cell type. Alternative spelling: eucaryote.

eukaryotic cell   A cell type having a true nucleus. Alternative spelling: eucaryotic cell.

Euramerica   A composite continent formed by collision of Laurentia with Avalonia and Baltica during the Devonian Period.

eustatic   The term applied to global sea level.

evaporite   A mineral or deposit of minerals formed under evaporative conditions (usually hot, dry conditions). Evaporite minerals include halite, gypsum, and anhydrite.

evolution   1. In a general sense, change through time; may involve biologic, physical, or chemical changes on Earth. 2. Descent with modification in biologic organisms (biologic evolution).

evolutionary theory   The scientific theory that explains processes by which biologic species give rise to other species, principally by way of genetic changes and natural selection.

exceptional preservation   The fossilization mode involving preservation of nonbiomineralized or lightly biomineralized tissues of organisms.

extrusive rock   An igneous rock that resulted from the cooling and solidification of magma erupted onto the Earth's surface. Extrusive rocks are commonly finely crystalline. Synonym: volcanic rock.

facies (pronounced "fay-shees")   As used for sedimentary deposits, a unit having a set of characteristics (such as lithology, color, or fossil assemblage) particular to a local environment, and usually reflecting conditions of its origin.

fault   A break in rock along which significant movement has taken place.

foliation   In a metamorphic rock, a planar arrangement of texture caused by alignment of crystals that grew perpendicular to the direction of stress applied to the rock during metamorphism. See cleavage.

foraminifera   An aqueous protozoan group characterized by a test of one or more chambers, usually calcareous or made of agglutinated particles.

foreland basin   A linear sedimentary basin that subsides in response to thrust loading of the crust.

formation   As used in stratigraphy, the fundamental unit of lithostratigraphy; it has a definable top and bottom, and is mappable across geographic space.

fossil   Any evidence of ancient life. It may include any remains, traces of activity, or chemical marker left by an organism that lived prior to the Holocene Epoch.

freshwater   Water having very low salinity values, typically less than 1‰.

Fungi (pronounced "fun-jí")   In biologic classification, the eukaryotic kingdom containing fungi, which are conjugating osmotrophs that develop from nonmotile spores, and that lack undulipodia.

gene   A unit of information about a heritable trait that is passed on from parents to offspring. A gene originates in the DNA component of a chromosome and is located at a specific position on a chromosome.

genetic classification   As used in geology, the classification of rock types according to their inferred history (for example, classification as igneous, sedimentary, metamorphic, biogenic, marine, or eolian rocks).

genetics   The branch of biology concerned with heredity and variation.

genus (plural, **genera)**   In biologic classification, a group of species. The taxonomic category ranking above species and below families.

geochronologic unit   A geologic time unit. Geologic time is intangible, so geochronologic units are spans of time represented materially (in stratigraphy) by chronostratigraphic units. A period is the fundamental geochronologic unit.

geochronology   The science of dating and determining the time sequence of events in Earth history.

geologic time scale The chronology (or "calendar") of Earth history.

geology The science of the Earth, including its composition, structure, origin, life forms, physical and chemical processes that act upon it, and its history.

geopetal structure Any sedimentary feature that shows the original stratigraphic younging direction. Examples of geopetal structures are ripples, cross-bedding, mudcracks, raindrop impressions, stromatolites, flute casts, and footprints.

glacial drift Sediment deposited by glaciers or icebergs, either directly or indirectly (as in stream meltwater, in lakes, or in the ocean).

glacial striations A series of parallel linear furrows, usually narrow, inscribed on bedrock by the rasping action of rock fragments embedded at the base of a moving glacier. Glacial striations are oriented in the direction of glacial movement.

***Glossopteris* flora** An assemblage of plants, dominated by fossil leaves of the progymnosperm *Glossopteris*, occurring in Carboniferous-Permian strata of the Southern Hemisphere.

gnathostome A vertebrate animal having true jaws.

Gondwana The large Paleozoic to mid-Mesozoic continent that included much of the present-day Southern Hemisphere landmasses: South America, the Falkland Islands, Africa, Madagascar, India, Australia, Antarctica. The term, as used today, has been adapted from an earlier usage (Gondwanaland), which originally referred to a hypothesized (but evidently nonexistent) land bridge that connected Australia with India, Madagascar, and Africa from the Cambrian to the Jurassic. Synonym: Gondwanaland.

graben An elongate basin formed through downdropping of a fault block, and bounded on both sides by a normal fault.

graded bedding A sedimentary bed resulting from one depositional event in which there is a progressive vertical change in grain size (usually fining upward).

grain size The general dimensions (such as diameter or volume) of the particles in a sediment or rock.

granite A felsic, intrusive igneous rock having coarse crystal (coarse grain) size. Granite is the intrusive equivalent of rhyolite.

graptolite A member of an extinct group of colonial organisms, referred to as hemichordates, having organic skeletons. Stratigraphic range: Cambrian to Carboniferous.

gravel Sediment having a grain size larger than 2 mm; includes granules, pebbles, cobbles, and boulders.

greenhouse gases Atmospheric gases, such as carbon dioxide, water vapor, and nitrous oxide, that trap the heat from solar radiation near the Earth's surface. Greenhouse gases act in a manner analogous to glass windows on a greenhouse: solar radiation passes into the enclosure but the glass limits reradiation of heat (infrared radiation) from escaping, ultimately causing temperature within the enclosure to rise.

greenstone belt An elongate area within an Archean shield that contains metamorphosed and deformed volcanic rocks (ultramafic plus mafic to felsic rocks) and sedimentary rocks, and that is characterized by abundant chlorite-rich greenstone.

Grenville orogenic belt An arcuate orogenic region that developed 1.3 to 1.0 billion years ago and that affected an extensive area of present-day North America and adjacent regions.

GSSA Global Standard Stratotype Age, an internationally ratified point in geologic time marking the boundary between two time-rock (chronostratigraphic) units.

GSSP Global boundary Stratotype Section and Point, an internationally ratified point in strata marking the boundary between two time-rock (chronostratigraphic) units, and their equivalent time units. A GSSP is placed at the base of a time-rock unit such as a system (equivalent to a period), series (equivalent to an epoch), or stage (equivalent to an age), and its position automatically defines the end of the previous unit.

guide fossil A fossil useful in biostratigraphic correlation as a guide to the relative age of strata. Synonym: index fossil.

guyot A sunken seamount, or undersea volcanic island, having a flat top due to erosion at sea level.

gymnosperm Any seed plant in which the seeds are not covered by a carpel wall. Gymnosperms include conifers, cycads, ginkgos, and seed ferns.

half-life The amount of time it takes for one-half of a radioactive parent isotope to transform, or decay, to a daughter product.

halide A mineral having positive ions of such elements as sodium and potassium attached to negative ions of such elements as chlorine and bromine. One important halide mineral is halite.

herbivore An organism that feeds on plant, algal, fungal, or bacterial material.

Hercynian orogeny Late Paleozoic orogeny that affected Europe and northwestern Africa during the Carboniferous and Permian; equivalent to the Alleghanian orogeny that affected North America. Synonym: Variscan orogeny.

heterotrophy A means of obtaining nutrients by ingesting or breaking down organic matter.

heterozygous The condition of having two different alleles for a trait at the same location on a pair of homologous chromosomes.

hexapod A six-legged terrestrial arthropod. Hexapods include insects (primitively winged hexapods) and non-winged forms such as springtails.

historical geology See Earth history.

hominid An advanced primate; the group embraces australopithecines and species in the genus *Homo*.

homologous structures Morphological features in biologic organisms that have a similar position and evolutionary origin, but not necessarily identical structure or the same function.

homoplasy　The separate development of similar characters in two or more taxa by different evolutionary pathways. Synonym: convergent evolution.

homozygous　The condition of having two identical alleles for a trait at the same location on a pair of homologous chromosomes.

horst　An elongate fault block, bounded on both sides by normal faults, that has remained relatively high standing compared to a downthrown block, or graben, in an area experiencing extension, or pulling apart, of the lithosphere.

hot spot　A volcanic center, often in the interior of a plate, caused by a plume of magma rising from the mantle.

hydrothermal vent　An opening in the Earth's crust, usually associated with magmatic activity, where hot water, often enriched in ions, is released.

hypothesis　A scientific idea, usually an explanation of a process, that is subject to modification, abandonment, or provisional acceptance after testing.

hypsometric curve　A graph indicating the proportions of the Earth's surface above and below sea level.

Iapetus　The proto-Atlantic ocean existing between Laurentia, Baltica, and Gondwana during much of the Paleozoic.

icehouse condition　A condition of relatively cool or cold global temperature, and normally characterized by the expansion of polar glaciers.

ichnofossil (pronounced "ick-no-fahs-sill)　A trace fossil.

ichthyosaur (pronounced "ick-thee-oh-sore")　A marine reptile with a dolphin-like body. Stratigraphic range: Triassic to Cretaceous.

igneous arc　See volcanic arc.

igneous rock　Any rock formed from the cooling and crystallization of magma.

index fossil　See guide fossil.

infaunal　A term applied to an animal that lives within the sediment in an aqueous setting.

ingroup　In a phylogenetic analysis, the group of taxa whose relationships are of interest (or under investigation).

insect　A member of a group of hexapods, exclusively terrestrial, that at least primitively has wings. Insects are the most diverse animals.

interglacial　The term applied to a relatively warm time interval between successive glacial episodes.

intermontane basin　A basin situated between mountains.

intrusive rock　An igneous rock that resulted from the cooling and solidification of magma within the Earth's crust. Intrusive rocks are commonly coarsely crystalline. Synonym: plutonic rock.

invertebrate　An animal lacking a backbone. Examples are arthropods, echinoderms, mollusks, brachiopods, bryozoans, cnidarians, and sponges.

ion　An atom that is either positively or negatively charged.

isostasy　A condition of equilibrium, similar to floating, of lithospheric units above the asthenosphere. Loading of the crust by thrust sheets or glacial ice leads to isostatic depression or downwarping, and removal of the load leads to rebound or upwarping.

isotope　One of two or more variations of the same chemical element, differentiated by atomic weight (related to the number of neutrons). Isotopes of the same element have slightly different physical and chemical properties.

isotopic excursion　A positive or negative shift in the isotopic ratio of an element recorded through a succession of stratigraphic layers.

Kaskaskia sequence　The name applied to a large scale, unconformity-bounded sequence of sediments deposited on the craton of Laurentia during the Early Devonian through mid-Carboniferous periods.

kingdom　In biologic classification, the rank above phylum (or divisions, if plants) and below domain. The kingdom represents a fundamental type of body plan and trophic habit, as reflected in animals, plants, fungi, protoctists, bacteria, and archeans.

komatiite　A group of ultramafic rocks having very high cooling temperatures (above 1100°C).

lamination　Layering or bedding, usually in sedimentary rocks. Often lamination is applied to fine-scale layering.

Laramide orogeny　An episode of deformation that occurred from the Late Cretaceous to the Paleocene and that resulted in many of the structural features of the eastern Rocky Mountains.

Laurasia　A large ancient continent composed of many continental and large island areas of the present-day Northern Hemisphere (North America, Greenland, and Eurasia); used to differentiate the Northern Hemisphere landmass from the Southern Hemisphere landmass (Gondwana).

Laurentia　The Paleozoic continent that included North America, Greenland, and parts of Northern Ireland, Scotland, and western Norway.

lepidosaur　A type of reptile characterized by diapsid skull type but lacking derived traits of the archosaurs. Lizards, snakes, tuataras, and mosasaurs are lepidosaurs.

limestone　A sedimentary rock composed predominantly of calcium carbonate.

Linnaean classification system　The hierarchical classification of living organisms. Categories representing progressively smaller groups of organisms are: kingdom (plural, kingdoms), phylum (plural, phyla), class (plural, classes), order (plural, orders), family (plural, families), genus (plural, genera), and species (plural, species). Often the term "domain" is used above the rank of kingdom.

lithification　The processes involved in changing sediments to rock, including compaction, dewatering (desiccation), cementation, and crystallization.

lithofacies A mappable sedimentary unit distinguished from other units on the basis of lithology.

lithology A rock type; distinguishing characteristics include color, mineralogic composition, grain size, and grain type.

lithosphere The outer, relatively rigid layer of the Earth, approximately 75 to125 km thick, overlying the more plastic asthenosphere. The lithosphere includes the entire crust plus the upper part of the mantle.

lithostratigraphic unit A stratigraphic unit defined on the basis of lithologic criteria. A formation is the fundamental lithostratigraphic unit.

lithostratigraphy Stratigraphic description and correlation on the basis of lithology (rock type).

lycopod A member of the vascular land plant group that includes club mosses.

macroevolution 1. Biologic evolution at the species level. 2. Large-scale evolutionary patterns, trends, and rates of change among groups of species.

magma Molten rock, including any suspended crystals (mineral grains) and dissolved gases; normally formed from the melting of rock at high temperatures.

magma ocean A condition in which molten magma covers the outer surface of a planet or its moon.

magmatic arc See volcanic arc.

mammal A vertebrate animal characterized by one jaw bone, three inner ear bones (malleus, incus, and stapes), endothermy (warm-bloodedness), hair or fur; most living species are viviparous (bearing live young). Among extant animals, mammals are the only ones that nourish their young with milk produced by mammary glands in the females.

marine Water having high salinity values, normally about 35‰ (35 parts per thousand), but often ranging between 28 and 37‰.

mass extinction When numerous species become extinct within a geologically short time interval.

meiosis The division of chromosomes to produce two haploid (1n) cells in the production of gametes (sperm and eggs, or spores in plants) necessary for sexual reproduction.

metazoan A multicellular animal; characterized by multiple cell types and cells organized into tissues and organs.

meteorite A relatively small rock, either silicate or metallic, or some combination of the two, that falls to a planetary surface from interplanetary space.

microbe A microscopic organism. Microbes include archaebacteria (archeans), eubacteria (bacteria or monerans), some protoctists, and some fungi.

microbial mat A layer of microscopic bacteria and fungi growing at the sediment surface.

microevolution 1. Changes in allele frequency caused by mutation, gene flow, genetic drift, and natural selection. 2. Changes that occur within species, usually between generations.

Milankovitch cycle A periodic change in one of Earth's orbital patterns (precession, obliquity, eccentricity) with respect to the Sun.

mineral A naturally occurring crystalline solid or a synthetic, chemically identical equivalent.

mitosis The division of cell nuclei in which the parental chromosomal number is maintained. Mitosis is the basis for somatic (bodily) growth, and in many eukaryotic species, asexual reproduction.

mobile crust phase An interval of Earth history during which amalgamation of the continental crust occurred; corresponds to much of the Archean Eon.

mold As used for a mode of fossilization, an impression made in sediment or its lithified equivalent by the body of an organism.

molecular clock The hypothesis that the mutation rate in DNA or RNA remains essentially constant over time, and can be used to measure the amount of time elapsed since separate species diverged.

molecular evolution Changes through time in organisms at the DNA level.

mollusk A solitary invertebrate having a tissue fold (mantle) draped around a soft, fleshy body; may have bilateral symmetry or coiled asymmetry, and may be with or without a calcium carbonate shell. Mollusks include clams, oysters, snails, and cephalopods (squids, belemnoids, ammonoids, and octopuses). Alternative spelling: mollusc.

monophyletic group See clade.

moraine A mound, ridge, blanket, or other buildup of unstratified, unsorted glacial drift (till).

mosasaur A type of large, marine Mesozoic lepidosaur.

mud Sediment consisting of silt and clay; grain size ranges up to 1/16 mm.

mudcracks Irregular fractures in a roughly polygonal pattern formed by the drying and shrinkage of mud.

mutation A heritable change in DNA or chromosomal structure. Mutation is the source of most alleles (molecular version of genes), and ultimately, of life's diversity.

natural selection A microevolutionary process by which individuals best adapted to their environment survive and reproduce, and less well adapted individuals are eliminated from the population.

nektonic A term referring to a swimming pelagic organism.

Nevadan orogeny Deformation, metamorphism, and plutonism that occurred in western North America during the Jurassic and Early Cretaceous. Emplacement of the Sierra Nevada batholiths began during this orogenic episode.

niche As pertaining to species, the full range of physical, chemical, and biologic conditions under which its members live and reproduce; it includes requirements for food, living space, and interactions with other species. Synonym: ecologic niche.

nonconformity An unconformity in which sedimentary strata overlie crystalline (igneous or metamorphic) rocks.

nonfoliated A monominerallic metamorphic rock, such as marble or quartzite, that lacks foliation.

normal fault A fault in which the hanging wall has moved downward relative to the footwall.

nothosaur A type of relatively small, extinct marine reptile. Stratigraphic range: Triassic to Jurassic.

nucleotide A small organic compound having a five-carbon sugar (deooxyribose), a nitrogen-forming base, and a phosphate group. Nucleotides are the structural units of nucleic acids (DNA and RNA), adenosine phosphates (such as ATP), and nucleotide coenzymes.

numerical age dating The technique of establishing when events occurred according to how much time has elapsed since their occurrence. Geologic time that has elapsed is measured in thousands, millions, or billions of years. Synonym: absolute age dating.

obliquity The tilt of the Earth's axis.

oceanic crust The solid, outer part of the Earth underlying the ocean basins; consists largely of mafic igneous rocks (mostly basalt).

oil shale Kerogen-bearing, finely laminated shale or siltstone, black or brown in color; can be distilled to produce hydrocarbons.

ontogeny The development of an individual organism from conception or other initiation through maturity.

ophiolite An assemblage of ultramafic and mafic igneous rocks representing oceanic crust.

ornithischian dinosaur An archosaur characterized by a bifurcate or reduced pelvic structure (bird-like hip); many species show elaborate batteries of teeth adapted for herbivory, and a beak. Stratigraphic range: Triassic to Cretaceous.

orogenic belt A linear or arcuate region subjected to folding and other deformation during a mountain building cycle. Synonym: orogen.

orogenesis The process of mountain building, including thrusting, folding, and faulting in outer layers of the Earth, and plastic folding, metamorphism, and plutonism (intrusion) in lower layers.

Ouachita orogeny Mostly Carboniferous (Pennsylvanian) mountain building along the southern margin of Euramerica (extending from Mississippi through Oklahoma and Texas to Utah) related to collision of Eauramerica with Gondwana.

outgroup In a phylogenetic analysis, a taxon that is related to, but not part of, the ingroup.

oxic The condition in which oxygen is present.

oxide A mineral having a metallic ion combined with oxygen. Important oxides include hematite and magnetite.

oxygen-minimum zone An interval in a water body in which the amount of dissolved oxygen is less than that above or below it; often characterized by the accumulation of laminated, organic-rich muds.

paleoecology The study of the relationships between ancient organisms and their environment, including the factors controlling the distribution and abundance of species.

paleomagnetism The study of natural remnant magnetism in rocks to determine properties of the Earth's magnetic field, especially the direction and intensity of Earth's magnetic field, in the geologic past

paleontological species concept The concept that the limits of ancient species may be inferred from their preserved physical traits.

paleosol An ancient soil horizon or profile.

Pangea (pronounced "pan-jee-ah") The late Paleozoic to early Mesozoic supercontinent comprising most of the world's continental crust. Alternative spelling: Pangaea.

Pannotia (pronounced "pan-oh-she-ah") A hypothesized supercontinental assembly of many cratonic areas in the late Neoproterozoic.

paraphyletic group In phylogenetics, a group that contains some, but not all, species descended from an ancestral species.

parent (nuclide) In a radioactive decay series, an unstable isotope that decays, or transforms, into a daughter product.

passive margin The trailing edge of a tectonic plate, where active tectonic interaction with another plates is not occurring.

pelagic 1. A term used for an organism whose habitat is the open water (either a swimming or floating organism). 2. A term used for an open aqueous environment.

pelycosaur A synapsid having a "sail" supported by rodlike extensions of the vertebrae extending from the back. Stratigraphic range: Carboniferous to Permian.

period 1. As used formally in stratigraphy, the geochronologic unit (geologic time unit) lower in rank than era and higher than epoch; the unit of geologic time equivalent to a system. 2. An informal term used to indicate a geologic interval of some duration.

perissodactyl An odd-toed ungulate mammal such as a horse, rhinoceros, hippopotamus, or tapir.

photic zone The depth in water to which sunlight will penetrate sufficiently for photosynthesis to take place.

photosynthesis The process by which plants, algae, and some bacteria trap energy from sunlight and convert it to chemical energy (ATP or NADPH, a phosphorylated nucleotide coenzyme), then synthesize sugar phosphates that become converted to organic molecules (sucrose, cellulose, starch, and other compounds). CO_2 is converted to organic matter, and O_2 is produced. It is the way most energy and carbon enters the web of life.

phyletic gradualism A speciation process in which morphologic change occurs gradually and continuously through an

...onary lineage so that an ancestral species appears to grade imperceptibly into its descendant species.

phylogeny The line or lines of descent in an evolutionary series.

phylogenetic analysis See cladistics.

phylum (plural, **phyla**) In biologic classification, the rank above class and below kingdom.

piercing point A geologic feature, such as an orogenic belt, that shows evidence of having once been continuous between two or more areas that were formerly adjacent parts of the same continental (or oceanic) block but are now separated.

placental mammal A mammal that nourishes its developing fetus with a placenta and gives lives birth.

placoderm A jawed fish characterized by jointed dermal armor (bony plates) covering the head and trunk. Stratigraphic range: Devonian-Carboniferous.

planktonic A term used for aquatic organisms that drift or weakly swim.

Plantae In biologic classification, the plant kingdom, which consists of haploid eukaryotes that grow from spores produced by meiosis. Plants are usually multicellular, photosynthetic autotrophs.

plate tectonics theory The scientific theory that the Earth's outer shell, or lithosphere, is cracked and composed of pieces that interact with each other as they float on a hot, deformable asthenosphere. Interactions at plate boundaries cause seismic and tectonic activity.

platform The part of a continent covered by flat-lying or gently tilted, mostly sedimentary strata.

platform phase An interval of Earth history characterized by relatively stable, amalgamated continental cores; begins with the Proterozoic Eon.

playa lake A shallow, ephemeral lake in an arid or semiarid region that dries up through evaporation, leaving a playa (a flat area at the lowest point of an undrained desert basin).

plesiosaur A Mesozoic marine reptile characterized by euryapsid skull type, a broad body, and large paddlelike limbs.

pluton An intrusive igneous rock body.

plutonic rock An igneous rock that resulted from the cooling and solidification of magma within the lithosphere. Synonym: intrusive rock.

pluvial lake A lake formed during an interval of exceptionally heavy rainfall.

polyphyletic group An artificial group of organisms in which the common ancestor is classified in another taxon; artificial classification of species from different clades in the same group.

preadaptation A feature already present, but seemingly nonadaptive, in an ancestral species that later becomes adaptive, providing an evolutionary advantage to its descendant species, in a different or changed environment.

Precambrian An informal term referring to the Archean and Proterozoic eons.

precession A slight wobble of the Earth's axis over time.

predator An organism that feeds on other organisms; it may or may not kill other organisms (prey). Unlike parasites, predators do not live in or on their prey.

primate A placental mammal belonging to the group that includes lemurs, monkeys, apes, and hominids.

principle of biotic succession The principle that body fossils occur in strata in a definite, determinable order.

principle of cross-cutting relationships The principle that a rock unit, sediment body, or fault that cuts another geologic unit is younger than the unit that was cut.

principle of included fragments The principle that fragments of a rock or sediment body contained within another rock or sediment are from a preexisting (older) rock or sediment than the one in which they are contained.

principle of original horizontality The principle that sedimentary strata were originally deposited nearly horizontally and parallel to the Earth's surface.

principle of original lateral continuity The principle that, at the time of deposition, a sedimentary unit extended laterally and continuously in all directions until it thinned out or otherwise reached the limits of its depositional range.

principle of superposition The principle that, in an undisturbed succession of strata, the oldest strata are at the base of the succession, and the youngest strata are at the top of the succession.

progymnosperm An extinct vascular plant with fernlike leaves and coniferlike wood.

Prokarya A name applied to prokaryotic organisms; it embraces the domains Archaea and Bacteria.

prokaryote An organism that lacks nucleus-bearing cells. Alternative spelling: procaryote.

Protoctista (pronounced "proh-toh-tis-tah") In biologic classification, the kingdom consisting of eukaryotic microorganisms and their descendants other than fungi, plants, and animals; presumed to have evolved through integration of former microbial symbionts.

pteridophyte (pronounced "tehr-id-oh-fite") A fern; a type of vascular plant that produces spores.

pteridospermophyte (pronounced "tehr-id-oh-sperm-oh-fite") A seed fern, the first known gymnosperm. Synonym: seed fern.

pterosaur (pronounced '"tehr-oh-sore") A type of extinct archosaur having wings adapted for flight. Stratigraphic range: Triassic to Cretaceous.

punctuated equilibrium A macroevolutionary pattern in which species arise quickly and undergo rapid morphologic change, then endure long intervals of relatively little morphologic change (morphologic stasis).

radioactive isotope   An isotope of an element that is unstable over time; it will decay to eventually form a stable daughter product.

recrystallization   1. The formation, mostly in a solid-state condition, of new crystals in a rock. 2. A fossilization mode in which crystals of the original biominerals are replaced by other crystals.

redbed   Reddish or reddish-brown sediments; the color is imparted by oxidation of iron-bearing minerals.

reef   As used in geology, a wave-resistant structure constructed by organisms, often corals, sponges, mollusks, or brachiopods. Stromatolites and thrombolites can be considered reefs in the broad sense. As used in biology, a reef community should show evidence of ecologic interaction among organisms. The original, nautical meaning of reef referred to any impediment to shipping.

regional metamorphism   A type of metamorphism (changes in texture and mineralogy of a rock) that affects large areas of the lithosphere.

regression   As used in stratigraphy, a drop of sea level and withdrawal of water from the land. Nonmarine facies shift seaward.

relative age dating   The technique of establishing a chronology of events arranged in sequential order (without reference to numerical age).

replacement   As used for a fossilization mode, substitution of the original organic body parts of an organism by new substances. Often, original biominerals are dissolved and replaced by other minerals or amorphous solids.

rhyniophyte   A leafless, rootless, extinct vascular plant having simple dichotomous (forked) branches and bearing terminal sporangia; a member of the group of first true vascular land plants.

ripple   1. A sedimentary bed form having a roughly triangular transverse cross section formed by the interaction of a moving air or water current with a mobile sediment. 2. The light ruffling of the surface of water by a breeze. Synonym: ripple mark.

RNA   Ribonucleic acid. A nucleic acid characterized by the sugar ribose and the pyramidine (nucleotide base) uracil; usually a single-stranded polynucleotide.

rock   1. An aggregate of minerals. Examples include granite, sandstone, shale, limestone, and marble. Often the definition is extended to include amorphous solids such as obsidian and opal, and to include lithified organic matter such as coal. 2. A colloquial term for a gemstone.

rock cycle   A conceptual model that describes the origin, alteration, destruction, and reformation of rocks through the action of Earth processes.

Rodinia   An early supercontinent, assembled in the Mesoproterozoic and separated in the Neoproterozoic.

roundness   The degree to which a sedimentary particle's original edges and corners have been smoothed off.

rudist   A type of clam (bivalve) characterized by unequal valve sizes that lived solitarily, gregariously, or in reefal masses. Stratigraphic range: Jurassic to Cretaceous.

ruminant   A cud-chewing ungulate mammal such as a bison, cow, or giraffe.

sand   Sediment having a grain size between 1/16 mm and 2 mm. Sand is commonly composed of quartz grains, but may be calcium carbonate skeletal debris or any other mineral.

sandstone   The lithified equivalent of sand; sedimentary rock composed of sand-size grains.

Sauk sequence   The name applied to a large scale, unconformity-bounded sequence of sediments deposited on the craton of Laurentia during the Ediacaran through Early Ordovician periods.

saurischian dinosaur   An archosaur characterized, at least primitively, by a triradiate pelvic structure (lizard-like hip). The clade includes sauropodomorphs, theropods, and birds (descendants of the Mesozoic theropods).

sauropod   An extinct saurischian dinosaur characterized by quadrupedal (four-legged) gait, a long neck and tail, teeth adapted for herbivory, and typically, large adult size. Stratigraphic range: Jurassic to Cretaceous.

scavenger   An organism that feeds on dead carcasses, decaying organic matter, or other nonliving organic matter.

science   A philosophical discipline based on empirical observation, hypothesis formulation, prediction or retrodiction, and testing.

scientific method   A general term for a scientific investigation involving an iterative process of empirical observation, hypothesis building (with a predictive or retrodictive component), and testing.

scientific theory   A scientific concept that is tantamount to fact. See theory.

sclerite   A hard, chitinous or biomineralized skeletal element of an invertebrate animal.

seafloor spreading   The hypothesis proposed by Harry Hess that ocean basins expand by the addition of new rock from spreading centers, and that older rock is destroyed near the basin margins. This concept was later amalgamated with aspects of continental drift to develop the theory of plate tectonics.

seamount   An elevation of the seafloor, 1000 m or higher, usually an undersea volcanic island.

sediment   Unconsolidated particles of broken rock that have been transported by agents of erosion and unconsolidated particles formed as skeletal material through biomineralization.

sedimentary rock   A rock formed from sediments (particles of preexisting rock or skeletal remains of organisms), usually together with minerals precipitated under aqueous conditions, and normally deposited in layers.

seed fern   See pteridospermophyte.

sequence 1. As used in sequence stratigraphy, a relatively conformable succession of genetically related sedimentary strata bounded by unconformities (erosion surfaces) or their correlative conformities. Synonym: depositional sequence. 2. Any formal or informal succession of sedimentary strata.

sequence-stratigraphic unit A genetically related sedimentary package delimited below and above by disconformities or their lateral conformities. A depositional sequence (sequence) is the fundamental sequence-stratigraphic unit.

sequence stratigraphy The study of stratigraphic relationships within a chronostratigraphic framework of repetitive, genetically related packages of sediments or sedimentary rocks bounded by disconformities or their lateral conformities. Genetic sedimentary packages are called depositional sequences (sequences).

series As used in stratigraphy, the chronostratigraphic equivalent of an epoch.

Sevier orogeny Folding and eastward thrusting in the eastern Great Basin (Utah) that culminated in the Late Cretaceous; the mountain building occurred in time between the Nevadan orogeny (to the west) and the Laramide orogeny (to the east).

shale A sedimentary rock composed of clay-size grains (smaller than 1/256 mm), and showing fine laminations and fissility.

shear boundary See transform plate boundary.

shelf break the edge of the continental shelf, separating the continental shelf from the continental slope.

shield As used in geology, a large area of exposed Precambrian (mostly Archean) basement rocks on a continent; commonly gently convex, and covered around the sides by platform sediments.

shocked quartz A quartz grain showing distinctive parallel sets of fused microscopic planes, called shock lamellae, formed under explosive or extreme impact conditions.

silicate 1. A mineral that has a silicate tetrahedron (SiO_4) as the basic chemical property. Silicates are the dominant mineral group in igneous, sedimentary, and metamorphic rocks. Common silicate minerals include quartz, feldspars, micas, and clay minerals. 2. A rock composed of silicate minerals.

siliciclastic Clastic (or detrital) particles composed of such silica-bearing minerals as quartz, feldspars, and clay minerals.

silt Sediment having a grain size between 1/256 mm and 1/16 mm.

siltstone The lithified equivalent of silt.

sima Basaltic-type rock comprising much of the oceanic crust. The word is formed from silicon and magnesium, two major components of basalts.

sister group In a phylogenetic ananlysis, the outgroup that is most closely related to the ingroup.

slate A low-grade metamorphic rock that has slaty cleavage, breaks into flat, plate-like pieces, and was formed from metamorphism of sedimentary mudrocks.

small shelly fossil (SSF) One of many skeletal fossils (sclerites), usually pieces of skeletons having diverse shapes, 1 or 2 millimeters in size, and calcitic or phosphatic in composition, common to deposits of the lower half of the Cambrian System.

Snowball Earth hypothesis The concept that during the Proterozoic Eon, the entire surface of the Earth was repeatedly enveloped in ice.

Sonoman orogeny Late Permian to Early Triassic mountain building activity recognized from Nevada that resulted in great westward expansion of the North American continent.

sorting A measure of the range of grain sizes in a sedimentary deposit.

speciation A macroevolutionary process resulting in the development of a new species from a preexisting one.

species (plural, **species**) 1. A group of organisms that can interbreed and produce fertile offspring (biological species definition). 2. Fossils showing sufficiently close similarity that they are inferred to represent remains of organisms that were, when they were living, a biological species (paleontological species definition).

spicule A small biomineralized skeletal element of an invertebrate animal, an alga, or other organism.

sponge A multicellular invertebrate animal lacking true tissue layers.

spore A tiny, unicellular reproductive cell, typical of plants or algae.

spreading center See divergent plate boundary.

stage The chronostratigraphic (time-rock) equivalent of age.

stoma A pore flanked by guard cells in the leaf of a photosynthetic green plant.

strata (singular, **stratum**) Layers of rocks, usually sedimentary rocks, but also including such extrusive igneous rocks as pyroclastic flows and volcanic ash beds.

stratigraphy (pronounced "strah-tig-ruh-fee") The study of layered rocks (or strata), including their compositions, origins, geometric relationships, and ages.

stromatolite A thinly layered biogenic-sedimentary structure resulting from the trapping and binding of fine sediment in layers by photosynthetic cyanobacteria.

subduction A tectonic process in which old, cold, dense lithosphere sinks into the asthenosphere and is remelted.

subduction zone A long, narrow belt, normally including a deep-sea trench, along which subduction occurs.

subfossil Evidence of past, but post-Pleistocene, life.

sulfate A mineral containing a sulfate ion (SO_4^{-2}). Important sulfate minerals include gypsum and anhydrite.

sulfide A mineral usually containing metallic elements combined with sulfur. An example is pyrite (FeS_2), in which iron combines with sulfur.

supercontinent cycle A tectonically driven cycle defined by the assembly of a supercontinent and later fragmentation and dispersal of its pieces.

suture 1. In tectonics, where two crustal blocks have become welded together through metamorphic processes, usually associated with collision. 2. In cephalopods, where internal partitions, called septa, meet the inside surface of the outer wall of the shell.

symbiosis A condition in which two or more dissimilar organisms live together in close association.

synapomorphy A shared derived character; used as key evidence of a common ancestry among two or more species.

synapsid 1. A vertebrate skull type having one opening (temporal fenestra) located behind and below the orbit (eye socket). 2. A vertebrate animal belonging to the clade characterized by a synapsid skull type. Primitive synapsids and mammals have synapsid skulls.

system The chronostratigraphic equivalent of a period, and the fundamental unit of chronostratigraphy.

systematics See taxonomy.

Taconic orogeny Orogenic activity during the Ordovician Period along the Appalachian margin of Laurentia.

taphonomy The study of the processes of fossilization.

taxon (plural, **taxa**) Any unit of biologic classification, such as a species, genus, family, order, class, phylum, kingdom, or domain.

taxonomy The theory and practice of classifying biologic organisms.

tectonosedimentary cycle The record of continental collision in basinal sedimentation patterns: passive margin sedimentation gives way to foreland basin development (deep water flysch deposits followed by shallow water and marginal-marine to nonmarine molasse deposits).

tectonostratigraphic terrane A rock body having an internally consistent geologic makeup (structural style and genetically related stratigraphy), and distinct from a continental block or an adjacent terrane by bounding faults. Synonym: terrane.

Tejas sequence The name applied to a large scale, unconformity-bounded sequence of sediments deposited on the craton of North America during the Cenozoic Era.

teleost A derived form of bony fish, characterized by a bony skeleton, thin, rounded scales, and a symmetrical tail.

terrestrial 1. A term referring to the Earth. 2. A nonmarine environment other than a lake, stream, or pond.

Tertiary Until the Paleogene, Neogene, and Quaternary were ratified as the periods comprising the Cenozoic Era, "Tertiary" was used as a term embracing the Paleocene through Pliocene epochs. The last age (Gelasian Age) of the Pliocene Epoch as defined prior to 2007 is now included in the Quaternary Period, making the "Tertiary" as used historically equivalent to the Paleogene, Neogene, and the early part of the Quaternary as used today.

Tethys Ocean A Paleozoic-Mesozoic ocean that occupied the area east of Pangea, and largely obliterated by the Alpine-Himalayan orogenic belt.

tetrapod Any vertebrate animal with four limbs. Amphibians, reptiles, birds and mammals are tetrapods.

theory 1. In popular parlance: an idea, often one not fully tested or necessarily accepted (equivalent to a hypothesis). 2. In scientific parlance: a concept that is tantamount to fact, well-tested, and accepted by practicing scientists. In this book, the term scientific theory is used to differentiate a well-accepted scientific concept from a popular "theory."

theropod A saurischian dinosaur characterized by bipedal (two-legged) gait and, at least primitively, teeth adapted for carnivory. The clade includes birds, which descended from Mesozoic theropods.

thrombolite A nonlaminated biogenic-sedimentary structure constructed by cyanobacterial consortia, characterized by a vague clotted texture internally.

thrust fault A low-angle reverse fault, in which the hanging wall has moved up relative to the footwall; generally forms an angle less than 45° compared to the Earth's surface over most of its length.

tidal flat An extensive, nearly horizontal, relatively barren, coastal area alternately covered and uncovered by tides, and consisting of unconsolidated sediments (mostly mud and sand).

till A glacial deposit consisting of unsorted, unstratified particles.

Tippecanoe sequence The name applied to a large scale, unconformity-bounded sequence of sediments deposited on the craton of Laurentia during the Middle Ordovician through Early Devonian periods.

trace fossil Any evidence of the activity of an ancient organism.

transform fault A strike-slip fault that links two other faults or two other plate boundaries. Separate segments of a mid-ocean ridge are commonly linked by a transform fault.

transform fault boundary A boundary between two crustal plates characterized by a transform fault, and where crust is neither created nor destroyed.

transgression A rise of sea level and submergence of the continent under seawater. Marine facies shift landward.

trilobite (pronounced "try-loh-bite") A type of extinct marine arthropod characterized by a calcified exoskeleton divided lengthwise into three lobes; most forms have a distinctive head shield, a thorax composed of multiple, articulating segments, and a tail shield. Stratigraphic range: Cambrian to Permian.

triple junction A junction of three spreading edges of plates. The arms of a triple junction often meet at angles approximating 120°.

turbidite A submarine landslide deposit formed by downslope movement of a turbid mix of sediment and water (a turbidity current).

unconformity   A surface of erosion or non-deposition.

ungulate   A hoofed placental mammal, or a mammal evolved from one with hooves.

uniformitarianism   The principle that processes acting upon the Earth today have also operated in the geologic past.

Variscan orogeny   See Hercynian orogeny.

vascular plant   A plant having internal vascular tissues as well as true stems, roots, and leaves. Club mosses, rhyniophytes, horsetails, ferns, and angiosperms are vascular plants.

vertebrate   An animal possessing a skeleton of bone or cartilage; the skeleton includes a backbone.

vestigial structure   A structure in a biologic organism that is usually reduced in size or function compared to the homologous structure in the ancestral organism. A vestigial structure is commonly in the process of disappearing.

virus   A microscopic "living" thing that does not form cells, but is composed of DNA or RNA enclosed in a coat of protein; a virus does not exist independently and must enter a cell and use its biologic machinery to replicate.

volatile   An element or a compound that is transformed to the gaseous phase.

volcanic arc   An arcuate line of active volcanoes and igneous plutons formed at a convergent plate margin where subduction is occurring. Magma forms by melting of the downgoing plate. Synonyms: igneous arc, magmatic arc.

Walther's Law of Facies   The concept that in an unbroken sequence, vertically superimposed lithofacies were laterally adjacent to each other at the time of deposition.

Wilson cycle   The successive recurrence of plate-tectonic spreading and convergence.

weathering   Chemical alteration or mechanical breakdown of Earth materials.

zone   See biozone.

zooid   A tiny individual animal that lives as part of a colony. Graptolites, bryozoans, and other colonial animals comprise large integrated associations of zooids.

Zuni sequence   The name applied to a large scale, unconformity-bounded sequence of sediments deposited on the craton of North America during the Middle Jurassic to Cretaceous periods.

TABLE AND LINE ART CREDITS

Chapter 1
Figures 1.3, 1.4, 1.5: From Murck, Barbara, Brian J. Skinner, and Dana Mackenzie. *Visualizing Geology*. Copyright 2007 John Wiley & Sons, Inc. Reprinted with permission of John Wiley & Sons, Inc.

Chapter 2
Figures 2.2, 2.3E, 2.3G, 2.10, 2.27: From Murck, Barbara, Brian J. Skinner, and Dana Mackenzie. *Visualizing Geology*. Copyright 2007 John Wiley & Sons, Inc. Reprinted with permission of John Wiley & Sons, Inc.; Figures 2.4B, 2.4C: From Skinner, Brian J. and Stephen C. Porter. *The Dynamic Earth: An Introduction to Physical Geology, 5th Edition*. Copyright 2003 John Wiley & Sons, Inc. Reprinted with permission of John Wiley & Sons, Inc.; Figures 2.4D, 2.20: From Levin, Harold L. *The Earth Through Time, 8th Edition*. Copyright 2005 John Wiley & Sons, Inc. Reprinted with permission of John Wiley & Sons, Inc.

Chapter 3
Figure 3.6: Babcock, Loren E., Richard A. Robison, Margaret N. Rees, Shanchi Peng, and Matthew R. Saltzman. "The Global boundary Stratotype Section and Point (GSSP) of the Drumian Stage (Cambrian) in the Drum Mountains, Utah, USA." *Episodes* Vol. 30, no. 2, June 2007. Figure 9, p. 92. Reprinted with permission of the International Union of Geological Sciences; Figure 3.7: Modified from Zhu, Mao-Yan, Loren E. Babcock, and Shan-Chi Peng. Nov 2006. "Advances in Cambrian stratigraphy and paleontology: integrating correlation techniques, paleobiology, taphonomy and paleoenvironmental reconstruction." *Palaeoworld*, Vol. 15, Issues 3–4, pp. 217–222; Figures 3.11A, 3.11B, Table 3.1: From Skinner, Brian J. and Stephen C. Porter. *The Dynamic Earth: An Introduction to Physical Geology, 5th Edition*. Copyright 2003 John Wiley & Sons, Inc. Reprinted with permission of John Wiley & Sons, Inc.

Chapter 4
Figure 4.4: From Salyers, Abigail A. and Dixie D. Whitt. *Microbiology: Diversity, Disease, and The Environment*. Copyright 2000 John Wiley & Sons, Inc. Reprinted with permission of John Wiley & Sons, Inc.; Figure 4.13: From Raven, Peter H. and Linda R. Berg. *Environment, 5th Edition*. Copyright 2005 John Wiley & Sons, Inc. Reprinted with permission of John Wiley & Sons, Inc.

Chapter 5
Figure 5.2: Adapted from Gould, Stephen Jay. *The Book of Life: An Illustrated History of the Evolution of Life on Earth*. W.W. Norton & Co. Inc: January 2001. p. 199; Figures 5.3, 5.4A, 5.5B: From Levin, Harold L. *The Earth Through Time, 8th Edition*. Copyright 2005 John Wiley & Sons, Inc. Reprinted with permission of John Wiley & Sons, Inc.; Figure 5.9A: Adapted from Figs. 13.5, p. 445 and 13.6, p. 448 from *Evolutionary Analysis*, 2nd edition by Scott Freeman and Jon C. Herron. Copyright 2001 by Prentice-Hall, Inc. Reprinted by permission of Pearson Education, Inc.; Modified from Figures 5.10A, 5.10B: From Fitch, W. M. and E. Margoliash. "Construction of Phylogenetic Trees." *Science 279: 279–84* (1967). Figure 2. Reprinted with permission from AAAS.

Chapter 6
Figure 6.17: Adapted from Strahler, Alan H. and Arthur Strahler. *Introducing Physical Geography, 4th Edition*. Copyright 2005, John Wiley & Sons, Inc. Reprinted with permission of Alan H. Strahler; Figure 6.24: Adapted from Gradstein, F., Ogg, J. & Smith, A. "A Geologic Time Scale 2004." *Geological Magazine*, September 2005, v. 142, no. 5, p. 633; Figure 6.25: Modified from Peng, S. C., L. E. Babcock, R.A. Robison, H. L. Lin, M. N. Rees, and M. R. Saltzman. 2004. "Global Standard Stratotype-section and Point (GSSP) of the Furongian Series and Paibian Stage (Cambrian)." *Lethaia*, 37:365–379, figure 7, p. 374. Reprinted with permission of Wiley-Blackwell, Inc.

Chapter 7
Figures 7.1A, 7.1C, 7.2E, 7.5B, 7.9, What A Geologist Sees Parts A, B, F (pp. 212–213): From Murck, Barbara, Brian J. Skinner, and Dana Mackenzie. *Visualizing Geology*. Copyright 2007 John Wiley & Sons, Inc. Reprinted with permission of John Wiley & Sons, Inc.; Figures 7.2C, 7.2D, 7.3B, 7.3C, 7.6A, 7.10 A, B, C, D, 7.12, 7.13, 7.14, 7.16: From Skinner, Brian J. and Stephen C. Porter. *The Dynamic Earth: An Introduction to Physical Geology, 5th Edition*. Copyright 2003 John Wiley & Sons, Inc. Reprinted with permission of John Wiley & Sons, Inc.; Figures 7.5A, 7.17: From Levin, Harold L. *The Earth Through Time, 8th Edition*. Copyright 2005 John Wiley & Sons, Inc. Reprinted with permission of John Wiley & Sons, Inc.

Chapter 8
Figure 8.5C: From Murck, Barbara, Brian J. Skinner, and Dana Mackenzie. *Visualizing Geology*. Copyright 2007 John Wiley & Sons, Inc. Reprinted with permission of John Wiley & Sons, Inc.; Figure 8.7: From Skinner, Brian J. and Stephen C. Porter. *The Dynamic Earth: An Introduction to Physical Geology, 5th Edition*. Copyright 2003 John Wiley & Sons, Inc. Reprinted with permission of John Wiley & Sons, Inc.; Figures 8.9A, 8.14A: From Levin, Harold L. *The Earth Through Time, 8th Edition*. Copyright 2005 John Wiley & Sons, Inc. Reprinted with permission of John Wiley & Sons, Inc.

Chapter 9
Figure 9.2: From Skinner, Brian J. and Stephen C. Porter. *The Dynamic Earth: An Introduction to Physical Geology, 5th Edition*. Copyright 2003 John Wiley & Sons, Inc. Reprinted with permission of John Wiley & Sons, Inc.; Figures 9.3A and 9.3B: Adapted from Scotese, C. R., 2001. Atlas of Earth History, Volume 1, Paleogeography, PALEOMAP Project, Arlington, Texas, 52 pp; Figure 9.5A: Modified from James A. Drahovzal, David C. Harris, Lawrence H. Wickstrom, Dan Walker, Mark T. Baranoski, Brian D. Keith, Lloyd C. Furer, compiled and edited by David C. Harris. "The East Continent Rift Basin: a New Discovery." Ohio Division of Geological Survey Information Circular 57. 1992: Figure 8.; Figure 9.10: From Levin, Harold L. *The Earth Through Time, 8th Edition*. Copyright 2005 John Wiley & Sons, Inc. Reprinted with permission of John Wiley & Sons, Inc.

Chapter 10
Figure 10.2: Modified from Blakey, Ronald C. Companion Poster to *A Geologic Time Scale 2004*. Early Cambrian Map. Reprinted with permission of Ronald C. Blakey, Professor of Geology; Figure 10.3: Modified from Babcock, L. E., Zhang, W. T. and Leslie, S. A. 2001. "The Chengjiang Biota: record of the early Cambrian diversification of life and clues to exceptional preservation." *GSA Today*, Vol. 11, no. 2, pp. 4–9. p. 5, Figure 2; Figure 10.5A: Adapted from *The Emergence of Animals*, by Mark A. S. McMenamin and Dianna L.S. McMenamin. p. 54, Figure 4.6. Copyright 1990, Columbia University Press. Reprinted with permission of the publisher; Figure 10.13: Adapted from J.H. Stitt, *Oklahoma Geological Survey Bulletin*. 124: 1–79, 1977; Figure 10.14: Modified from Blakey, Ronald C. Companion Poster to *A Geologic Time Scale 2004*. Middle Ordovician Map. Reprinted with permission of Ronald C. Blakey, Professor of Geology; Figure 10.15: Modified from Williams, Harold and Robert D. Hatcher, Jr. "Suspect terranes and accretionary history of the Appalachian orogen." *Geology*, 1982, v. 10, p. 530, Figure 1; Figure 10.20B: Modified from Blakey, Ronald C. Companion Poster to *A Geologic Time Scale 2004*. Early Middle Devonian Map. Reprinted with permission of Ronald C. Blakey, Professor

of Geology; Figure 10.21A: Modified from Hellstrom, L. W. and L. E. Babcock, 2000. "High-resolution stratigraphy of the Ohio Shale (Upper Devonian), Ohio." *Northeastern Geology and Environmental Sciences*, 22: 202–226, fig. 1; What A Geologist Sees Figure A (pp. 308–309), Figures 10.28A, 10.28B: From, Dunbar, Carl Owen and Karl M. Waage. *Historical Geology*. Wiley: 1969. Fig. 11-31. Reprinted with permission of John Wiley & Sons, Inc.; Figure 10.28C (Left-hand image: Pectoral fin of *Eusthenopteron*): Reproduced by permission of the Royal Society of Edinburgh from *Transactions of the Royal Society of Edinburgh*: vol. 68 (1970), pp. 207–329.; Figure 10.28C (Right-hand image: Hind limb of *Ichthyostega*). Modified from Jarvik, E., 1964. "Specializations in early vertebrates," Annals Society Royal Zoological Belgium, 94, 11–95; Figures 10.29A, 10.29B: Modified from Blakey, Ronald C. Companion Poster to *A Geologic Time Scale 2004*. Middle Mississippian Map. Reprinted with permission of Ronald C. Blakey, Professor of Geology; Figure 10.30.B: Modified from Heckel, Philip H: "Origin of the Phosphatic Black Shale Facies in Pennsylvanian Cyclothems of Mid-Continent North America." *The American Association of Petroleum Geologists Bulletin*, V 61, No. 7 (July 1977), p. 1045–1068. Figure on p. 1047, "Black Shale in Pennsylvanian cyclothems".; Figure 10.35A: Modified from Sepkoski, J. J., Jr., and Raup, D.M., 1986. "Periodicity of marine extinction events," pp. 3–36. In: Elliot, D.K., ed., *Dynamics of Extinction*. John Wiley & Sons, NY; Critical and Creative Thinking Questions #10: Modified from Seposki, John J. Jr. "A factor analytic description of the Phanerozoic marine fossil record." *Paleobiology*, 7(1), 1981, pp. 36–53. p. 49, Figure 5.

Chapter 11

Figure 11.2: Modified from Blakey, Ronald C. Companion Poster to *A Geologic Time Scale 2004*. Late Triassic Map. Reprinted with permission of Ronald C. Blakey, Professor of Geology; Figure 11.7: Modified from Dr. David Normal and Ron Sibbick. *Illustrated Encyclopedia Of Dinosaurs—Original And Compelling Insight Into Life In The Dinosaur Kingdom*. Crescent Books: 1985. p.21; Figure 11.10: Modified from Blakey, Ronald C. Companion Poster to *A Geologic Time Scale 2004*. Early Jurassic Map. Reprinted with permission of Ronald C. Blakey, Professor of Geology; Figures 11.11A, 11.11B: From Dunbar, Carl Owen and Karl M. Waage. *Historical Geology*. Copyright 1969 John Wiley & Sons, Inc. Reprinted with permission of John Wiley & Sons, Inc.; Figures 11.18B, 11.18C: Modified from Spencer George Lucas. *Dinosaurs: The Textbook*. McGraw Hill: 1994, (illustrator William C. Brown.) Figures 5.13, 6.5, 9.5. Reprinted with permission of The McGraw-Hill Companies, Inc.; Figure 11.24C: Modified from Spencer George Lucas. *Dinosaurs: The Textbook*. McGraw Hill: 1994, (illustrator William C. Brown.) Figure 16.8. Reprinted with permission of The McGraw-Hill Companies, Inc.; Figure 11.24C: Modified from Chatterjee, Sankar. *The Rise of Birds: 225 Million Years of Evolution*. Fig. 5.4, p.92. Copyright 1997: The Johns Hopkins University Press. Reprinted with permission of The Johns Hopkins University Press.; Figures 11.25A, 11.25B: Modified from Blakey, Ronald C. Companion Poster to *A Geologic Time Scale 2004*. Early Creta-ceous Map. Reprinted with permission of Ronald C. Blakey, Professor of Geology; Figure 11.26A: Modified from Levin, Harold L. *The Earth Through Time, 8th Edition*. Copyright 2005 John Wiley & Sons, Inc. Reprinted with permission of John Wiley & Sons, Inc.; Figure 11.33: Modified from Sepkoski, J. J., Jr., and Raup, D.M., 1986. "Periodicity of marine extinction events", p. 3–36. In: Elliot, D.K., ed., *Dynamics of Extinction*. John Wiley & Sons, Inc.; Figure 11.34E: Modified from Cowen, Richard: *History of Life, 3e* Blackwell Science, Inc: 2001. p. 284, fig 18.1. Reprinted with permission of Wiley-Blackwell, Inc.

Chapter 12

Figure 12.3: Modified from Cole, Mark R. and John N. Armentrout, "Neogene Paleogeography of the Western United States." In *Cenozoic Paleography of the Western United States, SEPM Pacific Coast Section, Third Paleography Symposium*, Anaheim, California. (1979). Figure 1; Figure 12.5: Modified from Blakey, Ronald C. Companion Poster to *A Geologic Time Scale 2004*. 44 ma (mid-Paleogene) Map. Reprinted with permission of Ronald C. Blakey, Professor of Geology; Figure 12.6: Modified from Barker, P.F. and Thomas, E. Origin, "Signature and palaeoclimatic influence of the Antarctic Circumpolar Current" *Earth Science Reviews 66 (2004) 143–162* © 2003. Elsevier B.V. Figure 1a and 1b; Figure 12.7A: Modified from Ernst, W.G. (Ed). *The Geotectonic Development of California*. Robey, Volume 1. Prentice-Hall. p. 308; Figure 12.14: Modified from McFadden, Bruce J. "Patterns of phylogeny and rates of evolution in fossil horses: hipparions from the Miocene and Pliocene of North America". *Paleobiology*, 11(3), 1985, pp. 245–257. p. 247, Figure 1; Based on Simpson, G. G. 1951. *Horses: The Story of the Horse Family in the Modern World and Through Sixty Million Years of History*. Oxford University Press, and Marsh, O. C. 1879. "Polydactyle horses, recent and extinct." *American Journal of Science*, 17:499–505; Figure 12.18: Modified from Blakey, Ronald C. Companion Poster to *A Geologic Time Scale 2004*. 11.5 ma (mid-Neogene) Map. Reprinted with permission of Ronald C. Blakey, Professor of Geology; Figures 12.20A, 12.20B: Modified from Culver S. J., and Rawson, P. F. (eds.), "Biotic Response to Global Change." *The Last 145 Million Years* © The Natural History Museum, London, 2000. Cambridge University Press. p. 86, Fig. 6.3 and p. 89, Fig. 6.4; Figure 12.22: Modified from T.E., Wang, Yang, and Quade, Jay. "Expansion of C_4 ecosystems as an indicator of global ecological change in the late Miocene." *Nature*, Vol. 361, 28 January 1993 © Nature Publishing Group. p. 345, Figure 1; Figures 12.30A, 12.30B, 12.31A, 12.33A, 12.33B, 12.33C: Modified from Flint, R. F. *Glacial and Quaternary Geology*, Copyright 1971 John Wiley & Sons, Inc. Reprinted with permission of John Wiley & Sons, Inc.; Figure 12.35: Adapted from Nilsson, T. *The Pleistocene, Geology and Life in the Quaternary Ice Age*. Boston: D. Reidel Publishing Company, 1983. p. 408, Fig. 16.21. Reprinted with kind permission of Springer Science and Business Media; Critical and Creative Thinking Question #3: Based on Crowley, T, J.: "Causes of climate change of the past 1000 years". *Science*, 14 July 2000 Vol. 289.

PHOTO CREDITS

Chapter 1

Pages 2–3: NASA/NG Image Collection; page 4: (left) Courtesy Loren Babcock; page 4: (right) Maria Stenzel/NG Image Collection; page 5: (bottom left) ©Georges Antoni/Hemis/©Corbis; page 5: (bottom right) Reza/NG Image Collection; page 7: (top left) O. Louis Mazzatenta/NG Image Collection; page 7: (top center) Jonathan Blair/NG Image Collection; page 7: (top right) Jonathan Blair/NG Image Collection; page 7: (bottom inset) Courtesy Loren Babcock; pages 8–9 NG Maps; page 8: (inset) NG Maps; page 10: Karen Kasmauski/NG Image Collection; page 10: (inset) NG Maps; page 13: (center) Raymond Gehman/NG Image Collection; page 13: (bottom right) Courtesy Loren Babcock; page 15: (center left) Ira Block/NG Image Collection; page 15: (center right) Courtesy Loren Babcock; page 15: (center right inset) NG Maps; page 15: (bottom right) Courtesy Loren Babcock; page 17: (top) Courtesy Loren Babcock; page 17: (bottom) Courtesy Loren Babcock; page 18: Courtesy Loren Babcock; page 20: (left) Medford Taylor/NG Image Collection; page 20: (right) Kenneth Garrett/NG Image Collection; page 21: (left) Courtesy Loren Babcock; page 21: (right) John Eastcott and Yva Momatiuk/NG Image Collection; page 22: (top) Jonathan Blair/NG Image Collection; page 22: (center) Karen Kasmauski/NG Image Collection; page 22: (bottom) Courtesy Loren Babcock; page 23: (center right) Courtesy Loren Babcock; page 23: (bottom left) Courtesy Loren Babcock; page 24: (top) Courtesy Loren Babcock; page 24: (top inset) NG Maps; page 24: (bottom) Courtesy Loren Babcock; page 25: (top right) Michael Nichols/NG Image Collection; page 25: (top left) NG Maps.

Chapter 2

Pages 26–27: Courtesy Loren Babcock; page 28: (right) Courtesy Loren Babcock; page 28: (bottom left) C.D. Winters/Photo Researchers, Inc.; page 29: (left) Courtesy Loren Babcock; page 29: (right) Wolfgang Poelzer/Peter Arnold, Inc.; page 31: (top) C.D. Winters/Photo Researchers, Inc.; page 31: (just below top) Courtesy Loren Babcock; page 31: (center) Mark A. Schneider/Photo Researchers, Inc.; page 31: (bottom) Mark A. Schneider/Photo Researchers, Inc.; page 34: (top) Courtesy Loren Babcock; page 34: (bottom) Neil Barks/Alamy; page 35: (top left) Roberto de Gugliemo/Photo Researchers, Inc.; page 35: (top center) Breck Kent; page 35: (top right) Harold Levin; page 35: (center left) Ward's Natural Science Establishment; page 35: (center) Neil Barks/Alamy; page 35: (center right) Charles D. Winters/Photo Researchers, Inc.; page 35: (bottom far left) Andrew J. Martinez/Photo Researchers, Inc.; page 35: (bottom left) Photo by William Sacco, provided courtesy of Brian J. Skinner; page 35: (bottom center) Visuals Unlimited/©Corbis; page 35: (bottom right) Breck Kent; page 38: (left) Robert madden/NG Image Collection; page 38: (left inset) NG Maps; page 38: (right) Courtesy Loren Babcock; page 38: (right inset) NG Maps; page 38: (bottom) Courtesy Loren Babcock; page 39: (top left) Norbert Rosing/NG Image Collection; page 39: (top right) Courtesy Loren Babcock; page 39: (center left) James L. Stanfield/NG Image Collection; page 39: (center right) Courtesy Loren Babcock; page 40: Courtesy Brian J.Skinner; page 41: Courtesy Brian J. Skinner; page 42: Courtesy Loren Babcock; page 43: Courtesy Stephen C. Porter; page 44: (top) William E. Ferguson; page 44: (bottom) Philip Richardson/©Corbis; page 44: (bottom inset) NG Image Collection; page 45: (bottom left) Cary Wolinsky/NG Image Collection; page 45: (bottom right) Cary Wolinksi/NG Image Collection; page 46: (top) Marc Moritsch/NG Image Collection; page 46: (bottom left) Marc Moritsch/NG Image Collection; page 46: (bottom right) Courtesy Loren Babcock; page 47: (left) Panoramic Images/Getty Images; page 47: (right) Courtesy Loren Babcock; page 48: Courtesy Loren Babcock; page 48: Courtesy Loren Babcock; page 50: (left) Courtesy Loren Babcock; page 50: (right) Courtesy Loren Babcock; page 51: (top) Darlyne A. Murawski/NG Image Collection; page 51: (bottom) Courtesy Loren Babcock; page 52: Courtesy Loren Babcock; page 53: Courtesy Brian J. Skinner; page 54: (left) ©Breck Kent; page 54: (right) A,J. Copley/Visuals Unlimited; page 56:

Courtesy Loren Babcock; page 57: (top) Courtesy Loren Babcock; page 57: (bottom right) Marc Moritsch/NG Image Collection; page 58: (top) Courtesy Loren Babcock; page 58: (bottom) Courtesy Loren Babcock; page 59: (top right) Wolfgang Poelzer/Peter Arnold, Inc.; page 59: (bottom) Courtesy Brian J. Skinner.

Chapter 3

Pages 60-61: Ralph Lee Hopkins/NG Image Collection; page 62: ©picture-dimensions/Alamy; page 62: (inset) NG Image Collection; page 63: Courtesy John Isbell, University of Wisconsin, Milwaukee; page 65: Courtesy Loren Babcock; page 66: Courtesy Loren Babcock; page 67: (top) Michael Nichols/NG Image Collection; page 67: (bottom) ©1967 Grand Canyon Natural History Association; page 68: Courtesy Loren Babcock; page 70: Courtesy Loren Babcock; page 71: Courtesy Loren Babcock; page 72: Courtesy Loren Babcock; page 73: (center) From Bally, A.W. (editor), Atlas of Seismic Stratigraphy, Vol. 1, page 8, 1987. AAPG ©2008. Reprinted by permission of the American Association of Petroleum Geologists whose permission is required for further use; page 73: (bottom) From Bally, A.W. (editor), Atlas of Seismic Stratigraphy, Vol. 1, page 10:, 1987. AAPG ©2008. Reprinted by permission of the American Association of Petroleum Geologists whose permission is required for further use.; page 80: Courtesy Loren Babcock; page 82: Courtesy Loren Babcock; page 83: ©picturedimensions/Alamy; page 85: (top) Courtesy Loren Babcock; page 85: (bottom) ©Marli Bryant Miller.

Chapter 4

Pages 86-87: O. Louis Mazzatenta/NG Image Collection; page 90: (top) James L. Amos/NG Image Collection; page 90: (bottom left) Dr. Jeremy Burgess/SPL/Photo Researchers, Inc.; page 90: (bottom right) CNRI/ SPL/Photo Researchers, Inc.; page 92: (left) Paul Nicklen/NG Image Collection; page 92 (right) O. Louis Mazzatenta/NG Image Collection; page 93: (center left) Karen Kuehn/NG Image Collection; page 93: (center left inset) NG Maps; page 93: (bottom) Ken MacDonald/Photo Researchers, Inc.; page 93: (bottom inset) NG Maps; page 94: (top left) Scimat/Photo Researchers, Inc.; page 94: (top right) Bob Krist/NG Image Collection; page 94: (center left) Courtesy Loren Babcock; page 94: (center right) Jonathan Blair/NG Image Collection; page 95: Rick Smolan/NG Image Collection; page 96: (left) Brian J. Skerry/NG Image Collection; page 96: (center) Dee Breger/Photo Researchers, Inc.; page 96: (right) Darlyne A. Murawski/NG Image Collection; page 97: (top left) Darlyne Am Murawski/NG Image Collection; page 97: (top right) Darlyne A. Murawski/NG Image Collection; page 97: (center left) William Douthitt/NG Image Collection; page 97: (center right) Courtesy Loren Babcock; page 98: (left) Courtesy Loren Babcock; page 98: (right) Phil Schermeister/NG Image Collection; page 99: (center left) Charles Kogod/NG Image Collection; page 99: (center right) Courtesy Loren Babcock; page 99: (bottom) Courtesy Loren Babcock; page 100: (top left) Paul Nicklen/NG Image Collection; page 100: (top center) Andrew J. Martinez /Photo Researchers, Inc.; page 100: (top right) Barry Tessman/NG Image Collection; page 100: (bottom left) Robert Sisson/NG Image Collection; page 101: (top left) Janice Healey/NG Image Collection; page 101: (top right) Tim Laman/NG Image Collection; page 101: (bottom left) Courtesy Loren Babcock; page 101: (bottom right) Nicole Duplaix/NG Image Collection; page 102: Phil Schermeister/NG Image Collection; page 106: (top left) Nick Caloyianis/NG Image Collection; page 106: (top right) Brian J. Skerry/NG Image Collection; page 107: (left) Jonathan Blair/NG Image Collection; page 107: (right) Jonathan Blair/NG Image Collection; page 108: (top right) Courtesy Loren Babcock; page 108: (center left) Tom Bean/©Corbis; page 108: (bottom left) Courtesy Loren Babcock; page 108: (bottom right) Courtesy Loren Babcock; page 110: (top) Beverly Joubert/NG Image Collection; page 110: (bottom) Courtesy Loren Babcock; page 111: (top) Calvin Larsen/Photo Researchers, Inc.; page 111: (bottom left) Courtesy Loren Babcock and Philip Borkow; page 111: (bottom center)

Courtesy Loren Babcock and Philip Borkow; page 111: (bottom right) Courtesy Loren Babcock and Philip Borkow; page 113: (top left) Karen Kuehn/NG Image Collection; page 113: (top right) Ken Lucas/Visuals Unlimited; page 113: (center left) Courtesy Loren Babcock; page 113: (center right) James L. Amos/NG Image Collection; page 113: (bottom left) Phil Schermeister/NG Image Collection; page 113: (bottom right) John Cancalosi/NG Image Collection; page 114: (center left) Paul Zahl/NG Image Collection; page 114: (center right) Smithsonian Museum of Natural History, photo by Chip Clark; page 114: (bottom right) Jonathan Blair/NG Image Collection; page 115: (center) Courtesy Loren Babcock; page 115: (bottom) Courtesy Loren Babcock and Alycia L. Stigall; page 116: (top left) Courtesy Loren Babcock; page 116: (top right) Courtesy Loren Babcock and Alycia L. Stigall ; page 116: (bottom left) Andrew Syred/Photo Researchers, Inc.; page 116: (bottom right) Courtesy of W.D. Hamilton, U.S. Geological Survey; page 117: (left) O. Louis Mazzatenta/NG Image Collection; page 117: (right) ©Roy Larimer; page 118: (top left) O. Louis Mazzatenta/NG Image Collection; page 118: (top right) James L. Amos/NG Image Collection; page 118: (bottom left) James L. Amos/NG Image Collection; page 118: (bottom right) O. Louis Mazzatenta/NG Image Collection; page 119: (center) Rick Smolan/NG Image Collection; page 119: (top right) Darlyne A. Murawski/NG Image Collection; page 119: (bottom) Janice Healey/NG Image Collection; page 120: (top left) Tim Laman/NG Image Collection; page 120: (bottom left) Barry Tessman/NG Image Collection; page 120: (top right) Courtesy Loren Babcock; page 120: (bottom right) O. Louis Mazzatenta/NG Image Collection; page 121: James L. Amos/NG Image Collection; page 122: (top) Joel Sartore/NG Image Collection; page 122: (bottom) Tim Laman/NG Image Collection; page 123: (bottom right) Jonathan Blair/NG Image Collection.

Chapter 5

Pages 124–125: Kirk Moldoff/NG Image Collection; page 126: David Young-Wolff/Stone/Getty Images; page 129: (left) Paul Nicklen/NG Image Collection; page 129: (center) Tim Laman/NG Image Collection; page 129: (right) Phil Schermeister/NG Image Collection; page 130: Ralph Hutchings/Visuals Unlimited; page 131: Robert Clark/NG Image Collection; page 132: Courtesy Loren Babcock; page 133: Phil Schermeister/NG Image Collection; page 135: (left) James King-Holmes/Photo Researchers, Inc.; page 135: (right) Alan & Linda Detrick/Photo Researchers, Inc.; page 136: (left) Kirk Moldoff/NG Image Collection; page 136: (right) Bruce Dale/NG Image Collection; page 151: (top left) John Cancalosi/NG Image Collection; page 151: (top right) Darlyne A. Murawski/NG Image Collection; page 151: (center left) James L. Stanfield/NG Image Collection; page 151: (center right) Rich Reid/NG Image Collection; page 154: (top) Courtesy Loren Babcock; page 154: (bottom) Kirk Moldoff/NG Image Collection; page 155: (center left) John Cancalosi/NG Image Collection; page 155: (center right) Darlyne A. Murawski/NG Image Collection; page 156: (center) Flip Nicklin/Minden Pictures; page 156: (bottom) Raul Touzon/NG Image Collection; page 157: Kirk Moldoff/NG Image Collection.

Chapter 6

Pages 158–159: Courtesy Loren Babcock; page 160: Frans Lanting/NG Image Collection; page 161: Peter Carsten/NG Image Collection; page 163: (top) ©Yann Arthus-Bertrand/©Corbis; page 163: (bottom left) Courtesy Loren Babcock; page 163: (bottom right) Courtesy Loren Babcock; page 164: Courtesy Loren Babcock; page 165: (top left) Courtesy Loren Babcock; page 165: (top right) Courtesy Loren Babcock; page 165: (bottom) Courtesy Loren Babcock; page 166: (center) Courtesy Loren Babcock; page 166: (bottom) ©epa/©Corbis; page 168: (top left) ©Marli Bryant Miller; page 168: (top center) Jim Richardson/NG Image Collection; page 168: (top right) Gerald & Buff Corsi/Visuals Unlimited; page 168: (bottom) Courtesy Sergio Longhitano, Dept. of Geological Science, University of Basilicata, Campus Universitario di Macchia Romana, Potenza, Italy; page 169: (top) Georgette Douwma/Photo Researchers, Inc.; page 169: (bottom) Courtesy Loren Babcock; page 170: Courtesy Loren Babcock; page 171: (left) Gilbert Twiest/Visuals Unlimited; page 171: (right) Courtesy Loren Babcock; page 171: (center) Courtesy Loren Babcock; page 175: (bottom left) ©Steven Vidler/Eurasia Press/©Corbis; page 175: (right) Harald Sund/Getty Images; page 176: (top right) Emory Kristof/NG Image Collection;

page 176: (center left) Courtesy Stephen C. Porter; page 176: (center right) O. Louis Mazzatenta/NG Image Collection; page 177: (top left) ©Marli Bryant Miller; page 177: (top right) ©Marli Bryant Miller; page 177: (bottom left) Peter Carsten/NG Image Collection; page 177: (bottom right) Melissa Farlow/NG Image Collection; page 179: (top left) Courtesy Loren Babcock; page 179: (top right) Courtesy Loren Babcock; page 179: (bottom left) Courtesy Loren Babcock; page 179: (bottom right) Courtesy Loren Babcock; page 179: (center) NG Maps; page 180: (bottom left) Science Source/Photo Researchers, Inc.; page 180: (bottom right) Kevin Fleming/©Corbis; page 181: (top left), page 181: (top right), page 181: (center), page 182: (bottom left) NASA/Photo Researchers, Inc.; page 182: (bottom right) M-Sat, Ltd/Photo Researchers, Inc.; page 183: (center) Marli Miller/Visuals Unlimited; page 183: (bottom) L. Sat, Ltd./Photo Researchers, Inc.; page 184: (bottom left) Courtesy NASA; page 184: (bottom right) Steve Allen/Brand X/©Corbis; page 185: Courtesy NASA; page 190: (top left) Frans Lanting/NG Image Collection; page 190: (center) Courtesy Loren Babcock; page 190: (bottom) Courtesy Loren Babcock; page 191: (top) Courtesy Loren Babcock; page 191: (bottom) Courtesy Loren Babcock; page 192: (top left) ©Stephen Frink/Corbis; page 192: (top right) SUPERSTOCK; page 192: (bottom) William A. Bake/©Corbis; page 193: Peter Carsten/NG Image Collection.

Chapter 7

Pages 194–195: Peter Carsten/NG Image Collection; page 197: Rendered by Robert Stacey, WorldSat International Inc. for National Geographic. Atlas of the World Eighth Edition, (Washington, D.C.: National Geographic Society), 2005. Image is a blend of Landsat and NOAA AVHRR data. NOAA/NGDC and DMSP lights at night data was also used (human settlements, white; fires, yellow; fishing fleets, blue; natural gas flares, red). ; page 198: Courtesy Loren Babcock; page 199: (left) Courtesy Loren Babcock; page 199: (right) A.J. Copley/Visuals Unlimited; page 201: (top) Rendered by Robert Stacey, WorldSat International Inc. for National Geographic. Atlas of the World Eighth Edition, (Washington, D.C.: National Geographic Society), 2005. Image is a blend of Landsat and NOAA AVHRR data. NOAA/NGDC and DMSP lights at night data was also used (human settlements, white; fires, yellow; fishing fleets, blue; natural gas flares, red). ; page 201: (center) Gregory G. Dimijian, M.D./Photo Researchers, Inc.; page 201: (bottom) David Hosking/©Corbis; page 203: (left) Marie Tharp and Bruce C. Heezen; page 203: (right) NOAA/SPL/Photo Researchers, Inc.; page 206: Courtesy Loren Babcock; page 208–209: NG Maps; page 209: (bottom right) Tomonari Tsuji/Getty Images; page 212: Marli Miller/Visuals Unlimited; page 213: (top) Radius Images/Alamy; page 213: (bottom) W.A. Rogers/NG Image Collection; page 214: Courtesy NASA; page 215: Breck Kent; page 217: William E. Ferguson; page 222: (top right) Rendered by Robert Stacey, WorldSat International Inc. for National Geographic. Atlas of the World Eighth Edition, (Washington, D.C.: National Geographic Society), 2005. Image is a blend of Landsat and NOAA AVHRR data. NOAA/NGDC and DMSP lights at night data was also used (human settlements, white; fires, yellow; fishing fleets, blue; natural gas flares, red). ; page 222: (bottom) Tomonari Tsuji/Getty Images; page 225: Rendered by Robert Stacey, WorldSat International Inc. for National Geographic. Atlas of the World Eighth Edition, (Washington, D.C.: National Geographic Society), 2005. Image is a blend of Landsat and NOAA AVHRR data. NOAA/NGDC and DMSP lights at night data was also used (human settlements, white; fires, yellow; fishing fleets, blue; natural gas flares, red); page 224: Courtesy Pennsylvania Department of Conservation and Natural Resources.

Chapter 8

Pages 226–227: Marli Miller/Visuals Unlimited; page 230: Courtesy D.A. Walker, Alaska Geobotany Center, University of Alaska Fairbanks; page 231: James L. Amos/NG Image Collection; page 232: (top right) ©NASA/Roger Ressmeyer/©Corbis; page 232: (center) Francois Gohier/Photo Researchers, Inc.; page 233: John R. Foster/Photo Researchers, Inc.; page 234: (bottom left) Mark Garlick/Photo Researchers, Inc.; page 234: (bottom right) Shigemi Numazawa/Atlas Photo Bank/Photo Researchers, Inc.; page 236: Bonestell Space Art; page 237: ©NASA/Roger Ressmeyer/©Corbis; page 238: Gerard Lodriguss/Photo Researchers, Inc.; page 240: ©METI and NASA. All rights reserved. Processed and distributed by ERSDAC, Japan; page 241: Courtesy Russell Mapes; page 242: (top and center) James L. Amos/NG Image Collection; page 242: (top right inset) NG Maps; page 242: (bottom left) James L. Amos/NG Image Collection; page 242: (bottom right) James L. Amos/NG Image Collection; page 244: (top left) Sinclair Stammers/Photo Researchers, Inc.; page 244: (top right) Cary Wolinsky/NG Image Collection; page 245: (left) O. Louis Mazzatenta/NG Image Collection; page 245: (right) Dr. Stanley L. Miller; page 247: (bottom left) Ken MacDonald/Photo Researchers, Inc.; page 247: (bottom right) Photo courtesy of Colleen M. Cavanaugh, Department of Organismic and Evolutionary Biology, Harvard University; page 249: (bottom) James L. Amos/NG Image Collection; page 249: (top) Gerard Lodriguss /Photo Researchers, Inc.; page 251: (top) Courtesy Loren Babcock; page 251: (bottom) Cary Wolinsky/ NG Image Collection.

Chapter 9

Pages 252–253: Courtesy Loren Babcock; page 258: Courtesy Loren Babcock; page 261: Courtesy Loren Babcock; page 261: (top inset) NG Maps; page 262: Courtesy Loren Babcock; page 263: (left) Courtesy Dr. Robert Darling, SUNY Cortland; page 263: (right) Courtesy Loren Babcock; page 265: (left) Courtesy Loren Babcock; page 265: (right) Courtesy Loren Babcock; page 267: Sinclair Stammers/Photo Researchers, Inc.; page 268: Courtesy Loren Babcock; page 269: O. Louis Mazzatenta/NG Image Collection; page 270: Courtesy G. Vidal; page 271: (top left) Courtesy Loren Babcock; page 271: (top center) Courtesy Loren Babcock; page 271: (top right) Courtesy Loren Babcock; page 271: (center) Courtesy Loren Babcock; page 271: (bottom) John Dawson/NG Image Collection; page 272: Inga Spence/Visuals Unlimited; page 273: (left) Courtesy Loren Babcock; page 273: (right) O. Louis Mazzatenta/NG Image Collection; page 274: Courtesy Loren Babcock; page 275: Courtesy Loren Babcock; page 276: (top) Courtesy Loren Babcock; page 276: (bottom) Courtesy Shuhai Xiao; page 277: (left) Courtesy Dr. Robert Darling, SUNY Cortland; page 277: (top right) Sinclair Stammers/Photo Researchers, Inc.; page 277: (bottom right) Courtesy Loren Babcock.

Chapter 10

Pages 278–279: Courtesy Loren Babcock; page 283: Courtesy Loren Babcock; page 284: (left) Courtesy Jean Vannier, CNRS; page 284: (right) Courtesy Simon Conway Morris, University of Cambridge; page 285: (top) Courtesy Loren Babcock; page 285: (bottom left) Courtesy Loren Babcock; page 285: (bottom right) Courtesy Loren Babcock; page 286: Courtesy Loren Babcock; page 287: (top left) Smithsonian Museum of Natural History, photo by Chip Clark; page 287: (top right) Smithsonian Museum of Natural History, photo by Chip Clark; page 287: (center left) Smithsonian Museum of Natural History, photo by Chip Clark; page 287: (center right) Smithsonian Museum of Natural History, photo by Chip Clark; page 287: (bottom left) Smithsonian Museum of Natural History, photo by Chip Clark; page 287: (bottom right) Courtesy Gabriella Bagnoli, University of Pisa, Italy; page 288: (center left) Courtesy Loren Babcock; page 288: (center right) Courtesy Loren Babcock; page 288: (bottom right) Courtesy Loren Babcock; page 289: (top left) Smithsonian Museum of Natural History, photo by Chip Clark; page 289: (top center) Courtesy Loren Babcock; page 290: (center) Alan Sirunlnikoff/Photo Researchers, Inc.; page 290: (bottom right) Smithsonian Museum of Natural History, photo by Chip Clark; page 290: (top inset) NG Maps; page 291: (center) Courtesy Dr. Derek J. Siveter; page 291: (bottom left) Courtesy Loren Babcock; page 291: Courtesy Loren Babcock and Wengang Zhang; page 291: (top left inset) NG Maps; page 292: Tom McHugh/Photo Researchers, Inc.; page 295: NG Maps; page 296: Smithsonian Museum of Natural History, photo by Chip Clark; page 297: Martin Land/Photo Researchers, Inc.; page 298: Courtesy Paul K. Strother; page 299: Courtesy Loren Babcock; page 303: (top) Courtesy Loren Babcock; page 303: (bottom left) Courtesy Loren Babcock; page 303: (right) Courtesy Pennsylvania Department of Conservation and Natural Resources; page 304: (bottom left) Smithsonian Museum of Natural History, photo by Chip Clark; page 304: (bottom right) Courtesy Loren Babcock; page 305: (top right) Courtesy The Cleveland Museum of Natural History; page 305: (bottom left) Courtesy Loren Babcock; page 305: (bottom right) Courtesy

Loren Babcock; page 306: (top left) Scott Camazine/Photo Researchers, Inc.; page 306: (top right) Courtesy Loren Babcock; page 306: (bottom left) Ted Kinsman/Photo Researchers, Inc.; page 306: (bottom right) C. Permian Creations/Alamy; page 308: (bottom left) Courtesy Loren Babcock; page 308: (bottom right) Philip Scalia/Alamy; page 310: The Natural History Museum; page 311: (top right) Jonathan Blair/NG Image Collection; page 315: Courtesy Loren Babcock; page 316: Courtesy Loren Babcock; page 318: (bottom left) Courtesy Loren Babcock; page 318: (bottom right) Smithsonian Museum of Natural History, photo by Chip Clark; page 319: (bottom left) Courtesy Loren Babcock; page 319: (center right) Courtesy Loren Babcock; page 319: (bottom right) Courtesy Loren Babcock; page 320: Ken Lucas/Visuals Unlimited; page 321: (left) Jonathan Blair/NG Image Collection; page 321: (right) Francois Gohier/Photo Researchers, Inc.; page 323: Smithsonian Museum of Natural History, photo by Chip Clark; page 324: (top) Courtesy Loren Babcock; page 324: (bottom) Courtesy The Cleveland Museum of Natural History; page 325: Courtesy Loren Babcock; page 326: Courtesy Loren Babcock; page 327: (left) Smithsonian Museum of Natural History, photo by Chip Clark; page 327: (right) Courtesy Loren Babcock.

Chapter 11

Pages 328–329: Todd Gipstein/NG Image Collection; page 332: Courtesy Loren Babcock; page 333: (top left) Jonathan Blair/NG Image Collection; page 333: (top right) ©The Field Museum, #CK34T; page 333: (bottom left) Chris Butler/Photo Researchers, Inc.; page 333: (bottom right) Courtesy Forschungsinstitut Senckenberg; page 334: Phil Schermeister/NG Image Collection; page 335: (left) Ken Lucas/Visuals Unlimited; page 335: (right) Mike Danton/Alamy; page 336: (top left) Science Photo Library/Photo Researchers, Inc.; page 336: (top right) O. Louis Mazzatenta/NG Image Collection; page 336: (bottom) Courtesy Forschungsinstitut Senckenberg; page 339: Superstock; page 340: (top left) ©tompiodesign.com/Alamy; page 340: (top right) Mark A. Schneider/Visuals Unlimited; page 340: (bottom) Courtesy of the Sternberg Museum of Natural History, Hays, Kansas; page 341: (top left) E.R. Degginger/Bruce Coleman, Inc.; page 341: (top center) Heather Angel/Alamy; page 341: (top right) Jonathan Blair/NG Image Collection; page 341: (bottom left) Smithsonian Museum of Natural History, photo by Chip Clark; page 341: (bottom right) Smithsonian Museum of Natural History, photo by Chip Clark; page 342: Courtesy Loren Babcock; page 343: (right) Yuji Higashida/Sebun Photo/Getty Images; page 343: (left) Smithsonian Museum of Natural History, photo by Chip Clark; page 344: Courtesy Loren Babcock; page 346: (center left) ©2000 The Field Museum, GN89777_12c; page 346: (center right) Courtesy Loren Babcock; page 346: (bottom) Tom Bean/Getty Images; page 347: (top) O. Louis Mazzatenta/NG Image Collection; page 347: (center) O. Louis Mazzatenta/NG Image Collection; page 348: (center left) Louie Psihoyos/Science Faction/Getty Images; page 348: (center right) Louie Psihoyos/Science Faction/Getty Images; page 348: (bottom) Joe Tucciarone/Photo Researchers, Inc.; page 349: (left) Louie Psihoyos/Science Faction/Getty Images; page 349: (right) Louie Psihoyos/Science Faction; page 350: (center left) Francois Gohier/Photo Researchers, Inc.; page 350: (center right) Francois Gohier/Photo Researchers, Inc.; page 350: (bottom left) Louie Psihoyos/Science Faction; page 350: (bottom right) Kevin Schafer/Stone/Getty Images; page 351: (left) Francois Gohier/Photo Researchers, Inc.; page 351: (right) Science Photo Library/Photo Researchers, Inc.; page 352: (left) Courtesy Loren Babcock; page 352: (right) J.C. Carton/Bruce Coleman, Inc.; page 357: (top left) Steve Gschmeissner/Photo Researchers, Inc.; page 357: (top right) Norbert Rossing/NG Image Collection; page 357: (bottom) Smithsonian Museum of Natural History, photo by Chip Clark; page 358: (left) Courtesy Loren Babcock; page 358: (right) Courtesy Loren Babcock; page 359: (top left) Richard T. Nowitz/Photo Researchers, Inc.; page 359: (top right) Photo by Daniel Miller ©University of Michigan Museum of Paleontology. Used by permission; page 359: (bottom left) O. Louis Mazzatenta/NG Image Collection; page 359: (bottom right) Jonathan Blair/NG Image Collection; page 362: (top) Courtesy V.L. Sharpton, LPI/NASA; page 362: (center right) Jonathan Blair/NG Image Collection; page 363: Courtesy Loren Babcock; page 364: (bottom left) Yuji Higashida/Sebun Photo/Getty Images; page 364: (center right) Louie Psihoyos/Science Faction/Getty Images; page 365: (top) Courtesy Loren Babcock; page 365: (center) Francois Gohier/Photo Researchers, Inc.; page 366: (center left) Willam P. Leonard/DRK Photo; page 366: (center right) Kathy Merrifield/Photo Researchers, Inc.; page 366: (bottom) Courtesy Forschungsinstitut Senckenberg; page 367: (left) O. Louis Mazzatenta/NG Image Collection; page 367: (right) J.C. Carton/Bruce Coleman, Inc.

Chapter 12

Pages 368–369: Thomas Ernsting/Bilderberg/Aurora Photos; page 372: Ashvin Mehta/Dinodia Images/Alamy; page 374: Harvey Lloyd/Getty Images; page 375: Courtesy Loren Babcock; page 378: Marc Adamus/Getty Images; page 379: (top left) Jeff Foott/Getty Images; page 379: (center right) James L. Amos/NG Image Collection; page 380: Courtesy Loren Babcock; page 381: (top) Kevin Schafer/Alamy; page 381: (bottom left) Imagebroker/Alamy; page 381: (bottom right) CMSP/NewsCom; page 384: (bottom left) Smithsonian Museum of Natural History, photo by Chip Clark; page 384: (bottom right) Alex Kerstitch/Visuals Unlimited; page 385: (top) Phil Degginer/Mira.com/Digital Railroad, Inc.; page 385: (bottom) Courtesy Skulls Unlimited International, Inc., www.skullsunlimited.com; page 386: (top left) Jonathan Blair/NG Image Collection; page 386: (top right) Courtesy Loren Babcock; page 386: (bottom left) Tam C. Nguyen/Phototake; page 386: (bottom right) Kaj. Svensson/SPL/Photo Researchers, Inc.; page 387: James Dean, Smithsonian Institution; page 389: (far left) Adrian Davies/Bruce Coleman, Inc.; page 389: (left of center) Courtesy Loren Babcock; page 389: (center) ISM/Phototake; page 389: (right of center) ISM/Phototake; page 389: (far right) T.A. Wiewandt/DRK Photo; page 391: Alfred Pasieka/Photo Researchers, Inc.; page 392: (top) Courtesy Loren Bab-

cock; page 392: (bottom) Jonathan Blair/NG Image Collection; page 393: Marc Moritsch/NG Image Collection; page 394: Robert Landau/Digital Railroad, Inc.; page 394: (bottom left inset) NG Maps; page 395: (bottom left) Tom McHugh/Photo Researchers, Inc.; page 395: (center right) Karen Kuehn/NG Image Collection; page 395: (bottom right) Imagebroker/Alamy; page 397: (center left) Pascal Goetgeluck/Photo Researchers, Inc.; page 397: (top left) Eduardo Contreras/Zuma Press; page 397: (bottom left) ISM/Phototake; page 397: (bottom center) NewsCom; page 397: (bottom right) Pascal Goetgheluck/Photo Researchers, Inc.; page 397: (center right) Pascal Goetgheluck/Photo Researchers, Inc.; page 398: John Reader/Photo Researchers, Inc.; page 399: Robert Harding Picture Library/Alamy; page 402: Courtesy Loren Babcock; page 403: (center left) Courtesy Loren Babcock; page 403: (bottom left) Photodisc/Getty Images; page 403: (center right) Courtesy Loren Babcock; page 403: (bottom right) Courtesy J. Manchester, New York State Geological Survey; page 405: (left) Getty Images; page 405: (right) Courtesy Loren Babcock; page 406: (bottom left) Peter Menzel/Digital Railroad, Inc.; page 406: (bottom right) Chase Studio/Photo Researchers, Inc.; page 407: (bottom left) Courtesy Loren Babcock; page 407: (bottom right) Courtesy Loren Babcock; page 407: (right inset) NG Maps; page 408: (top) Ashvin Mehta/Dinodia Images/Alamy; page 408: (bottom) Courtesy Loren Babcock; page 409: Pascal Goetgeluck/Photo Researchers, Inc.; page 410: (top right) Courtesy Loren Babcock; page 410: (top left) Courtesy Loren Babcock; page 410: (center) Courtesy Loren Babcock; page 410: (bottom) Maria Stenzel/NG Image Collection; page 411: (top left) Marc Adamus/Getty Images; page 411: (bottom left) Tom McHugh/Photo Researchers, Inc.; page 411: (center right) John Reader/Photo Researchers, Inc.

INDEX

Note to the reader: *italicized* entries refer to figures, locators with a "t" refer to tables.